BIOMEMBRANES
Volume 9

BIOMEMBRANES

A series edited by Lionel A. Manson
The Wistar Institute, Philadelphia, Pennsylvania

Recent Volumes in this Series

1972 . Biomembranes . Volume 3
Passive Permeability of Cell Membranes
Edited by F. Kreuzer and J. F. G. Slegers

1974 . Biomembranes . Volume 4A
Intestinal Absorption
Edited by D. H. Smyth

1974 . Biomembranes . Volume 4B
Intestinal Absorption
Edited by D. H. Smyth

1974 . Biomembranes . Volume 5
Articles by Richard W. Hendler, Stuart A. Kauffman, Dale L. Oxender, Henry C. Pitot, David L. Rosenstreich, Alan S. Rosenthal, Thomas K. Shires, and Donald F. Hoelzl Wallach

1975 . Biomembranes . Volume 6
Bacterial Membranes and the Respiratory Chain
By N. S. Gel'man, M. A. Lukoyanova, and D. N. Ostrovskii

1975 . Biomembranes . Volume 7
Aharon Katzir Memorial Volume
Edited by Henryk Eisenberg, Ephraim Katchalski-Katzir, and Lionel A. Manson

1976 . Biomembranes . Volume 8
Articles by Robert W. Baldwin, William C. Davis, Paul H. DeFoor, Carl G. Gahmberg, Sen-itiroh Hakomori, Reinhard Kurth, Lionel A. Manson, Michael R. Price, and Howard E. Sandberg

1977 . Biomembranes . Volume 9
Membrane Transport—An Interdisciplinary Approach
By Arnošt Kotyk and Karel Janáček

A Continuation Order Plan is available for this series. A continuation order will bring delivery of each new volume immediately upon publication. Volumes are billed only upon actual shipment. For further information please contact the publisher.

BIOMEMBRANES, Volume 9

MEMBRANE TRANSPORT
An Interdisciplinary Approach

Arnošt Kotyk and Karel Janáček
Institute of Microbiology
Czechoslovak Academy of Sciences, Prague

PLENUM PRESS • NEW YORK AND LONDON

ISBN: 0-306-39809-5
Library of Congress Catalog Card Number: 75-18942

Scientific Editor: Prof. Dr. J. Koštíř, DrSc.
Scientific Advisers: Prof. Dr. J. Koryta, DrSc.
Doc. Dr. Ľ. Drobnica, CSc.

Published in coedition by ACADEMIA
Publishing House of the Czechoslovak
Academy of Sciences, Prague

and outside of the Socialist countries by
PLENUM PRESS
A Division of Plenum Publishing Corporation
227 West 17th Street, New York, N.Y. 10011

© A. Kotyk, K. Janáček, Prague 1977

All rights reserved

No part of this book may be reproduced, stored in a retrieval system, or transmitted, in any form or by any means, electronic, mechanical, photocopying, microfilming, recording, or otherwise, without written permission from the Publishers

Printed in Czechoslovakia

PREFACE

Not many years ago, problems of membranes and transport attracted the attention of but a few dozen enthusiasts, mainly physiologists who recognized the significance of membranes for the stabilization of the general steady state of organisms. The first symposium organized some fifteen years ago could boast of the attendance of perhaps fifty scientists (the remaining fifty were not yet sure that membranes was the topic of their choice), ranging in specialization from physical chemistry to bacterial genetics, who clairvoyantly decided to study what now has become the number one subject at most congresses of biophysics, physiology, and even biochemistry and microbiology.

As is the case with many rapidly developing fields, the interest in membranes and transport seems to be growing out of bounds and the whole field of membranology, interdisciplinary as it is, has penetrated into the realms of a number of branches of physics, chemistry, and biology. Its subject is primarily biological and, although much has been done in the world to increase the "exactness" of biology over the past thirty years, one cannot strive for a rigorous mathematical description of biological phenomena since, as M. H. Jacobs wrote appropriately back in 1935, "In the first place, it is utterly hopeless for the biologist with the means at present at his disposal, to reduce the variables that enter into his problem to the small number usually encountered in physical investigations. He is compelled, therefore, regretfully but of necessity, to be content with a lesser degree of precision in his results than that attainable in the so-called 'exact sciences'. It follows that in dealing with most biological problems it is not only useless, but actually unscientific, to carry mathematical refinements beyond a certain point, just as it would be both useless and unscientific to employ an analytical balance of the highest precision for obtaining the growth-curve of a rat."

The present book is rather an attempt at bringing together data on membranes and their primary function, viz. transport of substances, in a relatively small volume. Recent literature abounds with compendia, highly specialized monographs and conference proceedings dealing with various aspects of membranes, but it is our feeling that there is a need for an introductory text that would integrate at least most of the available information about membranes and transport.

Several years ago, it was possible to include in a single volume devoted exclusively to transport studies (Kotyk and Janáček, Plenum Press, New York, 1970; 2nd edition 1975) a large section of comparative aspects, including various microorganisms, plants and animal tissues. At present, there are so many pieces of information available from various biological objects that they deserve specialized reviews. For this reason, only generally valid and most important observations are included in the present book while, on the other hand, due attention is being paid to membrane functions other than transport, including their physical chemistry and biochemistry.

We are aware of the fact that some chapters of the present book are better than others because some of the authors' personal experience and perhaps even hobbies are reflected in them. We hope that the reader will not fail to peruse the more erudite monographic works that we refer to at the appropriate places in the text if he wishes to obtain more enlightenment.

In preparing the manuscript we were greatly aided, both through consultations and through technical assistance, by some of our co-workers, in particular Dr. K. Sigler, Dr. J. Horák, and Dr. R. Rybová of this laboratory. The most competent photography of Dr. A. Wolf of this Institute is gratefully acknowledged. When mentioning these collaborators, we do not overlook the general stimulating atmosphere created by the whole staff of the department of membrane transport. Without it, this book could not have been written.

Arnošt Kotyk
Karel Janáček

Laboratory for Cell Membrane Transport
Institute of Microbiology
Czechoslovak Academy of Sciences

CONTENTS

Preface

1.	**Introduction (K. J.)**	11
1.1.	Important events in the history of membranology	12
1.2.	Evolution and transport	14
1.3.	Transport in space and time	16
2.	**Membranes (A. K.)**	21
2.1.	Surfaces and interfaces	22
2.2.	Chemical composition	36
2.2.1.	Lipids	36
2.2.1.1.	Chemistry	36
2.2.1.2.	Distribution	45
2.2.1.3.	Fatty acids	49
2.2.1.4.	Solubilization	51
2.2.2.	Proteins	53
2.2.2.1.	Types and occurrence	53
2.2.2.2.	Solubilization	57
2.2.3.	Carbohydrates	60
2.3.	Structure of membranes	61
2.3.1.	Physico-chemical techniques	61
2.3.1.1.	X-ray diffraction	61
2.3.1.2.	Infrared spectroscopy	63
2.3.1.3.	Ultraviolet spectroscopy	64
2.3.1.3.1.	Absorption spectra	65
2.3.1.3.2.	Optical rotatory dispersion	66

2.3.1.3.3.	Circular dichroism	67
2.3.1.4.	Nuclear magnetic resonance	70
2.3.1.5.	Electron spin resonance	75
2.3.1.6.	Fluorescence	79
2.3.1.7.	Calorimetry and related techniques	82
2.3.1.8.	Other techniques	85
2.3.2.	Organization of lipids	86
2.3.2.1.	Biological membranes	86
2.3.2.2.	Artificial membranes	95
2.3.3.	Organization of proteins	99
2.3.4.	Lipid−protein interactions and membrane structure.	101
2.4.	Assembly of membranes	107
2.5.	Electron microscopy	117
2.6.	Isolation of membranes	119
2.7.	Morphology and function of different biological membranes	126
2.7.1.	Plasma membrane	126
2.7.1.1.	Morphology	126
2.7.1.2.	Functional properties	130
2.7.1.2.1.	Antigenicity	130
2.7.1.2.2.	Enzyme content	132
2.7.1.2.3.	Cell walls	133
2.7.1.2.4.	Binding and receptor properties	145
2.7.2.	Mitochondrion	147
2.7.3.	Chloroplast	152
2.7.4.	Mesosome	156
2.7.5.	Endoplasmic reticulum	156
2.7.6.	Golgi apparatus	158
2.7.7.	Lysosome	161
2.7.8.	Tonoplast	163
2.7.9.	Nucleus	163
2.7.10.	Other membranes	165
Synopsis		165
3.	**Thermodynamics of transport (K. J.)**	169
3.1.	Thermodynamic equilibrium, passive and active transport processes	169
3.2.	Thermodynamics of the steady state	172
3.3.	Network thermodynamics	177
Synopsis		181

4.	Transport of nonelectrolytes (A. K.)	183
4.1.	Principles of diffusion (K. J.)	184
4.2.	Diffusion across membranes (A. K. + K. J.)	192
4.3.	Kinetics of mediated transport	207
4.3.1.	Mediated or facilitated diffusion	207
4.3.1.1.	Steady-state approach	208
4.3.1.2.	Equilibrium approach	211
4.3.1.3.	Two-site carrier	220
4.3.2.	Primary active transport	222
4.3.2.1.	Kinetics	222
4.3.2.2.	Combined systems	226
4.3.2.3.	Energetics	228
4.3.3.	Coupled transport	229
4.4.	Chemical nature of nonelectrolyte transport systems	233
4.4.1.	Group-translocation systems	233
4.4.2.	Oxidoreductive systems	235
4.4.3.	Binding proteins	238
4.5.	Distribution and role of nonelectrolyte transport	239
Synopsis		240
5.	Transport of ions (K. J.)	243
5.1.	Equilibria of ions	243
5.1.1.	A simple membrane equilibrium and membrane potentials	243
5.1.2.	Gibbs-Donnan equilibrium	247
5.1.3.	Diffuse electrical double layer	250
5.2.	Electrodiffusion and membrane potentials	256
5.2.1.	Introduction	256
5.2.2.	The electrodiffusion equation—general considerations	259
5.2.3.	Schlögl's (1964) derivation of the general differential equation of electrodiffusion	262
5.2.4.	Henderson's equation—potential difference across a continuous layer with constant concentration gradients of individual ions	269
5.2.5.	Planck's procedure—potential difference across a microscopically electroneutral continuous layer	271
5.2.6.	Goldman's procedure—potential difference across a continuous layer with constant field	279
5.2.7.	Constant-field equation for potential difference across the whole membrane	282

5.2.8.	Constant-field equation for steady-state membrane potential in the presence of an electrogenic sodium pump	284
5.2.9.	The Hodgkin-Horowicz equation	286
5.3.	Chemical nature of ion-translocating systems (A. K.)	288
5.3.1.	Na,K-Adenosinetriphosphatase	288
5.3.2.	Ca-Adenosinetriphosphatase	293
5.3.3.	Other adenosinetriphosphatases	294
5.3.4.	Ion-binding proteins	294
5.3.5.	Transport of ferric ions	295
5.3.6.	Ionophores	296
5.4	Distribution and role of ion transport	301
Synopsis		304
6.	**Transport of water (K. J.)**	307
6.1.	Steady-state thermodynamics of water permeation	307
6.2.	The state of water in cells	314
Synopsis		316
7.	**Transport by special mechanisms (A. K.)**	317
7.1.	Oligopeptide permeases	317
7.2.	Pinocytosis	318
7.3.	Uptake of nucleic acids and special proteins	320
Synopsis		323
References		325
Subject index		339

1. INTRODUCTION

"Whence is it that Nature doth nothing in vain; and whence arises all that Order and Beauty which we see in the World?"

Isaac Newton, Opticks

The principal aim of the present book is to show how the interdisciplinary scientific branch of steadily growing importance, which may be called biological membranology, explains phenomena observed on biological membranes in terms of physics, chemistry and biology. This introduction follows a similar purpose, limiting itself, however, to considerations of the most general nature; its second section represents a speculative attempt to evaluate the biological survival value of membrane transport and the third considers the transport as a physical phenomenon taking place in space and time, whereby certain general characteristics of its physical description by mathematical formulae are predetermined. Still, although most diverse developments in the whole realm called by our ancestors natural history and natural philosophy are thus seen to contribute directly or indirectly to membranological theories, some of the scientific exploits of the past are especially closely related to them. The first section of this introduction mentions briefly these milestones in the early history of membranology.

1.1. IMPORTANT EVENTS IN THE HISTORY OF MEMBRANOLOGY

It may be a relatively fair estimate to state that the science of membranology began with German physiologists and botanists at about the middle of the last century. In the forties, one of the celebrated pupils of the great physiologist Johannes Peter Müller (1801 to 1858), Emil Heinrich Du Bois - Reymond (1818 – 1896) describes the electrical potential difference across surviving frog skin (Untersuchungen über tierische Elektricität, 1. Band, Berlin, 1848) and in 1851 the German physiologist Hugo von Mohl (1805 – 1872) describes plasmolysis of plant cells, assuming that their cell wall functions as a membrane (Grundzüge der Anatomie und Physiologie der vegetabilischen Zelle, Braunschweig). In 1855, the Swiss-born botanist Karl Wilhelm von Nägeli (1817 – 1891) explains the osmotic behavior of cells by the presence of a semipermeable cell membrane (Nägeli, K. und Cramer, K.: Pflanzenphysiologische Untersuchungen, Zürich) and in the same year the physiologist Adolf Eugen Fick (1829 – 1901) derives the phenomenological laws of diffusion (Über Diffusion, *Poggendorffs Annalen*, **94**, 59). In 1877, the botanist Wilhelm Pfeffer (1845 – 1920) publishes his "Osmotische Untersuchungen" (Leipzig) in which he postulates the existence of the cell membrane on the basis of similarities in the behavior of cells and osmometers with artificial semipermeable membranes (from deposited copper ferrocyanide, prepared shortly before by Moritz Traube).

In the eighties, the Dutch botanist Hugo Maria de Vries (1848 to 1935) continued with osmotic studies on plant cells, believing that the whole protoplasmic layer between the plasmalemma and the tonoplast functions as a membrane (e.g., Plasmolytische Studien über die Wand der Vacuolen, *Jahrbuch wiss. Botanik*, **16**, 465). His research served as a basis of the physicochemical theories of osmotic pressure and electrolyte dissociation by the Dutchman, Jacobus Hendricus van't Hoff (1852 – 1911), and the Swede, Svante August Arrhenius (1859 – 1927). In 1888, an equation for the liquid junction potential was derived by the German physicist and chemist Walther Hermann Nernst (1864 – 1941); in 1890, the German physical chemist and philosopher Wilhelm Ostwald (1853 – 1932) stressed the probable role of membranes in the generation of bioelectric phenomena. Between the years 1895 and 1902, Ernst Overton (1865 – 1933) measured the permeability of cell surfaces to a great number of substances and demonstrated the relation between their permeability and lipid solubility (e.g., Über die osmotischen Eigenschaften der lebenden Pflanzen- und Tierzelle, *Vierteljahrsschr. Naturforsch. Ges. Zürich*, **40**, 159, 1895; Beiträge zur allgemeinen Muskel- und Nervenphysiologie, *Pflügers Arch.* **92**, 115, 1902.

Edward Waymouth Reid (1862—1948) appears to have been the first to introduce a split chamber to study transepithelial transport (Transport of fluid by certain epithelia, *J. Physiol.* 26, 436, 1901). In 1902, Julius Bernstein (1839—1917) explains the electrical phenomena in living cells by a membrane hypothesis (Untersuchungen zur Thermodynamik der bioelektrischen Ströme, *Pflügers Arch.* **92**, 521). The existence of the cell membrane was later proved beyond any doubt by experiments of Rudolf Höber (1873—1953), who demonstrated a high electrical resistance of whole erythrocytes and a low one of their interior (e.g., Physikalische Chemie der Zelle und der Gewebe, 6th edition, Leipzig, 1926) and by experiments of Robert Chambers (1881—1957), who demonstrated by microinjections that substances not permeating easily across the cell surface diffuse freely in the cell interior (A microinjection study on the permeability of the starfish egg, *J. Gen. Physiol.* **5**, 189, 1922). Studies on artificial membranes date back to the pioneering work of Leonor Michaelis (1875—1949) who used collodion membranes during his stay in Nagoya, Japan, in the early twenties (see, e.g., *Kolloid Z.*, **10**, 575, 1933) to demonstrate their selective permeability. Likewise, the work of Karl Sollner on membranes started as early as 1930 (*Z. Elektrochem.* **36**, 36, 234, 1930).

With P. J. Boyle and Edward J. Conway, who observed the similarity between distribution of potassium ions in cells and the Gibbs—Donnan equilibrium (Potassium accumulation in muscle and associated changes, *J. Physiol.* **100**, 1, 1941), with H. B. Steinbach who showed that the cell membrane is permeable to sodium ions (Sodium and potassium in frog muscle, *J. Biol. Chem.* **133**, 695, 1940), and with the concept of a sodium pump extruding sodium ions from muscle fibers, as proposed by R. B. Dean (Theories of electrolyte equilibrium in muscle, *Biol. Symp.* **3**, 331, 1941) we are approaching the discovery of active transport and the stage of biological membranology, which forms the subject matter of the present book.

1.2. EVOLUTION AND TRANSPORT

There is little doubt that the transport faculties of living cells and organisms developed in the same way as any other of their useful properties, i.e. as a result of spontaneous random mutations which have afterwards proved their survival (adaptive) value in the process of natural selection. A rather powerful argument in favor of this proposition can be recognized in the fact that in a number of cases elaborate transport mechanisms capable of concentrating nutrients in cells can be seen to disappear when they are of little or no survival value, a satisfactory inflow of these substrates being ensured for long evolutionary periods by some other means. This is obviously the case with a number of organ cells and with anucleate erythrocytes as well as with some species of yeast cells cultivated for millenia.

In this connection we would like to draw the attention to two rather common features of the evolutionary process.
(*1*) The first of these principles, even if its occurrence may not be a logical necessity, is obviously often encountered in the process of evolution: *a single problem of adaptation is often solved in a number of different ways.* We can cite, e.g., the diversity of devices used by various organisms to solve the problem of locomotion or of direct as well as indirect utilization of the sun radiation energy.
(*2*) The second principle appears to be an obvious corollary of the combination of random mutations with natural selection: *There may be a number of adaptive values of a given single mutation and not all of them may be apparent at the same time.* Thus a mutation which was seemingly useless at the time of its appearance may (provided that it was not lethal) prove advantageous in a number of ways when the organism itself and/or its environment is changed.

The lack of uniformity in the mechanisms responsible for transport of nonelectrolyte nutrients across cell membranes may well serve as an illustration of the first principle – simple diffusion, mediated diffusion as well as active transport may be encountered in this field (see, e.g., Kotyk, 1973, and section 4.5. of the present book). Already the simple diffusion across a membrane of lipoid character seems to be an important factor rendering the basic biochemical processes much more efficient. Most important intermediate metabolites (like the organic acids of Krebs' cycle) are very polar compounds and hence they diffuse only slowly across lipoid membranes and do not readily escape from cells, so that sequences of biochemical reaction steps inside the cells are possible (Kubišta, 1974). Concerning the more sophisticated mechanisms, it has been proposed (Holden, 1968) that evolution of specific and even active (i.e., uphill-accumulating) transport mechanisms possibly preceded the evolution of biosynthetic pathways, "or occurred at the same time as an alternative solution to the stress of nutrient depletion". Indeed, an increase in the rate of passive transport of the required nutrient, achieved by the mechanism of a specific equilibrating transport, or even the capability of an active transport system to maintain the intracellular concentration of the nutrient at a higher level than that in the environment, certainly helps to sustain a higher rate of growth than in cells which are less endowed. The abundant growth of organism so developed then might result in a still more pronounced depletion of the environment of the given substrate, with the result that mutants possessing more efficient transport mechanisms (e.g., those with a higher affinity for the substrate) or, alternatively, capable of converting a more abundant precursor into the growth-limiting nutrient, will have a selective advantage. If, on the other hand, the transport mechanisms did not develop at this early stage of evolution, they were likely to develop only much later. For, it is argued, if the biosynthetic enzymes have a lower level of structural complexity than the transporting mechanisms and hence appeared earlier in evolution, a selective advantage may be obtained only by sparing the energy required by a complex biosynthetic pathway under conditions where the biosynthesis is already highly developed and the environment partly re-enriched in organic substrates (Holden, 1968). Be it as it may, "there are more advanced cell types where most unnecessary properties have been weeded out (like the plurality of carrier systems for a single compound in many unicellular organisms), and there are the less advanced cell types which do not yet possess the obviously efficient coupled transports of higher organisms and particularly of their specialized tissues" (Kotyk, 1973).

The second principle, that of a single mechanism of mutational origin serving a multiple biological purpose is perhaps best apparent in the case of the most wide-spread active ion-transporting mechanism, the so-called sodium pump. The hypothesis that the active sodium transport is primarily a volume regulating device appears to be ill-founded now. Not only are sodium pumps operative in plant cells in which the problem of volume control is entirely solved by the presence of a rigid cell wall, but also in animal cells a sodium-independent regulation of the cell volume was demonstrated (Rorive *et al.*, 1972). There are, however, a number of other ways in which sodium pumps are useful for cells and organisms. The gradients of the sodium electrochemical potential serve as energy reservoirs in excitability phenomena as well as in some of the so-called secondary active transports of sugars and amino acids (see section 5.4.). Several intracellular enzymes are activated by potassium ions; sodium ion activation of such enzymes is much rarer or even negative (see, e.g., Ussing, 1960). Although it may not be easy to decide which of the two properties developed first, whether the sodium extrusion from cells or the potassium activation of intracellular enzymes, their combination could represent a useful device of metabolic control. May it not be that the basal metabolism of some cells in the cold is reduced not only by the cold itself but also due to a cold-induced sodium inflow into cells? Finally, homeostasis of whole organisms with respect to salt and water is maintained by specialized tissues in which full use is made of the ability of cells to transport sodium actively – without an active sodium transport the migration of life from seas to fresh water and on dry land would be hardly possible. Thus not only are the transport mechanisms a product of evolution, they are also its integral part; without membrane transport the whole evolution as we know it would be unthinkable.

1.3. TRANSPORT IN SPACE AND TIME

Physical, chemical and biological phenomena of membrane transport take place in space and time. A description of these phenomena is hence bound to include the fundamental physical concepts of space and time explicitly or implicitly. The aim of the present section of the introduction is to characterize the mathematical means used to describe transport phenomena and to point out how they reflect the properties of space and time variables.

As in other branches of natural science, statements of qualitative or semiquantitative nature (like "the substance does or does not permeate a given membrane" or "the substance is accumulated on one side of the membrane") are only an initial step toward a quantitative description of membrane transport phenomena. From the time of Newton and Leibniz, scientists have attempted to derive quantitative laws by establishing, from fundamental principles, such relations between states as are only infinitesimally distant in space and time. Thus, instead of trying to comprehend in one step the laws governing various processes as a whole, they are looking first for microscopic laws of evolution known as differential equations. By establishing relations between states neighbouring in time, the differential equations are an expression of the deterministic character of natural phenomena, of causality in nature.

There are two different types of differential equations, partial differential equations, containing more independent variables, and ordinary differential equations, containing only one independent variable. It might seem that the behavior of quantities which are of interest in transport studies, e.g., of concentrations, is properly described only by partial differential equations, quantities of this type having a definite value at each point of space (given by three independent variables corresponding to the coordinates of Euclidean space) and being, moreover, dependent on a fourth independent variable, on time. Still, in the description of membrane transport we often encounter ordinary differential equations which are genetically related to the description of movement of indestructible particles, the coordinates of which are solely functions of time. Actually, they will appear as plausible approximations of the appropriate partial differential equations, the dependence of, say, concentration, on space variables being replaced with the qualitative notions of "inside" and "outside" of the membrane-surrounded compartment. The plausibility of such an approach rests in this case on the assumption that equilibration in the media adjacent to the membrane proceeds at a much faster rate than the transport across the membrane itself.

A new and possibly very promising approach, alternative to the straightforward description of the membrane transport by ordinary differential equations, has emerged quite recently in the formalism of network thermodynamics (Kedem, 1972; Oster *et al.*, 1973). Using a suitable algorithm, differential equations describing a system can be obtained directly from the network graphs; the graphs, at the same time, reveal the topology of the system (the way in which individual parts of the system are interconnected) and thus they contain more information than the differential equations themselves (Oster *et al.*, 1973).

It will be seen that in a number of cases the differential equations need not be actually integrated to obtain valuable mathematical formulae governing membrane phenomena. At least in two important instances are the differential equations solved by other means.

(*1*) The time derivative is approximated by the quotient of a finite change of the quantity and of the elapsed time. Typically, the so-called initial velocities of transport are measured under various conditions and then tested graphically in order to see whether they satisfy this or that differential equation.

(*2*) In biological systems, the time-independent steady states play as important a role as thermodynamic equilibria do in other physical systems. For both these stationary states, the time derivatives are zero and the differential equations become algebraic equations.

Ordinary differential equations commonly encountered in the field of membrane transport show the property of being autonomous, i.e., the time derivatives occurring in them are not, or need not be, explicit functions of time. This property is both convenient and interesting. Convenient, for it allows the nature of the solutions to be studied even when they are not accessible in an analytical form, using the qualitative theory of differential equations. The time differential is easily excluded from the sets of such equations and relations between the remaining variables may be studied in a phase plane or phase space. It is interesting, too, since it points to an important characteristic of time, its homogeneity. Any process described by an autonomous differential equation can be reproduced at any time, the only prerequisite being that all other conditions be reproduced — the law expressed by an autonomous differential equation is eternally valid and applicable. The existence of laws of this kind is somewhat less obvious in biology than in physics; it may be argued that, due to the evolution of the organic world and to the development of individual organisms, the biological and physiological time is less homogeneous than the physical one. However, the need to account for such phenomena by nonautonomous equations is probably quite remote.

Whereas the time used in the description of membrane phenomena may be homogeneous to a very good approximation, it certainly lacks another of the symmetry properties, the isotropy, i.e. the property of being equivalent in all directions. Time flows in one direction only; transport phenomena are accompanied by dissipation of energy and once they have taken place they can never be returned completely, without leaving some changes on the surroundings of the system studied. It is for this reason that thermodynamics is of great value in considerations of membrane transport.

Space, too, is homogeneous and the same observations on membrane transport are found to be reproducible all over the world (or, at least, one wishes them to be). As for the other symmetry property, empty space is obviously isotropic but space filled with matter (e.g., a membrane) need not be isotropic and may thus show different properties in different directions. This consideration is of importance when the so-called active transport, a flow of substance across a membrane coupled to a chemical reaction in the membrane or in its proximity, is considered. According to the principle called the Curie–Prigogine principle (see, e.g., Katchalsky and Curran, 1965), in an isotropic medium there cannot be any direct coupling between vectorial and scalar quantities so that a vectorial flow of a substance cannot be directly driven by a scalar affinity of a chemical reaction In an anisotropic membrane, however, such coupling is permissible.

2. MEMBRANES

> *"And where it is of a less thickness the Attraction may be proportionally greater, and continue to increase, until the thickness do not exceed that of a single Particle of the Oil. There are therefore Agents in Nature able to make the Particles of Bodies stick together by very strong Attractions. And it is the Business of experimental Philosophy to find them out."*
>
> Isaac Newton, Opticks

The view that all phenomena of life are more or less intimately associated with the existence of cell membranes has become a powerful candidate for superseding the often quoted dictum of F. Engels that life is a form of existence of proteins. Indeed, various types of membranes serve as the locale of most enzyme activities of a cell, provide the physical basis of locomotion, intercellular contacts, secretion and uptake, are involved in protein synthesis and cell division. To explore the manifold functions of cell membranes would mean to write a book on biochemistry, physiology and biophysics combined, something that no serious scholar could attempt at this age of specialization. Still, while bearing in mind the title of the book and concentrating on one particular function of cell membranes, *viz.* the transport of ions and molecules, we shall first briefly review the existing knowledge of the structure and function of biological membranes in general.

The word "membrane" has been used for over a century (*cf.* von

Mohl, 1851; Nägeli, 1855) to designate the boundary of a cell serving both as a mechanical barrier between the cell interior and its surroundings and as a semipermeable partition permitting the passage of water and of some but not all solutes. While the last-named function is at the core of most investigations into the transport of substances, the first, mechanical attribute is now recognized to belong rather to structures that lie externally to the cell membrane proper, namely the various types of wall-like structures to be discussed later (*cf.* p. 133). On the other hand, membranes of one kind or another have been discovered to form the structural framework of the whole cell and give its architecture a unique common denominator.

There are perhaps two principal features that permit to distinguish a membrane from other morphological cell constituents: (*1*) The typical trilaminar, "railroad-track" appearance seen after negative staining in an electron microscope, (*2*) the high lipid (particularly phospholipid) content, accompanied by varying amounts of protein and only occasionally by rather minute amounts of other types of compounds, especially polysaccharides and, in some special membranes, possibly by nucleic acids.

2.1. SURFACES AND INTERFACES

The striking stability of membranes in aqueous solutions and their spontaneous formation from suitable mixtures of lipids with water has been at one time considered analogous to the spreading of monomolecular films of an immiscible substance lighter than water (typically a lipid) on the surface of water. It is well known (e.g. Langmuir, 1933) that such films can exist in several different states of molecular freedom, that can be likened to gaseous, liquid and solid states of matter. If the surface area of the solvent (usually water) is large enough, the second-component molecules tend to disperse randomly, with little or no interaction between them. Using a larger-size fatty acid, the mean area at the disposal of a "floating" molecule will be above 0.5 nm^2. These monolayers correspond to low surface pressures, such as exerted in a Langmuir trough, a device employed to spread or compress surface films and measure the force exerted by the film in a given state to oppose compression. The pressures in this type of spreading are generally less than 0.001 N m^{-1} and the situation may be considered as the "gaseous" state. It is properly described by a "two-dimensional" gas law which may be written as

$$\pi(A_m - A_0/N) = 10^5 \, kT \qquad (2.1)$$

where π is the surface pressure in $N\,m^{-1}$, A_m is the molecular area in nm^2, A_0 is the limiting molar area, N is the Avogadro constant ($6.023 \cdot 10^{23}\,mol^{-1}$), k is the Boltzmann constant ($1.38 \cdot 10^{-23}\,J\,K^{-1}$) and T is the absolute temperature. Hence a plot of π versus A_m will yield an equilateral hyperbola (part **A** of Fig. 2.1.)

It the surface pressure is increased, interactions between the floating molecules will occur and a coherent film will be formed on the surface, the area taken up by a molecule corresponding roughly to $0.5\,nm^2$ at the highest compression, i.e. several times the smallest cross section of the molecule. This situation, corresponding to the liquid three-dimensional state, is termed expanded monolayer and exists at surface pressures of $0.001-0.01\,N\,m^{-1}$. This state is properly described by

$$(\pi - \pi_0)(A_m - A'_0/N) = 10^5 \, k\mathrm{T} \qquad (2.2)$$

where π_0 is the surface pressure at the transition from the gaseous state and A'_0 the limiting molar area under the new pressure conditions (part **B** of Fig. 2.1).

If the pressure is increased further (beyond $0.01\,N\,m^{-1}$) the molecules on the surface will tend to arrange themselves in a pattern of closest packing, in the case of fatty acids, aldehydes, alcohols, phospholipids, etc., in such a way that the polar heads of the molecules will be immersed in water while the hydrocarbon chains will stick out into the surrounding air. This packing pattern, termed condensed monolayer, still has some degree of translational motion since the heads are hydrated and have a larger cross-sectional area than the hydrophilic tails. The equation describing the situation is then

$$\pi = ask(A''_0/N - A_m) \qquad (2.3)$$

where a is the activity of the solute ($a = f_a c$, c being concentration, f_a the activity coefficient), s the surface area of the solution and k a constant (part **C** of Fig. 2.1).

A further increase in pressure will result in stripping the polar heads of their solvation water molecules and in the formation of a condensed monolayer which is solid and practically noncompressible. The equation describing this situation is like (2.3) with a different constant replacing k (part **D** of Fig. 2.1). That this particular type of

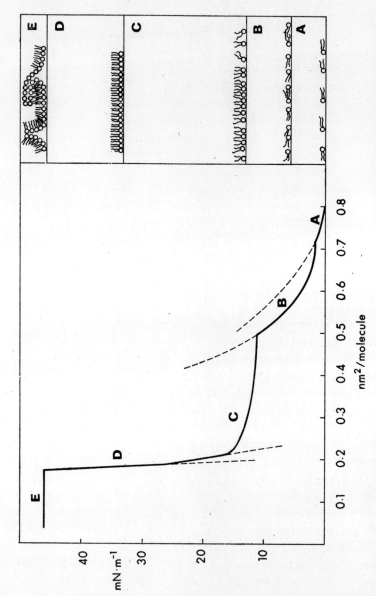

Fig. 2.1. A generalized force-area curve for a long-chain fatty acid. The right-hand panel shows schematically the position of the lipid molecules at the air–water interface. Explanation of the curve segments is given in the text.

packing occurs is supported by the fact that in a series of homologues of amphiphiles the cross-sectional area per molecule (about 0.2 nm^2) is practically independent of the hydrocarbon chain length.

If the limiting pressure is exceeded, the monolayer will gradually collapse, giving finally rise to a multilayered structure as in part **E** of Fig. 2.1. The equation describing the situation is

$$\pi = ask''. \qquad (2.4)$$

The ability to form surface monolayers is not limited to lipids, proteins being known to spread in films on the surface of solutions (especially ammonium sulfate). However, the spreading of protein (achieved artificially) often results in a collapse of the tertiary structure or at least in partial reversible unfolding. In fact, mixed monolayers, both of miscible and immiscible components, can be prepared, the latter forming larger or smaller patches on the liquid surface, corresponding to separate phases.

It is of practical interest that surface monolayers have been used successfully for the determination of molecular weight both of proteins and lipids. If πA is plotted against π (A being the total area of a Langmuir trough), one obtains for the "gaseous" phase, i.e., at low solute concentrations, a straight line with a slope of nA_0, intersecting the vertical axis at nRT; this yields the number of moles n, as well as the limiting molar area A_0.

Monolayers need not form only on the surface of a liquid. If a suitable proportion of lipid, organic solvent and water (e.g., 50 mg dipalmitoyllecithin, 2 mg steroid, 0.5 ml benzene-chloroform of density equal to one, 50 ml 2 mM CsCl) is sonicated (e.g., 10 min at 45 °C with 70 W power) spherical vesicles about 50 nm in diameter are formed which enclose the organic solvent (Sackmann and Träuble, 1972).

The tendency to form monolayers on the surface is associated with the degree of immiscibility of the solute in the solvent and the ability to alter the surface tension of the solution. The surface pressure exerted by the bar in a Langmuir through is not to be confused with surface tension which is a qualitative property of liquids at the boundary with a gaseous phase and is due to the tendency of liquids toward cohesion. An analogous, interfacial, tension exists at the boundary between two immiscible liquids or a liquid and a solid. Every liquid tends to take up the smallest possible volume and hence form spheres but when dealing with larger amounts of liquids this trend is overcome by the pull of gravity which causes the liquid to distribute itself flat. The two quantities are related by $\pi = \sigma_0 - \sigma$, so that the surface pressure π is equal to the difference in the surface tension of the pure solvent σ_0 and of the solution σ. The surface tension, like the surface pressure, is expressed in dynes per cm or preferably in newtons per m (Table 2.1).

TABLE 2.1. **Values of interfacial tension σ of some liquids at 20 °C (in N m^{-1})**

		Liquid: air	
Liquid	σ	Liquid	σ
Mercury	0.484	Toluene	0.028
Water	0.073	Carbon tetrachloride	0.027
Glycerol	0.064	Acetone	0.023
Aniline	0.043	Ethanol	0.022
Carbon disulfide	0.034	Hexane	0.018
Benzene	0.029	Diethyl ether	0.017

		Liquid: water	
Liquid	σ	Liquid	σ
n-Octane	0.051	Benzene	0.035
Carbon tetrachloride	0.045	Aniline	0.006

The surface tension decreases linearly with increasing temperature to a point lying close to the critical temperature of the liquid to disappear completely at the critical temperature (the temperature above which a given compound cannot exist as liquid no matter what pressure be applied — e.g., it is 647 K for water and 304 K for carbon dioxide). What is more important in the context of film formation is the consequence of the Gibbs surface isotherm derived for the relationship between surface concentration and changes in surface tension with the bulk concentration of solute. The internal energy change dU is expressed in terms of change of entropy dS, change of volume dV at a given pressure p, surface tension σ, surface area A

and chemical potentials μ of the two components, n'_1 and n'_2, thus:

$$dU = T\,dS - p\,dV + \sigma\,dA + \mu_1\,dn'_1 + \mu_2\,dn'_2. \quad (2.5)$$

The primed n's refer to concentration in a column of solution including the surface layer and are related to concentration in the same volume of solution excluding the surface layer in the following manner:

$$\Gamma = \frac{n' - n}{A} \quad (2.6)$$

Γ being called the surface excess of moles of solute per unit surface area.

Equation (2.5) can be integrated and then differentiated completely to yield

$$dU = T\,dS + S\,dT - p\,dV - V\,dp + \sigma\,dA + A\,d\sigma + \\ + \mu_1\,dn'_1 + n'_1\,d\mu_1 + \mu_2\,dn'_2 + n'_2\,d\mu_2$$

whence (by comparison with eq. 2.5)

$$S\,dT - V\,dp + A\,d\sigma + n'_1\,d\mu_1 + n'_2\,d\mu_2 = 0$$

so that, at constant temperature and pressure,

$$A\,d\sigma + n'_1\,d\mu_1 + n'_2\,d\mu_2 = 0. \quad (2.7)$$

Inside the solution, the Gibbs–Duhem equation will hold:

$$n_1\,d\mu_1 + n_2\,d\mu_2 = 0. \quad (2.8)$$

The equilibrium condition is obtained by subtracting eq. (2.8) after multiplying it with a suitable constant such that $n'_1 = kn_1$, from eq. (2.7):

$$A\,d\sigma + (n'_2 - kn_2)\,d\mu_2 = 0. \quad (2.9)$$

Rearrangement of this equation gives

$$-\frac{d\sigma}{d\mu_2} = \frac{n'_2 - kn_2}{A} = \Gamma_2. \quad (2.10)$$

Since n_1 does not enter the equation, the surface concentration of component 2 ($= \Gamma_2$) is independent of volume. The chemical potential change $d\mu_2$ being equal to $RT\,da/a$, eq. (2.10) becomes

$$\Gamma = -\frac{a\,d\sigma}{RT\,da} \quad (2.11)$$

where a is the activity of solute in the given solution.

Equation (2.11) is the Gibbs surface isotherm (or adsorption equation) and shows that compounds decreasing the surface tension will concentrate at the surface since if $d\sigma/da$ (sometimes called the surface activity of a substance) is negative, $\Gamma(=\Delta a)$ will be positive. Thus, surface-active agents (soaps, detergents, etc.) will decrease the surface tension by accumulating on the surface of a liquid. The actual value of Γ can be arrived at from an empirical equation due to von Szyszkowski (1908)

$$\sigma_0 - \sigma = \alpha \ln(1 + \beta a) \tag{2.12}$$

where σ_0 is the surface tension of pure solvent, σ that of the solution, α and β are constants relating to the solute. Derivation of eq. (2.12) yields

$$-d\sigma/da = \alpha\beta/(1 + \beta a)$$

which may be substituted in (2.11) to give

$$\Gamma = \alpha\beta a/RT(1 + \beta a). \tag{2.13}$$

(The similarity with a Langmuir adsorption isotherm is obvious on rearranging to $\Gamma = (\alpha/RT) \cdot a/(1/\beta + a)$.) The value of α is identical for a given homologous series (e.g., 12.95 for monocarboxylic fatty acids) while the value of β depends strikingly on the molecular species (e.g., it is 6.1 for propionic acid but 2 327 for caproic acid). It may be seen that Γ tends toward a limiting value equal to α/RT as concentration is raised and that it is reached the sooner the greater the value of β which may be compared with the surface activity of the given compound.

It should be realized that the surface monolayer film of a lipid on water, although superficially resembling a biological "half-membrane" and although being identical with one half of an artificial black membrane (see p. 95), is a far cry from a true biological membrane. One arrives somewhat closer at what appears like a typical railroad-track membrane if one increases vastly the concentration of lipid in the lipid-water system, particularly when working again with amphipathic lipids, i.e., those that contain both hydrophilic and hydrophobic portions. As the "concentration" of lipid in the system rises, there will be not only a superficial film formed corresponding to eq. (2.13) but the lipid molecules will begin to form aggregates within the solution, the so-called micelles. These clusters may contain tens or hundreds of lipid molecules arranged as shown in Fig. 2.2. The solution may still appear transparent but light scattering and particularly X-ray scattering will reveal the presence and size of the aggregates formed. Their shape may range from spheres to ellipsoids to rods and generally the higher the concentration the greater the participation of rods (but there are examples where spheres persist at all concentrations, e.g., with palmityltrimethylammonium chloride; or where rods are found at all concentrations, e.g. with sodium oleate).

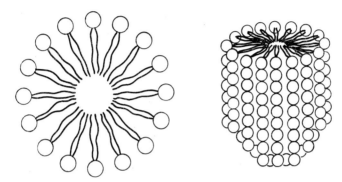

Fig. 2.2. A schematic drawing of micelles of amphiphilic molecules in an aqueous solution.

If the concentration of lipid is increased further (generally above 50% v/v) an abrupt change in viscosity occurs and the system becomes anisotropic. The organization of lipids is now one of several patterns, containing periodic repeats in one or two dimensions and is called mesomorphic or liquid-crystalline (in contrast with crystalline structure where the periodic repeats occur in three dimensions). The word smectic is occasionally used to describe this semicrystalline state (from Greek smektikos = cleansing, purifying, since the structure was first described in detergent mixtures); the smectic phase contains parallel planes or sheets of aggregated molecules (in contrast with the nematic state, rather rare indeed, where parallel chains or strands are formed). The sequence of semicrystalline phases, in the order of increasing lipid concentration, is hexagonal − deformed hexagonal − rectangular − complex hexagonal − cubic − lamellar but rarely do all the patterns occur. Thus, sodium stearate will pass from hexagonal to deformed hexagonal, complex hexagonal and lamellar, while sodium oleate from hexagonal to rectangular, complex hexagonal and lamellar. The arrangement of lipid molecules in the various patterns is shown in Fig. 2.3. The structure designated as deformed hexagonal is surmised from a particular set of reflections in X-ray diffraction, the one called cubic is postulated from analogy with the structure of anhydrous soaps. Distinction between hexagonal type I and hexagonal type II is possible on the basis of intensity of X-ray reflections and also from the fact that in type I the area taken up by a given molecule is almost independent of the hydrocarbon chain length and lipid concentration

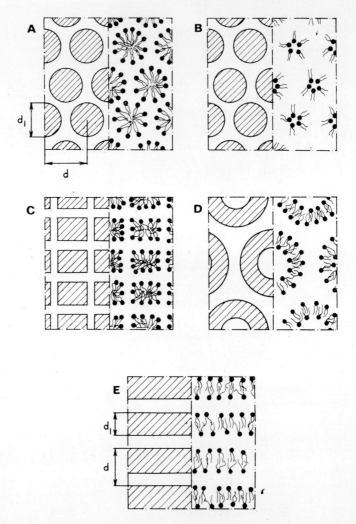

Fig. 2.3. Schematic structure of some liquid-crystalline phases of lipid−water systems. **A** Hexagonal I, **B** hexagonal II, **C** rectangular, **D** complex hexagonal, **E** lamellar. d Repeat distance, d_1 diameter of lipid cylinder or lamella. (Adapted from Luzzati, V. and Husson, F. (1962). *J. Cell Biol.* **12**, 210.)

whereas in type II the area is a complex function of both. Type I is found in water mixtures of lysolecithin or soaps, type II in water mixtures of phosphatidyl ethanolamine, phosphatidyl choline, mitochondrial lipids and other more biological substances.

At very low water contents (about 10−15%) lipids tend to form gels or coagels which may also have a lamellar structure, the coagel containing two phases, liquid water and crystalline lipid, unlike the gel which is homogeneous but usually metastable (Vincent and Skoulios, 1966).

It is to be noted that transitions similar to those described for lipid-water mixtures occur in anhydrous lipids, particularly soaps, as temperature is decreased and an isotropic (possibly micellar) liquid passes over to a liquid crystalline state and finally to a solid crystalline state (Luzzati, 1968).

The phase diagrams of a number of lipid-water mixtures have been obtained, some of them rather complicated. As an example, based on the work of Abrahamsson and co-workers (1972), the system of sodium sulfatide and water is shown in Fig. 2.4.

In many cases, the existence of a given repeating structure is detectable without the aid of X-ray diffraction or other sophisticated techniques. Electron microscopy of a phosphotungstate-fixed mixture

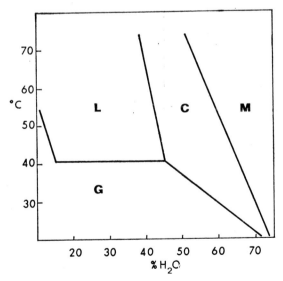

Fig. 2.4. Phase diagram of sodium sulfatide in water. L Lamellar phase, C cubic phase, M micellar phase, G gel phase. (Adapted from Abrahamsson et al., 1972.)

of lecithin with water or an alkylphosphoric acid will reveal stacked or concentric lamellae, their repeat distance being 5—10 nm, depending on the amount of aqueous phase present. These "membranes" have been called myelin figures because they resemble the multilayered myelin sheath of a nerve axon (Fig. 2.5; cf. Fig. 2.35E). In other cases, hexagonal arrays are formed (Fig. 2.6).

In all the patterns assumed by lipids in water the obvious tendency is to ensure the maximum polar-to-polar and nonpolar-to-nonpolar interactions, this being dictated by the necessity of forming patterns with the lowest free energy content. The principal factor active in steering the molecules into their positions are the hydro-

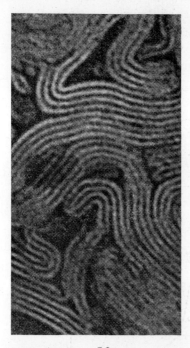

Fig. 2.5. Myelin figures—a negatively stained preparation of lecithin and cholesterol. (Taken with kind permission from Glauert, A. M. and Lucy, J. A. (1968). In: *The Membranes* (ed. by Dalton, A. J. and Haguenau, F.), p. 1. Academic Press, New York—London.)

——— 50 nm

Fig. 2.6. **A** Hexagonal phospholipid structure in 3 % water. The dark spots are probably aqueous cylinders surrounded by lipid. (Taken with kind permission from Stoeckenius, W. (1962). *J. Cell Biol.* **12**, 221.)
B The hexagonal phase of a lecithin-cholesterol-saponin complex. The light rings appear to be composed of subunits (arrow). (Taken with permission from Glauert and Lucy, as in Fig. 2.5.)

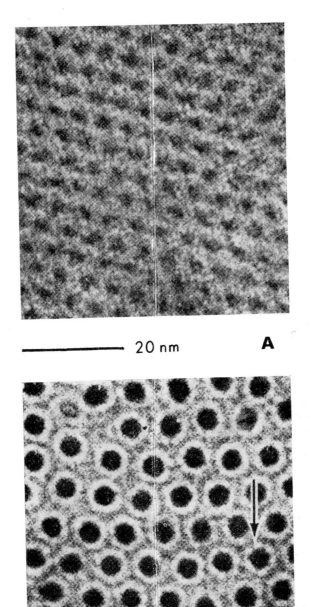

20 nm A

20 nm B

phobic interactions (Kauzmann, 1959). To demonstrate the nature of these interactions, let us take the simple example of methane distributed between benzene and water. The change in free energy (properly, the unitary free energy, i.e. the total free energy minus translational entropy, the so-called cratic entropy) on going from benzene to water is 10.9 kJ mol^{-1}, reflecting the higher solubility of methane in benzene than in water. However, the change in enthalpy is negative, -11.7 kJ mol^{-1}, so that it is the negative entropy change, -75 J K^{-1} mol^{-1} at 25 °C, which makes the methane reside rather in benzene than in water. The entropy decrease is apparently associated with an ordered arrangement of water molecules about the methane, either in a cage form (clathrate) or not. The same consideration holds for the "dissolving" of lipids (or proteins or any kind of high-molecular compound).

With phospholipids, carrying both positive and negative charges at different parts of their polar heads, electrostatic interaction acts as an additional factor in the orientation of their molecules in an aqueous medium. In fact, of naturally occurring lipids, it is only the phospholipids that can form extensive bilayer structures oriented with their polar heads toward water and their hydrocarbon tails toward each other.

Van der Waals forces make a relatively minor contribution to the structural patterns of lipids in water, perhaps most importantly by ensuring the closest possible packing of the hydrocarbon chains of lipids with no gaps or spaces between them.

We have thus arrived at an artificial system containing only lipid and water but resembling the biological membrane in its appearance under the electron microscope as well as in its dimensions. However, it was pointed out at the beginning of this chapter that perhaps a half of the weight of biological membranes is due to proteins. How would then a protein—lipid—water mixture behave as compared with a lipid–water mixture? Especially X-ray diffraction studies have shown that two principal structural arrangements appear, just like in the protein-free mixtures: a lamellar one and a hexagonal one (Shipley *et al.*, 1969; Gulik-Krzywicki *et al.*, 1969).

Depending on the lipid-protein ratio, the spacing between repeating patterns will change, the assumption being that when more protein is present, there will be an increasing number of protein layers intercalated between layers of lipids. Analogously, the situation will develop in the hexagonal array (Fig. 2.7). Pertinent studies were

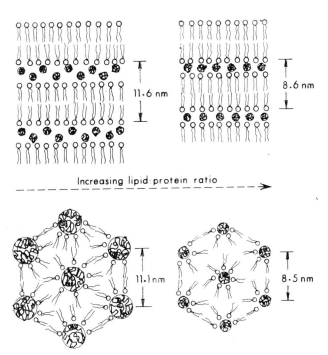

Fig. 2.7. Schematic representation of the lamellar and hexagonal phases formed in a mixture of cytochrome c, phospholipids and water. The repeat distance is shown to decrease as the relative amount of phospholipids increases. (Redrawn with permission from Shipley, G. G. (1973). In: *Biological Membranes* (ed. by Chapman, D. and Wallach, D. F. H.), p. 1. Academic Press, London—New York.)

undertaken with oxidized cytochrome c, lysozyme and serum albumin, and a variety of phospholipids. As demonstrated by the example shown in Table 2.2, both pH and temperature influence the lattice spacing but the range of lattice dimensions tallies with the dimensions of biological membranes estimated both by X-ray diffraction and by electron microscopy.

In the next chapter it will be seen that, although the range of biological membrane functions precludes a simple reconstitution from dispersed components, many of the characteristics described so far for purely artificial membrane-like structures are retained in cell membranes and may be vital for the maintenance of membrane structure as a whole. However, before making such comparisons, we should know more about the chemical composition and supramoecular organization of actual biological membranes.

TABLE 2.2. Characteristics of artificial mixtures of lysozyme—phosphatidyl inositol—water (from Gulik-Krzywicki et al., 1969)

pH	°C	Lattice type	Spacing nm
4	0—35	two-dimensional square	6.35
5.5—7	0—35	two-dimensional hexagonal	8.0—8.25
7	0	two-dimensional hexagonal	9.6
7	30	lamellar	10.6
7	40	lamellar	8.3
8	25	lamellar	10.6
8	40	lamellar	8.3

2.2. CHEMICAL COMPOSITION

The variety of chemical species occurring in biological membranes is impressive but more than 95% of the bulk in practically all membranes falls into two groups, lipids* and proteins. We shall now describe in some detail the constituents of the two groups and proceed from the static chemical composition to the structural arrangement in biological membranes.

The percent content of lipids and proteins in various membranes is shown in Table 2.3.

2.2.1. Lipids

2.2.1.1. Chemistry

Although not necessarily all lipid compounds found in nature occur in membranes (leaf waxes and fatty alcohols, insect cuticle waxes; reserve triglycerides in adipose tissue; steroids used as means of defense or as hormones, pheromones, isoprenoids in latex) the

---

* The term lipids is used here in the more physical sense of the word, designating hydrophobic, oil-soluble compounds. From the chemist's point of view, lipids should be probably restricted to esters of glycerol and sphingosine.

spectrum of lipids actually found in membranes is rather broad and includes practically all groups of lipids plus a number of complex substances of lipid-carbohydrate-protein nature.

There exist different types of classification of lipids: here we shall use a simple system, recognizing glycerol-substituted lipids, sphingosine-substituted lipids, sterols and other lipids.

TABLE 2.3. **Protein and lipid content of various membrane preparations (from Dewey and Barr, 1970; Suomalainen** et al.**, 1973; Wallach, 1972; Wolfe, 1964)**

Membrane	% Protein	% Lipid
Bovine myelin	22	78
Human erythrocyte[a]	49	44
Ehrlich ascites cell membrane	70	30
Liver plasma membrane	60	40
Intestinal villi	85[b]	15
Inner mitochondrial membrane	78	22
Outer mitochondrial membrane	55	45
Retinal rod outer segment	59	41
Brain synaptic vesicles	66	34
Rat liver microsomes[c]	62	32
Rabbit muscle microsomes[c]	54	22
Chloroplast lamellae	45	55
Micrococcus lysodeikticus[d]	50	28
Bacillus megaterium[e]	68	20
Pseudomonas aeruginosa	60	35
Halophilic bacteria	65	35
Acholeplasma	70	30
Baker's yeast plasma membrane[f]	46 (27, 35)	38 (46, 38)
Candida utilis plasma membrane[f]	39	40
Amoeba[f]	43	32
Myxoviruses[g]	73	20
Leukoviruses[h]	64	31

[a] Also about 7 % carbohydrate; [b] a great part of this may be nonmembrane protein; [c] the remaining material is apparently RNA from attached ribosomes; [d] contains 15—20 % carbohydrate; [e] contains 6—10 % carbohydrate; [f] the remaining material is mainly carbohydrate from adhering contamination by the cell wall; [g] also some 6 % carbohydrate and 1 % RNA; [h] also some 6 % carbohydrate and 2 % RNA.

A. Based on substituted glycerol

The backbone of these lipids is formed by the substituted trihydroxy-alcohol glycerol

$$\begin{array}{c} CH_2OR^1 \\ | \\ R^2OCH \\ | \\ CH_2OR^3 \end{array}$$

the three substituents R^1, R^2, and R^3 being generally different.

a. *Containing only* C, H, O

If R^1 is a fatty ester, $CH_3(CH_2)_nC=O$ ($n = 10-20$), R^2 and $R^3 = H$, the compound is a monoglyceride; if R^1 and R^3 are fatty esters, the compound is a diglyceride; if all three R's are fatty esters, it is a triglyceride.

If R^1 or R^2 is a fatty ether, $CH_3(CH_2)_nCH_2-$ ($n > 10$), the compound is a glycerol ether.

If R^1 and R^2 are fatty esters and R^3 is a galactoside, the compound

is a monogalactosyl diglyceride. If R^3 is a digalactoside,

the compound is a digalactosyl diglyceride (both are neutral glycolipids). Analogously, a monoglucosyl and a diglucosyl diglycerides have been described, as well as other, rather rare, compounds, such as 6-sulfo-6-deoxyglucosyl diglyceride (Okaya, 1964).

b. *Containing* C, H, O, P

R^1 and R^2 are fatty esters. If R^3 is a phosphate ester, $-\overset{\overset{O}{\|}}{\underset{\underset{OH}{|}}{P}}-O^-$, the compound is phosphatidic acid. If R^3 is an inositol phosphate ester,

$$-\overset{\overset{O}{\|}}{\underset{\underset{O^-}{|}}{P}}-O-\underset{\underset{OH}{}}{\overset{\overset{OH\;OH}{}}{\left\langle\;HO\;\right\rangle}}-OH$$

the compound is phosphatidyl inositol. The inositol moiety may be further phosphorylated in position 4 (or 4 and 5).

If R^3 is a glycerol phosphate ester,

$$-\overset{\overset{O}{\|}}{\underset{\underset{OH}{|}}{P}}OCH_2\underset{\underset{OH}{|}}{C}HCH_2OH$$

the compound is called phosphatidyl glycerol. If it is a phosphatidylglycerol phosphate ester,

$$-\overset{\overset{O}{\|}}{\underset{\underset{O^-}{|}}{P}}OCH_2\underset{\underset{OH}{|}}{C}HCH_2O\overset{\overset{O}{\|}}{\underset{\underset{O^-}{|}}{P}}OCH_2\underset{\underset{OR}{|}}{C}HCH_2OR'$$

the compound is diphosphatidyl glycerol or cardiolipin.

c. *Containing* C, H, O, P, N

R^1 and R^2 are fatty esters. If R^3 is ethanolamine phosphate ester,

$$-\overset{\overset{O}{\|}}{\underset{\underset{O^-}{|}}{P}}OCH_2CH_2N^+H_3$$

the lipid is phosphatidyl ethanolamine (one of the cephalins). If R^3 is serine phosphate ester,

$$-\overset{\overset{O}{\|}}{\underset{\underset{O^-}{|}}{P}}OCH_2\underset{\underset{O^-}{|}}{C}HC\overset{\overset{N^+H_3}{|}}{=}O$$

the lipid is phosphatidyl serine (another of the cephalins). If R^3 is choline phosphate ester,

$$\begin{array}{c} \text{O} \\ \parallel \\ -\text{POCH}_2\text{CH}_2\text{N}^+(\text{CH}_3)_3 \\ | \\ \text{O}^- \end{array}$$

the lipid is phosphatidyl choline or lecithin.

Occasionally, the occurrence of lysolecithin or lysophosphatidyl ethanolamine has been reported, these compounds being formed from the parent ones by splitting off the fatty acyl residue R^2.

If R^3 is aminoethyl phosphonate ester,

$$\begin{array}{c} \text{O} \\ \parallel \\ -\text{PCH}_2\text{CH}_2\text{N}^+\text{H}_3 \\ | \\ \text{O}^- \end{array}$$

the compound is a phosphonolipid. Some protozoans contain also phosphatidyl-N-(2-hydroxyethyl)alanine, the R^3 being here

$$\begin{array}{c} \text{O} \\ \parallel \\ -\text{POCH}_2\text{CH}_2\text{NHCHCH}_3 \\ | \qquad\qquad\qquad | \\ \text{O}^- \qquad\qquad \text{COO}^- \end{array}$$

Many bacteria contain the so-called lipoamino acids in which the R^3 substituent is

$$\begin{array}{c} \text{O} \qquad\qquad \text{O} \\ \parallel \qquad\qquad \parallel \\ -\text{POCH}_2\text{CHCH}_2\text{OCCHR} \\ | \qquad | \qquad\quad | \\ \text{O}^- \quad \text{OH} \quad\;\; \text{NH}_2 \end{array}$$

the R representing an amino acid residue (Arg, Phe, Thr, Leu, Orn, Ala, Lys, Glu, Asp, His having been found so far).

R^1 is a vinyl ether $CH_3(CH_2)_n CH = CH-$ ($n > 9$). The compounds, usually substituted at R^3 with ethanolamine phosphate, are called plasmalogens. A phosphonoplasmalogen is also known.

B. Based on substituted sphingosine

The backbone of the following lipids is the substituted aminoalcohol sphingosine

$$\begin{array}{c} \text{CH}_2\text{OR}^3 \\ | \\ \text{R}^1\text{CH} \\ | \\ \text{NHR}^2 \end{array}$$

in which substituent R^1 is known to occur in three different forms: with $CH_3(CH_2)_{14}CH(OH)-$ it is dihydrosphingosine; with $CH_3(CH_2)_{12}CH=CHCH(OH)-$ it is sphingosine proper; with $CH_3(CH_2)_{13}CH(OH)CH(OH)-$ it is phytosphingosine (in all cases $R^2 = R^3 = H$). Substituent R^2 is either an acyl group $CH_3(CH_2)_nC=O$ with $n = 14-24$, or an α-hydroxy acyl group $CH_3(CH_2)_nCHC=O$.
$\quad\quad\quad\quad\quad\quad\quad\quad\quad\quad\quad\quad\quad\quad\quad\quad\quad\quad\quad\quad\quad$ OH

The main distinctions arise from the type of substituent R^3. If $R^3 = H$, the resulting amide is called ceramide. If it is a choline phosphate ester, the lipid is sphingomyelin (R^3 may also be an ethanolamine phosphate ester or even an aminoethyl phosphate ester).

If R^3 is a glucoside

(or infrequently galactoside), the lipid belongs to the cerebrosides (a class of *N*-containing glycolipids). An important group of compounds, the gangliosides, contain an oligosaccharide as R^3 and are comprised of hematosides, globosides and blood-group substances. The common components of the oligosaccharide substituents are glucose, galactose and particularly *N*-acetylgalactosamine and *N*-acetylneuraminyl residues. *N*-Acetylneuraminic or sialic acid is a pyranose derivative of the following formula:

(*cf.* with *N*-acetylmuramic acid of bacterial walls; p. 135).
Some of the gangliosides are listed below.

Monosialoganglioside (GM_1)	Galactosyl-*N*-acetylgalactosaminyl--(*N*-acetylneuraminyl)-galactosyl-glucosyl ceramide
Disialoganglioside	*N*-Acetylneuraminyl-galactosyl--*N*-acetylgalactosaminyl-(*N*-acetylneuraminyl)-galactosyl-glucosyl ceramide

CTH	α-Galactosyl-galactosyl-glucosyl ceramide
Hematoside	N-Acetylneuraminyl-galactosyl-glucosyl ceramide
Disialohematoside	N-Acetylneuraminyl-N-acetylneuraminyl-galactosyl-glucosyl ceramide
Globoside 1 (cytolipin)	N-Acetylgalactosaminyl-galactosyl-galactosyl-glucosyl ceramide (if second to third galactose is linked 1 → 4, it is cytolipin K of humans; if the link is 1 → 3, it is cytolipin R of the rat)
Forssman glycolipid	N-Acetylgalactosaminyl (α1→3)-N-acetyl--acetylgalactosaminyl (β1→3)-galactosyl(α1→4)-galactosyl(β1→4)-glucosyl ceramide

The blood group substances are glycolipids of the ganglioside type present on the surface of red cells as well as in tissue fluids and secretions. The sugar terminal units defined in these substances are the following: O-α-N-acetyl-D-galactosaminyl(1→3)-O-β-D-galactosyl-(1→4)-N-acetyl-D-glucosaminyl (group A), 3-O-α-D-galactosyl(1→3)-D-galactosyl (group B), O-α-L-fucosyl(1→2) and O-β-N-acetyl-D-glucosaminyl (group O or H), N-acetylneuraminyl (group M and N), 3-O-β-D-galactosyl-N-acetylglucosaminyl or 4-O-α-L-fucosyl-N-acetylglucosaminyl (Le^a).

A special group of glycolipids are the so-called phytoglycolipids where the ceramide is substituted with an oligosaccharide phosphate ester (= R^3 of the sphingosine backbone).

Quite recently (Lester *et al.*, 1974), three sphingolipids containing a sugar residue have been discovered in the fungus *Neurospora crassa* and in baker's yeast. The major one in yeast is an inositol phosphorylceramide with R^1 being a dihydroxyaliphatic chain with 16 or 18 carbon atoms, R^2 an α-hydroxy acyl group with 26 carbon atoms.

The last group of sphingosine (or ceramide) derived lipids are the sulfatides where R^3 is a glucosyl sulfate.

C. Sterols

Only a few sterols have been found in major amounts in the membranes of eukaryotic (but not prokaryotic) cells:

cholesterol

ergosterol

β-sitosterol

zymosterol

Other sterols have been reported from various sources, particularly from lower plants (Goodwin, 1973). They include campestrol, brassicasterol, cycloartenol, desmosterol, fucosterol, spinasterol, ascosterol, lanosterol and a number of others, as well as several sterol glycosides.

D. Minor lipids

Although of restricted occurrence and insignificant abundance, some of these lipids may play very important roles in special metabolic reactions (derivatives of β-carotene in the vision process; vitamin K in the terminal respiration shunt and blood clotting initiation, etc.). They may be classified as follows (basically according to Law and Snyder, 1972).

a. *Terpenoid compounds*, e.g. β-carotene

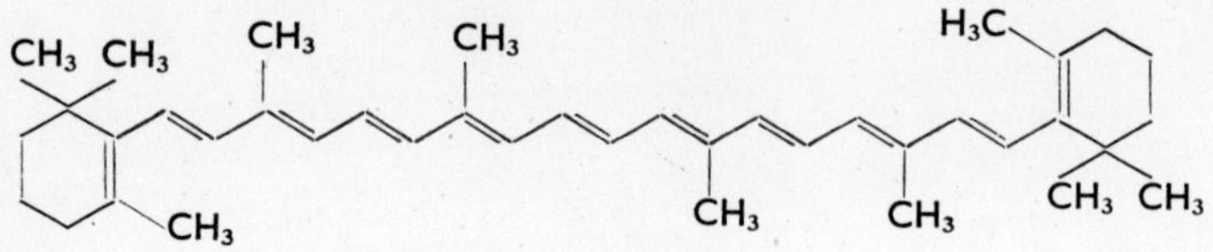

vitamin K_1

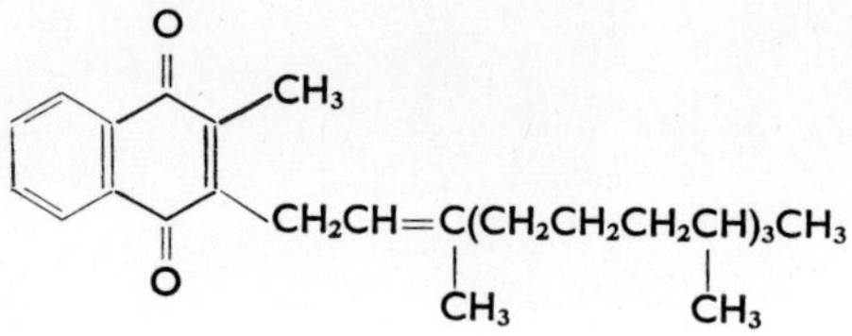

polyisoprenoid alcohols

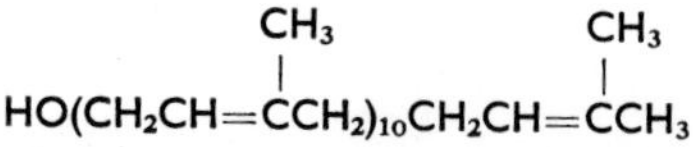

$HO(CH_2CH=CCH_2)_{10}CH_2CH=CCH_3$
(with CH_3 branches)

squalene

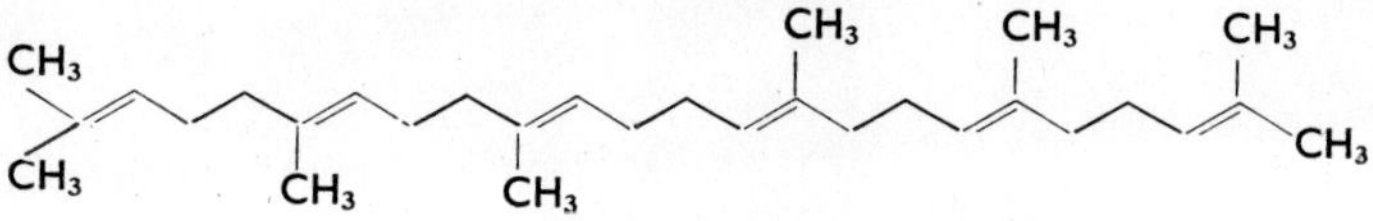

b. *Hydrocarbons*, particularly odd-numbered ones, such as $C_{31}H_{64}$.

c. *Wax esters* of the type

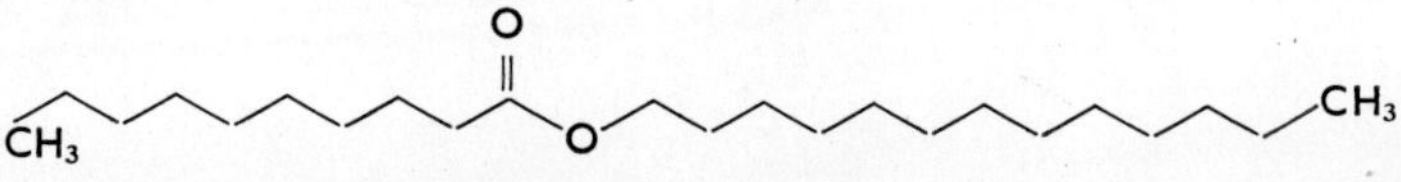

d. *Sugar esters* of the type

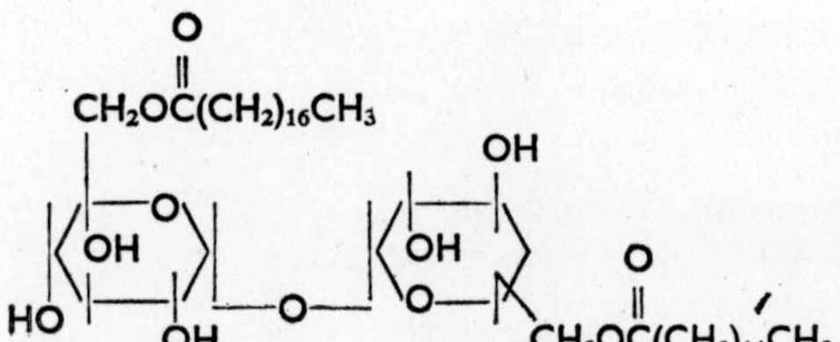

e. *Esters of diols* of the type

$$\begin{array}{l} \text{CH}_2\text{OC(CH}_2)_{14}\text{CH}_3 \\ | \\ \text{CH}_2\text{OC(CH}_2)_{16}\text{CH}_3 \end{array}$$

(with C=O groups on the ester carbons)

f. *Esters of monohydroxyalcohols* of the type

$$\text{CH}_3\text{OC(CH}_2)_{14}\text{CH}_3$$

(with C=O on the ester carbon)

g. *Sulfolipids*, such as

$$\begin{array}{l} \text{CH}_3(\text{CH}_2)_7\text{CH(CH}_2)_{12}\text{CH}_2\text{OSO}_3^- \\ | \\ \text{OSO}_3^- \end{array}$$

as well as halosulfolipids, such as

$$\begin{array}{cccc} \text{Cl} & \text{OSO}_3^- & & \text{Cl} \\ | & | & & | \\ \text{CH}_3(\text{CH}_2)_5\text{CHCHCHCHCH}_2\text{CH(CH}_2)_8\text{CCH}_2\text{OSO}_3^- \\ | & | & | & | \\ \text{Cl} & \text{Cl} & \text{Cl} & \text{Cl} \end{array}$$

h. *Lipopolysaccharides.* These represent large molecules of the outer layer of Gram-negative bacterial cell wall and will be discussed on p. 139.

i. *Other components* extractable with organic solvents, although generally not considered as lipids, include quinones and chlorophyll, occurring almost exclusively in plant and algal chloroplasts.

2.2.1.2. Distribution

Law and Snyder (1972) compiled an informative table on the occurrence of various lipid groups in nature. In a modified and extended form, it is reproduced here as Table 2.4. The table includes compounds that may not occur in membranes in major amounts (wax esters, hydrocarbons). It shows some significant differences between prokaryotic (bacterial) and eukaryotic (other) cells, most striking being the occurrence of sterols and sphingomyelins.

The relative representation of the various lipid groups variés from organ to organ but there is a considerable lack of variability for a given membrane among related species as well as taxonomically

TABLE 2.4. Distribution of lipid groups in nature

Lipid group	Viruses	Eubacteria	Myco-bacteria	Fungi	Plants	Protozoa	Inverteb-rates	Vertebrates
Triglycerides	?	traces	rare	+	+	+	+	+
Phospholipids	+	+	+	+	+	+	+	+
Phosphonolipids	?	—	rare	—	—	+	+	+
Plasmalogens	?	—[a]	?	?	?	?	+	+
Sphingomyelins	+	—[a]	—	+	+	+	+	+
Glycolipids	+	+	+	+	+	+	+	+
Sterols	+	traces	—	+	+	+	+	+
Hydrocarbons	—	—	?	?	+	?	+	+
Wax esters	—	?	?	?	+	?	+	+
Sulfolipids	—	?	?	?	+	+	?	+

[a] Only in some anaerobes.

distant groups. An exception to this rule is represented by erythrocyte membranes where the variety is overwhelming. Some examples of this variability are shown in Tables 2.5, 2.6 and 2.7.

TABLE 2.5. Percent content of phospholipids in the cell membranes of various organs of the rat (adapted from Rouser *et al.*, 1968)

Phospholipid[a]	Liver	Spleen	Heart	Lung	Kidney	Erythrocyte
PC	52.3	47.2	41.2	48.2	36.6	47.5
PE	25.2	27.2	33.8	23.3	28.8	21.5
PS	3.9	9.0	3.2	9.7	8.1	10.8
PI	9.0	6.3	4.0	4.2	6.1	3.5
PG	—	—	1.1	2.3	0.2	3.8[b]
DPG	4.4	2.4	13.0	1.1	7.1	
PA	0.4	0.6	0.2	0.3	0.3	0.3
SP	4.5	7.3	3.5	10.8	12.9	12.8

[a] Abbreviations used: PC, phosphatidyl choline; PE, phosphatidyl ethanolamine; PS, phosphatidyl serine; PI, phosphatidyl inositol; PG, phosphatidyl glycerol; DPG, cardiolipin; PA, phosphatidic acid; SP, sphingomyelin. [b] Lysophosphatidyl choline.

TABLE 2.6. Weight percent composition of phospholipids from various sources (adapted from Galliard, 1973; Jakovcic et al., 1971; Suomalainen et al., 1973; Wolfe, 1964)

Phospholipid	Brain						Kidney				Saccharomyces cerevisiae	Potato microsomes	Sugar beet protoplasts	Apple mitochondria	Euglena gracilis	Micrococcus lysodeikticus
	Man	Cow	Rattlesnake	Frog	Goldfish	Octopus	Fly	Lobster	Man	Rat	Mouse	Frog						
PC	29.2	29.6	39.0	43.2	49.6	38.8	18.6	58.4	37.9	36.6	38.5	36.7	45	45	38	45	47	—
PE	35.0	35.7	36.7	39.0	33.6	37.7	64.5	25.7	30.8	28.8	28.4	32.1	15	33	15	35	15	—
PS	17.6	16.9	9.1	10.2	8.8	11.8	5.3	5.7	7.0	8.1	7.6	6.5	7.3[a]	1	—	3	4	5
PI	2.0	2.7	3.9	2.4	3.6	2.6	2.7	1.2	6.1	6.1	6.3	6.1	20.3	16	9	5	9	—
PG	—	—	—	—	—	—	—	—	0.6	0.2	0.5	0.3	0.3	1	38	7	10	—
DPG	0.4	0.8	0.5	1.1	0.6	2.3	1.8	0.6	4.2	7.1	7.5	5.2	3.3	1	—	trace	14	68
PA	0.5	1.0	0.6	1.5	2.0	?	—	1.7	0.6	0.3	0.3	0.3	0.8	3	—	5	—	—
SP	13.6	13.2	10.1	2.7	1.7	1.0	?	6.7	12.8	12.9	11.0	13.0	2[b]	—	—	—	—	24[c]

Abbreviations as in Table 2.5. [a] According to Baraud et al. (1970) baker's yeast contains no phosphatidyl serine; [b] unidentified; [c] glycolipid.

TABLE 2.7. Percent content of lipids in erythrocyte membranes of various animals (adapted from Dawson et al., 1960; Dewey and Barr, 1970; Rouser et al., 1968)

Lipid	Cat	Cow	Dog	Guinea pig	Horse	Pig	Rabbit	Rat	Sheep	Man
Phospholipids	61.3	64.8	52.6	55.6	52.0	59.8	65.8	67.0	63.2	65.9
PC[a]	30.5	0	46.9	41.1	42.4	23.3	33.9	47.5	0	28.9
PE	22.2	25.1	22.4	24.6	24.3	29.7	31.9	21.5	26.2	27.2
PS	13.2	15.3	15.4	16.8	18.0	17.8	12.2	10.8	14.1	13.0
PI	7.4	3.7	2.2	2.4	0.3	1.8	1.6	3.5	2.9	1.3
PA	0.8	0.3	0.5	4.2	0.3	0.3	1.6	0.3	0.3	2.2
SP	26.1	46.2	10.8	11.1	13.5	26.5	19.0	12.8	51.0	26.9
Others	0.3	1.7	1.8	0.3	1.7	0.9	0.3	3.8	4.8	—
Cholesterol	26.8	27.5	24.7	27.0	24.5	26.8	28.9	24.7	26.5	22.2
Gangliosides	8.8	5.5	11.8	2.2	15.5	3.3	4.5	6.3	7.8	
Other glycolipids	3.1	2.2	10.9	15.2	8.0	10.1	0.8	2.0	2.5	10

[a] The various phospholipid fractions are expressed as percent of total phospholipids. Abbreviations as in Table 2.5.

Unfortunately, there are no comparative data regarding the occurrence in membranes of various other lipids although some regularities have already emerged, among them:

1. Animal membranes contain cholesterol, plant membranes contain β-sitosterol, yeast and fungal membranes ergosterol and zymosterol as major sterols.

2. Glycolipids in the broad sense do not occur in mitochondria and probably are restricted to plasma membranes.

3. Only mitochondria contain greater amounts of cardiolipin.

4. The composition of phospholipids in bacterial membranes is somewhat peculiar; the major component appears to be phosphatidyl glycerol and its amino acyl derivative. Phosphatidyl ethanolamine and phosphatidyl serine occur sporadically while sphingolipids and phosphatidyl choline are very rare.

Various glycolipids are of importance, e.g., monoglucosyl and diglucosyl diglycerides.

5. *Acholeplasma* cells, although prokaryotic like bacteria and although containing a great of phosphatidyl glycerol (36%) and

glucosyl diglycerides (46%) among its polar lipids, as bacteria do, show a striking content of cholesterol (10−30% of total lipid).

6. Virus lipids are apparently derived from their host cells and are thus of corresponding composition.

One may consider the distribution of various lipids from the aspect of occurrence of polar groups in their molecules. Classifying lipids as uncharged (cholesterol, triglycerides, polyhydroxy compounds), zwitterionic (phosphatidyl ethanolamine, phosphatidyl choline), weak acids (carboxylic acids and phosphatidyl serine) and strong acids (phosphatidic, sulfo acids) we arrive at a predominance of the uncharged class in myelin, erythrocytes and plant tissues and a prevalence of zwitterionic and weakly acidic substances in mitochondrial membranes.

2.2.1.3. Fatty acids

In all lipids based on glycerol and sphingosine one or more hydrocarbon chains derived from a fatty acid are present. It may be of interest to compare the pattern of fatty acid representation (1) in different lipids and (2) in different species. The variability is shown in Tables 2.8 and 2.9. The predominance of 16 : 0, 18 : 0, 18 : 1,

TABLE 2.8. Percent fatty acid content in various phospholipids of human erythrocytes (adapted from Dewey and Barr, 1970; Rouser et al., 1968)

Source of fatty acids	Carbon atoms:double bonds									
	16:0	16:1	18:0	18:1	18:2	20:4	22:0	22:5	22:6	24:0 24:1
PC	33.0	1.0	11.7	20.6	18.2	5.0	—	5.4	1.1	— —
PE	18.9	0.6	8.0	25.2	7.0	21.9	4.7	3.1	2.9	— —
PS	7.1	0.4	41.6	13.0	2.8	19.7	—	2.9	4.2	— —
SP	41.3	0.1	9.1	5.2	3.7	0.1	8.0	—	—	15.0 15.5
Whole membrane	28.2	0.7	15.1	18.3	10.6	10.8	—	4.0	2.1	? ?

Some authors have detected the following additional fatty acids; PC 14:0, 20:0, 20:2; PE 12:0, 14:0, 20:3; PS 20:3; SP 12:0, 14:0, 20:0, 22:4, 22:5.

TABLE 2.9. Percent fatty acid content in the total erythrocyte lipids from various species (compiled from several sources)

Animal	Carbon atoms: double bonds													
	12:0	14:0	16:0	16:1	18:0	18:1	18:2	18:3	20:0	20:3	20:4	22:5	22:6	24:1
Rat	—	0.3	31.1	2.2	13.7	18.5	7.2	—	—	2.3	24.0	—	—	—
Rabbit	—	—	20.1	2.6	17.8	11.0	21.5	—	—	—	18.5	—	—	—
Guinea pig	—	—	16.1	2.6	26.6	12.3	15.7	—	—	—	16.1	—	—	—
Sheep	0.6	1.6	16.0	2.3	9.7	49.6	11.5	0.7	0.9	1.9	1.4	—	—	—
Cow	—	—	12.1	2.7	14.1	34.5	21.1	—	—	—	4.8	—	—	—
Pig	—	—	21.4	2.4	10.4	32.1	23.2	—	—	—	6.4	—	—	—
Dog	—	—	16.9	1.7	19.0	14.2	12.9	—	—	—	30.8	—	—	—
Cat	—	—	20.1	2.6	17.8	11.0	21.5	—	—	—	18.5	—	—	—
Monkey	—	0.3	15.6	0.5	15.2	15.7	11.0	—	—	1.0	12.3	4.2	10.8	7.5?
Chicken	—	0.2	28.7	0.8	8.8	21.7	31.8	0.9	—	0.6	4.6	1.2	3.2	—

It should be noted that the analyses vary somewhat depending on the technique and degree of refinement applied by the various authors.

18 : 2 and 20 : 4 acids is obvious in most of the species tested, palmitic and oleic acids leading the field. However, in microorganisms and plants there appears to be a much greater variety in fatty acid patterns.

To begin with, in addition to saturated and unsaturated fatty acids, there exist cyclopropane fatty acids with one or more of the three-membered rings along the hydrocarbon chain in various bacteria and both cyclopropane and cyclopropene fatty acids in some plants (*cf.* Gunstone, 1970). Another rarity found practically only in bacteria are the branched-chain fatty acids (mainly with 15 and 17 carbon atoms). The importance of all these acids for membrane fluidity will be discussed later in the chapter. The differences between microbial and plant species and the animal kingdom may be grasped from a comparison of Tables 2.8 and 2.9 with Table 2.10.

2.2.1.4. Solubilization

Extraction of lipid material from membranes presents no particular problem in cases when it is of no concern whether lipid protein interactions are disrupted in the process. Lipids can be extracted from cells or tissues with a number of organic solvents or their mixtures, such as aqueous acetone, tertiary amyl alcohol, isobutyl alcohol, *n*-butanol, ether, etc. The usual procedure is homogenization with a mixture of chloroform and methanol (2 : 1, v/v), using a 20-fold excess of solvent over tissue with subsequent washing of the extract with water (one-fifth of the volume is recommended) to remove the water-soluble components (some sugars and most amino acids are extracted by the chloroform-methanol treatment). The only lipid that remains unextracted is phosphatidyl inositol (particularly in the polyphosphorylated form) and this must be removed by using chloroform-methanol plus hydrochloric acid or, as suggested by Folch-Pi and Stoffyn (1972) for brain proteolipids, by applying five volumes of chloroform-methanol (1 : 1, v/v) and 0.5 volume $2M$ KCl in water. The lower phase now contains all phospholipids, cholesterol, polyphosphoinositol, as well as lipoprotein, the upper phase contains gangliosides (plus protein and low-molecular weight components).

Since the extraction of lipids occurs almost as well at subzero as at room temperature it is advisable to work at the lowest practicable temperature, there being less risk of chemical changes of the lipids. Likewise, some lipid components being rapidly degraded by oxygen,

TABLE 2.10. Percent fatty acid content in the cell envelopes or external membranes of some plants and microorganisms (adapted from Cho and Salton, 1966; Galliard, 1973; Nurminen and Suomalainen, 1971; Tourtellotte, 1972)

Source of fatty acid	Carbon atoms: double bonds								
	10—13:0	14:0	14:1	16:0	16:1	17:0	18:0	18:1	18:2,3
Saccharomyces cerevisiae[a]	7	6	6	6	51	1	4	18	<1
Acholeplasma laidlawii[b]	7.5	24.8	—	53.5	—	—	3.0	6.6	4.4
Bacillus licheniformis[c]	—	4.2	—	3.9	12	—	0.2	11.1	—
Micrococcus lysodeikticus[d]	0.4	4.4	—	0.2	—	—	—	—	—
Aerobacter aerogenes	—	5.7	—	56.4	5.7	3.8	1.7	9.6	18.8
Escherichia coli B	—	6.8	2.1	42.0	2.3	—	6.1	25.7	—
Serratia marcescens[e]	—	10.5	—	55.5	0.9	—	4	3.1	2.6
Apple fruit	—	—	—	25	—	—	2	7	65
Moss (*Hypnum*)[f]	—	1	—	14	5	—	+	7	44
Chlorella pyrenoidosa[g]	—	+	—	+	3	trace		14	43

[a] There are a number of fatty acids with more than 20 carbon atoms present, albeit in minute amounts, the order of decreasing occurrence being 26:OH, 26:0, 20:0, 22:0, etc. Differences depending on strain character and type of cultivation may be substantial (*cf.* Longley *et al.*, 1968). According to Baraud *et al.* (1970) the plasma membrane phospholipids contain appreciable amounts of 3,7,11-trimethyldodecanoic acid. [b] Like in many bacteria, the fatty acid content of *A. laidlawii* membranes can be influenced by growth on a fatty acid. Thus growth on pentadecanoic acid will produce membranes containing 83% of this acid, growth on isopalmitic acid will yield 79% of this acid in membranes. [c] Contains also 50.4 % branched 15:0 acid and 28.2 % branched 17:0 acid. [d] Contains 85.4 % branched 15:0 acid. [e] Contains nearly 20 % 9,10-methylenehexadecanoic acid (a cyclopropane fatty acid). [f] Also 12 % of 20:4, 7 % of 20:5, and 5 % of 22:0. [g] Total saturated acids represent 17 %; also some 18:4, 16:2 and 16:3 acids are present.

the extraction procedure should be done in the absence of air (under argon or nitrogen).

Once extracted, the lipid mixture can be resolved into classes and subclasses by a number of chromatographic techniques, among them chromatography on silicic acid, on Sephadex LH (a lipophilic variety), thin layers of alumina or other adsorbents. For further analysis, various techniques are available, including the application of phospholipases.

Fatty acyl esters can be hydrolyzed either in acid or in alkaline media and the fatty acids liberated then esterified and as esters analyzed elegantly by gas-liquid chromatography.

There is a wealth of methods applicable to the definitive analysis and much can be learned from compendia such as those of Johnson and Davenport (1971) or Lowenstein (1969).

2.2.2. Proteins

2.2.2.1. *Types and occurrence*

Unlike the lipids which (*1*) have a single major role in membranes, *viz.* that of maintaining their mechanical stability and overall hydrophobicity and (*2*) are easily extracted and analyzed as chemical individuals, membrane proteins have proved to be singularly difficult both to extract and to analyze structurally as well as functionally.

It will be recalled from Table 2.3 that proteins usually represent more than one-half of the membrane dry weight; still, no universal structural protein or even a subunit has been established. In fact, in cases where the protein content is relatively high (mitochondrial inner membrane, bacterial cytoplasmic membranes) the membranes are endowed with numerous enzymic functions, the inference being that membrane proteins are largely enzymes. Although this is apparently not so, most membrane proteins do play a more or less specific role: in enzyme catalysis, as receptors of hormonal or antigenic signals, as recognition elements in membrane transport as well as pinocytosis and chemotaxis, as transmembrane carriers for low-molecular substances, etc. There are thus few proteins with no apparent physiological function left.

The rather crude methods of extraction and separation of membrane proteins generally do not permit to analyze the whole spectrum of proteins present, the highest number extracted from a given membrane being about 30. However, it must be obvious that an average cell, both a eukaryotic organ cell or a unicellular organism, contains probably 30—50 different proteins whose sole role is to select solutes for transport, plus a variety of proteins with other functions.

Specific proteins, mostly of enzyme character, of various membranes will be described in the appropriate section. Here we shall consider some characteristic protein types occurring in membranes and examine whether any of the protein may be considered as a structural backbone of a membrane.

The molecular weights of polypeptides extracted from membranes vary from about 10 000 to 240 000. (The "miniproteins" of mol. wt. of 6000 reported by Laico and associates (1970) were later shown by the same group headed by Dreyer (1972) to be artifactual.)

Erythrocytes. Depending on the solubilization procedure, 8—20 polypeptides have been resolved, the major ones being (according to Bretscher, 1973) peptide *a* (100 000), a glycoprotein (30 000) and tektin A (220 000 + 240 000), this last component being probably identical with the spectrin isolated by Clarke (1971). Steck (1974) reports 8 major polypeptides and 2 glycoproteins in the same material. After treatment with 2-chloroethanol, the molecular weights of the peptides all lie between 10 000 and 50 000 (Zahler, 1969), the inference being that the large molecular weights reported above may be due to association of subunits. Apparently, none of the components is a structural protein.

Mitochondria. Apart from a multitude of enzymes (*cf.* Table 2.26) these organelles were reported to contain a structural protein (Richardson *et al.*, 1963) with mol. wt. of about 23 000, the protein showing an affinity for phospholipids and ATP. Analogous proteins were detected in liver microsomes, bovine erythrocytes and spinach protoplasts and assumed to form a part of the integral membrane structure. However, no universal occurrence of these proteins could be established and the concept was eventually abandoned (Green *et al.*, 1968). In fact, even the inner and outer mitochondrial membranes differ completely in their individual polypeptides (Schnaitman, 1969).

Bacteria. In *Escherichia coli*, the outer envelope (see p. 134) appears to contain six polypeptides while the plasma membrane proper (with its array of enzymes) has 27 major polypeptides plus many minor ones, none of them in comparable amounts in the two layers (Schnaitman, 1970). Recently, many more were detected by using two-dimensional gel electrophoresis (Johnson et al., 1975). In Acholeplasma laidlawii, disc-gel electrophoresis revealed 20−30 polypeptides with molecular weights from 15 000 to 100 000, none of them predominating in any way (Morowitz and Terry, 1969).

Interesting observations exist on membrane protein mutants of various bacteria (for a review see Machtiger and Fox, 1973), the deletions affecting either the function of a membrane-located enzyme or the number of proteins eluted from the membrane. An extreme case in this connection is the strain Bacillus megaterium PP where more than 90% of membrane proteins is formed by a single type of polypeptide (Patterson and Lennarz, 1970).

Other membranes. In several instances, the distribution of extracted proteins was rather asymmetric, there being a single predominant polypeptide present, although by no means a "structural" protein. Thus, rhodopsin is the single important polypeptide in the rod outer segments, brain myelin contains three major polypeptides, similarly to sarcoplasmic reticulum. The purple membrane of *Halobacterium halobium* was found to contain a single protein. In these cases it is rather likely that the protein or few proteins present in the membrane do in fact play a structure-preserving role but it is not their primary function. Still, the existence of a functional structural protein is supported by experiments where thorough washing of liver microsomes with sodium chloride, sodium carbonate and sodium bicarbonate removes practically all proteins but a few with a predominant 52 000 polypeptide with no apparent function (Hinman and Phillips, 1970).

In their amino acid composition, membrane proteins do not exhibit any peculiarities in comparison with the composition of soluble enzymes. The hydrophobic amino acids represent about one-third of the total; acidic amino acids amount to about 20%, basic ones to about 12% but the carboxylic groups may be more strongly amidated so that no net charge arises from the apparent preponderance of the acidic over the basic amino acids (Table 2.11). The only difference distinguishing the membrane proteins from other cell proteins is the low amount of cysteine (apparently no disulfide bridges are formed in membrane proteins). Although in their average the membrane proteins are no more hydrophobic than soluble proteins, an important distinction arises when extrinsic and intrinsic membrane proteins are considered. Dividing amino acids into polar (Asp, Asn,

TABLE 2.11. Amino acid composition (in molar %) of proteins of various cell membranes (from Engelman and Morowitz, 1968; Haurowitz, 1963; Longley et al., 1968; Steck and Fox, 1972; Wallach, 1972)

Amino acid	Human erythrocyte	Ehrlich ascites tumour plasma membrane	Ehrlich ascites endoplasmic reticulum	Rat liver bile fronts	Baker's yeast[a] plasma membrane	Acholeplasma laidlawii[b]	Human serum albumin	Insulin
Lys	5.2	6.3	6.5	7.2	7.4	6.37	10.7	2.0
His	2.4	2.6	2.1	2.6	2.5	1.46	2.9	3.6
Arg	4.5	4.7	5.2	5.2	4.2	2.95	4.5	2.3
amide NH$_3$	6.9	14.7	10.8	12.4	?	?	8.0	14.2
Asp	8.5	8.8	8.7	9.3	10.8	11.4	9.9	5.7
Glu	12.2	10.1	10.6	12.0	10.2	8.23	15.0	15.5
Thr	5.9	5.5	5.4	5.3	10.9	6.77	5.4	3.1
Ser	6.3	6.6	6.2	6.0		6.40	4.4	6.2
Pro	4.3	5.2	5.4	4.9	5.1	3.60	5.6	2.8
Cys	1.1	trace	trace	0.9	1.0	0.19	6.6	11.0
Met	2.0	2.7	2.5	2.3	1.5	2.33	1.1	—
Gly	6.7	8.5	7.7	7.8	8.3	6.95	2.7	6.9
Ala	8.2	7.8	7.6	8.0	7.7	8.23	8.9	3.7
Val	7.1	6.6	6.7	6.6	6.1	7.55	8.4	9.5
Ile	5.3	6.1	5.1	5.1	5.5	7.39	1.7	14.2
Leu	11.3	10.1	10.3	9.6	9.6	9.79	11.6	
Tyr	2.4	3.1	3.4	2.7	7.8	4.81	3.3	7.7
Phe	4.2	4.8	4.8	4.5		5.40	6.0	5.4
Trp	2.5	1.5	1.5	—	1.2	?	0.1	—

[a] Ornithine was also detected. [b] Also 1.73% glucosamine and 5.75% galactosamine.

Glu, Gln, Lys, Arg), intermediate (Ser, Thr, Tyr, His, Gly) and nonpolar (Ala, Val, Leu, Ile, Cys, Met, Pro, Phe, Trp) ones and assigning to the first group the polarity of unity, to the second of one-half and to the third of zero, and expressing them in molar percent, Vanderkooi and Capaldi (1972) arrived at an interesting distribution of polarities among proteins in general. While 206 nonmembrane proteins had polarities of 46 ± 6%, intrinsic proteins had polarities ranging from 30 to 40%, extrinsic ones from 41 to 53%, the difference being apparently associated with the localization of the former within the membrane, of the latter on the membrane surface.

2.2.2.2. Solubilization

The task of solubilizing proteins in a native form is incomparably more difficult than analogous work with lipids. The ease of denaturation and of possible subtle changes in the tertiary and particularly quaternary (subunit) structure have been frequently used by critics as arguments against the validity of deductions drawn from analyses of membrane proteins made soluble by one means or another; still, the natural way for an investigator at this stage of experimental capabilities is to start from a thorough analysis of constituents before proceeding to a synthetic approach that might give a more definitive picture of what a membrane is like. Hence, proteins must be released from membranes and sorted out. The applicable techniques are numerous, practically all of them based on disruption of weak bonds between individual proteins as well as between proteins and lipids. It is an obvious disadvantage of the methods that, depending on the potency of the solubilizing agent, proteins may be broken down to smaller units in the process, these units either being able to reaggregate or not if the solubilizing agent is removed. Most estimates of molecular weights of membrane proteins are thus hampered and one may encounter values that refer obviously to a single polypeptide chain, as well as others that would embrace most of the membrane protein in a few dozen supermolecules.

The solubilizing procedures may be divided as follows (*cf.* Steck and Fox, 1972; Coleman, 1973; Jakoby, 1971).

1. Proteins that are merely adsorbed on the membrane surface are readily washed out at a low ionic strength, even with distilled water or with e.g. $10^{-5}\,M$ NH$_4$OH.

2. Periplasmic proteins are immediately released by removing the external obstacle of the cell wall by treatment with lysozyme or penicillin (in bacteria), or with snail-gut or analogous enzymes (in yeasts and fungi). Occasionally, it is difficult to distinguish between a true periplasmic protein and an extracellular enzyme which can cross the wall without any special processing. A procedure frequently employed for the release of periplasmic proteins has been the osmotic shock, pioneered by Heppel (e.g., Nossal and Heppel, 1966) in bacteria.

In a frequently used modification, intact cells are suspended in cold 0.5 M sucrose (buffered with Tris-HCl and with some EDTA added) and after a few minutes centrifuged and rapidly dispersed in ice-cold 10^{-4} M MgCl$_2$; the supernatant liquid is collected and used for further work. A partial list of proteins thus released from *Escherichia coli* is given in Table 2.12. Sucrose has sometimes been

TABLE 2.12. **Proteins known to be released from *Escherichia coli* by an osmotic shock (adapted from Heppel, 1971)**

Alkaline phosphatase	Asparaginase (II)
5'-Nucleotidase	Penicillinase
2' : 3'-Cyclic-nucleoside monophosphate phosphodiesterase	Deoxyribose-phosphate aldolase
Acid phosphatases (of three kinds)	Purine-deoxyribonucleoside phosphorylase
Ribonuclease I	
Deoxyribonuclease I	Phosphopentomutase
Glucose-1-phosphate adenylyltransferase	Various binding proteins

replaced with saturated mannitol (e.g., Schwencke *et al.*, 1971) or even concentrated salt solutions. There are indications that some periplasmic proteins may be released by simple transfer of cells to distilled water without previous exposure to high osmolarities (Opekarová *et al.*, 1975).

3. Removal of cations can occasionally release a membrane protein if it is held by ionic bonds or if ions are essential for its integrity. The case in point is the application of ethylenediamine tetraacetic acid for solubilization of adenosinetriphosphatase from *Streptococcus faecalis* protoplasts, from the mitochondrial inner membrane and from thylakoid membranes.

4. High ionic strength may be effective in removing electrostatically bound proteins, 0.5 – 1.2 M NaCl being employed to this end.

5. Similarly to the preceding case, various denaturing agents in high concentrations are occasionally used under the name of chaotropic agents. These substances disturb the ionic milieu in the membrane, break the weak bonds and, moreover, decrease substantially the actual concentration of water at the membrane. They include particularly salts of thiocyanate perchlorate, as well as guanidine and urea. The concentrations of these last-named compounds

are usually 6 M. A drawback of the technique is the fact that the solubilized peptide chains are frequently devoid of activity and, in the case of urea application at neutral pH, may be heavily carbamylated.

6. Treatment with hydrolytic enzymes is frequently used as a preliminary step in membrane breakdown. Papain and trypsin have been employed widely both for the release of enzymes (e.g., β-fructosidase, L-β-naphthylamidase) and for disruption of larger complexes, even of tissue. Phospholipases have been used for releasing numerous lipid-associated proteins (e.g., 3-hydroxybutyrate dehydrogenase, mitochondrial NADH dehydrogenase).

7. Sonication can be used to break membranes by mechanical shearing into small fragments but there is little hope that proteins separated from lipids will be obtained by this technique.

8. Organic solvents, by extracting lipids from membranes, may in favourable cases release some of the proteins even in a native (or rather active) state. An effective solvent in this context is 2-chloroethanol in water (9 : 1) at pH below 2 or, alternatively, N,N-dimethylformamide in water. Some authors prefer mixtures of phenol−formic acid−water (e.g. 14 : 3 : 3). n-Butanol or n-pentanol, or even a mixture of n-butanol with urea, will release proteins in an association with lipids, in some cases as the active complexes.

9. Detergents, by being amphiphilic or amphipathic, will bind hydrophobically to membrane proteins while interacting through their polar moieties with the aqueous medium, thus pulling the proteins out of the membrane matrix. Like in the preceding case, lipid-protein complexes are likely to be extracted in this way. The nonionic detergents of the Triton type, Lubrols and various Tweens have been used most widely. The cationic detergents include in particular cetyltrimethylammonium bromide (used successfully for solubilization of rhodopsin); the anionic ones are sodium cholate and deoxycholate and, most important, sodium dodecyl sulfate which dissociates proteins into monomeric units by forming regular complexes with them. Such complexes are readily separated by gel electrophoresis and their molecular weight can be estimated.

10. Various other compounds, such as the organic mercurial mersalyl, have been used for special purposes (Cantrell, 1973).

Further processing and analysis of released proteins can be effected in a number of ways, various chromatographic and electrophoretic techniques being used (for detailed information see, e.g., Jakoby, 1971).

2.2.3. Carbohydrates

In many reports on membrane composition, the content of carbohydrates looms relatively large side by side with proteins and lipids. However, there is probably very little free carbohydrate in cell membranes, most of the sugar residues being constituents of glycolipids and glycoproteins. This is in sharp contrast with the composition of cell walls, particularly of plant or fungal origin, where true carbohydrates, such as cellulose, glucan, mannan, chitin and the like, are found as the major structural components (cf. p. 141).

For the sake of completeness we shall only briefly enumerate the carbohydrate parts of the complex membrane molecules.

The sugar moieties of glycolipids are described on p. 41. The structure of bacterial lipopolysaccharides and of peptidoglycans will be mentioned on p. 139 and p. 134, respectively.

The principal monosaccharides of glycoproteins are similar to those of glycolipids, viz. D-galactose, D-glucose, N-acetylglucosamine, N-acetylgalactosamine, sialic acid, D-fucose, but also D-mannose and D-xylose, plus a spattering of less usual sugars, such as are found in lipopolysaccharides (the question remains whether this is not a contamination caused by inadequate separation of membranes).

In glycoproteins, the monosaccharides usually form heterologous sequences of several units, the oligosaccharide being linked either O-glycosidically to the hydroxyl group of serine and threonine or N-glycosidically to asparagine.

The glycoproteins are found in many membranes but may be completely absent in some. An important glycoprotein of human erythrocyte membranes was already mentioned previously; it can be resolved into three sialoglycopeptides, their composition being characterized by tetrasaccharide groupings, but larger oligosaccharides are also present. There may be over 60% carbohydrate in this material (cf. Winzler, 1969).

An outstanding representative of the glycoprotein family is the retinal rod component rhodopsin—its species from bovine retina contains about 4% carbohydrate, arranged as an oligosaccharide chain attached to a 28 000 polypeptide (Heller and Lawrence, 1970).

The role of glycoproteins as cell surface receptors of signals (hormones, antigens, phytohemagglutinins, etc.) has been recognized in many instances (e.g., Kornfeld and Kornfeld, 1970).

Glycoproteins represent major constituents of viral envelopes (as much as 40% of total virus proteins) their molecular weights being rather alike. Thus, that of the glycoprotein of group A arboviruses is 52 000, that of rhabdoviruses 67 – 80 000, those of paramyxoviruses 53 – 56 000 and 65 – 74 000, those of orthomyxoviruses then 24 – 32 000, 45 – 51 000, 49 – 58 000 and 74 – 78 000 (for more details see Klenk, 1973).

2.3. STRUCTURE OF MEMBRANES

In this section, the organization of membrane components, including their interactions, will be considered. It was during the last few years that considerable insight into the membrane structure has been gained, particularly through the use of previously unavailable techniques of physical chemistry. Before presenting what has been deduced about membrane structure, let us first survey the methodological approaches that assisted in these deductions.

2.3.1. Physico-chemical techniques

2.3.1.1. X-Ray diffraction

Crystalline solids can act as a three-dimensional diffraction grating for X-rays if the interatomic or intermolecular distances of the lattice are of the same order of magnitude as the wavelength of the X-rays, i.e. about 10^{-10} m (0.1 nm). Ordered constructive and destructive interference occurs as the wavefronts are diffracted, obeying the Bragg equation

$$2d \sin \theta = n\lambda \quad (2.14)$$

d being the distance between repeating lattice planes, 2θ the angle between the incident and the diffracted beams, n the sequential number of a constructive maximum and λ the wavelength of the incident X-ray. With greater values of d (as in macromolecular systems), the angle will be rather small – hence we often speak of low-angle diffraction spectroscopy in the context of lipid or membrane studies.

Every crystalline sample can be envisaged as an array of the smallest repeating units, the so-called unit cells. It is the properties of this unit cell that determine the features of the diffracted waves. Thus, the positions of intensity maxima are determined by the unit cell dimensions; the intensity of the maximum depends on the distribution of electrons in the unit cell; the angular length of the maximum depends on the degree of relative orientation between the unit cells; the angular width of the maximum depends on the number of unit cells in the crystal examined.

The intensity of the diffraction pattern obtained is related to the amplitude F of a given wavefront by

$$I = F^2 \qquad (2.15)$$

so that the absolute amplitude but not its phase can be computed. Using the Fourier relationship for the vectorial electron density in real space (in the x direction in this particular case) it is defined by

$$\varrho(x) = \int F(r)\, e^{-2\pi\, irx}\, dr \qquad (2.16)$$

where $F(r)$ is the so-called structure factor, r being the reciprocal spacing (in nm^{-1}), i then $(-1)^{0.5}$. To obtain the correct phase of each $F(r)$, several ingenious techniques are in use. The complete solution

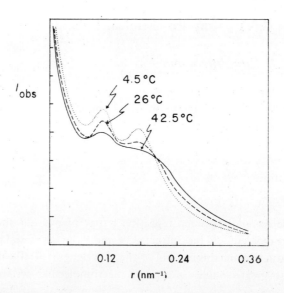

Fig. 2.8. Microdensitometer tracing of X-ray diffraction patterns from frog retinal rod membranes at different temperatures. The high intensity recorded near zero corresponds to the incident beam. I_{obs}, relative observed intensity; r, reciprocal space distance. (Taken with kind permission from Blasie and Worthington, 1969).

for $\varrho(x)$ yields the number of electrons at different intervals of the repeat spacing and, with the use of computers, can be interpreted to reveal the mutual positions of the component atoms or molecules.

In membrane studies, the usual procedure is to read the diffraction pattern intensity with a microdensitometer (a plot of observed intensity reciprocal spacing, such as in Fig. 2.8) from which an electron density profile (density vs. radial distance d in nm) is calculated.

X-ray diffraction studies have been applied both to artificial and native phospholipid membranes where, for a hexagonal packing, a sharp peak at $d = 0.42$ nm is obtained. At higher temperatures (when the hydrocarbon chains lose their rigidity) the peak becomes diffuse and is shifted to 0.46 nm. Refined techniques have shown that (at least in myelin) the packing distance of hydrocarbon chains is 0.46 to 0.48 nm, that there is a lipid bilayer 5–7 nm thick (depending on hydration) and that the bilayer is coated with protein. The membranes are apparently symmetrical with respect to dimensions but asymmetrical with respect to electron density, probably due to different participation of cholesterol at the two membrane faces (*cf.* Caspar and Kirschner, 1971).

2.3.1.2. Infrared spectroscopy

This technique is widely used in structural analysis particularly in organic chemistry and has found only limited application in the analysis of membrane lipid and protein structure. The absorption spectrum of a sample in the 2–15 μm region is due to absorption by vibrating atoms in molecules, various bands being ascribed empirically to certain functional groups. Since water obscures some of the important sections of the spectrum, organic chemists dissolve their samples in organic solvents or disperse then in KBr pellets; for membrane studies, ordinary water may be replaced with D_2O to achieve comparable results. Alternatively, air-dried films of lipids in AgCl can be analyzed. Examples of infrared spectra of membrane material are shown in Fig. 2.9.

In protein analysis, the amide bands are of particular importance, there being absorption maxima due to the amide grouping at 1652, 1630 and 1535 cm^{-1}, the first of these supporting the α-helical structure, the second the pleated sheet (β-conformation) structure. With the exception of some mitochondrial "structural-type" protein, all membranes were shown to possess a predominance of the α-helix.

Only recently spectroscopy based on Raman scattering of IR laser light has been applied to the study of membrane constituents and even whole membranes, the technique showing promise in the definition of various interactions (e.g., Lutz and Breton, 1973).

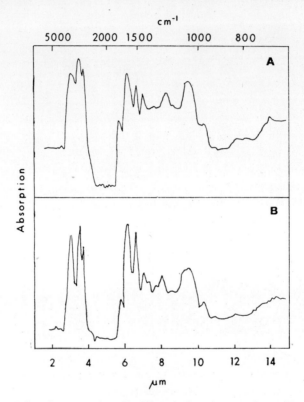

Fig. 2.9. Infrared spectra of myelin (**A**) and of erythrocyte ghosts (**B**). (Adapted from Chapman and Wallach, 1968.)

2.3.1.3. *Ultraviolet spectroscopy*

Electron transitions between the ground state and the electronically excited states underlie the absorptions observed in the spectral region from 150 to 600 nm, as well as the optical activity of various groupings.

Both classical absorption spectroscopy in the ultraviolet region and the more recent optical rotatory dispersion (ORD) and circular dichroism (CD) have been exploited in the study of membrane composition and structure, but restricted mostly to the investigation of proteins.

2.3.1.3.1. Absorption spectra

The UV absorption spectra are usually expressed in terms of absorbance A, which is defined as $-\log(I/I_0)$, I_0 being the intensity of incident light, I that of transmitted light. To provide a more standard basis for comparison between samples, the molar absorptivity ε (previously called the molar extinction coefficient) has been defined as A/bc where b is the path in cm and c the solute concentration in mol dm^{-3}, the overall dimension of ε thus being dm^3 mol^{-1} cm^{-1}. The use of the term optical density (O.D.) for absorbance should be restricted to cases where not only absorption but also light scattering contribute to the attenuation of the beam as it passes through the sample (for turbidimetry of cell suspensions, for instance).

In the case of membrane proteins, two contributions to the UV absorption may be recognized. The first is the $\pi° \to \pi^-$ (nonbonding to antibonding; in the transition of the peptide N—C—O atomic group) bond (absorption at about 190 nm, $\log \varepsilon$ being 3.7–4.0) accompanied by the $n \to \pi^-$ (nonbonding to antibonding, in the O atom) transition (absorption at 220 nm, $\log \varepsilon < 2.5$).

The second contribution arises from the presence of amino acid side chains, the highest absorptivities being displayed at the following wavelengths: $\log \varepsilon$ 4.78 (188 nm, phenylalanine, $\pi \to \pi^*$), 4.68 (193 nm, tyrosine, $\pi \to \pi^*$). 4.67 (219 nm, tryptophan, $\pi \to \pi^*$), 4.30 (197 nm, tryptophan, $\pi \to \pi^*$), 4.15 (185 nm, arginine, $n \to \sigma^*$), 3.75 (280 nm, tryptophan, $\pi \to \pi^*$), 3.15 (274 nm, tyrosine, $\pi \to \pi^*$).

In spite of their high absorptivities, these absorptions contribute relatively little to the overall protein spectrum because of the preponderance of the —NH—CO— group in the sample.

Among lipids, it is particularly the conjugated diene ($\log \varepsilon$ 3.47–4.0, λ_{max} 230–260 nm), the ketone ($\log \varepsilon$ 3.30–3.40, λ_{max} 190–220 nm) and the conjugated enone ($\log \varepsilon$ 3.95–4.18, λ_{max} 220–240 nm) that contribute to the diagnostic nature of UV absorption spectrum.

In addition to the UV-absorbing properties of natural membrane components, use has been made of external probes which attach to membranes and whose optical characteristics can be examined directly in the membrane material.

2.3.1.3.2. Optical rotatory dispersion

As a monochromatic, linearly polarized beam of light passes through an optically active sample (using UV wavelengths and hence quartz cuvettes), the plane of vibration of its electrical vector will be rotated. The magnitude of rotation can be evaluated by analyzing prisms. The molar unit of rotation Φ is given by $\alpha M'/100cl$, where α is the angle of rotation in degrees*, M' the molecular weight of the solute, expressed numerically as g/cmol (for proteins the average weight of one amino acid residue is taken), c the solute concentration in g cm^{-3} and l the path in dm, the overall dimension of Φ being rad cm^2 dmol^{-1}. Since the solvent itself may be polarizable, a Lorentz correction term is used to account for this. Then

$$\Phi' = \frac{3}{n^2 + 2} \Phi \qquad (2.17)$$

where n is the index of refraction of the solvent. The ORD spectrum has usually the form of a plot of Φ against λ but, obviously, if a preparation of undefined molecular weight is examined, one can plot α against λ. The wavelengths applied to ORD studies are usually in the region from 180 to 240 nm where $\pi° \to \pi^-$ and $n_1 \to \pi^-$ electron transitions occur.

In membrane studies it is mainly the protein which yields useful ORD spectra and in this connection, they have been used particularly for distinguishing between the α-helical and other structures of peptide chains (Fig. 2.10). The diagnostic wavelength is 233 nm, the empirically (as well as theoretically; cf. Yang, 1967) calculated $[\Phi']_{233}$ being $-15\,000$ for a perfect α-helix, $-2\,000$ for a disordered chain, and $-3\,000$ to $-4\,000$ for the pleated-sheet conformation, taking a mean residue molecular weight of 115. Thus, for $[\Phi']_{233}$ more negative than, say, $-6\,000$, a considerable amount of α-helix is indicated.

The ORD spectra have been generally interpreted to indicate a prevalence of the α-helix in membranes but there are factors that may influence the quality of the spectra and, at least partly, affect the deductions made on their basis. The four important ones are the pos-

---

* This angle in radians is defined by

$$\alpha = \frac{\pi l(n_L - n_R)}{\lambda} \qquad (2.17')$$

where l is the optical path length through the sample, λ the wavelength used and n_L and n_R the indices of refraction for the left and right circularly polarized light.

sible contributions by side chains or prosthetic groups; the lack of equivalence between synthetic polypeptides used as standard for random-coil conformation and nonrepeating secondary structure found in proteins; the differences in the effect of longer helical structures as compared with a sum of shorter ones; possible existence of distorted helical regions in native proteins.

2.3.1.3.3. Circular dichroism

The value of circular dichroism, usually designated as ellipticity θ is actually the difference in absorbance by a sample of left and right circularly polarized light. The ellipticity has the same units as optical rotation, viz. rad cm^2 dmol^{-1}. It is related to the difference in absorptivity $\Delta\varepsilon$ of the left and right circularly polarized components as follows:

$$[\theta] = (2.303 \times 18\,000/4\pi)\,\Delta\varepsilon = 3\,298\,\Delta\varepsilon \qquad (2.18)^*$$

$\Delta\varepsilon$ is obtained from

$$\Delta\varepsilon = \varepsilon_L - \varepsilon_R = \frac{A_L - A_R}{cl} \qquad (2.19)$$

where A_L and A_R are the actual measured absorbances, c is the molarity of the solution and l is the optical path length (in cm). Like with the optical rotatory dispersion, the molar ellipticity $[\theta]$ can be corrected for polarizability of the solvent by the Lorentz factor $3/(n^2 + 2)$.

Thus, while ORD reflects the difference in the indices of refraction (eq. 2.17'), the CD is related to the difference in absorbances. It follows from a fundamental consideration of polarizability that they are mutually related (see Moffitt and Moscowitz, 1959), a suitable expression for this fact being in the Kronig–Kramers transform

$$[\theta'(\lambda)] = -\frac{2}{\pi\lambda} \int_0^\infty [\Phi'(\lambda')] \frac{\lambda'^2}{\lambda^2 - \lambda'^2}\,d\lambda' \qquad (2.20a)$$

$$[\Phi'(\lambda)] = \frac{2}{\pi} \int_0^\infty [\theta'(\lambda')] \frac{\lambda'}{\lambda^2 - \lambda'^2}\,d\lambda' \qquad (2.20b)$$

---

* 2.303 is the conversion factor from natural to base-ten logarithms, the factor $4\,500\pi$ derives from the definition of absorbance $A = (4\pi\varepsilon/\lambda)\log e$ and from conversion of units of radians/cm to angular degrees/dm (*cf.* Djerassi, 1960, and Beychok, 1967).

where λ is the wavelength at which Φ or θ is determined and λ' is the current integration coordinate. Although the transformation is valid strictly only for the limits of 0 and ∞, even over finite regions of the spectrum useful information can be obtained. From this consideration it follows that both ORD and CD spectra yield

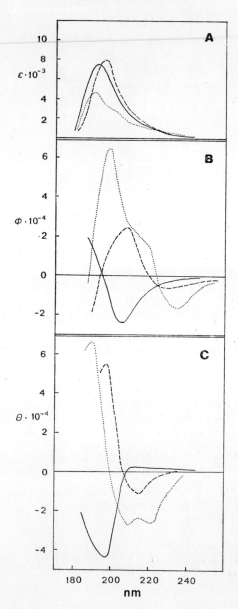

Fig. 2.10. Comparison of ultraviolet absorption (**A**), optical rotatory dispersion (**B**) and circular dichroism (**C**) spectra of poly-L-lysine with different types of secondary structure. (Adapted from Rosenheck and Doty, 1961; Holzwarth, 1972; Greenfield et al., 1967.)

much the same information. However, ORD can be measured even at wavelengths where no absorption band is present (unlike CD). On the other hand, the number of Cotton effects* is much more readily obtained from CD measurements.

Like the ORD spectra discussed above, the CD curves have been used mostly for supporting the prevalence of one protein secondary structure over another. The value of ellipticity at 225 nm, $[\theta]_{225}$, is highly negative for an α-helix ($-40\,000$) but only less so for a pleated sheet (-7000) and is positive for a completely random coil ($+1500$).

Fig. 2.10 compares the absorption, ORD and CD spectra of synthetic poly-L-lysine of three different secondary structures.

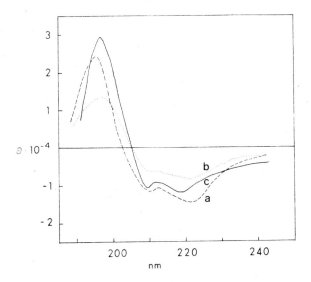

Fig. 2.11. Circular dichroism spectra of various membrane preparations. a Human red cell ghosts, b Ehrlich ascites cell plasma membranes, c supernatant at 20 000 g of sonicated mitochondrial fragments. (Adapted from Wallach, 1969; Ji and Urry, 1969.)

In working with membranes, several difficulties are encountered that complicate interpretation of CD and ORD spectra, mainly the rather low amplitudes and a red shift of the CD bands. The principal reason for this is the particle inhomogeneity of membrane suspensions.

* The Cotton effect is the term used to describe the deviations from monotony of the ORD and corresponding CD curves.

While the flattening of the curves is accounted for by the so-called absorption statistics (also Duysens' effect*, sieve effect, etc.), the red shift is probably due to light scattering on the particles (for a scholarly treatment of the subject consult Holzwarth, 1972).

Fig. 2.11 compares the CD spectra of several membrane preparations, showing the predominance of the α-helical structure, where the mentioned effects of particle size can be recognized.

2.3.1.4. *Nuclear magnetic resonance*

The technique of nuclear magnetic resonance is now in common use for a variety of analytical purposes. Their application to the investigation of lipids, both isolated and *in situ* in membranes, has yielded important information particularly on the fluidity of membrane lipid sheets.

The principle of the technique is the measurement of energy required to change the alignment of atomic nuclei in an external magnetic field. A nucleus with a nonzero spin (a paramagnetic nucleus) will absorb the energy of electromagnetic radiation applied at right angles to the external magnetic field, at a frequency defined by

$$v = \mu H_o / hI = \gamma H_o / 2\pi \qquad (2.21)$$

where μ is the nuclear magnetic moment, H_o is the external magnetic field strength, h is the Planck constant, I the nuclear spin and γ the so-called gyromagnetic ratio, specific for a given nucleus. Some of the characteristic values for biologically important isotopes are given in Table 2.13. The resonant frequency is usually found by varying the magnitude of the applied magnetic field at a constant frequency of electromagnetic radiation.

The value generally referred to in NMR spectrometry is the chemical shift δ defined by

$$\delta = \frac{v_S - v_R}{v_R} \cdot 10^6 \qquad (2.22a)$$

where v_S is the resonant frequency actually measured in a sample, v_R the resonant frequency of a reference compound (often tetramethyl-

---

* The absorption-flattening coefficient (Q_A), relevant to the Duysens effect is defined as the ratio of suspension absorbance to the true solution absorbance.

TABLE 2.13. Characteristics of some nuclei used in NMR studies (in a magnetic field of 1 T)

Nucleus	Spin I	NMR resonant frequency ν (MHz)	Magnetic moment μ (in nuclear magnetons[a])	Relative sensitivity at constant field	Relative sensitivity at constant frequency
^{1}H	1/2	42.57	2.793	1.00	1.00
^{2}H	1	6.54	0.857	0.009 64	0.409
^{7}Li	3/2	16.55	3.257	0.294	1.94
^{13}C	1/2	10.71	0.702	0.015 9	0.251
^{14}N	1	3.077	0.404	0.001 01	0.193
^{15}N	1/2	4.316	—0.283	0.001 04	0.101
^{17}O	5/2	5.772	—1.893	0.029 1	1.58
^{19}F	1/2	40.07	2.627	0.834	0.941
^{23}Na	3/2	11.262	2.217	0.092 7	1.32
^{31}P	1/2	17.24	1.131	0.066 4	0.405
^{33}S	3/2	3.267	0.643	0.002 26	0.384
^{35}Cl	3/2	4.173	0.821	0.004 71	0.490
^{43}Ca	7/2	2.864	—1.315	0.063 9	1.41
^{127}I	5/2	8.519	2.794	0.093 5	2.33

[a] A nuclear magneton is analogous to the Bohr magneton (referring to an electron) and is defined as $eh/4\pi mc$ (e = elementary electrical charge, h = Planck's constant, m = mass of the proton, c = speed of light). When expressed in the absolute electromagnetic cgs system, it has the value of $5.05 \cdot 10^{-24}$ erg/gauss; in the SI system it is $5.05 \cdot 10^{-27}$ J T^{-1}.

silane or sodium 2,2-dimethyl-2-silapentane-5-sulfonate). δ may also be defined as

$$\frac{H_S - H_R}{H_R} \cdot 10^6 \qquad (2.22b)$$

the H's referring to the field strength at which resonance is observed. The chemical shift is brought about by the fact that the local magnetic field around the nucleus is different from that applied externally due to various shielding effects, dipole interactions and through the bond coupling with different neighboring nuclei. Because of undesirable effects of dissociable solvent hydrogen, measurements are usually performed in D_2O solutions (or in $CDCl_3$ in organic chemistry).

Instead of the δ-scale, a τ-scale is sometimes used, the relationship between the two being $\delta = 10 - \tau$. The chemical shifts due to various groupings are shown schematically in Fig. 2.12.

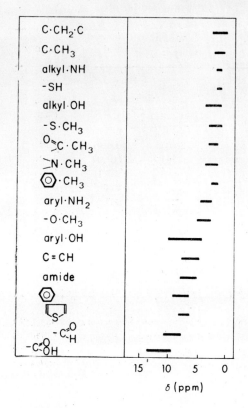

Fig. 2.12. Chemical shift values of various chemical groups in nuclear magnetic resonance.

The character of the NMR spectrum is determined by the life-time of the spin states, by the dipole interactions with neighboring nuclei, by magnetic field inhomogenities and by various minor factors.

In describing the lifetime of the spin states, let us consider a proton which can exist in two spin states: parallel and antiparallel to the magnetic field applied. The originally equal distribution in the two states will change if a radiation of proper frequency is operative (resonant frequency of the nucleus). The time required for attaining 63.2 % of the new equilibrium state ($= (e - 1)/e$) is called the spin-lattice relaxation time (for analogies with chemical relaxations see, e.g., Gutfreund, 1972). The frequency of the field surrounding the given nucleus is determined by fluctuating magnetic dipoles in the vicinity, the strength of interaction being inversely proportional to r^6 (r being the distance between interacting dipoles) so that intramolecular interactions are much greater than intermolecular ones. The effect on the lattice-spin relaxation is the greater the larger the magnetic dipole of the neighboring nucleus (paramagnetic molecules, for instance).

The line width in the NMR spectrum is also affected by the so-called low-frequency dipole interactions between nuclei with a nonzero magnetic moment, the underlying effect being a change in the field surrounding a given nucleus. Like the spin-lattice relaxation, it is more powerful in the presence of interaction of large magnetic dipole moments.

Another important factor in changing the line width is the magnetic field inhomogeneity. The total line width is proportional to the reciprocal of the transverse relaxation time T_2, thus:

$$\pi \Delta v = 1/T_2 = 1/2T_1 + 1/T_{2m} + 1/T_{2d} \qquad (2.23)$$

where T_1 is the relaxation time for the spin-lattice relaxation process, T_{2d} that for the dipole interaction between nuclei and T_{2m} that for the inhomogeneity of the magnetic field (these last two are computed using the uncertainty principle; cf. Abragam, 1961). Δv for a Gaussian-type line (bell-shaped) is the peak width at maximum slope, for a Lorentzian-type line (fine-tipped, with long segments of practically invariant slope) is the peak width at half-maximum height.

NMR spectra are usually obtained at suitable applied frequencies (40, 60, 100, 120, 220 MHz) and show the absorption of energy (as a peak) versus magnetic field strength H, in either of two ways: If the sample is exposed to the frequency for a longer period of time we speak of the continuous-wave method; if intense pulses of the field are applied we have to do with the pulse method. This second method has the advantage that it requires much less time but, on the other hand, requires computer assistance to achieve results comparable to the continuous-wave method. The sensitivity of the method (generally 1 ml sample of a 1 mM solution) can be increased by sweeping over the entire field strength many times and computing the averaged transients (the CAT method).

Depending on the homogeneity of the applied magnetic field we speak of wide-line NMR and of high-resolution NMR. The first of these requires a homogeneity of about $1 : 10^6$ and is suited for studies where motion is relatively slow and anisotropic. The second technique, with magnetic field homogeneity of at least 1 : 10^8 is technically more demanding. An advantage in resolution may be obtained by spinning the sample at the "magic" angle, defined as $\sec^{-1} \sqrt{3}$ (= 54.74°), to the magnetic field.

An example of NMR spectrum is shown in Fig. 2.13.

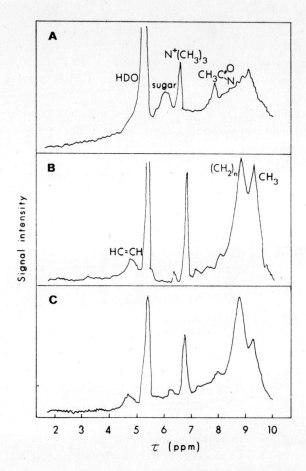

Fig. 2.13. Nuclear magnetic resonance spectra of 5 % sonicated dispersion in heavy water. **A** Erythrocyte membrane fragments, **B** total membrane lipid (phospholipids plus cholesterol), **C** total membrane phospholipid. (Redrawn from Chapman et al., 1968.)

2.3.1.5. *Electron spin resonance*

Whereas all the above-described methods studied or could study the membrane components either *in situ* or after suitable extraction without any alteration of their composition*, electron spin resonance is a technique which requires the introduction into the membrane of a probe, generally a molecule with an unpaired electron, a paramagnetic recorder. Practical application of the technique rests on the possibility to synthesize compounds containing a nitroxy group where, at the nitrogen atom, there is an unpaired electron in the $2p\pi$ orbital confined practically to the nitrogen atom. The basic structure of a nitroxide is

$$\begin{array}{c} H_3C \quad R^1 R^1 \quad CH_3 \\ C\text{—}N\text{—}C \\ H_3C \quad \quad CH_3 \\ \quad O \end{array}$$

and the following typical derivatives have been prepared:

Tempol

* Although UV and fluorescence external probes are also in use.

```
              CH₃
            |—CH₃
       O   N—O
CH₃                              OCH₂
 \/\/\/\/\/\/\/\/\/\/             |
                                  C=O
                                  |
                                  OC   O
CH₃                               ||   ||
 \/\/\/\/\/=\/\/\/\/\/            O   CH₂OPOCH₂CH₂N⁺CH₃
                                       |
                                       O⁻
```

The usefulness of the ESR method is in providing information about the viscosity and polarity of local areas of membranes, the fluidity of these areas, the degree of molecular ordering and rotational and translational freedom.

The underlying theory proceeds from the fact that a magnetic moment μ precesses in a magnetic field of intensity H with a characteristic frequency (the Larmor precession frequency) defined by

$$\omega = 2\pi v = 2\pi\mu H/h. \qquad (2.24)^*$$

For paramagnetic atoms, this equation may be written as

$$\omega = 2\pi g\beta H/h \qquad (2.25)$$

where β is the Bohr magneton (cf. Table 2.13), h is the Planck constant and g is a characteristic value of free radicals (2.0023 for pure electron spin). The value of g has a tensorial order, being 2.0088 in the x direction, 2.0062 in the y direction and 2.0027 in the z direction (the example is taken from di-tert-butyl nitroxide).

If the above frequency is applied externally (it lies in the microwave region) the magnetic moments are made to precess coherently so that a magnetic field is induced in the sample at right angles to the constant external field. At this frequency an absorption of energy is recorded.

Practically speaking, in an ESR spectrum where the magnetic field H (in gauss or fractions of tesla, such as mT) is varied at a fixed frequency (usually 9.5 or 35 GHz), a peak is observed when the condition of eq. (2.25) is reached. An example of ESR spectrum is shown in Fig. 2.14. This spectrum is often electronically converted to

* This principle is the same as that underlying the use of NMR; cf. eq. (2.21).

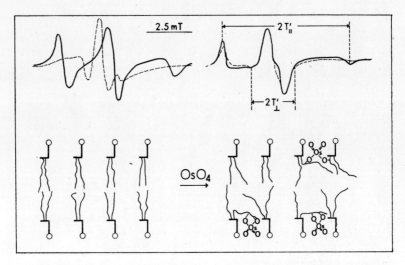

Fig. 2.14. Electron spin resonance spectrum of 5-[4',4'-dimethyloxazolidine-N-oxyl]stearic acid in a lecithin multilayer. At left, normal membrane; at right, after treatment with osmium tetroxide. The heavy lines in the spectra correspond to the supporting slide being perpendicular to the preparation, the dashed lines to the slide being parallel with the preparation. The values of $T'_\parallel$ and $T'_\perp$ refer to equation (2.27). The lower part of the figure shows schematically the arrangement of phospholipid molecules before and after treatment with OsO_4. (Adapted from Jost et al., 1971.)

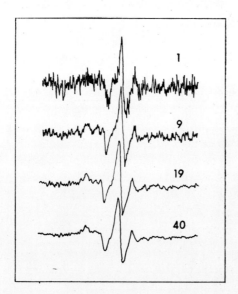

Fig. 2.15. Increasing the signal-to-noise ratio by repeated scanning (the number of sweeps is shown at each curve). The sample used was $3.75 \cdot 10^{-5}\,M$ sodium 5-[4',4'-dimethyloxazolidine-N-oxyl]stearate in an aqueous solution of 5 % egg lecithin. (Redrawn from Jost et al., 1971.)

its first or second derivative to provide further information. The interpretation (and simulation) of ESR spectra has been closely linked to the use of computers. There exist methods making it possible to predict the shape of an ESR spectrum on the basis of various assumptions about rotation of the spin-labelled molecule (e.g. Itzkowitz, 1967). Another computer application lies in the processing of repeated scans of the same spectrum to differentiate between background noise and true signal, the improvement obtained being proportional to $n^{0.5}$ where n is the number of sweeps (cf. Fig. 2.15).

The fine structure of the spectral lines contains information on the effects of rotation, interaction between molecules, transfer of spin labels across membranes and changes due to temperature (cf., e.g., Jost et al., 1971).

2.3.1.6. Fluorescence

Application of fluorescence techniques to membrane investigation has been scanty and probably more may be expected in the future. Still, some important information on subtle perturbations of conformation has been obtained by using both intrinsic protein fluorescence and the fluorescence of extrinsically added labels.

The principle of fluorescence is easy to grasp. A molecule in the ground state of energy may interact with light of an appropriate wavelength (its energy E is equal to hv, where h is the Planck constant and v ($= c/\lambda$) is the frequency) and be thus raised to a higher energy level, represented either by an excited singlet state (where all electrons in the molecule have paired spins) or by an excited triplet state (where one electron pair has unpaired spins). The triplet state energy level is somewhat lower than that of the corresponding singlet state, no matter whether the transition is of the type $\pi \to \pi^*$ or $n_1 \to \pi^*$. The energy of the excited state can then be released in various ways. (1) A very frequent one is the internal conversion which is not accompanied by any except thermal radiation; it may proceed with rate constants k_c of $10^2 - 10^{11}$ s^{-1}. (2) An important one is fluorescence characterized by a rate constant k_f of 10^8 s^{-1} where light of a lower frequency than that used for excitation is given off. (3) A rare one is phosphorescence, occurring from a triplet state where the rate constant k_p is of the order of 10^2 s^{-1} and which persists for appreciable periods after the excitatory radiation is turned off (cf. Fig. 2.16). The characteristics of some fluorescent molecules are shown in Table 2.14.

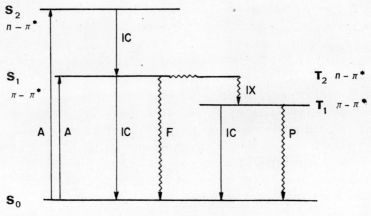

Fig. 2.16. Scheme of energy transitions during luminescence processes. The ground level is shown by S_0, the excited singlet levels by S_1 and S_2, the excited triplet levels by T_1 and T_2. A Absorption, IC internal conversion, IX intersystem crossing (with reversal of electron spin), F fluorescence, P phosphorescence.

TABLE 2.14. Wavelengths used in fluorescence assay

Wavelength (nm)	Energy (kJ mol^{-1})	Color	Region of typical excitation maxima	Region of typical emission maxima
200	598.7			
250	478.9		Phenylalanine	
300	399.0		Tryptophan, tyrosine	
350	342.1		NADH, ANS[a], vitamin A	Tryptophan, tyrosine
400	299.4	Visible limit		
450	265.9	Violet-blue	Flavins	NADH
500	239.5	Green		Flavins, vitamin A, protein-bound ANS
550	217.7	Yellow		Free ANS
600	199.7	Orange		
650	184.2		Chlorophyll	Chlorophyll
700	170.8	Red		

[a] 1-Anilinonaphthalene-8-sulfonate.

An important characteristic of the efficiency of fluorescence is the so-called quantum yield, defined as

$$Q_f = \frac{\text{number of quanta emitted}}{\text{number of quanta absorbed}} = \frac{k_f}{k_f + k_p + k_c} \quad (2.26)$$

Of the naturally occurring membrane components, only proteins show an intrinsic fluorescence which is due to the presence of phenyl-

TABLE 2.15. Fluorescence of amino acids

Amino acid	Absorption		Emission	
	λ_{max} (nm)	$\log \varepsilon$ (M^{-1} cm^{-1})	λ_{max} (nm)	Q_f
Phenylalanine	257	2.30	282	0.035
	206	3.95		
	187	4.76		
Tyrosine	275	3.08	303	0.21
	222	3.90		
	192	4.67		
Tryptophan	280	3.74	350	0.20
	220	4.51		
	196	4.32		

alanine, tyrosine and tryptophan. However, fluorescence quantum yield of these amino acids (as shown in Table 2.15) is markedly depressed in proteins, apparently due to quenching through collisions of the high-energy excited rings with protons from dissociable protein groups ($Q_f = 0.008 - 0.07$). Moreover, the fluorescence of proteins (except those without tryptophan) is due to the presence of tryptophan alone, the position as well as the height of the maximum depending both on the degree of hydrophobicity of the tryptophan milieu and on the quality of the solvent.

In membranes as such, the fluorescence behavior resembles that of proteins but there the various subtle changes in the emission spectrum may be interpreted in terms of interactions with other components (hydrophobic vs. polar) and of conformational changes. An example of this last-named phenomenon, described for membrane proteins *in vitro* is provided by the glutamine-binding protein from *Escherichia coli* (Weiner and Heppel, 1971).

A rather greater impact on our knowledge of membrane protein-lipid interactions has come from the use of so-called extrinsic probes which can be bound to the protein either covalently or, more frequently, adsorbed through noncovalent bonds.

Covalent probes include 4-acetamido-4′-isothiocyanostilbene-2,2′-disulfonic acid, diazosulfanilate, formylmethionylsulfone methyl phosphate and 1-dimethylaminonaphthalene-5-sulfonyl chloride.

The indicator of choice for the noncovalent type of binding has been sodium 1-anilinonaphthalene-8-sulfonate (ANS). In solution, it has a Q_f of 0.004 but when bound to a protein or a lipoprotein membrane the Q_f may rise to 0.8 and the emission λ_{max} is shifted (*cf.* Table 2.14). Moreover, it is highly sensitive to minute changes in mutual subunit positions, such as occur in allosteric proteins and possibly in biological membranes.

Other noncovalent fluorescent probes used are 1-toluidinonaphthalene-5-sulfonate (TNS), *N*-phenyl-1-naphthylamine (NPN), arsanilinochloromethoxyacridine and pyrene-3-sulfonate. Other fluorescent molecules come into use particularly as membrane markers and will be mentioned in the chapter on fractionation of membranes (2.6).

Fluorescent probes have also been frequently used for the observation of transport across cell or mitochondrial membranes (retinol, ostruthin, umbelliferone), a sensitive fluorescence microscope or microspectrofluorimeter being of essence for attaining useful data (e.g., Rotman and Papermaster, 1966).

2.3.1.7. *Calorimetry and related techniques*

There are two principal methods based on measuring the thermal properties of membranes or their components: the differential thermal analysis (DTA) and the differential scanning calorimetry (DSC). In the first of these, a sample is heated and then left to cool spontaneously while its temperature is being recorded. A generally smooth decrease is obtained in regions where no phase changes occur but a break or a kink is observed at the point of a thermal transition (solidification, for example). If another, inert reference sample is processed in parallel, it is possible to record the differences in the temperature of the two and these are then plotted against temperature (*cf.* Fig. 2.17).

In differential scanning calorimetry, which is superior to the preceding technique, both the experimental and the reference sample are kept at the same temperature (or a constant temperature difference) while being heated by a regulated heat flow. The differences in current

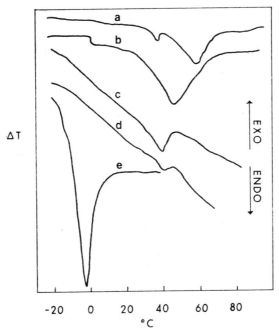

Fig. 2.17. Differential thermal analysis of myelin at different relative humidities. The percent water content was as follows: a 3 %, b 5 %, c 10 %, d 15 %, e 30 %. (Redrawn from Ladbrooke et al., 1968.)

required for maintaining the samples at the same temperature are then plotted against temperature as shown in Fig. 2.18. Whereas in DTA an endothermic transition is plotted downward, in DSC it is shown as an upward deflection.

The techniques have been used particularly to study the melting of membrane lipids, the effects of lipid-protein interactions, composition, presence of water, lipid-metal interactions, etc. Dependence of various secondary marker effects (fluorescence probes in particular) on gradually increasing temperature has been widely used as a diagnostic for membrane structure but this is not actually a thermal technique and it will be discussed in connection with the structure of membrane lipids.

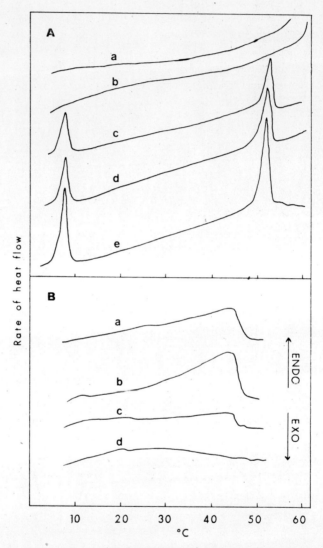

Fig. 2.18. Differential scanning calorimetry. **A** Mixtures of water with distearoyl-lecithin. Water content (% w/w): a 10, b 20, c 25, d 30, e 40. (Adapted from Chapman and Dodd, 1971.) **B** 50 % dispersions of lipids from *Acholeplasma* (*Mycoplasma*) *laidlawii* B membranes. a Total lipids, b glycolipids, c phospholipids, d neutral lipids. (Adapted from Chapman and Urbina, 1971.)

2.3.1.8. Other techniques

Dilatometry has been little used so far but it is a suitable auxiliary technique to fluorescence and calorimetric methods. In analogy with changes in the specific heat content of a sample, its volume will change (this is true particularly of lipids; see Träuble and Haynes, 1971). Using a dilatometer, a volumetric device that can be designed in

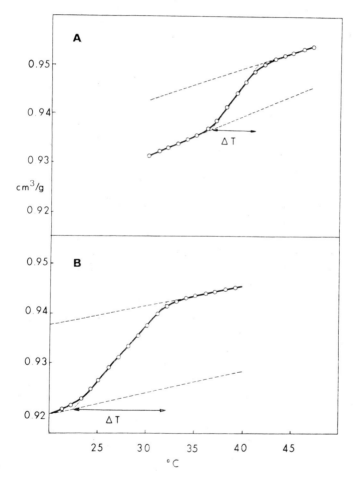

Fig. 2.19. Dilatometry of 18.5 % (v/v) mixtures of trans-18 : 1-phospholipids **A** and of trans-16 : 1-phospholipids **B** in water. The ordinate shows the specific volume of the lipid at increasing temperature after subtraction of water expansion. The transition temperature T_t lies half-way between the two extrapolated straight lines. (Redrawn from Overath and Träuble, 1973.)

different ways to measure small volume changes (of 0.1 % or so) one can compare the volume expansion of a mixture of lipid with water, with that of distilled water alone. It will be seen from Fig. 2.19 that during the gel−liquid transition of lipids a relatively greater increase in the specific volume of the dispersed lipid is found.

A basically identical information can be obtained from the measurement of light scattering of a lipid−water dispersion. Since the amount of light scattered from particles increases with their size, a plot of scattered light intensity *vs.* temperature will show steeper negative slopes in the region where volume suddenly increases.

2.3.2. Organization of lipids

2.3.2.1. Biological membranes

The amount of data pertaining to the structure and organization of membrane lipids is enormous and rather heterogeneous but the last few years have witnessed a certain convergence of opinion among lipid chemists and physical chemists dealing with membranes so that a fairly clear outline of the situation can be offered.

The majority of membrane lipids are probably in the bilayer arrangement, such as formed spontaneously in mixtures of phospholipids with water (p. 32). Thermodynamic considerations as well as physicochemical evidence show rather convincingly that the polar heads of the lipids face outward, the hydrophobic chains pointing into the membrane. Molecular models of lipids have been employed to bring analytical data into agreement with the steric requirements of such a bimolecular leaflet (Finean, 1958; Vandenheuvel, 1963; Jost *et al.*, 1971). One of the possibilities, based partly on X-ray diffraction data, is shown in Fig. 2.20.

As indicated in the lowest panel of the figure, there is a more or less stiff region of the hydrocarbon chains near the polar heads and a more flexible region in the center of the bilayer. The thickness of the double layer depends, among other things, on the presence of negatively charged phospholipids in the juxtaposed layers, mutual repulsion resulting in greater hydration. Addition of electrolytes with strongly binding positive counterions (Na^+, K^+) shrinks the membrane; addition of Ca^{2+} may result in bridges being formed between the neighboring phosphatides with the result of closer packing and decreased permeability.

The biological bilayers are far from being symmetrical in composition. Of the various examples available, the erythrocyte membrane is best understood (Bretscher, 1972; Verkleij et al., 1973). Phosphatidyl choline and sphingomyelin appear predominantly in the outer layer, phosphatidyl serine and phosphatidyl ethanolamine mainly in the inner one.

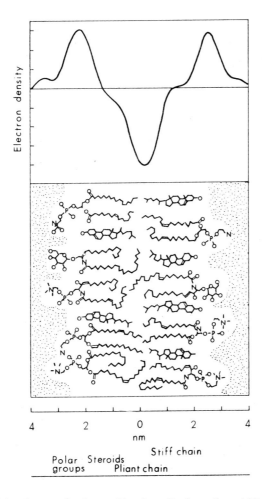

Fig. 2.20. The electron density profile of myelin from the rabbit sciatic nerve (upper panel) and a schematic representation of the myelin structure, showing the rigid and flexible parts of the phospholipid + cholesterol bilayer. (Adapted from Caspar and Kirschner, 1971.)

The closeness of packing of the phospholipids is greatly influenced by the quality of the fatty acyl residues and by the presence of nonpolar lipids, in particular of sterols. As shown in Table 2.16, several

Table 2.16. Packing areas of fatty acids and phospholipids (adapted from O'Brien, 1967) in nm^2/molecule

Compound	Native	After adding cholesterol 1 : 1
Myristic acid (14 : 0)	0.37	—
Palmitic acid (16 : 0)	0.24	—
Stearic acid (18 : 0)	0.23	—
Oleic acid (18 : 1)	0.48	—
Erucic acid (22 : 1)	0.40	—
16-Methylheptadecanoic acid	0.32	—
1,2-Dimyristoyllecithin	0.72	0.61
1,2-Distearoyllecithin	0.45	0.45
1,2-Dioleyllecithin	0.83	0.77
1,2-Dilinoleoyllecithin	0.99	0.99

effects may be recognized. Straight-chain fatty acids alone, as well as their corresponding lecithins, take up less area the longer the hydrophobic chain (due to stronger hydrophobic interactions); branching and particularly *cis*-unsaturated bonds cause a substantial increase in packing; cholesterol condenses the packing particularly of monounsaturated phospholipids. There is steric justification for this since phospholipids (especially PC and PS) have surface head groups greater than their hydrophobic ends while cholesterol is frayed so that they fit nicely (Israelachvili and Mitchell, 1975). In comparable membranes, such as those of various mammalian erythrocytes, the closeness of packing is directly proportional to the relative impermeability of the membrane. Thus, in the series of erythrocytes of rat, rabbit, man, pig, ox, the content of arachidonic (20 : 4) acid decreases (*cf.* Table 2.9) in the same order as the susceptibility to hemolysis (measured by the rate of bursting in glycol, glycerol or thiourea; Jacobs *et al.* 1950). The reducing effect of cholesterol on permeability observed both in *Mycoplasma* and in erythrocytes is in keeping with these arguments (McElhaney *et al.*, 1970; Bruckdorfer *et al.*, 1969).

A high rigidity of the packing, however, may not be the desirable property of lipid membranes. On the contrary, as shown by growing

Acholeplasma laidlawii in the presence of stearic acid plus small amounts of various loosely packing fatty acids, cell lysis ensuing in the presence of stearic acid alone can be prevented the more effectively the greater the spatial demands of the accompanying fatty acid (*cf.* Tourtellotte, 1972).

At "biological" temperatures, the membrane lipids (at least in the center of the bilayer) appear to be in a liquid-crystalline or rather defective ordered state which, on lowering the temperature, passes into a crystalline state. The transition point, often designated as the melting temperature T_m or transition temperature T_t, can be defined by various techniques (*cf.* Fig. 2.17, 2.18, 2.19) including extrinsic fluorescence, its value depending greatly on the composition of the lipid bilayer as well as on various external factors such as pH and, obviously, on the interactions with other membrane components.

As to the composition effects, the more unsaturated or branched or cyclized the fatty acyl residues, the lower the T_m. This may be of critical importance for psychrophilic organisms which, to be able to thrive at low temperatures, contain high amounts (up to one-third of the total) of *cis*-unsaturated acids.

An interesting example of a regulatory effect of lipid composition is provided by the isomerization of pinifolic acid present in relatively large amounts in conifer chloroplasts. In the frost-resistant state in winter it occurs as

while in summer the three-ring form is present, rendering the chloroplasts sensitive to a lowering of temperature (Bervaes and Kuiper, 1975)

(Position of the carboxyl group is not definitely established.)

Artificially prepared lecithins show the following transition temperatures in dependence on the type of acyl present:

Acid	T_t, °C
14 : 0	23
16 : 0	41
18 : 0	58
22 : 0	75
18 : 1 (cis)	−22

In qualitative agreement, *Escherichia coli* membranes enriched by growth on different fatty acids, show the following T_t:

Enriched with	T_t
16 : 1 (trans)	27
18 : 1 (trans)	35
18 : 1 (cis)	15

The effect of pH on the transition temperature is apparently associated with the degree of protonation of phospholipids and hence their electrostatic interaction with the surrounding milieu and with

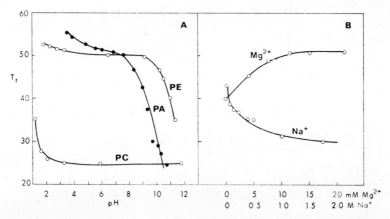

Fig. 2.21. Effect of pH and of cations on the transition temperature of different phospholipids, using fluorescence measurement of $2 \cdot 10^{-6}$ M N-phenylnaphthylamine. The transition temperatures shown were recorded at increasing temperature. **A** Effect of pH on the transition temperature of phosphatidyl ethanolamine (**PE**), phosphatidyl choline (**PC**) and phosphatidic acid (**PA**), all $2.5 \cdot 10^{-4}$ M, with myristic acid as the constituent acyl residues. **B** Effect of Na^+ and Mg^{2+} on the transition temperature of $2.5 \cdot 10^{-4}$ M C_{14}-phosphatidic acid at pH 9.7 (for Na^+) and 9.2 (for Mg^{2+}). (Adapted from Träuble and Eibl, 1974.)

neighboring molecules. Similarly, divalent cations Mg^{2+} and Ca^{2+} increase the transition temperature by binding to negative charges of phospholipids. Somewhat surprisingly, the effect of univalent cations Li^+, Na^+ and K^+ is to decrease the transition temperature. These ions do not interact with negatively charged phospholipids and their effect is probably due to an increase of ionic strength which reduces the surface potential and thus brings about further ionization of phospholipids (Fig. 2.21).

Events taking place at the transition temperature lend further support to the bilayer model. Above T_t the apparent cross-section area per lipid increases in the case of dipalmitoyllecithin from 0.48 to 0.58 nm² while the thickness of the bilayer decreases from 4.6 to 4.0 nm, in agreement with the view of kinks being formed in the array of lecithin molecules (Träuble and Haynes, 1971).

The importance of the "fluid" state of membrane lipids for various membrane processes can be documented by comparing the temperature dependence of various transports with the lipid composition of the membrane. The Arrhenius plot of a transport rate shows usually

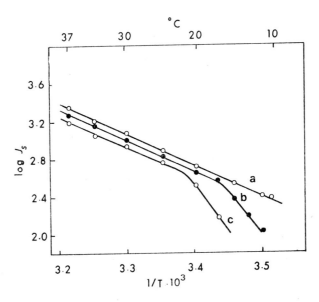

Fig. 2.22. Arrhenius plots of the rate of uptake of 2-deoxyglucose by *Acholeplasma laidlawii* grown in a medium containing extra oleic acid (a), without any supplement (b) and without supplement but cells containing exceptionally many saturated lipids (c). (Adapted from Tourtellotte, 1972).

a break at the transition temperature—this holds for thiomethyl-β-galactoside efflux, *o*-nitrophenyl-β-galactoside uptake in *Escherichia coli* (Overath *et al.*, 1971*a*), for Na,K-adenosinetriphosphatase of sheep kidney medulla (Grisham and Barnett, 1973), for the Mg^{2+}-adenosinetriphosphatase of *Acholeplasma laidlawii* (de Kruyff *et al.*, 1973) and for 2-deoxyglucose uptake by the same microorganism (Tourtellotte, 1972) as shown in Fig. 2.22.

The more general importance of the fluid state of the membrane for transport processes is best demonstrated by the fact that the transition temperatures for both β-galactoside and β-glucoside uptake by *Escherichia coli* are identically related to the membrane lipid composition (influenced by supplementing the medium with one or another fatty acid).

Supplement to medium	Break in Arrhenius plot °C
Elaidic acid (*trans*-Δ^9-18 : 1)	30
Oleic acid (*cis*-Δ^9-18 : 1)	13
Dihydrosterculic acid $(CH_3\text{—}(CH_2)_7\text{—}CH\text{——}CH\text{—}(CH_2)_7COOH$ $\diagdown CH_2 \diagup)$	11
cis-Vaccenic acid (*cis*-Δ^{11}-18 : 1)	10
Linoleic acid (*cis*, *cis*-$\Delta^{9,12}$-18 : 2)	7

The temperature characteristic of ion transport processes (studied mainly in artificial membranes) is an important tool for defining the action of various ionophore antibiotics (*cf.* section 5.3.6.). Thus, the abrupt cessation of the conductivity effect of nonactin, valinomycin, and the like, as the lipids "crystallize" at a certain temperature, indicates the mobility of the ionophores to be instrumental for their function. On the other hand, the channel-forming ionophores, such as gramicidin, remain operative even well below the transition temperature.

In a number of instances (*cf.* Linden *et al.*, 1973) two breaks were found in the temperature dependence both of transport rate and of spectral parameters of the Tempol spin label. The higher-temperature transition denotes the moment where solid patches of membrane lipids are first detected and when an increase in lateral compressibility is to be expected, the lower-temperature one denotes the end of lateral phase separation when all of the lipid is in a solid phase. The data suggest that the transport agency lies perpendicularly in the membrane and is sensitive to lateral compression (an example is shown in Fig. 2.23).

A major component of cell membranes that bears on the fluidity of cell lipids is cholesterol (and possibly ergosterol and other sterols) but not epicholesterol and other 3α-OH isomers. By interacting

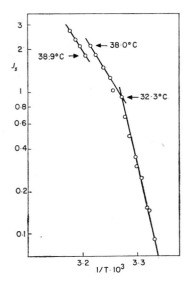

Fig. 2.23. Arrhenius plot of β-galactoside transport in *Escherichia coli*, using cells grown in elaidate-supplemented medium. (Redrawn from Linden et al. 1973.)

particularly with the liquid-crystalline lecithin, cholesterol affects the ordered arrangement of hydrocarbon chains and hence its increased concentration results in a lowering of T_t both in artificial mixtures (Chapman, 1969) and in biological membranes (de Kruyff et al., 1973, studying the membrane Mg^{2+}-adenosinetriphosphatase).

Using data of electron spin resonance, one can characterize the fluidity of a membrane by the so-called order parameter S (Seelig, 1970; Hubbell and McConnell, 1971) which may be defined as

$$S = (T'_\parallel - T'_\perp)/(T_z - T_x) \qquad (2.27)$$

$T_z - T_x$ being 2.5 mT, $T'_\parallel$ is half the distance between two outer side bands in the ESR spectrum, $T'_\perp$ is half the distance between two inner lines of the spectrum (cf. Fig. 2.14). S defines the average orientation of the long axis of the $2p\pi$ orbital of the spin-labelled molecule with respect to the direction normal to the membrane. It can be used for computing the flexibility of a lipid bilayer (cf. Seelig et al., 1973).

The fluidity of the lipid phase does not signify a completely random movement of lipid molecules in the membrane but still the value of lateral diffusion is remarkably high, being of the same order of magnitude in an artificial dipalmitoyllecithin membrane ($D = 3 \cdot 10^{-8}$ cm^2 s^{-1}) as in an *Escherichia coli* membrane ($D = 3.25 \cdot 10^{-8}$ cm^2 s^{-1}). This corresponds to an average distance traveled per second of 3 μm, on the basis of Einstein's equation (eq. 4.26). This corresponds to about 1/10 the distance around an average cell. The lateral diffusion coefficient was arrived at by an elegant method of electron spin resonance using N-oxyl-4′,4′-dimethyloxazolidine derivatives of stearic acid (Sackmann *et al.*, 1973). The exchange interaction between spin-labelled molecules is related to the rate of lateral (two-dimensional) diffusion and brings about a broadening of the ESR spectrum if the label-to-lipid ratio is greater than 0.025. From the broadening of the spectrum (ΔH_{ex}) which is a component of the total width of the central line

$$\Delta H = \Delta H_0 + \Delta H_{dipole} + \Delta H_{ex} \qquad (2.28)$$

(ΔH_0 is the width at an extremely low label concentration, ΔH_{dipole} is the broadening due to dipole-dipole interactions and can be neglected for ratios less than 0.1), the ΔH_{ex} can be obtained. This value is related to the exchange frequency W_{ex} ($W_{ex} = 1.4 \cdot 10^6 \, H_{ex}$; in Hz) and this, in its turn, to the lateral diffusion coefficient

$$D = 3/4 \frac{F \lambda W_{ex}}{d_c} \cdot \frac{1 + c}{c}. \qquad (2.29)$$

where F is the area per lipid molecule (0.5−0.6 nm^2), d_c the critical distance for the occurrence of spin exchange (about 2 nm) and λ the length of one diffusional jump in the lipid lattice (about 0.8 nm); c designates the label-to-lipid ratio. Now, in this case

$$D = 1.82 \cdot 10^{-15} W_{ex} \frac{1 + c}{c} \qquad (2.29')$$

The magnitude of lateral diffusion is of paramount importance in considering processes of biogenesis of membranes and immunological response.

Besides lateral diffusion, there is the possibility of a flip-flop movement of lipids in the membrane. However, in contrast with lateral diffusion, the flipping is extremely improbable, with half-times of the order of hours.

The significance of the phase transition of lipid membranes for the regulation of cellular processes remains to be explored. Although it is obvious that the operational state of membrane lipids is the fluid state there may be instances when a sudden decrease of a membrane function (generally a transport process) is required – a case in point is the transmission of the nerve impulse when a change in permeability for Na^+ vs. K^+ is held responsible for its propagation (cf. p. 284). One could perhaps speculate that local changes in cation concentrations could bring about a phase transition either of the whole membrane or, more likely, of an area surrounding the transport agency or governing the permeability to a given solute. This mechanism would be of general applicability and might account for some of the rapid permeability responses particularly of excitable membranes.

However, the fluidity of lipids in membranes is not a sufficient condition for transport processes to occur. It should be made clear that the presence of proteins in general and of specific proteins in particular is essential, as will be shown in the following sections of the book.

2.3.2.2. Artificial membranes

As was stated several times before, molecules of amphipathic lipids dispersed in a high enough concentration in an aqueous medium tend to aggregate to a bilayer structure with the hydrocarbon chains pointing inward and the polar heads interacting with the medium. Not only are the membranous structures thus formed stable but they can be fairly easily formed as a unit bilayer membrane that is amenable to experimental manipulation. There are two main forms of such membranes: (1) The classical (Mueller et al., 1962; Mueller and Rudin, 1968) *flat membrane* (often called the black lipid membrane because of negative interference effects causing it to appear black under the microscope). Its formation is shown in Fig. 2.24 and some of its properties appear in Table 2.17.

(2) The *spherical membrane*, either with the same or with different media outside and inside. These membranes range in size from lipid vesicles (e.g., Bangham, 1968) as little as several tens of nm across, to larger (hydrated and dehydrated) vesicles (e.g., Mueller and Rudin, 1969) about 1 µm across, to large "bubble" membranes as large as a centimeter in diameter (Mueller et al., 1964). The small-size vesicles can be formed in a multilayered version and this is the type usually

called liposomes. The formation of these types of membranes is shown in Fig. 2.25 and 2.26. A recent comprehensive treatise of liposomes may be found in Bangham *et al.* (1974).

Table 2.17. Physical characteristics of biological and lipid bilayer membranes (adapted from Wallach, 1972)

Property	Biological membrane	Lipid bilayer
Thickness (nm)	4—13	4.6—9.0
Resistance (Ω cm^2)	10^2—10^5	10^3—10^9
Capacitance (μF cm^{-2})	0.5—1.3	0.3—1.3
Resting potential difference (mV)	10—200	0—140
Index of refraction	1.55	1.37
Interfacial tension (mN m^{-1})	0.03—3.0	0.2—6.0
Permeability to water (10^{-4} cm s^{-1})	25—33	5—10
Dielectric breakdown (mV)	100	150—200

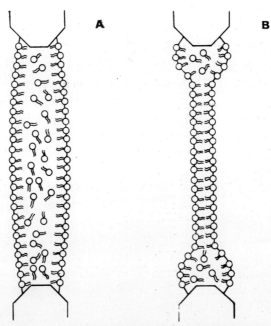

Fig. 2.24. Schematic representation of black lipid membrane formation. In **A**, the lipid layer is several molecules thick and shows various interference colors. In **B**, it has thinned spontaneously to a bilayer which appears black when viewed in a microscope.

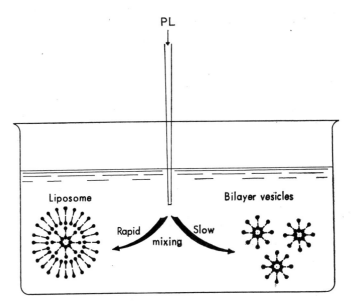

Fig. 2.25. Schematic representation of formation of multi-layer liposomes and bilayer vesicles, depending on the rate of stirring the phospholipid (PL) emulsion. Liposomes can also be formed from bilayers by sonication.

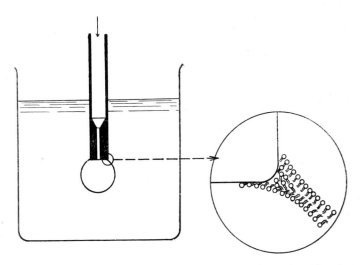

Fig. 2.26. Schematic representation of bubble membrane formation. The pipette at left carries a drop of phospholipid at its tip and a solution within. By gently blowing at arrow the solution is expelled from the pipette and a bubble containing it is formed, ideally composed of a lipid bilayer.

There is an analogy between these round vesicles and soap bubbles but bubbles in air have their hydrophobic parts exposed on the surface, with polar parts pointing into the bilayer film.

Spherical vesicles behave like small osmometers and transport of water across their membranes brings about volume changes which can be readily monitored, e.g., by light-scattering measurements.

Materials used for the preparation of artificial membranes include various phosphatides as well as neutral glycerides and combinations with cholesterol and various minor additives, such as α-tocopherol. Many lipid mixtures of biological origin have been used, such as extracts of erythrocytes, brain fractions, synaptic vesicles, mitochondria, chloroplasts, etc. The stability of the membranes formed (up to several hours) is difficult to predict and, besides the quality of the major components, it is subject to subtle effects due to oxidation (work in oxygen-free atmosphere is of fundamental importance), presence of contaminants in the solvents used (hydrocarbons completely free of other homologues are relatively costly and difficult to obtain) and some other poorly defined factors.

If cholesterol alone is used, it should be partly oxidized in air, whereupon black membranes are formed that are stable and show permeability properties like phospholipid membranes do (Tien et al., 1966).

In most cases, the symmetrical phospholipid bilayers were studied. However, asymmetrical membranes can also be constructed with, say, phosphatidyl choline in one film and phosphatidyl ethanolamine in the other (Montal and Mueller, 1972). This is done by using a Teflon partition partly immersed into water so as to separate two different lipid films on the surface. The partition is provided with an opening and, as it is immersed deeper, it carries the lipid films with it so that at the opening they are apposed and form a true black membrane.

In principle, it is also possible to prepare membranes with a single film of lipid and an adherent single film of protein (Frommhertz, 1970). The technique, consisting in transferring a lipid monolayer onto free medium, can be applied to the formation of multilayer structures of defined layer sequence and composition.

The usefulness of existing artificial lipid membranes is now beyond any doubt but their potential significance depends on the success of incorporating native proteins, particularly of transport function, into their structure, something that has been possible only in a few rather special cases, e.g., Na,K-adenosinetriphosphatase (Jain et al., 1969; Redwood et al., 1969) or sucrase-isomaltase involved in intestinal glucose transport (Storelli et al., 1972).

Perhaps the greatest value of artificial membrane models derives from the application of agents affecting their permeability to both ions and nonelectrolytes. The results will be treated in some detail in the respective chapters (p. 296 and p. 204).

2.3.3. Organization of proteins

Although proteins represent usually more than a half of total membrane weight no general pattern of their organization in membranes appears to exist. Various optical methods (in particular ORD and CD) have shown the proteins to be $30-50\%$ in the α-helical configuration, with a predominance of random-coil arrangement in the rest. Perhaps only the mitochondrial protein contains less, if any, α-helix. There is no known membrane protein with predominant β-configuration (the antiparallel pleated sheet), an exception being the proteins of the outer membrane of *Escherichia coli* (Nakamura et al., 1974).

The overall structure of a membrane is probably governed by the lipids but associations among proteins may play a role in maintaining the membrane shape. Thus, practically delipidated mitochondrial membranes, myelin and *Mycoplasma* membranes retain their appearance after OsO_4 staining in the electron microscope, this supporting the possibility of a structural framework of proteins.

In mitochondrial cristae, chloroplast lamellae and bacterial cytoplasmic membranes one would perforce expect some sort of defined mutual arrangement of proteins that ensures the highest efficiency of their function, namely the immediate transfer of the product of one reaction to act as substrate for the next enzyme in the functional sequence. (The dependence of these enzyme activities on the presence of lipids is another matter and will be dealt with on p.108).

Indirect support for a wide protein structuration comes from experiments indicating the lateral cooperativity of some membrane phenomena, such as the effect of a few molecules of hormones on the entire cell membrane (Sonnenberg, 1971) or a similar effect of colicin El on the character of the *Escherichia coli* envelope (Phillip and Cramer, 1973). Although treated theoretically for an array of lipoprotein subunits (for a review see Wallach, 1971), a cooperativity among lipids (in the form of a pressure wave) might account for these findings as well.

Membrane proteins are now generally assumed to be of two types, viz. the peripheral and the integral proteins. Peripheral proteins are those that are easily removed from the membrane by mild agents or even distilled water and are assumed to blanket the surfaces of the membrane, albeit asymmetrically. When solubilized, they generally form true solutions and are free of lipid. They include, for instance, cytochrome c, spectrin, α-lactalbumin, the heat-stable protein HPr (p. 234), aldolase, nectin, the oligomycin-sensitivity conferring protein (OSCP) and the periplasmic binding proteins.

Integral proteins, on the other hand, are embedded to a greater or lesser degree in the lipid continuum and may span the entire thickness of the membrane. They require detergents or chaotropic agents for solubilization and are usually associated with lipids, not forming true solutions. They include cytochrome b_5, various phycoproteins, hormone receptors, lectin receptors, some eukaryotic binding proteins, and others.

The two types of proteins must differ in the distribution of their hydrophobic amino acid residues. Whereas the peripheral proteins are to a large extent hydrophilic on their surface (they are mostly water-soluble) with their hydrophobic amino acids buried inside, the integral proteins have the hydrophobic residues exposed on their surface to allow for maximum interaction with the nonpolar milieu of the membrane. Still, parts of the integral proteins should be polar on the surface to permit interactions with the polar groups of lipids and with the peripheral proteins.

The best-known membrane proteins are globular: Rhodopsin from the rod outer segment in the retina has a molecular weight of 28 000 and is spherical with a diameter of over 4 nm; mitochondrial cytochrome c oxidase has a molecular weight of 20 000 – 25 000 and is cylindrical with dimensions of $9.8 \times 5.0 \times 5.0$ nm^3 (Vanderkooi and Sundaralingam, 1970).

The lateral diffusion of proteins in the membrane is slower than that of lipids, an estimate for the lateral diffusion coefficient of a protein of 100 000 mol. wt. being 3.10^{-10} cm^2 s^{-1} (Sackmann et al., 1973) with the corresponding average distance traveled per second being 0.3 µm.

If several hundred protein molecules are clustered together the diffusion in the plane becomes rather slow so that the cluster may not move much throughout the cell's life.

Beside the lateral translational movement some proteins are known to rotate in the plane of the membrane. Rhodopsin from the retinal rod outer segment rotates with a relaxation time of 0.3 ns, bacteriorhodopsin from *Halobacterium halobium* with $\tau = 10$ ms.

Proteins can apparently flip from one membrane face to the other (particularly those involved in transport) but so far no direct measurement of this movement exists. Judging from the maximum transport rates and from the number of binding protein molecule per cell one can estimate a frequency of movement of about $10^4 \, s^{-1}$.

2.3.4. Lipid-protein interactions and membrane structure

It will have become apparent from the two preceding sections that neither lipids nor proteins alone can account for the various membrane phenomena observed and that they must be considered in mutual context to arrive at a plausible membrane model.

Interactions between lipids and proteins must be considered in the first place. There can be covalent binding, electrostatic attraction, polar interaction, and hydrophobic interaction. While covalent binding is little probable, the other three factors may play their roles, the final outcome being the attainment of a thermodynamically most favorable structure, i.e., minimum free energy content of the system. This is achieved most plausibly by allowing for both polar and hydrophobic interactions. It should be noted that hydrogen bonds, although certainly playing a role in maintaining the structure of proteins, are probably of no special consequence for lipid-protein organization since a hydrogen bond is endowed with the same relative free energy in an aqueous and in a lipid environment. Some of the reasons behind this minimalization of free energy content have been described already on p. 32.

On the molecular level, this means that hydrophilic (polar) groups should be in contact with each other and with water while hydrophobic (nonpolar) groups should be in contact with each other but not with the polar groups or the external aqueous medium.

This requirement is most easily met by a simple lipid bilayer as described above (Gorter and Grendel, 1925), with proteins spread on its surfaces, such as suggested by Davson and Danielli (1943). This model, strongly supported and documented by electron microscopic evidence by Robertson (e.g., 1959), in its principles has survived for a long time and has been amended several times to greater sophistication (e.g., Hechter, 1965).

However, there are several properties of biological membranes that are not accounted for by a simple lipid bilayer (cf. Table 2.17), in particular the electrical resistance and water permeability. To explain the greater "leakiness" of natural membranes, hydrophilic pores were introduced by Stein and Danielli (1956), polar polypeptides permitting passage through the bilayer. A variant of the lipid bilayer, trying to account for the fact that there is not enough lipid in the membrane to form a continuous double sheet organized as in the classical Danielli models, is that by Hybl and Dorset (1970) or Fettiplace et al. (1971).

Electron-microscopic observations of lipid-water mixtures (cf. Fig. 2.6), as well as of ultrathin tangential sections of membranes (Fig. 2.27) and of freeze-etched plasma membrane preparations

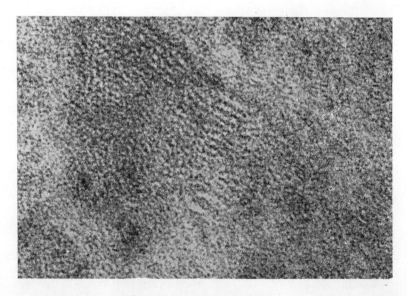

— 50 nm

Fig. 2.27. Globular structures seen in an ultrathin tangential section through the Golgi membranes of pancreatic acinar cells. (Reproduced with permission from Sjöstrand F. S. (1964). *Symp. Soc. Chem. Biol. Suppl.* **14**, 103.)

(*cf.* Fig. 2.37)* have brought in a new possibility of organization, *viz.* one of globules or micelles or lipoprotein units that would form the membrane sheet. To retain the optimum requirements for a membrane containing amphipathic components, the units could be arranged either as lipid globules in a protein matrix, the polar heads forming the periphery of the globules (the model of Lucy and Glauert, 1964) or, alternatively, as protein globules arranged in a lipid continuum (Vanderkooi and Green, 1970), this model retaining the lipid bilayer concept and being in fine agreement with electron-density distribution across membranes. The globular micelle models for proteins floating in a lipid bilayer were refined to accommodate some particular physicochemical data, e.g., that by Wallach and Gordon (1968) or by Zahler and Weibel (Zahler, 1969).

Another model, doing away with the bilayer concept altogether, is that due to Green and Perdue (1966) and to Benson (1968) where the membrane is composed entirely of lipoprotein units organized in a curved plane. An even more elaborate model along the same lines was proposed by Sjöstrand (1968).

All these models, although some of them rather complicated, failed to take account of the fact that the functions of membranes as well as their morphology are heterogeneous and hence a model including regularly repeating units in an array cannot meet the requirements placed on a universally acceptable model. Moreover, the now well recognized lateral mobility of membrane components and apparent phase transitions of lipids in membranes led to two types of membrane structure models.

The one by Singer and Nicolson (1972) assumes the existence of a lipid matrix broken up at places by individual protein molecules or their clusters (integral proteins), the whole sheet then covered with a layer of peripheral protein patches. The one by Green (for the latest version see Green *et al.*, 1972) assumes two continuous domains in the membrane, one of lipids (in which some proteins, particularly glycoproteins, may be immersed), one of proteins (the "intrinsic" proteins), the whole sheet again covered by an inhomogeneous sheet of extrinsic protein.

Each of these two models has its advantages and both must be viewed with these important aspects in mind: (1) the structures

---

* However, the size of the particles seen here is substantially greater than would be possible to accommodate within a native membrane.

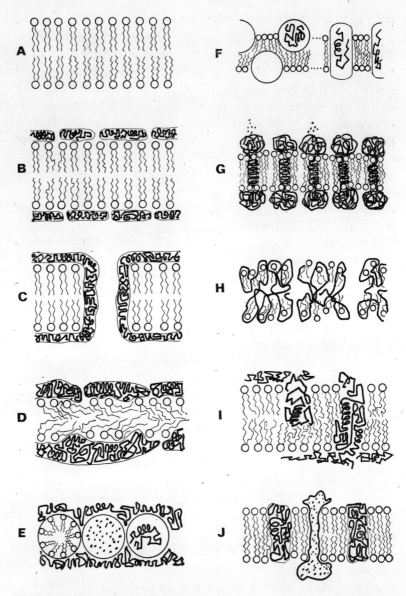

Fig. 2.28. A synopsis of various historical and modern models of cell membrane structure. An attempt was made to use the same graphical means for all the models, a circle with two wavy lines representing a phospholipid, heavy twisted line a polypeptide chain, dotted areas a glycoprotein. **A** Gorter and Grendel, 1925; **B** Davson and Danielli, 1943; **C** Stein and Danielli, 1956; **D** Hybl and Dorset, 1970; **E** Lucy and Glauert, 1964; **F** Vanderkooi and Green, 1970; **G** Zahler, 1969; **H** Benson, 1968; **I** Singer and Nicolson, 1972; **J** Green et al., 1972.

depicted are not rigid but are in a state of constant reorganization; (2) no two membranes are identical in respect of the relative participation of proteins and lipids and the extent of local domains. Extreme cases, such as the inner mitochondrial membrane or chloroplast thylakoids, may be highly ordered with a definite repeat distance between subunits (perhaps with alternating protein and lipid) allowing for efficient transmission of substrate material. At the other end of the scale, lymphocyte plasma membranes may be in a completely random state of (dis)organization, allowing for swift transmission of accidental signals. By the same token, the various structures, both seen and surmised to protrude from the membranes of mitochondria as well as plasma membranes (cf. Fig. 2.48 and 2.35**D**) must be viewed in the context of membrane heterogeneity, one of its aspects being asymmetry about the central plane, the other the ability to accomplish different functions at different areas (such as in polar cells of various epithelia, etc.). The models discussed are summarized in the composite Fig. 2.28 and an overall view, incorporating most of the known data, is presented in Fig. 2.29.

Whatever the actual local arrangement in the membrane, there exist well-founded estimates of the number of integral (intrinsic) proteins in relation to the number of lipids. In a brilliant sequence of

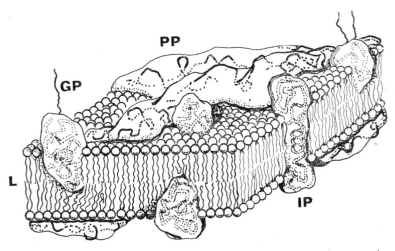

Fig. 2.29. A comprehensive view of a typical plasma membrane, incorporating most of the recently available information. **L** Lipid continuum, **IP** integral proteins, **PP** peripheral proteins, **GP** glycoproteins. The number of polar heads of phospholipids should be roughly 10 times greater with respect to the number of integral proteins.

deductions based on the use of fluorescent markers with *Escherichia coli* membranes, Träuble and Overath (1973) arrived at a value of one integral protein per 600 lipid molecules, 130 of these being closely associated with the protein molecule and not taking part in the phase transition described on p. 89. In terms of membrane area, the integral proteins take up roughly 10% of the area (the value is subject to variation with the cross-sectional area of the protein). Peripheral proteins, on the other hand, take up in this case 47% of the surface (an average for both membrane faces).

The association of some lipid molecules with particular proteins has its corollaries in the finding of (*a*) interaction between particular proteins and (*b*) with lipids at an air−water interface (London et al., 1974). While the so-called Folch-Lees protein interacts with various phospholipids but most powerfully with cholesterol, the other major protein, the *A* basic protein, interacts practically only with cholesterol sulfate.

The second argument has been demonstrated for a number of enzymes, the role of the lipid being of several types: (*1*) It can activate the substrate for the enzyme (in the case of lipopolysaccharide synthesis; p. 139); (*2*) it can activate the enzyme by stabilizing a certain conformation (probably quite frequent); (*3*) it can organize a multienzyme system (possibly in mitochondria); (*4*) it can act as a covalently linked cofactor in a reaction (the carrier lipids). Table 2.18 lists some of the lipidactivated enzymes. The effects of phospholipids may be surprisingly subtle. Thus, adenylate cyclase sensitivity to glucagon is restored by phosphatidyl serine, its sensitivity to noradrenaline by phosphatidyl inositol.

The direct effect of membrane fluidity on membrane-bound enzymes is nicely seen in the recent report by Bloj and co-workers (1973) who found that the Hill coefficient (an indicator of allosteric cooperativity) of the inhibition by fluoride of rat erythrocyte Na,K-adenosinetriphosphatase is raised from 2.07 in rats fed corn oil (containing a high proportion of saturated fatty acids) to 3.60 in animals on a cholesterol supplement (cholesterol decreases the T_t; see p. 93). Even if the transition temperature is not affected by diet or cultivation medium (oleic *vs.* linolenic acid in *Escherichia coli*), there may be effects on the Hill coefficient (1.60 *vs.* 2.12 for Ca^{2+}-ATPase (Siñeriz et al., 1973).

2.4. ASSEMBLY OF MEMBRANES

Whereas lipid components in a suitable aqueous medium tend to form bilayer structures spontaneously, the assembly of biological membranes is a rather complicated process that is only poorly understood. It may be viewed from two different aspects which we shall discuss briefly.

The first concerns the reconstitution of solubilized membrane components to a membrane that would be morphologically and functionally equivalent to the original, a feat that is relevant to the structure and strength of interactions of the components.

There are a number of proteins that can be bound again to membranes from which they have been extracted (cf. Razin, 1972). They include first of all various adenosinetriphosphatases that have been solubilized by washing with dilute Tris buffer or sonication of *Streptococcus faecalis*, *Micrococcus lysodeikticus*, *Bacillus megaterium* and the inner mitochondrial membrane. They can be reattached in the presence of Mg^{2+} or Ca^{2+} and usually several coupling factors. Other proteins thus reattached include mitochondrial flavin-containing amine oxidase, mitochondrial cytochrome c, mitochondrial hexokinase, hydrogen hydrogenase from *Vibrio succinogenes* (Razin, 1972), the phosphate-binding protein from *Escherichia coli* (Medweczky and Rosenberg, 1969, 1970), the glutamate-binding protein from *Escherichia coli* onto depleted cells (Willis and Furlong, 1974), the glutamate-binding protein from *Bacillus subtilis* to vesicles prepared from a transport-negative mutant (Diesterhaft and Freese, 1974).

Like proteins, extracted membrane lipids may be added to depleted membranes to restore enzyme activity, a number of examples being found in Table 2.18.

There are ways of reconstituting solubilized membranes simply by adding an extracted protein or lipid to the remaining membrane. Generally, this is achieved by removing the detergent that has been used for the solubilization. If sonication is used for membrane disruption (it usually does not lead to real solutions) the reconstitution is achieved by manipulating the ion content of the mixture. Likewise, if restoration is to be accomplished from separated proteins and lipids (which offers the possibility of forming "hybrid" membranes) the presence of divalent cations appears to be of essence. It is quite characteristic that many reconstituted systems form vesicles.

Among the most important achievements in membrane re-

Table 2.18. Membrane enzymes requiring lipids for activity (adapted from Coleman, 1973; Finean, 1973; Machtiger and Fox, 1973)

Enzyme	EC number	Source	Reactivating lipid
3-Hydroxybutyrate dehydrogenase	1.1.1.30	Beef heart mitochondria	PC
Malate oxidase	1.1.3.3	*Mycobacterium avium*	DPG
Pyruvate oxidase	1.2.3.3	*Escherichia coli*	LPE, PC, FA, TG
Succinate dehydrogenase	1.3.99.1	Beef heart mitochondria	DPG PE PC
Acyl-CoA dehydrogenase	1.3.99.3	Various	Mixed TG, FA
Cytochrome *c* oxidase	1.9.3.1	Beef heart mitochondria	PE DPG PC
UDP-glucuronosyltransferase	2.4.1.17	Guinea-pig liver microsomes	Mixed lipids
UDP-galactose-lipopolysaccharide galactosyltransferase	2.4.1.44	*Salmonella typhimurium*	Dioleyl PE PG PS[a]
UDP-glucose-lipopolysaccharide glucosyltransferase I	2.4.1.58	*Salmonella typhimurium*	PE PA, PG[a]
Isoprenoid-alcohol kinase	2.7.1.66	*Staphylococcus aureus*	PG, DPG LPG, LDPG PC, PE

Phospho-enol-pyruvate-HPr-phosphotransferase (EII)	2.7.3.9	*Escherichia coli*	PG PS, DPG
Cholinephosphate cytidylyltransferase	2.7.7.15		LPC PC
Phospho-*N*-acetylmuramoyl-pentapeptidetransferase	2.7.8.13		PG, PS, PE, PI
Phosphatidate phosphatase	3.1.3.4		Mixed lipids
Glucose-6-phosphatase	3.1.3.9	Rat liver microsomes	LPC PC
Alkenyl-glycerophosphinicocholine hydrolase	3.3.2.2	Rat liver microsomes	PC, SP PE
Ca^{2+}-adenosinetriphosphatase	3.6.1.3	Rat muscle microsomes	PI, mixed lipids
Na^+, K^+-adenosinetriphosphatase	3.6.1.3	Various sources	Cholesterol; or PS; or PC LPC PA
Adenylate cyclase	4.6.1.1	Rat liver plasma membrane	PS PC
Acyl-CoA synthetase (GDP-forming)	6.2.1.10	Rat liver mitochondria	

[a] For reconstruction, the protein must be combined first with substrate, then with lipid. Abbreviations used: PC, phosphatidyl choline; PE, phosphatidyl ethanolamine; PS, phosphatidyl serine; PI, phosphatidyl inositol; PG, phosphatidyl glycerol; DPG, cardiolipin; PA, phosphatidic acid; SP, sphingomyelin; LPC, lysophosphatidyl choline; LPE, lysophosphatidyl ethanolamine; LPG, lysophosphatidyl glycerol; LDPG, lysodiphosphatidyl glycerol; FA, fatty acids; TG, triglycerides.

constitution is the functional electron-transport chain put together from soluble succinate dehydrogenase, cytochromes b, c and c_1, cytochrome c oxidase, phospholipids and coenzyme Q_{10} (Yamashita and Racker, 1969) – the recovery of activity took several hours in this case. Another is the almost complete reconstitution of oxidative phosphorylation (e.g., Racker *et al.*, 1969; Kagawa and Racker, 1971) and another still the partial restoration of photosynthetic activity of chloroplast membranes (e.g., Takacs and Holt, 1971). A long way toward reconstituting all the functions of sarcoplasmic reticulum membrane has been made (e.g., Martonosi, 1968). A truly remarkable feat is the identification of all the components of the phosphotransferase transport system (see p. 233) and their joining together in a definite sequence (enzyme IIB + Ca^{2+} + phosphatidyl glycerol + + enzyme IIA) to form a functional system (Kundig and Roseman, 1971).

Functional systems can be constructed even from solubilized enzymes and artificial bilayer membranes or liposomes. Many proteins decrease the electrical resistance of these membranes without being enzymically active or endowing them with a specific transport function. The interactions may be merely electrostatic, as suggested by the necessity of low pH and/or a basic protein (e.g., lysozyme, cytochrome c, ribonuclease). Transport systems successfully attached so far include adenosinetriphosphatase (Redwood *et al.*, 1969; Jain *et al.*, 1969), sucrase-isomaltase (Storelli *et al.*, 1972) and the receptor protein for acetylcholine (De Robertis, 1971). Romeo *et al.* (1970*a, b*) reconstituted a functional galactosyltransferase in a phospholipid monolayer containing lipopolysaccharide.

The second aspect from which membrane assembly can be considered is the biogenesis of various cell membranes. Here the field abounds with hypotheses but already a number of rather definite points can be made.

The whole process of biogenesis can be split into three stages: (*1*) Biosynthesis of the component molecules; (*2*) assembly of the component molecules; (*3*) modification and migration of the components.

1. The first of these points is a matter of the enzyme equipment of the cell. Phosphatidic acid is synthesized in bacteria from glycerol by phosphorylation (ATP + glycerol kinase), monoacylation (acyl-carrier protein + glycerolphosphate acyltransferase) and diacylation (acyl-carrier protein + monoacylglycerolphosphate acyltransferase)

whereas in eukaryotic cells the usual pathway is from dihydroxyacetonephosphate which is reduced to glycerol phosphate and this then reacts with two molecules of acyl-CoA to the phosphatidic acid.

From here on, the pathways differ again in bacteria and in eukaryotes as indicated in the following scheme:

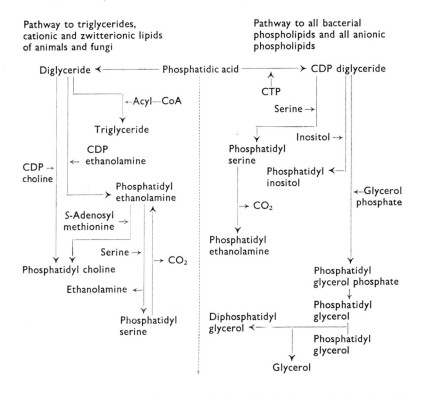

For the synthesis of sterols and minor lipids the reader is referred to specialized biochemical texts.

The site of lipid synthesis is presumably the endoplasmic reticulum (in bacteria, the plasma membrane carries all the enzymes required) although, obviously, not all cells are properly equipped so that in many cases the lipids formed on one membrane may have to travel through the cytoplasm to the locus of their insertion into another membrane.

The insertion is assisted by two mechanisms. (*1*) Action of lysophosphatides which solubilize parts of the phospholipid membrane by forming small micelles and thus permit the entry of another lipid into the vacated place. (*2*) Operation of the so-called lipid exchange proteins which can be isolated from various animal tissues (e.g., Wirtz et al., 1972; Helmkamp et al., 1974) and which are specific carriers for certain types of phospholipids, such as lecithin or phosphatidyl inositol, effectively assisting in an exchange of a membrane phospholipid for this particular lipid species.

Proteins are synthesized, as is well known, by the highly complex process taking place on the ribosomes. It is not clear at present what part of protein is synthesized on the endoplasmic reticulum ribosomes and what part of the total arises on membrane-associated ribosomes or ribosomes included in other organelles (mitochondria and chloroplasts). At any rate, some of the specific membrane proteins must similarly travel through the cytoplasm to the point of their destination.

2. The assembly step is the crucial one in the sequence. One may envisage several possibilities at the molecular level (cf. Overath et al., 1971a): Lipids are incorporated (a) in patches near the enzymes of their synthesis, or (b) in a random fashion. Proteins are incorporated (a') into new lipid areas, (b') into both new and old ones, (c') in a random fashion. Evidence listed below supports the possibilities under (b) and (c'). The main arguments in favor of a random insertion of lipids as well as proteins into the membrane framework derive from estimates of the turnover rate of the components.

The half-lives of phospholipids (using ^{32}P as a label) were found to differ from type to type. Thus, phosphatidyl inositol of rat brain had a half-life of 12.5 days, sphingomyelin of 40 days (Freysz et al., 1969). In bacteria, the same label shows the half-life of phosphatidyl glycerol to be 1 h, that of diphosphatidyl glycerol perhaps twice that while phosphatidyl ethanolamine shows practically no turnover at all (Kanemasa et al., 1967).

The turnover rate of some phospholipids can be markedly increased by specific stimulation of cells. Thus, Lapetina and Michell (1973) report enhanced phosphatidyl inositol turnover in cerebral cortex sympathetic ganglia, the vagus nerve, pancreas, submaxillary glands and other exocrine tissues on exposure to acetylcholine, propranolol or electrical shock.

The relatively slow turnover of some membrane lipids (it may come to a complete stop when bacteria undergo extensive growth; McElhaney and Tourtellotte, 1969) contrasts sharply with the capabilities of many cells to support fantastically rapid new membrane formation. The paradigm of this phenomenon is the behavior of amoebae which, when pierced or even severed in two pieces, surround the injured areas within tens of seconds with a membranous sheath appearing trilaminar under the microscope (e.g., Szubinska, 1971).

Similarly striking differences in the rates of turnover were found among membrane proteins. Although not enough comparative data are available on particular plasma membrane proteins, different enzymes of rat endoplasmic reticulum have the following half-lives: hydroxymethylglutaryl-CoA reductase 2−3 h, NADH dehydrogenase 3−4 days, cytochrome b_5 reductase 5 days, and NAD^+ nucleosidase 16 days.

The rate of turnover of membrane proteins is in a striking relationshop to their size. The larger the molecule, the faster it turns over−this being perhaps the single rather universal law valid not only for membranes but for soluble proteins as well (cf. Dehlinger and Schimke, 1971). The experimental protocol supporting this statement deserves to be described. Animals (rats) were fed ^{14}C-labeled amino acid; after 4 days, they were given the same, but ^3H-labeled, amino acid. Several hours after this administration they were killed, the plasma membranes of their liver were separated, extracted with 1% Triton X-100 ("soluble" proteins) and then with 0.5% sodium dodecyl sulfate ("insoluble" proteins). The extracts were chromatographed on Sephadex and the radioactivity due to ^{14}C and ^3H was assayed in the individual fractions (Fig. 2.30). In this experimental schedule, the $^3H/^{14}C$ ratio is seen to be twice as high for very large molecular-weight proteins as for the small molecules (such as cytochrome c).

Among other things, this relationship suggests that membrane proteins are degraded probably only after dissociation from the membrane matrix−what the underlying control mechanism is remains to be explored.

Another set of data supporting the rather random incorporation of membrane components derives from experiments with bacterial auxotrophs requiring fatty acids for full growth (cf. Cronan et al., 1969) or using mutants defective either in the reduction of dihydroxyacetone phosphate (Hsu and Fox, 1970) or in the acylation of either glycerol-3-phosphate (Cronan et al., 1970) or of monoacylglycerol-3-phosphate (Hechemy and Goldfine, 1971). In the fatty acid auxotrophs one can alter the fatty acid composition of the membrane lipids (and hence the transition temperature) at will, in the latter mutants one can stop and start lipid synthesis by removing or adding glycerol or glycerol phosphate to the medium. By growing cells first

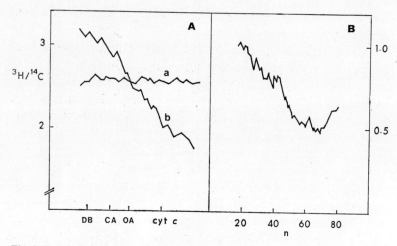

Fig. 2.30. **A** Relative rate of degradation of "soluble proteins" of rat liver as a function of molecular weight. ^{14}C-Labeled leucine was administered to rats four days before ^{3}H-labeled leucine. The animals were killed 4 h later and the proteins chromatographed on Sephadex (curve b). Control animals (curve a) received both isotopes at the same time. The chromatography was calibrated by dextran blue (DB), conalbumin (CA, 78 000), ovalbumin (OA, 45 000) and cytochrome c (13 000). **B** A similar pattern is obtained with double-labeled plasma membrane, the number of fractions (n) being plotted on the abscissa. (Adapted from Dehlinger and Schimke, 1971.)

in the presence of oleic acid (transition temperature for β-galactoside transport 14.9 °C) and shifting to palmitelaidic acid (*trans*-16 : 1), or first in palmitelaidic acid (transition temperature for the same process 33.4 °C) and then shifting to oleate, and by exposing the cell to a transport inducer at different intervals after the shift, one obtains a smooth gradual increase or decrease, respectively, of the transition temperature. Moreover, a completely analogous situation is observed when the inducer is added for a short period before the shift from one to the other fatty acid (Overath *et al.*, 1971*b*; Tsukagoshi and Fox, 1973*a, b*). Thus, under all conditions it is the average composition of membrane lipids which determines the total membrane transport characteristics, this militating against any localized, patchy assembly of either lipids or proteins.

Finally, evidence based on membrane growth supports the absence of localized and permanent domains of synthesis and incorporation. Using a density label for lipid synthesis (bromostearic acid) makes it possible by sucrose gradient centrifugation to distinguish between cells or subcellular bodies containing lipids synthesized in the presence and in the absence of bromostearic acid (Fox et al., 1970). If a rod-shaped bacterium synthesized its membrane only at one pole, within a single generation time (after one division) there should exist a group of heavy and a group of light cells. If the growth zone were at both poles or equatorially located, two divisions should produce an analogous result. However, a complete uniformity of weight was always found in *Escherichia coli* (Tsukagoshi et al., 1971). Likewise, in the *Escherichia coli* strain P678-54, forming tiny spherical cells apically during division, a localized growth zone should be reflected in unequal density of the daughter and mother cells but no such differences were found (Wilson and Fox, 1971b).

Thus the conclusion may be drawn that there are no localized sites of membrane assembly. However, one should be wary of making hasty judgments for several reasons.

Lateral diffusion of lipids is so rapid that even if there are special lipid-synthesizing domains we would be unable to detect them. The rapid spreading of membrane components over the entire surface is also documented by the distribution of antigens (mouse and human) as shown by cell fusion techniques (Frye and Edidin, 1970). However, the fact remains that replacement of an "old" phospholipid with a new one is specific and errors probably rarely occur.

On the other hand, in the case of proteins at least, there is much indirect evidence that, in spite of the large diffusion coefficients, there exist areas in the membrane that contain different proteins for various special purposes. In the case of polarized cells (e.g., in epithelial tissues) this is most obvious but there are indications that even in microorganisms, the entire membrane area is not statistically homogeneous. To cite one example, morphological cell division starts in the membrane at a particular site which is then probably equipped with different proteins (and possibly lipids) as compared with the rest of the membrane.

Moreover, highly specialized membranes, such as the mitochondrial inner membrane or chloroplast lamellae, are certainly not randomized since they could not fulfil their functions as efficiently as they do.

There is still room for speculation on how the heterogeneity of membrane proteins is maintained. A likely explanation is in an intimate contact between the plasma membrane and the intracellular organelles, in particular with the endoplasmic reticulum, whereby information and possibly even building material is transmitted to the plasma membrane. An interaction through microtubules would explain a great deal in this respect. Microtubules are filamentous cylindrical structures, 18−25 nm in diameter, now known to exist in practically all cell types, composed of globular subunits of the protein tubulin. The structures may carry side chains or bridges to other microtubules. The protein may exist in the soluble state and then undergo self-assembly whenever need arises. This process is prevented by colchicine, vinblastine, colcemid and other antimitotic agents. Microtubules are well known from the locomotion organs of cells (cilia, flagella) where they function under the energy input from ATP by sliding along each other like muscle fibrils do. Beside this function, they may be instrumental in maintaining cell structure, directing cytoplasmic transport, separating chromosomes during mitosis, governing secretion and many other vectorial processes. Their implication in the immobilizing of membrane proteins was documented by the action of colchicine which randomizes the distribution of, for instance, lectin and receptor proteins or transport proteins in lymphocytes which otherwise are held in place so as to prevent clustering during concanavalin A binding, on the one hand, and to protect transport proteins from digestion during phagocytosis, on the other.

3. Much has been said about the translocation of both lipids and proteins in the foregoing section. Let it be added that there is evidence for transmission of material from one membrane to another. Lipopolysaccharide synthesized at the inner cytoplasmic membrane of Gram-negative bacteria is rapidly transferred to the outer membrane. Lipid components are exchanged between mitochondria and microsomes (Wirtz and Zilversmit, 1968). Fusion of lysosomes with plasma membranes and analogous processes necessarily involve transmission of membrane material.

To conclude this section, it should be stressed that although the magnitude of lateral diffusion coefficients offers enough time for lipids and proteins to spread themselves homogeneously over the entire membrane surface so as to mask possible localized assembly lines there is much evidence on the nonhomogeneity of cell membranes, both with regard to their asymmetry about the inner hydrocarbon core and with regard to special features localized at special sites of the membrane plane.

2.5. ELECTRON MICROSCOPY

Our present knowledge of the morphology of subcellular organelles, most of which are of membranous character, is due to the extensive use of electron microscopy of various cell preparations. The now classical trilaminar appearance of membranes in ultrathin sections derives mainly from the use of electronic "stains", such as potassium permanganate, osmium tetroxide and uranyl acetate. The problem is not satisfactorily solved as to the chemical basis for the attachment of these compounds: Proteins are known to bind some of them but so are the polar heads of phospholipids and so are the double bonds of unsaturated fatty acids (cf. Korn, 1968).

One of the possibilities, fairly well documented for osmium tetroxide, is that during fixation the hydrocarbon chains are contorted and swung near the plane of the polar head groups where they are held by interaction with OsO_4 (cf. Fig. 2.14).

The thickness of the membranes apparent in ultrathin sections is not the same for different membranes (Table 2.19). The three component lines (two dense enclosing an electron-transparent one) themselves are of different thickness; proceeding outward the values in nm (after OsO_4 fixation) are $3.5-3.1-2.4$ (cat pancreas), $3.7-3.3-2.7$ (intestinal epithelium striated border) and $3.3-2.2-2.5$ (intestinal epithelium lateral cell surface).

TABLE 2.19. Thickness (in nm) of various membranes in ultrathin sections of mouse kidney proximal tubule cells (Sjöstrand, 1963)

Fixation	Plasma membrane	Smooth cytoplasmic membranes (Golgi apparatus)	Mitochondrial membranes
$KMnO_4$	9.6	7.7	6.2
OsO_4	9.3	6.2	5.1

Much effort has been expended to show the ultrastructure of the railroad-track pattern and in many cases the results have been rather convincing. Thus, mitochondrial membranes (cf. Fig. 2.48), chloroplast lamellae (cf. Fig. 2.51), retinal rod outer segments (Fig. 2.31) show a clear substructure. If radially symmetrical, the substructural elements may be visualized more clearly by using the technique of Markham *et al.* (1963). An enlarged microphotograph is carefully centered on a rotating disc which is then turned either manually by a certain angle (such that the total circle is divided into an integral number of angles n) and photographed in every position and the pictures then superimposed; or it is arranged like a fast-rotating stroboscope and exposed by an electronic device at suitable intervals corresponding to the above angles.

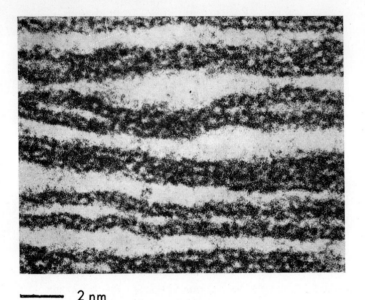

2 nm

Fig. 2.31. Globular structure in the membranes of the retinal rod outer segment of tadpole. (Taken with kind permission from Sjöstrand, F. S. (1968). In: *The Membranes* (ed. by Dalton, A. J. and Haguenau, F.), p. 151. Academic Press, New York—London.)

On the other hand, much of the granular structure purported to be present in electron micrographs of membranes is apparently due to (*a*) coarse grain of the film used, (*b*) peripheral, surface elements of the membrane rather than the intrinsic core of it which appears to be mostly planar.

The structure of the membrane surface is best visualized through the application of a more "native" technique, doing away with fixation and staining, *viz.* freeze-etching and -fracturing. Cells or their subfractions are rapidly frozen to below $-150\,°C$ in liquid freon (maintained at that temperature by liquid nitrogen), placed in a vacuum at $-100\,°C$ and fractured with a deeply cooled knife (Fit. 2.32**A**). Subsequently, some of the ice may be sublimed (etched away) for several minutes to expose the surrounding unfractured layers (Fig. 2.32**B**). Immediately after that, the object is shadowed with carbon and platinum to obtain a replica of the original sample. Examples of freeze-etched membranes may be seen in Fig. 2.37.

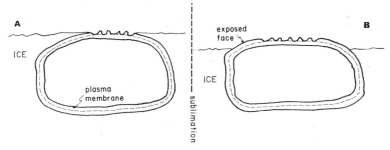

Fig. 2.32. Schematic drawing of the technique of freeze-fracturing (**A**) and freeze-etching (**B**), the latter technique including sublimation of the superficial ice layer and exposition of parts of unfractured material.

Various arrangements of globules are often found in the inner core of membrane split open by freeze-etching and often interpreted as evidence for integral proteins spanning the membrane (e.g., Tourtellotte, 1972).

2.6. ISOLATION OF MEMBRANES

The separation from one another of the various cell membranes is no simple task and, rigorously speaking, it has never been accomplished completely when starting from the whole cell. The situation is much simpler, say, with mammalian erythrocytes where osmotic bursting releases the entire cell content (with no intracellular membranes present) so that all one has to do is to centrifuge and wash the stromas (ghosts), which are in fact plasma membranes.

With other cells the procedure is usually to break up the cells by one of the following techniques.

1. Application of a hypotonic solution which effects an osmotic lysis of tissue cells with consequent emptying of the cell content into the medium.
2. Disruption of cells in a mechanical device. The one most commonly used is a glass or Teflon pestle fitting smoothly into a glass tube with the cell suspension in which it is rotated while the tube is moved up and down to ensure thorough mixing and shearing of the content (this is the Potter—Elvehjem type) (Fig. 2.33**A**).

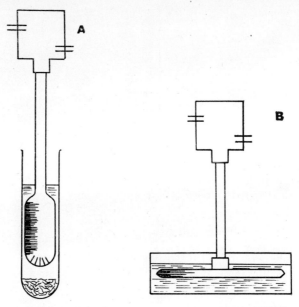

Fig. 2.33. Schematic drawing of a Potter—Elvehjem homogenizer (**A**) and a horizontal glass disc homogenizer (**B**). The former is drawn approximately half its size, the latter is reduced about five times.

Another rotating device used with excellent results for thick-walled microorganisms derives from a stirrer used in the paint and enamel industry. A smooth thick glass disc is fixed at the lower end of a vertical driving shaft and is immersed slightly into a cell suspension containing lead-free glass beads (Balottini) of the appropriate size (this is determined mostly empirically; a mixture of grade 12 and 14 will do well for yeast and fungal cells). With external cooling and without access of oxygen, cells are broken up in a few minutes at most (Fig. 2.33**B**).

3. A sudden change of pressure can be applied in different modifications for bursting cells. In a French or Hughes press a frozen cell pellet is pushed through a tiny opening, in the process of which it melts and the cells are disrupted.

In a procedure due to Wallach (1967) cells are compressed with a nonoxidant gas at 5−7 MPa and rapidly decompressed.

4. Freezing and thawing will reduce the cell to fragments but these are difficult to separate into any defined membrane types.

5. Sonication will break up some cells but here, again, the danger of progressive damage to individual membranes is high.

The technique generally employed for separation of the material obtained by cell disruption is centrifugation at different centrifugal forces. The rationale for the procedure is that particles of different size and different density sediment in the centrifuge at different rates, reflected in the average time t (in seconds) required for the particle to reach the bottom of the tube:

$$t = 843\eta K^{-1}(\text{r.p.m.})^{-2} D^{-2} (d_p - d_m)^{-1} \log(r_{max}/r_{min}) \quad (2.30)$$

where η is the viscosity of the liquid in poises (g cm^{-1} s^{-1}), which for water at 0 °C is 0.017 93, at 20 °C 0.010 09, at 40 °C 0.006 57; D is the particle diameter in cm, d_p the density of the particle, d_m that of the liquid in g cm^{-3}, r_{max} and r_{min} are the radii (cm) of rotation of the bottom and of the meniscus in the centrifuge tube, respectively; r.p.m. is the number of revolutions per min, K is the particle shape factor (0.222 for a sphere, 0.19 for a disc, 0.06 for a cylinder). Even if one can thus calculate the time required for a given particle to sediment, it is readily seen that a clear separation of related particles is impossible because the sedimentation time of a particle in a lower layer of the suspension is obviously shorter than that of a particle near the meniscus.

The above formula is related to the expression (for a spherical particle)

$$s = D^2(d_p - d_m)/18\eta \quad (2.31)$$

where s (in seconds) is the sedimentation coefficient per unit centrifugal field. This is usually expressed in Svedberg units $S(= 10^{-13}$ s).

For calculating the centrifugal field of a given centrifuge run, the well-known formula is used:

$$G = 1.12 \cdot 10^{-5} (\text{r.p.m.})^2 r \quad (2.32)$$

where G is the multiple of the value of acceleration due to gravity (g), r is the centrifuge radius in cm. There are two ways of indicating the centrifugal acceleration applied: (a) in multiples of g and the duration of run, thus 5000 g for 10 min or 10 000 g for 90 min; (b) in time-integrated units, thus $5 \cdot 10^4$ g min or $9 \cdot 10^6$ g min. Although both expressions are intrinsically equivalent, it should be appreciated that the time-integrated presentation must be understood within reasonable limits — it would be absurd to spin a sample either at 10^7 g for 1 min or at 10^3 g for 10^4 min.

The sedimentation properties of some important cell fractions of rat liver are shown in Table 2.20. The purity of a fraction obtained

TABLE 2.20. Sedimentation of rat liver homogenate (adapted from Steck and Fox, 1972)

Fraction	Diameter (μm)	Pelleting force (g min)	Sedimentation coefficient (S)	Equilibrium density (g cm^{-3})
Whole cells	15—20	10^3	10^7	$c. 1.20$
Nuclei	5—11	$5-10 \cdot 10^3$	10^6—10^7	$c. 1.32$
Golgi apparatus	1—3	$4 \cdot 10^4$	10^5	1.06—1.14
Mitochondria	0.7—1.1	10^5	10^4	1.18—1.21
Lysosomes	0.4—0.6	$4 \cdot 10^5$	$5 \cdot 10^3$	1.20—1.22
Peroxisomes	0.4—0.6	$4 \cdot 10^5$	$5 \cdot 10^3$	1.22—1.24
Rough microsomes	0.05—0.2	$3-10 \cdot 10^6$	$0.3-2 \cdot 10^3$	1.13—1.25
Smooth microsomes	0.05—0.3	$3-10 \cdot 10^6$	$0.3-2 \cdot 10^3$	1.10—1.20
Plasma membranes				
Bile fronts	2—10	10^4	10^6	1.16—1.18
Vesicles	0.1—0.7	10^5—10^7	$0.3-5 \cdot 10^3$	1.12—1.15
Soluble proteins	< 0.01	> 10^8	2—20	$c. 1.30$

in differential centrifugation is never very high and, to achieve better results, the desired fraction should be recentrifuged several times (cf. Cotman et al., 1970) and a suitable density gradient employed for final separation.

The density gradient to be used depends greatly on the objective of separation and on the type of membrane involved. Various inorganic and organic compounds are in use (Table 2.21).

There are two basic approaches to density gradient centrifugation. One uses a shallow gradient whose function is mainly to prevent

TABLE 2.21. Density of gradient solutions at 25 °C (adapted from Wallach, 1972)

Solute	Concentration (weight percent)					
	10	20	30	40	50	60
LiCl	1.054	1.113	1.178	1.250		
LiBr	1.073	1.160	1.261	1.281	1.529	1.716
KBr	1.072	1.158	1.257	1.371		
NaBr	1.078	1.172	1.281	1.410		
RbBr	1.079	1.174	1.285	1.419	1.582	
CsCl	1.079	1.174	1.286	1.420	1.582	1.785
CsBr	1.081	1.180	1.297	1.440	1.616	
Potasium acetate	1.048	1.100	1.155	1.213	1.242	1.333
Potassium citrate	1.066	1.140	1.221			
Potassium tartrate (20 °C)	1.066	1.139	1.218	1.305	1.400	
Glycerol	1.021	1.045	1.071	1.097	1.124	1.151
Sucrose	1.038	1.081	1.127	1.176	1.230	1.289
Polysucrose	1.034	1.068	1.102	1.136		
Polyglucose	1.038	1.076	1.114	1.152		
Urografin	1.043	1.087	1.130	1.173	1.217	1.260

convection during the centrifuge run. The homogenate is placed on the top and briefly spun – various particles will distribute themselves according to their sedimentation rates but they would all reach the bottom if centrifuged for a more extensive period of time. This is the technique known as rate zonal centrifugation.

The other approach used now more extensively is called equilibrium or isopycnic density gradient centrifugation. Here the gradient is steeper and its bottom layer is denser than any of the particles to be separated. After a sufficiently long run (generally 10^8 g min) the particles come to a stop at the gradient layer equal in density to their own. An example of such a separation in a discontinuous sucrose gradient is shown in Fig. 2.34.

With some cell organelles one can change their buoyant density artificially. If animals are fed with Triton, it accumulates in lysosomes and decreases their density quite substantially; likewise, polystyrene particles phagocytized by leukocytes are eventually enclosed in phagosomes which are in fact plasma membrane derivatives and can again be readily separated on the basis of altered density (*cf.* Wallach and Lin, 1973).

Although some membranes are readily recognized morphologically (e.g., mitochondria), with others (the "smooth microsomal fraction") the degree of purity of a given separation must be de-

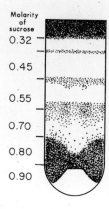

Fig. 2.34. Distribution of various subcellular fractions of a rat lung homogenate in a discontinuous sucrose gradient.

termined by a different method. The so-called enzyme markers are most often used, the rationale of the approach being in the fact that some cell enzymes are located almost exclusively in a given organelle or rather its membrane. More will be said about the enzyme equipment of the various organelles in the subsequent section; here we shall list only some of the most common enzyme markers used (Table 2.22).

Beside the intrinsic enzyme markers specific to the individual membrane types, plasma membrane lends itself to a different type of labeling. Intact cells are exposed to various fluorescent, paramagnetic or radioactive markers which attach themselves covalently to the cell surface and, after homogenization and fractionation, can be detected by suitable devices. There are two classes of such compounds, small and large.

The small ones (4-acetamido-4' isothiocyanostilbene-2,2'-sulfonic acid, diazonium salt of sulfanilate, formylmethioninesulfone methyl phosphate, 2,4,6-trinitrobenzene, 1-dimethylaminonaphthalene-5-sulfonyl chloride) are readily available but they penetrate after a longer exposure across the plasma membrane and can thus label even intracellular membranes.

The large compounds are superior because they do not penetrate across the plasma membrane. They include N-(3-mercuri-5-methoxy-

TABLE 2.22 Enzyme markers of various cell membranes

Membrane	Enzyme
Plasma membrane	Apyrase
	Na, K-adenosinetriphosphatase
	5'-Nucleotidase
Endoplasmic reticulum	Nucleoside phosphatase (acting on IDP)
	Glucose-6-phosphatase
	NADH dehydrogenase
Mitochondria	Choline dehydrogenase
	Glutamate dehydrogenase
	Amine oxidase (flavin-containing)
	Cytochrome c oxidase
	Succinate dehydrogenase
	Isocitrate dehydrogenase
Thylakoids	$NAD(P)^+$ transhydrogenase
	Ferredoxin-$NADP^+$ reductase
Lysosomes	Acid phosphatase
	Ribonuclease
	Deoxyribonuclease
	Arylsulfatase
Golgi apparatus	UDPgalactose-N-acetylglucosaminegalactosyl transferase
Peroxisomes	Catalase
	D-Amino-acid oxidase

propyl)poly-D,L-alanyl amide-^{203}Hg (Yariv et al., 1969), p-chloromercuribenzoate coupled with aminoethyl dextran of molecular weight of 250 000 (Ohta et al., 1971) and substituted anilinoflavazole attached to an isomaltose chain (Himmelspach et al., 1971) which can be diazotized to tyrosyl residues in the membrane

Furthermore, plasma membranes can be labeled in the presence of peroxidase, NaI and H_2O_2 whereby surface tyrosine residues are iodinated (Phillips and Morrison, 1971; Hubbard and Cohn, 1972).

Plasma membranes also can be tagged with antibodies (radioactive or fluorescent) to membrane antigens and with so-called lectins which are compounds of plant origin known to attach to surface glycoproteins and glycolipids (concanavalin A, phytagglutinins from beans, etc.).

2.7. MORPHOLOGY AND FUNCTION OF DIFFERENT BIOLOGICAL MEMBRANES

Although this is a book primarily on membrane transport, it may be useful to give an overview of the broad-ranging functions of various membranes in general. There may be more emphasis placed here on plasma membranes than on organelle membranes but this is due merely to the fact that, from the cell's point of view, plasma membrane is the locus of decisive transport functions which, for a limited period at least, can perform this role even in the absence of other cell organelles (in some cells, like anucleate erythrocytes, there is actually no other membrane present).

2.7.1. Plasma membrane

2.7.1.1. Morphology

The exterior surface of practically all animal cells and of some microorganisms that live in body fluids is formed by a relatively thick (about 10 nm at the least) membrane which may be quite flat (in its classical form in Fig. 2.35**A**), or corrugated (Fig. 2.35**B**), or deeply invaginated (Fig. 2.35**C**) or form finger-like projections (Fig. 2.35**A**). A special kind of plasma membrane is the myelin sheath of various

Fig. 2.35. Different modifications of plasma membrane. **A**. Human red blood cell membrane, fixed with $KMnO_4$. (Taken from Robertson, J. D. (1964). In: *Intracellular membraneous structures* (ed. by Seno, S. and Cowdry, E. V.), p.

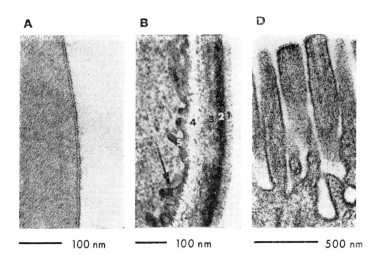

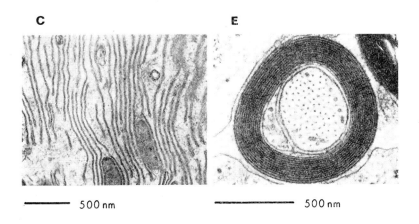

379. Japan Soc. Cell Biol., Okayama.) **B**. Cross section through the periphery of *Candida albicans* showing corrugated plasma membrane (arrow) and several layers of the cell wall (1—5). The cell was fixed with acrolein-tris-(1-aziridinyl) phosphine-OsO_4 and stained with lead and uranyl acetate. (Reproduced with kind permission from Djaczenko, W. and Cassone, A. (1972). *J. Cell Biol.* **52**, 186.) **C**. Ultrathin section of a cell from mosquito larva anal papilla. The membranes running vertically through the photograph are projections of basal plasma membrane. (Reproduced with permission from Copeland, F. (1964). *J. Cell Biol.* **23**, 253.) **D**. Intestinal brush border modification of the plasma membrane showing also the fuzzy glycocalyx. The preparation is from mouse jejunum (Reproduced with permission from Sjöstrand, 1963.) **E**. Membrane sheaths of a nerve central axon. (Taken with permission from Dean, G. (1970). *Scient. Amer.* **223**, 40.)

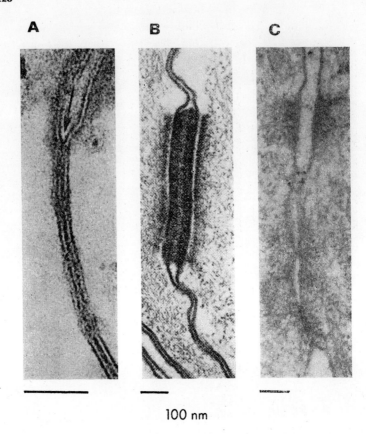

100 nm

Fig. 2.36. **A.** A tight junction (arrow) in isolated rat liver plasma membranes. (Taken with permission from Benedetti, E. L. and Emmelot, P. (1968). In: *The Membranes* (ed. by Dalton, A. J. and Haguenau, F.), p. 33. Academic Press, New York—London.) **B.** A desmosome in muscle tissue. (Taken from Curtis, A. S. G. (1967). *The Cell Surface: Its Molecular Role in Morphogenesis.* Logos Press and Academic Press, Woking and London.) **C.** A desmosome-like structure at the adjacent cell membranes of mouse jejunum. (Courtesy of Dr. J. Ludvík, Institute of Microbiology, ČSAV, Prague.)

Fig. 2.37. Surface views of various plasma membranes. **A.** Fractured outer membrane of *Mycoplasma* grown on oleate-supplemented medium (the particles are assumed to lie in the hydrophobic plane of the plasma membrane). (Taken with permission from Tourtellotte, M. E., Branton, D. and Keith, A. (1970). *Biochemistry* **66**, 909.) **B.** The plasma membrane of *Saccharomyces cerevisiae* with

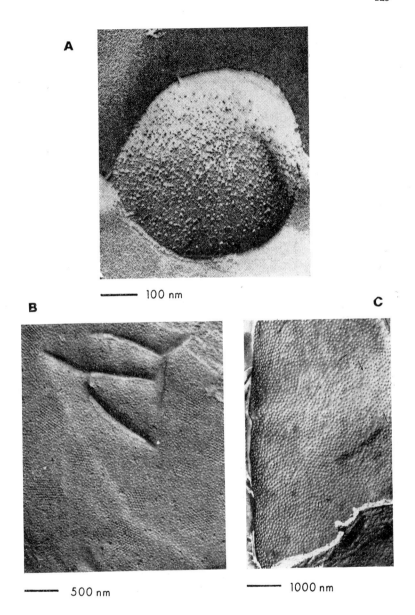

typical grooves and hexagonally arranged globules. (Courtesy of Prof. O. Nečas, J. E. Purkyně University, Brno, Czechoslovakia.) **C**. Plasmalemma of a myocardium capillary with numerous openings often of crater-like appearance. Note the much greater size of these surface features than of the globules or pits in **A** and **B**. (Taken with permission from Simionescu, M., Simionescu, N. and Palade, G. E. (1974). *J. Cell Biol.* **60**, 128.)

nerve fibers, arising from the enveloping Schwann cell which winds itself several times about the axon (Fig. 2.35**E**).

Where two adjacent organ cells come into close contact, the juxtaposed plasma membranes can form junctions of three types: (*1*) the tight junction, (*2*) the intermediate junction, and (*3*) the desmosome. The tight junction appears to represent a continuous belt surrounding a part of the cell surface and separating the lumen of an organ from intercellular spaces between the cells lining it (Fig. 2.36**A**).

The intermediate junction and the desmosome are less tightly joined, appearing rather as two trilaminar membranes cemented together. In the case of the desmosome, there is much cytoplasmic material adhering from both sides even in isolated membrane preparations. It is surmised that at the desmosomes special communication channels between cells exist (Fig. 2.36**B**).

Viewed in a freeze-fractured preparation, the plasma membrane appears to be spattered with protein particles in *Mycoplasma* (Fig. 2.37**A**), deeply grooved at places in *Saccharomyces cerevisiae* (Fig. 2.37**B**) and broken up by fenestrae in a capillary membrane (Fig. 2.37**C**).

Plasma membranes are more difficult to isolate in a native state than most other membranes, an exception being the erythrocyte stroma. They tend to break up and form small vesicles spontaneously, the vesicles appearing in the microsomal fraction during differential centrifugation. They can be separated by their buoyant density (Steck and Wallach, 1970) which for the liver bile fronts in sucrose is 1.17 and for the yeast plasma membranes in Urografin about 1.15.

2.7.1.2. *Functional properties*

The functions of plasma membranes are manifold indeed (Table 2.23). The questions of transport of various compounds will be dealt with in more detail in the following chapters of the book; here we shall restrict ourselves to a survey of the functions that are distinct from transport.

2.7.1.2.1. *Antigenicity*

On its outer surface, the plasma membrane of many cells carries its antigenic determinants in the form of glycolipids and glycoproteins. These components also play a major role in the phenomena of cell

TABLE 2.23. Functions of the plasma membrans in various cells

Function	Comment
Selective diffusion of small molecules and ions	p. 192 and p. 256
Passive and active carrier transport	p. 207 and p. 222
Pinocytosis and phagocytosis	p. 318
Cell adhesion and fusion	in wall-less cells
Mechanical protection of cell interior	especially amoebas, mycoplasmas
Electrical insulation	myelin sheath
Site of surface antigens	in wall-less cells
Generation of nerve impulses	nerve plasma membrane
Site of various metabolic enzymes converting oxidative to phosphatebond energy	especially in bacteria
Site of enzymes involved in cell-wall synthesis	bacteria, fungi, plants (?)
Site of hormone receptors	p. 146
Site of absorption of light quanta	photosynthetic bacteria
Locale of piezoelectric effects	motile bacteria and protozoans
Biological clock	p. 147

recognition, contact inhibition and transfer of surface material. It has been known for some time that normal tissue cells coming into contact are suddenly immobilized by a "freezing" of cytoplasmic streaming and thus can develop in alignment with one another (Abercrombie and Heaysman, 1953). This property is absent in cancer cells. The alignment of normal cells can come about either through structural fit of the surface glycoproteins and glycolipids (Fig. 2.38**A**) or through an enzyme-substrate link where a surface galactosyl transferase is involved (Fig. 2.38**B**).

The sugar residues exposed on the surface of the plasma membranes are also responsible for a number of agglutination reactions, themselves caused by glycoproteins of plant origin, e.g., wheat germ agglutinin, soybean agglutinin and the jack bean agglutinin with two binding sites on the molecule, termed concanavalin A (the binding site appears to contain two tyrosines, one aspartic acid and one arginine; Hardman and Ainsworth, 1973). They combine specifically with N-acetyl-glucosamine, N-acetylgalactosamine and α-methyl-D-glucoside, respectively, the reaction being generally cryptic in non-transformed cells but very powerful in cells altered by viruses.

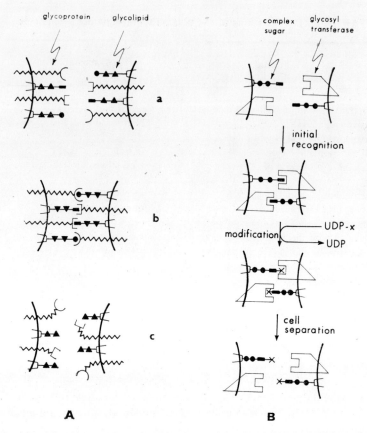

Fig. 2.38. **A.** Intercellular interaction through surface glycolipids and glycoproteins. **a** Separated cells, **b** confluent cells displaying complementary linkage through noncovalent bonds or possibly Ca^{2+} bridges, **c** malignant cells with incomplete carbohydrates and hence no complementarity. **B.** The role of cell surface glycosyl transferase in cell-cell recognition and cell contact modification.

2.7.1.2.2. Enzyme content

The enzyme equipment of plasma membranes can be very extensive, there being substantial differences between various plasma membranes. A surprisingly varied assortment is found in liver plasma membranes, erythrocyte membranes being much poorer. Few enzymic activities were detected in yeast membranes while the bacterial membrane contains practically all the mitochondrial enzymes of eukaryotic cells plus a number of others (Table 2.24). It is noteworthy that even in the enzyme equipment of plasma membranes a definite

TABLE 2.24. Major enzyme activities found in various plasma membranes

Rat liver	Erythrocyte	Baker's yeast	Mitrococcus lysodeikticus
Mg-ATPase	Mg-ATPase	Mg-ATPase	Mg-ATPase
Na, K-ATPase	Na, K-ATPase	Acid phosphatase	Cytochrome c oxidase
5'-Nucleotidase	Cathepsin I	Phospholipase	Catalase
Acetate kinase	Cathepsin II	Phosphatidylinositol kinase	NADH dehydrogenase
NAD^+ pyrophosphatase	Ribosephosphate isomerase	(β-Fructofuranosidase)	Lactate dehydrogenase
Phosphodiesterase	Acetylcholinesterase		Succinate dehydrogenase
Nucleosidediphosphatase (acting on IDP)			Oxoglutarate dehydrogenase
Leucine aminopeptidase			

asymmetry is apparent. Erythrocyte membranes contain acetylcholinesterase and NAD^+ nucleosidase on the outer surface but lipoamide dehydrogenase, protein kinase and adenosinetriphosphatase on the inner surface (Steck, 1974).

2.7.1.2.3. *Cell walls*

An important function of membranes of unicellular organisms of plants is the secretion of cell-wall material and, possibly, the function as a template or primer for some of the wall components.

All the rigid cell walls of plants and microorganisms are mainly polysaccharide plus varying amounts of peptide. Starting at the lower end of the phylogenetic tree, we shall deal first with bacteria.

Bacteria

There appear to be two fundamental types with regard to the structure of the cell envelope, the Gram-positive and the Gram-negative bacteria, the latter being distinguished from the former by the fact that they possess two membranous structures, an outer membrane and an inner, cytoplasmic, membrane (Fig. 2.39).

The peptidoglycan or mucopeptide or murein layer appears practically in all bacteria and is built on a skeleton of alternating N-acetylmuramic acid and N-acetylglucosamine

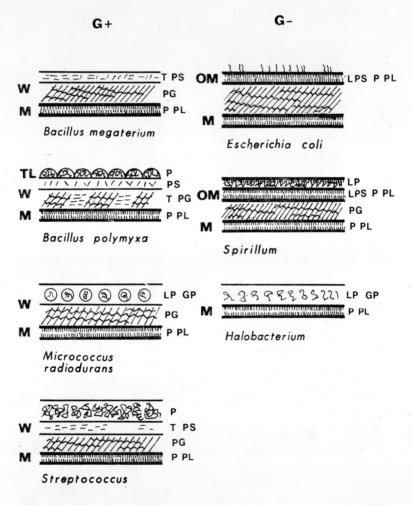

Fig. 2.39. Schematic models of bacterial cell envelopes both in Gram-positive and in Gram-negative species. **M** Plasma (cytoplasmic) membrane, **OM** outer membrane, **W** cell wall, **TL** T-layer; T teichoic acid, PS polysaccharide, PG peptidoglycan, P protein, PL phospholipid, LP lipoprotein, GP glycoprotein, LPS lipopolysaccharide.

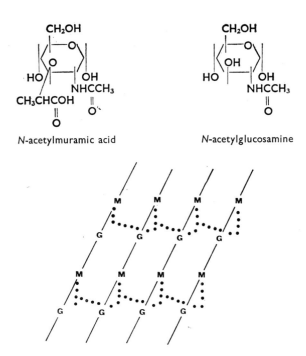

Fig. 2.40. Schematic representation of cell wall peptidoglycan, based on data from *Staphylococcus aureus*. The black dots represent amino acids of the peptide cross-links, **M** stands for *N*-acetylmuramic acid, **G** for *N*-acetylglucosamine. (Redrawn with permission from Ghuysen and Shockman, 1973.)

The *N*-acetylmuramic acid can be phosphorylated in position 6 or acetyl-ether linked or even cyclized between the carboxyl of the lactyl ether in position 3 and the nitrogen in position 2. Strands of *N*-acetylglucosamine and *N*-acetylmuramic acid are joined by oligopeptide links as shown in Fig. 2.40, the links being of several kinds:

$$\downarrow$$
L-ala→D-gln
$$\quad\quad\gamma\,|\!\rightarrow\text{L-lys}\rightarrow\text{D-ala}\rightarrow[\text{Gly}\rightarrow\text{Gly}\rightarrow\text{Gly}\rightarrow\text{Gly}\rightarrow\text{Gly}]\rightarrow\text{L-lys}$$
$$\quad\quad\quad\uparrow$$

(*Staphylococcus aureus*)

$$\cdots\cdots\rightarrow[\text{L-ala}\rightarrow\text{L-ala}\rightarrow\text{L-ala}\rightarrow\text{L-thr}]\rightarrow$$
(*Micrococcus roseus*)

$$\cdots\cdots\rightarrow[\text{Gly}_3, \text{L-ser}_2]\rightarrow$$
(*Staphylococcus epidermidis*)

$$\cdots\cdots\cdots\cdots\to [\text{L-ser}\to\text{L-ala}]\to$$
(*Lactobacillus viridescens*)

$$\cdots\cdots\cdots\cdots\to [\text{L-ala}\to\text{L-ala}]\to$$
(*Streptococcus pyogenes*)

$$\cdots\cdots\cdots\cdots\to [\text{L-ala}]\to$$
(*Arthrobacter crystallopoietes*)

$$\cdots\cdots\cdots\cdots\to [\text{D-asn}]\to$$
$$\beta\,|\!\to$$
(*Streptococcus faecalis, Lactobacillus casei*)

↓
L-ala→D-glu   DAP
  γ |→DAP*→D-ala ─────↑
  ┄┄┄↑
(*Bacillus megaterium*)

↓
L-ala→D-glu   ┄┄┄L-lys┄┄┄
  γ |→L-lys→D-ala ─────↑
  ┄┄┄↑
(*Aerococcus, Gaffkya*)

↓
L-ala→D-gln   DAP
  γ |→DAP→D-ala→Gly ─────↑
  ┄┄┄↑
(*Clostridium perfringens, Streptomyces albus*)

↓
L-ala→D-glu→Gly   ┄┄┄L-lys┄┄┄
  γ |→L-lys→D-ala→[L-ala→D-glu→Gly   ↑
  ┄┄┄↑   γ |→L-lys→D-ala$]_n$ ─────┘
(*Micrococcus lysodeikticus*)

* DAP = diaminopimelic acid

$$\begin{array}{l} \downarrow \\ \text{Gly} \rightarrow \text{D-glu} \cdots \cdots \\ \quad \gamma \mid \rightarrow \text{Homoser} \rightarrow \text{D-ala} \rightarrow \text{D-orn} \\ \quad \uparrow \qquad \qquad \qquad \qquad \uparrow \\ \quad \text{Gly} \rightarrow \text{D-glu} \cdots \cdots \cdots \cdots \cdot \mid \\ \qquad \quad \gamma \mid \rightarrow \\ \qquad \qquad (\textit{Corynebacterium}) \end{array}$$

$$\begin{array}{l} \downarrow \\ \text{L-ser} \rightarrow \text{D-glu} \cdots \cdots \\ \quad \gamma \mid \rightarrow \text{L-orn} \rightarrow \text{D-ala} \rightarrow \text{D-lys} \\ \qquad \qquad \qquad \qquad \uparrow \\ \qquad \qquad (\textit{Butyribacterium rettgeri}) \end{array}$$

The biosynthesis of peptidoglycans proceeds from cytoplasmic uridine triphosphate and N-acetylglucosamine-1-phosphate to uridine-diphospho-N-acetylglucosamine, followed by reactions with phosphoenolpyruvate, NADPH and the various amino acids, each reaction being catalyzed by a special enzyme. The UDP-oligopeptide formed then reacts in the membrane with a 55-carbon isoprenoid alcohol phosphate (undecaprenyl-P) to undecaprenyldiphospho-oligopeptide. This alcohol phosphate is called the glycosyl carrier lipid and appears to be a rather wide-spread membrane entity, known to be active also in the transport of teichoic acid precursor in *Staphylococcus* (Baddiley et al., 1968), of the lipopolysaccharide O-antigen of *Salmonella* (Osborn, 1969) and even in the glycosylation of glycoproteins (Caccam et al., 1969).

The undecaprenyldiphospho-oligopeptide reacts with UDP-N-acetylglucosamine to an undecaprenyldiphospho-disaccharide-oligopeptide and this is then translocated from the cytoplasmic membrane to the wall acceptor. The whole synthesis is shown in Fig. 2.41. together with the probable sites of action of several antibiotics. For a most up-to-date review see Ghuysen and Shockman (1973). It is rather likely that the peptidoglycan coat which maintains the shape of bacteria (rods, spheres, etc.) is not the determinant of gross cell morphology but rather a consequence of other factors in play (Henning and Schwarz, 1973).

The other major component found in the walls of Gram-positive bacteria is teichoic acid, a mixture of polymers of ribitol 5-phosphate and glycerol 1-phosphate, the polyols being linked phosphodiesteric-

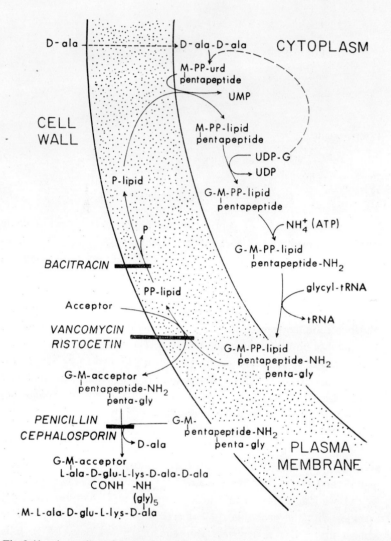

Fig. 2.41. An outline of the synthesis of peptidoglycan of type I (e.g. in *Aerococcus*) with the points of attack of various antibiotics. The dashed line at upper right represents the sequential building up of the UDP-M-pentapeptide from uridine-diphospho-*N*-acetylglucosamine. M *N*-acetylmuramic acid, G *N*-acetylglucosamine, lipid the undecaprenyl residue.

ally and frequently substituted with glucosyl, N-acetylglucosaminyl or ester-linked D-alanine residues.

$$\underset{O}{\overset{O^-}{HO\overset{|}{P}OCH_2}}-\left|-\right|-\underset{\underset{C=O}{\underset{|}{RO\ O\ O}}}{CH_2O}-\left[-\underset{O}{\overset{O^-}{\overset{|}{P}OCH_2}}-\left|-\right|-\underset{\underset{C=O}{\underset{|}{RO\ O\ O}}}{CH_2O}\right]_n-\underset{O}{\overset{O^-}{\overset{|}{P}-OCH_2}}-\left|-\right|-\underset{\underset{C=O}{\underset{|}{RO\ O\ O}}}{CH_2OH}$$

$$\underset{CH_3}{\underset{|}{HCNH_2}} \qquad \underset{CH_3}{\underset{|}{HCNH_2}} \qquad \underset{CH_3}{\underset{|}{HCNH_2}}$$

The chains are synthesized by sequential transfer of polyol phosphate from cytidinediphosphoribitol or CDP-glycerol to the acceptor (apparently preexisting teichoic acids). N-Acetylglucosamine attachment requires UDP. In some cases, the glycosyl carrier lipid is also involved.

An important component of the outer envelope of Gram-negative bacteria is the lipopolysaccharide. An example of its structure is shown in Fig. 2.42. It will be seen that various unusual sugars are present in its molecule. In fact, the O-antigen from other species of *Salmonella* was found to contain also paratose (3,6-dideoxy-D-glucose) and tyvelose (3,6-dideoxy-D-mannose). The lipid A region contains C_{12}, C_{14}, C_{16} and C_{14}-hydroxy fatty acids in a 1 : 1 : 1 : 3 ratio.

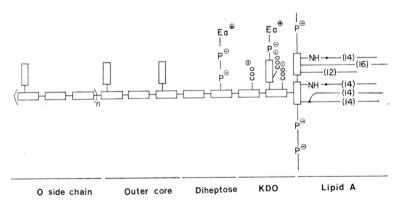

Fig. 2.42. A schematic outline of the lipopolysaccharide molecule. Ea Ethanolamine, P phosphate, KDO three molecules of 3-deoxy-D-manno-octulosonic acid. (Redrawn from Nikaido, 1973.)

Lipid A appears to be synthesized by the action of fatty acyl transferases on the diglucosamine skeleton. Once formed, it attaches the KDO residue, ATP being required for this step. Practically nothing is known about the attachment of the mannoheptoses but it is known that the glucose and galactose residues are joined through the action of UDPglucosyltransferases under activation of Mg^{2+}, phospholipids (particularly phosphatidyl ethanolamine) being required for the process (see Table 2.18).

The O-antigen part is synthesized separately under participation of the glycosyl carrier lipid and of diphospho derivatives of various nucleosides, *viz.* UDPgalactose, TDPrhamnose, GDPmannose and CDPabequose. The oligosaccharides are then transferred onto the

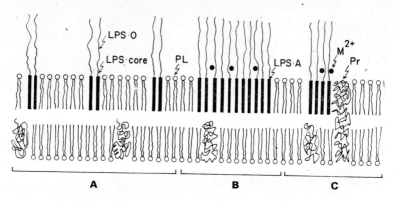

Fig. 2.43. Possible arrangement of molecules in the outer membrane layer of a Gram-negative bacterium. In **A**, lipopolysaccharide molecules are interspersed among phospholipids. In **B**, bivalent cations reduce the repulsion between lipopolysaccharide molecules so that these can lie next to each other. In C, proteins interact both with phospholipids and with lipopolysaccharide. LPS lipopolysacharide, PL phospholipid, O the O side chain, A lipid A, Pr protein. (Redrawn after Nikaido, 1973.)

core lipopolysaccharide again by mediation of the isoprenoid carrier lipid (all these facts, and more, are elegantly summarized by Nikaido, 1973).

The completed lipopolysaccharide must then be transported from the cytoplasmic membrane (*cf.* Fig. 2.39) to the outer membrane, apparently via the connections between the two membranes through the peptidoglycan layer. A model of the outer membrane is shown in Fig. 2.43.

The outer membrane is invested with several functions. Thus, the O-antigen chain has a survival value for the bacterium in a host animal, due to the possibility of enormous diversification of the sugar residues and hence relatively low levels of antibodies capable of coping with them.

The somatic O-antigen is not the only antigen carried by the surface of Gram-negative bacteria. Various mutants have been prepared with defects in the side-chain synthesis of the O-antigen or in their transfer to the lipopolysaccharide core – their antigens are usually termed R-antigens. There exist also transient forms designated as T-antigens (mainly in *Salmonella* and *Shigella*).

The wall contains also the common antigen CA, the C-antigen and the bacterial agglutinogen BA.

Moreover, in addition to the above somatic (or cellular) antigens, capsule-forming bacteria contain a variety of polysaccharide antigens in the outer capsule. These polysaccharides are pronouncedly negatively charged (due to phosphate groups, hexuronic acid and sialic acid) and often contain improbably high amounts of phosphorus. Thus, in the *Haemophilus* capsular antigen the sugar : P ratio is 1 : 1. The sugar composition of these polysaccharide antigens is not as varied as in the lipopolysaccharide, the predominant components being galactose, glucose, mannose, fucose, rhamnose, glucuronic acid and galacturonic acid.

The outer membrane as a whole apparently acts as a permeability barrier to various antibiotics and obviously other compounds although, generally, it is the inner, cytoplasmic membrane which carries specific transporting mechanisms.

Bacterial walls also contain polysaccharides without any admixture of either peptides or lipids but they are relatively of minor occurrence and importance.

Yeasts

In comparison with bacteria, the exterior coat of yeast cells is rather less complex although it may reach a thickness of 0.2 µm, comparable to the diameter of a small bacterium.

In *Saccharomyces*, the cell wall is over 80 % polysaccharide, one-half of it glucan, the other half mannan. It is generally assumed that mannan predominates in the outer part of the wall, together with some amorphous glucan, while fibrillar $\beta(1\rightarrow3)$glucan forms the innermost covering of the plasma membrane; the cross section through the wall is shown schematically in Fig. 2.44.

Other yeast genera, however, may contain no mannan at all but,

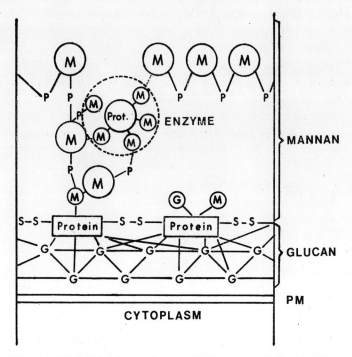

Fig. 2.44 A schematic representation of the structure of yeast cell wall. M Mannan, G glucan, P phosphate, ENZYME invertase, PM plasma membrane. (Adapted from Lampen, J. O. (1968). *Antonie van Leeuwenhoek, J. Microbiol. Serol.* **34**, 1.)

in compensation, large amounts of chitin (*Nadsonia, Rhodotorula, Endomyces*).

The *Saccharomyces* glucan contains mostly $\beta(1\rightarrow3)$ linkages in the fibrillar part but also $\beta(1\rightarrow6)$ linkages in the remaining part. The structure of the mannan is rather more varied, there being structural units of the following types present:

$$\begin{array}{cc}
\rightarrow\text{Man}^1\rightarrow{}^6\text{Man}^1\rightarrow & \rightarrow\text{Man}^1\rightarrow{}^6\text{Man}^1\rightarrow{}^6\text{Man}\rightarrow \\
\uparrow^2_1 & \uparrow^2_1 \\
\text{Man} & \text{Man} \\
\uparrow^2_1 & \uparrow^2_1 \\
\text{Man} & \text{Man} \\
 & \uparrow^3_1 \\
 & \text{Man}
\end{array}$$

The cell walls of yeast contain some protein (*c.* 10 %), possibly in a complex with glucan, and a small amount of lipid (1−10 %).

Both mannan and glucan are synthesized under participation of uridinediphosphosugar derivatives.

Like some yeast species, filamentous fungi contain major amounts of chitin which is a polymer of $\beta(1\rightarrow 4)$-N-acetyl-D-glucosamine. The walls are organized usually as closely packed microfibrils, parallel to the underlying plasma membrane. Sugar residues found in various fungal polysaccharides are (besides glucosamine) mainly glucose, mannose and galactose.

Plants

Especially heavy cell envelopes are known to occur in algae and in higher plants. Little is known about the architecture of algal cell walls but they are known to contain, as major fibrous components, the following polysaccharides: $\beta(1\rightarrow 4)$-mannan, $\beta(1\rightarrow 3)$-xylan, two types of cellulose and glucomannan. Hemicelluloses which are also present have been found to contain also galactose, arabinose, rhamnose and glucose.

Higher plants are characterized by the overwhelming content of cellulose in their cell walls, cellulose being a polymer of $\beta(1\rightarrow 4)$-linked molecules of D-glucose. However, only few cell walls are practically speaking pure cellulose, other polysaccharides being usually present. A common one is $\beta(1\rightarrow 4)$-linked xylan, often substituted with L-arabinose linked usually $(1\rightarrow 3)$. While it has been extremely difficult to arrive at a molecular weight of cellulose (owing ot the fact that the native state in a wall may represent a continuum of fibers, microfibers and ultramicrofibers) some reasonable estimates have been obtained for xylans (about $10-15\,000$, in general). Another polysaccharide of plant walls is a glucomannan with usually 2 or 3 mannose residues per glucose, linked $\beta(1\rightarrow 4)$. Another polysaccharide is arabinogalactan, probably a branched galactan (linked $\beta(1\rightarrow 3)$ in the backbone and $\beta(1\rightarrow 6)$ in the side chains) with mostly terminal arabinose residues.

Finally, there are the pectins, i.e. $(1\rightarrow 4)$-linked polygalacturonic acids.

A special group of plant cell wall polymers are the lignins, made up of networks of p-hydroxyphenylpropane, vanillin, syringaldehyde, derivatives of guaiacol and coniferol.

The architecture of the plant cell wall is rather intricate and no universal pattern has been arrived at but there is general agreement on the existence of a primary wall, formed during longitudinal growth of the cell, and of a secondary wall, formed by gradual layering of material from the cell interior moved through the plasmalemma onto the primary wall (Fig. 2.45).

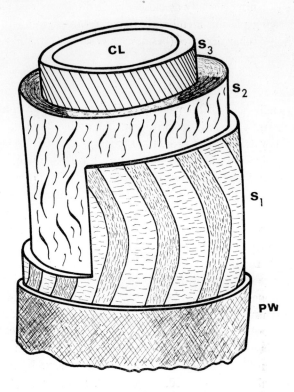

Fig. 2.45. A diagrammatic representation of the secondary wall of a mature cotton hair. **CL** Cell lumen, S_1, S_2, S_3 various layers of the secondary wall with different arrangement of the microfibrils, **PW** primary wall. (Redrawn from Roelofsen, P. A. (1959). *The Plant Cell Wall*. Gebr. Borntraeger, Berlin.)

There is little doubt that most (and perhaps all) wall polysaccharides are synthesized by attaching sugar moieties from uridinediphospho derivatives, there being specific enzymes to catalyze the attachment by a $(1\to3)$, $(1\to4)$ or $(1\to6)$ link. However, GDP derivatives are also known to operate under certain conditions. It is not clear by what means the UDP-sugar molecule is transported from the inside of the plasmalemma to the external polysaccharide primer.

Protozoans

An example of a nonplant extramembrane cell coat is the external envelope of amoebae, These protozoans have a thick (20 nm) unit membrane covered with filaments (100–200 nm long) extending into the surrounding medium (Fig. 2.46) and serving to adsorb nutrients which are then engulfed by pinocytosis. The composition of the whole coat is 45 % lipid, 34 % protein and 21 % carbohydrate. A rather surprising feature of this membrane is that it protects the cell against the strongly hypotonic external medium although it contains none of the generally necessary fibrillar meshwork.

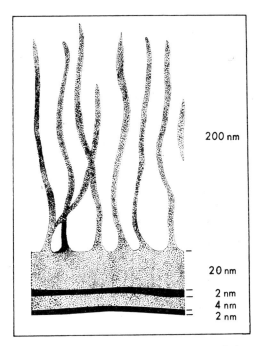

Fig. 2.46. A diagrammatic drawing of the surface coat of the amoeba *Chaos chaos*. The plasma membrane proper is represented by the three lowermost lines. (Adapted from Brandt, P. W. and Pappas, G. D. (1962). *J. Cell Biol.* **15**, 55.)

2.7.1.2.4. Binding and receptor properties

The surface of plasma membranes shows a distinct binding affinity for a number of ions, small molecules and polymers.

The cation-binding affinity of erythrocyte surface has been shown to decrease as follows: $Th^{4+} > UO_2^{2+} > La^{3+} > Cu^{2+}, Ni^{2+} > Ca^{2+}$,

Sr^{2+}, Ba^{2+}, Mn^{2+}. On the other hand, agglutination of erythrocytes is caused with highest efficiency by Ca^{2+}, Cu^{2+}, Th^{4+}, Ti^{3+}, Al^{3+}, less by Ag^+, Cd^{2+} and much less by Mn^{2+}, Co^{2+} and Hg^{2+}.

There is practically no binding of anions on the plasma membrane surface.

The cations bound to the surface reflect the magnitude of the surface charge, usually measured electrophoretically. The charge density has been measured for a variety of cells and found to range from about $7 \cdot 10^{-3}$ C m^{-2} in a hamster kidney cell to $2 \cdot 10^{-2}$ Cm^{-2} in a sheep erythrocyte (Curtis, 1967). This relatively high negative charge results in the isoelectric points of most single cells lying between pH 2 and 4. In mammalian cells, the surface charge is composed of $10^7 - 10^8$ negative charges per cell and of $10^6 - 10^7$ positive charges per cell, depending on the surface area, the density being as indicated above.

One of the important functions residing in some plasma membranes is the susceptibility to hormone stimulation. It appears that the hormone recognizing or initiator systems contain three parts: the receptor which specifically recognizes the hormone, the transducer which couples the function of the receptor with the effector, and the effector which then brings about appropriate change in the cell function. Some of the hormone receptors have been partly purified. The insulin receptor from adipose plasma membrane has a molecular weight of 300 000 and has two types of sites with dissociation constants of 10^{-10} M and 10^{-9} M (there is competition with insulin binding by plant lectins, both of them inhibiting adenylate cyclase); there appears to be 1 site per µm². The glucagon receptor has a K_D of $4 \cdot 10^{-9}$ M and a molecular weight near 50 000 (it stimulates adenylate cyclase). The catecholamines are bound to fat cells and to erythrocytes with K_D's of about 10^{-6} M, the erythrocytes containing about 5 sites per µm². Acetylcholine is bound with a K_D of 10^{-5} M and another one of 10^{-7} M in the *Electrophorus* eel electric organ, and with K_D's of 10^{-7} and 10^{-8} M in the *Torpedo* electroplax. The electric eel acetylcholine receptor has a molecular weight of about 360 000, and is composed of 50 000 subunits. The site density is extraordinarily high: about 33 000/µm² in the subsynaptic area, 100 000 per µm² in frog muscle, and 12 000/µm² in mouse diaphragm. Oxytocin is bound to frog-skin receptors in several ways, the most important one having a K_D of $5 \cdot 10^{-3}$ M, the maximum capacity being $1-2$ pmol/g tissue.

The character of the transducer remains unknown but physical changes in the lipid organization may be involved. The effector system is most often adenylate cyclase (certainly for peptide hormones, such as insulin, glucagon, oxytocin, vasopressin, thyreocalcitonin, secretin, adrenocorticotropin; *cf.* Jost and Rickenberg, 1971). The specificity of action is generally ensured by the receptor but, at least in the case of insulin, appears to be rather complicated.

A plasma membrane function perhaps related to the hormone sensitivity is the ability of specialized cells (lymphocytes, in particular of the bone-marrow or B type, or plasma cells) to respond to antigen binding by starting the synthesis of protein antibodies that inactivate the antigen either by neutralizing it (*antitoxins*), by causing the antigen-producing organisms to clump together (*agglutinins*), by causing the invading cell to disintegrate (*lysins*) or by making it more susceptible to attack by phagocyting macrophages (*opsonins*).

A special susceptibility carried by some bacteria is toward bacteriocins, such as the various colicins produced by *Escherichia coli* strains. Colicins are single-chain proteins, ranging in molecular weight from 45 000 to 80 000 and causing the death of various bacteria. They are coded by extrachromosomal genetic elements (the so-called Col factor plasmids) and they bring about cessation of rather varied cell functions: Protein synthesis is blocked by E3, DNA structure is broken down by E2, the integrity of the plasma membrane is disrupted by K and E1. The mechanism of their action may resemble the three-step action of hormones.

A rather peculiar function has been recently proposed for the plasma membrane by Njus and co-workers (1974), *viz.* that of maintaining circadian rhythms of organisms, involving lateral diffusion of proteins. The function is temperature-insensitive due to compensation in fatty acid composition of the membrane.

2.7.2. Mitochondrion

These are ubiquitous organelles of eukaryotic cells, about 1 μm thick and $1-3$ μm long, of a typical architecture (Fig. 2.47), containing an outer and an inner unit membrane. From a disrupted cell, they can be separated after removal of nuclei by centrifuging at $5-20 \cdot 10^4 \, g \, \text{min}$

and then purified in isopycnic density gradients, forming bands at $1.18-1.21$ g cm^{-3}.

In many respects they resemble bacteria (they contain circular DNA, small ribosomes and protein synthesis sensitive to chloramphenicol; they divide autonomously, etc.) and a popular theory has it that they have invaded primitive eukaryotic cells early in evolution and provided them symbiotically with an efficient system of storing oxidation energy. (For an excellent treatment of the data, see Getz, 1972.)

Be it as it may, the two mitochondrial membranes are the sites of various enzymes and have characteristic properties (Table 2.25).

TABLE 2.25. Some properties of mitochondrial membranes

Feature	Inner membrane	Outer membrane
Thickness (nm)	6.5	6.5
Fine structure	globular projections, 9 nm in diameter	flat, with 3nm pits
Buoyant density (g cm^{-3})	1.21	1.13
Protein-lipid ratio	3.6	1.2
% Cardiolipin	22	3
% Phosphatidyl inositol	4	14
Permeability	highly selective	lets larger molecules across
Osmotic response	+	—
Contraction after ATP	+	—

The mitochondrion is literally packed with enzymes which show a very clearly defined distribution in the various morphological compartments (Table 2.26).

The outer membrane resembles the endoplasmic reticulum of the cell, both containing cytochrome b_5, a rotenone-insensitive cytochrome c reductase and there being a phospholipid exchange between the

Fig. 2.47. The mitochondrion. **A** An ultrathin section of a mitochondrion from mouse kidney. **O** Outer mitochondrial membrane, **I** inner mitochondrial membrane. (Courtesy of Dr. J. Ludvik, Institute of Microbiology, ČSAV, Prague.)

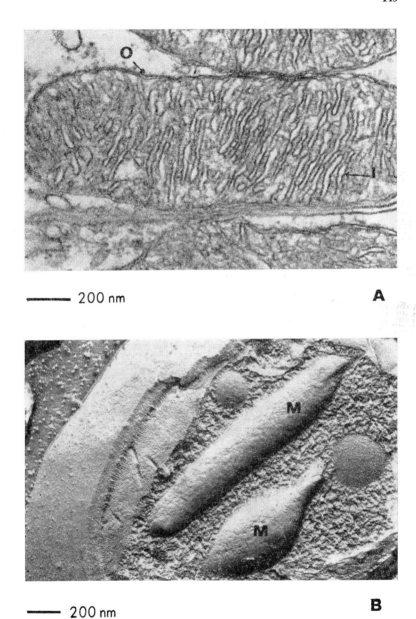

B A freeze-etched preparation of *Saccharomycodes ludwigii* showing two mitochondria (**M**) with their outer surface exposed. Parts of typical plasma membranes with grooves are seen on the left. (Courtesy of Dr. E. Streiblová, Institute of Microbiology, ČSAV, Prague.)

TABLE 2.26. Distribution of enzymes in liver mitochondria (adapted from Ernster and Kuylenstierna, 1970; Kroger and Klingenberg, 1970)

Outer membrane	Intermembrane space	Inner membrane	Matrix
Cytochrome b_5 reductase	Adenylate kinase	Cytochrome c oxidase (1)	Malate dehydrogenase (0.4)
Amine oxidase (flavin-containing)	Nucleosidediphosphate kinase	Cytochrome a (1)	Isocitrate dehydrogenase (NAD) (0.03)
Kynureninase	D-Xylulose reductase	Cytochrome c (1)	Isocitrate dehydrogenase (NADP) (0.25)
Acyl-CoA synthetase		Cytochrome c_1 (0.5)	
Glycerolphosphate acyltransferase		Cytochrome b (1)	Glutamate dehydrogenase (0.01)
Lysolecithin acyltransferase		Ubiquinone (11)	
Cholinephosphotransferase		Succinate dehydrogenase (0.16)	Oxoglutarate dehydrogenase
Phosphatidate phosphatase		NADH dehydrogenase (0.1)	Citrate (si)-synthase
Phospholipase A_1		NAD (11)	Aconitate hydratase
Nucleosidediphosphate kinase		NADP (3)	Fumarate hydratase
D-Xylulose reductase		3-Hydroxybutyrate dehydrogenase	Pyruvate carboxylase
Fatty acid elongation system		Ferrochelatase	Phosphoenolpyruvate carboxylase
		Carnitine palmitoyltransferase	Aspartate aminotransferase
		L-Xylulose reductase	Ornithine carbamoyltransferase
		Fatty acid elongation chain	Acyl-CoA synthetase
			L-Xylulose reductase

The figures in parentheses show the relative molar amounts of enzymes and coenzymes.

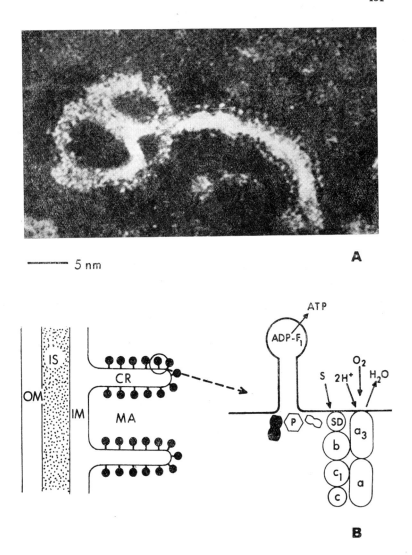

Fig. 2.48. Details of intramitochondrial architecture. **A.** A negatively stained part of the inner mitochondrial membrane of rat liver, showing the typical knobs. (Reproduced with permission from Ernster, L. and Kuylenstierna, B. (1970). In: *Membranes of Mitochondria and Chloroplasts* (ed. by Racker, E.), p. 172. Van Nostrand—Reinhold, New York.) **B.** Schematic drawing of mitochondrial membranes and an enlarged section of the vicinity of a cristal knob, together with the postulated functional components. OM Outer mitochondrial membrane, IS intermembrane space, IM inner mitochondrial membrane, CR crista, MA matrix, F_1 one of the coupling factors, P phosphate, S succinate, SD succinate dehydrogenase, b, c_1, c, a and a_3 various cytochromes.

outer membrane and the endoplasmic reticulum. The inner membrane, on the other hand, resembles very much a bacterial cytoplasmic membrane. It is the inner membrane that has been investigated by some of the best minds in the fields of biochemistry and membrane research, the most provocative feature being its ability to carry out oxidative phosphorylation, the detailed mechanisms of which is still a matter of dispute.

Three more or less antagonistic hypotheses exist, the chemical one, mainly due to Slater (e.g., 1958) and Racker (e.g., 1970), the chemiosmotic one due to Mitchell (e.g., 1966) and the electromechanochemical one due to Green and Ji (1972). The first of these envisages the function of various high-energy intermediates on the way from the oxidative chain to ATP, the second ascribes the mediating role to the translocation of protons across the inner membrane, the protonmotive force thus created generating ATP from ADP and inorganic phosphate. The third model endeavors probably most successfully to combine the merits of several previous hypotheses by introducing several configurations of the basepiece of the mitochondrial inner membrane. For thorough, healthily biased, reviews, the reader is referred to Greville (1969), Mitchell (1973) and Green (1974).

The various factors active in oxidation and oxidative phosphorylation have apparently a definite localization in the inner membrane, the latest information being summarized in Fig. 2.48.

The inner membrane is also the locale of various specific transport systems for anions, cations, organic acids (particularly those of the Krebs cycle), and nucleotides (see, e.g., Pressman, 1970).

2.7.3. Chloroplast

Like mitochondria, chloroplasts are cell organelles of considerable autonomy, containing ribosomes and a complete protein-synthesizing apparatus, starting from DNA to tRNA's. Like mitochondria, chloroplasts are suspected by some authors to represent early bacterial invaders of primitive plants with no photosynthetic apparatus.

Chloroplasts are enclosed in an outer envelope, the outer chloroplast membrane and are formed from lamellar vesicles (thylakoids) stacked in various ways, from very simple superposition in blue-green and red algae to irregular arrays several lamellae thick in green and brown algae to a more elaborate architecture in most green algae and higher plants (Fig. 2.49) where grana are seen at places where the lamellae overlap and partly fuse.

The obvious major function of chloroplast lamellae is to provide for conversion of radiation energy to chemical-bond energy, the key component of this photosynthetic process being chlorophyll of one kind or another. The postulated embedding of chlorophyll molecules

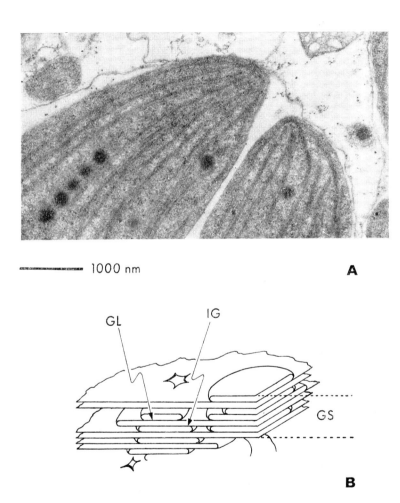

Fig. 2.49. The chloroplast. **A**. An ultrathin section through a cell of the green alga *Vaucheria* with strand-like chloroplasts. (Courtesy of Dr. S. Janda and Dr. V. Pokorný, Institute of Microbiology, ČSAV, Prague.) **B**. A schematic representation of the lamellar structure of a well developed chloroplast. GL Granum lamella, IG intergranum lamella, GS stack of grana.

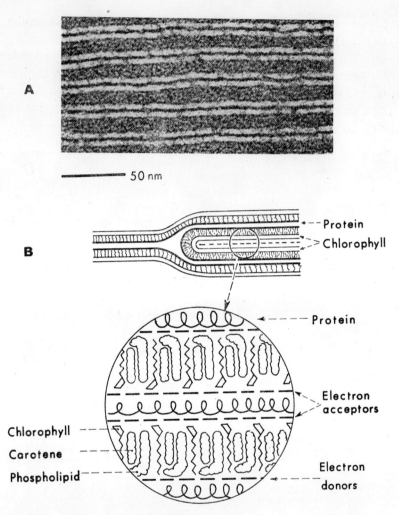

Fig. 2.50. Details of intrachloroplast architecture. **A**. An ultrathin section of a spinach chloroplast granum, stained with potassium phosphotungstate. **B**. A schematic cross section through the thylakoid membrane and an enlarged view of its part. The thylakoid here is a part of the granum stack shown in Fig. 2.49.

in the lamellar structure is shown in Fig. 2.50, together with a high magnification of a chloroplast lamella.

The chlorophyll is almost exclusively associated with galactosyl-diglyceride which has a characteristic spectrum of fatty acyl residues which is taxonomically defined; in photosynthetic bacteria it contains almost exclusively 16 : 0 and 16 : 1 acids, in blue-green algae also

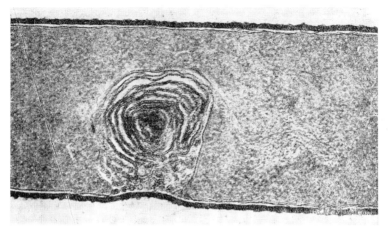

—— 100 nm   **A**

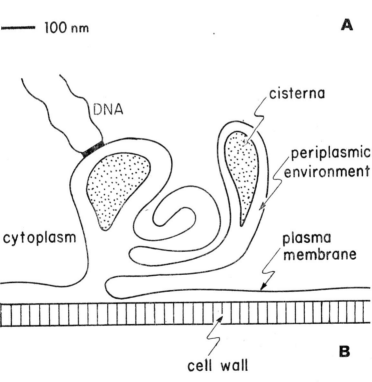

Fig. 2.51. **A**. A well developed mesosome in *Bacillus subtilis*. (Reproduced with permission from Weibull, C. (1968). In: *Microbial Protoplasts, Spheroplasts and L-Forms* (ed. by Guze, L. B.), p. 62. The Williams and Wilkins Co., Baltimore.) **B**. A schematic drawing of the mesosome showing its interconnection with the plasma membrane.

a great proportion of 18 : 2 and 18 : 3 fatty acids while in algae it has major amounts of higher unsaturated acids, such as 20 : 4, 20 : 5 and 18 : 4. In higher plants, the practically sole component of this glycolipid is the 18 : 3 acid.

The thylakoid membrane apparently possesses certain transport functions that have been investigated using cations and anions but very little is known about the translocation of molecules into and out of chloroplasts across the outer membrane. The outer membrane, like that of mitochondria, is apparently permeable to small as well as larger molecules. In contrast with the mitochondrial inner membrane, the thylakoid membrane carries a potential which is positive inside.

2.7.4. Mesosome

A number of bacteria, both Gram-negative (*Spirillum serpens*, *Escherichia coli*) but mainly Gram-positive (*Bacillus subtilis*, *B. licheniformis*, *B. medusa*, *B. megaterium*, *B. cereus*, *B. thuringiensis*, *Micrococcus lysodeikticus*, *Mycobacterium phlei*, *M. smegmatis*, *Lactobacillus casei*, *L. corinoides*, *Staphylococcus aureus*, *Streptococcus faecalis*) have been found to contain membranous structures, apparently derived from the cytoplasmic (plasma) membrane, of characteristic appearance (Fig. 2.51) that are called mesosomes.

Of the many postulated functions (energy metabolism, cell division, cell-wall synthesis, secretion of enzymes, site of episome attachment to membrane) none has been unequivocally confirmed. It appears most likely that the mesosomes are the site of DNA penetration into bacterial cells (*cf.* p. 320) (Reusch and Burger, 1973).

2.7.5. Endoplasmic reticulum

A common feature of the internal architecture of practically all eukaryotic cells is a more or less extensive network of membrane elements studded with particles now known to be ribosomes (Fig. 2.52). The primary function of the ribosomes is the synthesis of proteins but

---

Fig. 2.52. Endoplasmic reticulum. **A.** An ultrathin section through the tubular and cisternal elements of rough endoplasmic reticulum studded with ribosomes,

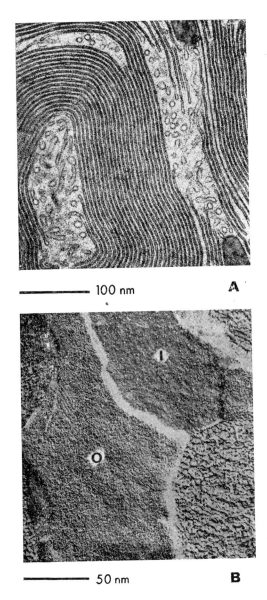

from a bat pancreas acinar cell. (Taken with permission from Fawcett, D. W. (1964). In: *Intracellular Membraneous Structure* (ed. by Seno, S. and Cowdry, E. V.), p. 15. Japan Soc. Cell Biol. Okayama.) **B**. A freeze-etched preparation of the yeast *Wickerhamia fluorescens* showing the outer surface (O) of the endoplasmic reticulum studded with 5 nm particles and the inner surface (I) of the endoplasmic reticulum with few 10 nm units. (Reproduced with permission from Bauer, H. (1970). *Can. J. Microbiol.* **16,** 219.)

this process, once the ribosomes carry the information-containing messenger RNA, probably takes place on ribosomal aggregates, the so-called polysomes whose association with the reticular membrane is not clear.

In a cell homogenate, the endoplasmic reticulum forms a fraction of "rough" microsomes which is separated from the general microsomal fraction by equilibrium density gradient centrifugation or selective aggregation in 15 mM CsCl and rate zonal sedimentation in a continuous sucrose gradient.

The membranes of the endoplasmic reticulum carry a number of enzymes, such as NADPH-cytochrome reductase, cytochrome P_{450}, cytochrome b_5, glucose-6-phosphatase, a mixed function oxidase and various esterases.

It is in the endoplasmic reticulum that a hypothetical unit, the membron, has been postulated to act in the regulation and stabilization of protein synthesis (e.g., Pitot et al., 1969).

The transport functions of endoplasmic reticulum (sometimes called the ergastoplasm) remain unexplored but it is known that even the large molecules of proteins synthesized on the ribosomes appear inside the lumen of the cisternae formed by the reticulum.

The endoplasmic reticulum is probably directly attached at places to the nuclear membrane. At the same time, it is probably functionally related to the Golgi apparatus, particularly in secretory cells.

2.7.6. Golgi apparatus

For some time, the Golgi apparatus has been included in the microsomal fraction as the "smooth" microsomes. However, it is apparently a separate system of membrane elements or cisternae (Fig. 2.53). The membranes of the apparatus or the dictyosomes are difficult to isolate without breakdown to small vesicles. The purest fraction is obtained at $1.05 - 1.09$ g cm^{-3} from a pellet sedimented at 10^6 g min after previous separation of a $2.6 \cdot 10^5$ g min sediment.

The function of the apparatus is to provide a vehicle for transport of large proteins but also the site of synthesis of carbohydrates and lipoproteins. The proteins synthesized on the ribosomes of the endoplasmic reticulum are translocated inward and then released in the form of vesicles pinched off toward the Golgi apparatus. There they fuse with some of the Golgi membranes, may be further supplemented

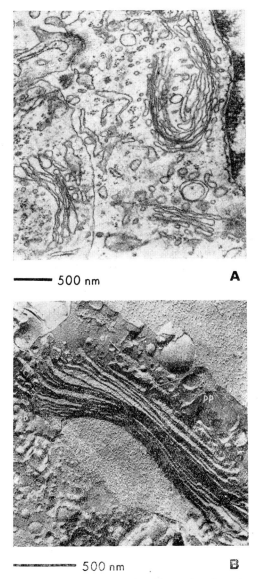

Fig. 2.53. The Golgi apparatus. **A**. An ultrathin section through a mouse tumor cell with several Golgi zones. A part of the nuclear membrane is also visible at right (Courtesy of Dr. J. Ludvik, Institute of Microbiology, ČSAV, Prague.) **B**. A freeze-etched Golgi apparatus (dictyosome) with individual cisternae seen in cross section (toward upper right) and in surface view (toward center and lower left). The membranes are associated with the endoplasmic reticulum at pp. (Reproduced with permission from Staehelin, L. A. and Kiermayer, O. (1970). *J. Cell Sci.* **7**, 787.)

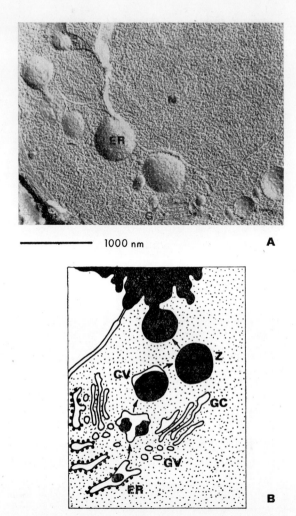

Fig. 2.54. Association of endoplasmic reticulum, Golgi lamellae and the exocrine process. **A**. A freeze-etched preparation of *Saccharomyces pombé* showing cisternae of endoplasmic reticulum (**ER**) associated with the nuclear membrane and moving toward the Golgi zone (**G**). (Courtesy of Dr. E. Streiblová, Institute of Microbiology, ČSAV, Prague.) **B**. A diagrammatic representation of the pancreatic exocrine cell function. Cisternae of the endoplasmic reticulum (**ER**) with and without ribosomes merge into the smooth vesicles of the Golgi peripheral region (**GV**) which arise from the Golgi cisternae proper (**GC**) to form the condensing vacuoles (**CV**) and finally the zymogen granules (**Z**) which are excreted into the lumen (black at top). (Compare with Fig. 7.3.) (Adapted from Jamieson, J. D. (1972.) In: *Current Topics in Membranes and Transport* (ed. by Bronner, F. and Kleinzeller, A.), vol. **3**, p. 273. Academic Press, New York—London.)

with carbohydrate moieties and stored in storage granules. These can then be expelled from cells by a process of reverse pinocytosis (p. 319). The train of events is shown schematically in Fig. 2.54.

The Golgi membranes are sometimes credited with the role of synthesizing new membrane elements.

2.7.7. Lysosome

Lysosomes are roughly spherical structure-less particles, about the size of a small mitochondrion, enclosed in a trilaminar membrane which may be variously corrugated (Fig. 2.55). Their apparent function is to wall off hydrolytic, digestive, enzymes. They may represent pinocytotic vacuoles but, more likely, they are special digestive

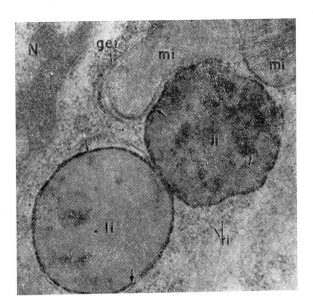

200 nm

Fig. 2.55. A thin section of a proximal convoluted tubule cell of mouse kidney, after injection of crystalline bovine hemoglobin. The most prominent features are the lysosomes (li) showing a positive acid phosphatase reaction on their surface (arrows). Other organelles to be seen include the nucleus (N), mitochondria (mi) and ribosomes (ri) as well as the rough endoplasmic reticulum (ger). (Reproduced with permission from De Robertis, E. D. P., Nowinski, W. N., and Saez, F. A. (1965). *Cell Biology*. W. B. Saunders Co., Philadelphia—London.)

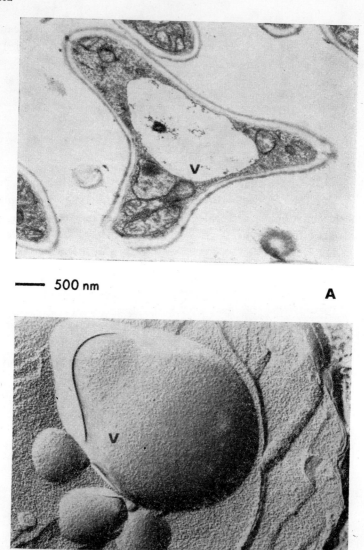

Fig. 2.56. The vacuole. **A**. A thin section through the triangular form of the dimorphic yeast *Trigonopsis variabilis* showing the vacuole (**V**) with various inclusions. (Courtesy of Dr. V. Šnejdar, Institute of Microbiology, ČSAV, Prague.) **B**. A freeze-etched preparation of the vacuole (**V**) of *Saccharomyces cerevisiae* with typical surface particles. (Courtesy of Dr. E. Streiblová, Institute of Microbiology, ČSAV, Prague.)

organelles of the cell. In damaged or dying cells they lyze and release their sanitary enzymes into the cell cytoplasm.

2.7.8. Tonoplast

Many fungal and plant cells contain large transparent structureless inclusions called vacuoles which are surrounded by a typical membrane called the tonoplast, shown in surface view in Fig. 2.56. Vacuoles can be isolated from lyzed protoplasts in the presence of 8 % Ficoll when they float on top (Matile and Wiemken, 1967). Like the lysosome of animal cells, the vacuole contains various hydrolytic enzymes (two proteases, an esterase, ribonuclease, aminopeptidase) as well as lipoamide dehydrogenase (NADH). It is, however, doubtful that these enzymes are associated with the tonoplast. The vacuole also contains metaphosphate deposits and is the principal storage pool for some amino acids (arginine, ornithine) in yeasts (Wiemken and Nurse, 1973).

If the vacuole has the lysosome function it is probably involved in protein degradation and thus in the regulation of protein synthesis but it is not known how proteins cross the tonoplast to come into contact with the proteolytic enzymes.

The transport properties of the tonoplast still require elucidation but this membrane might offer a simple model for transport studies.

2.7.9. Nucleus

The nuclear membrane is a typical structure found in all eukaryotic cells. It is actually a double membrane, the outer one being in close association with endoplasmic reticulum and covered with ribosomes. These two membranes form a continuum through large-size pores (Fig. 2.57) often plugged with electron-dense "blebs". In spite of these pores, the nuclei of cells can be isolated as such, sedimenting readily at 10^4 g min.

The two membranes are difficult to separate and usually collapse into small fragments at a density of about $1.18 - 1.22$ g cm^{-3}.

Very little is known about the enzyme equipment or permeability of nuclear membranes although it is readily appreciated that there must be channels permitting the passage of ribonucleic

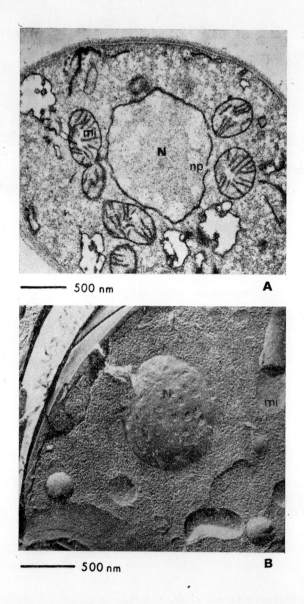

Fig. 2.57. The nucleus. **A**. A thin section through a cell of the yeast *Rhodotorula glutinis*, fixed with $KMnO_4$, showing a prominent nucleus (**N**) with several pores in the membrane (np) and several mitochondria (mi). (Courtesy of Dr. V. Šnejdar, Institute of Microbiology, ČSAV, Prague.) **B**. A freeze-etched preparation of the yeast *Saccharomyces cerevisiae* showing the nucleus (**N**) with numerous large pores. Several mitochondria are also visible (mi). (Courtesy of Dr. E. Streiblová, Institute of Microbiology, ČSAV, Prague.)

acids as well as proteins from and into the nucleus. There are some electron-microscopic indications that the transport may take place through the pores when the plugs are removed or dissolved after a suitable signal.

2.7.10. Other membranes

There is a variety of membranous organelles specific either for certain cells or certain organisms (the catalase-containing vesicular peroxisomes, the synaptic vesicles, hydrogenosomes etc.), some of their membranes possessing an informative architecture. A much studied example is the membrane of the retinal rod outer segment which is instrumental in the visual process. It is a regularly arranged membrane of globules about 4 nm in diameter, apparently the protein complexes including rhodopsin, retinal, vitamin A and the appropriate enzymes (Fig. 2.58).

SYNOPSIS

Biological membranes are recognized as trilaminar structures after negative staining in the electron microscope. They are the basic architectural framework of almost all cells and organelles (plasma membrane, nuclear membrane, mitochondria, chloroplasts, Golgi apparatus, endoplasmic reticulum, tonoplasts, lysosomes, mesosomes, etc.).

They contain mainly lipids and proteins with admixtures of carbohydrates. The lipids are either glycerol-based or sphingosine-based, phospholipids and glycolipids, plus sterols and minor lipids. Their fatty acid content is variable, the highest amount being found with palmitic and oleic acids. Membrane proteins are of two types, integral and peripheral, the integral ones being slightly more hydrophobic than the others and containing no cysteine.

The lipids are organized mainly in a bimolecular film with polar heads pointing outward, interspersed with isolated or clustered proteins, immersed in or spanning the entire thickness of the membrane. The lipids are present in several physical states, at biological temperatures in a liquid or liquid crystal form, below $12-18$ °C

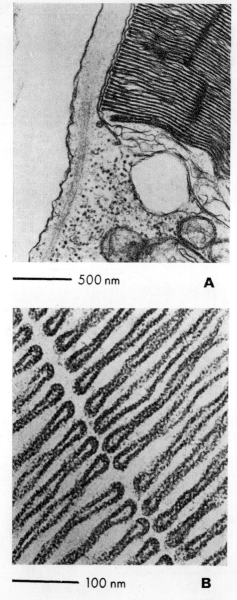

Fig. 2.58. Retinal rod outer membranes. **A**. Part of a retinal rod of frog, stained with osmium tetroxide. The lamellae are stacked at upper right. **B**. A greater magnification of the same object showing the granular structure of the lamellae. (Taken with permission from Robertson, J. D. (1964). In: *Intracellular Membraneous Structure* (ed. by Seno, S. and Cowdry, E. V.), p. 379. Japan Soc. Cell Biol., Okayama.)

in a crystalline, less mobile state. There is considerable lateral diffusion of all membrane components.

The membranes are put together by more or less random insertion of presynthesized proteins and lipids but a distinct functional pattern is rigorously maintained.

The functions of membranes, besides transporting molecules and ions, are varied: They act as receptors for signals, antigenic and hormonal, they are the site of many enzyme activities, they aid in the synthesis of external wall components, they serve as mechanical support and as electrical insulators.

3. THERMODYNAMICS OF TRANSPORT

"For Nature is very consonant and comformable to her self."
Isaac Newton, Opticks

Since thermodynamics deals with properties of matter and laws which can be understood without knowledge of the inner structure of matter, it is of great utility in the exploration of unknown physicochemical mechanisms of transport phenomena. Although thermodynamics itself does not describe the microscopic details of transport mechanisms, it shows which of the various hypothetical mechanisms are thermodynamically permissible. An admirable "Introduction to Thermodynamics" was written by Spanner (1964) and is recommended in this connection. Only a brief summary of the most important implications of thermodynamics in the field of transport is presented here.

3.1. THERMODYNAMIC EQUILIBRIUM, PASSIVE AND ACTIVE TRANSPORT PROCESSES

Thermodynamic equilibrium of a system is characterized by the absence of spontaneous processes. Since work can be performed only by a system which tends to a spontaneous change, an equilibrium system is one that is not capable of doing work.

The typical system encountered in membrane transport is both isothermal and isobaric. The constancy of the temperature is ensured by high thermal conductivity of the surrounding tissues or media and the constancy of pressure by the ambient atmosphere. The ability of an isothermal and isobaric system to perform work over the inevitable volume work at constant pressure is expressed by its Gibbs free energy (free enthalpy according to older nomenclature and used especially in German literature). In an equilibrium no changes of the Gibbs free energy occur. It may be shown (see, e.g., Edsall and Wyman, 1958, p. 161 – 162) that the same criterion of equilibrium holds also when the system consists of two or more compartments, all at the same temperature, but each at a different constant pressure. Such situations are encountered when studying transport phenomena in microorganisms or plant cells with rigid and elastic cell walls.

Compartments separated by a membrane represent open systems and their Gibbs free energy can be changed by transport of substances across the membrane. When the transfer of a substance across the membrane brings about a decrease in the Gibbs free energy of the system, the process proceeds spontaneously. Such transport processes are termed "passive" in order to distinguish them from those which can, as a result of a decrease of free energy in the course of coupled chemical reactions or in coupled downhill transport, proceed uphill and are called "active" and "secondary active", respectively.

The change in the Gibbs free energy, G, of a system, resulting from an inflow of dn_j moles of substance j may be written as

$$dG = \left(\frac{\partial G}{\partial n_j}\right)_{T,p} dn_j \qquad (3.1)$$

where the partial derivative is the partial molal Gibbs free energy of substance j at constant temperature and pressure and is called the chemical potential of the substance j, μ_j. There are other possible definitions of the chemical potential: it is also equal to the partial molal Helmholtz free energy at a constant temperature and volume, or to the partial molal internal energy at a constant volume and entropy, or to the partial molal heat content or enthalpy at a constant pressure and entropy. The latter two definitions, however, are less convenient; there are no obvious means of controlling the constancy of entropy in an experiment. With charged substances, ions, the term electrochemical potential with the symbol $\tilde{\mu}_j$ is used for the partial molal free energy.

Let us assume that a membrane separates two solutions of substance j and that its chemical potential in the outer solution is μ_{jo} and that in the inner one μ_{ji}, Transfer of dn_j moles of substance j from the inner to the outer compartment involves a change in the Gibbs free energy of the whole system equal to

$$dG = (\mu_{jo} - \mu_{ji})\, dn_j \qquad (3.2)$$

and the process is spontaneous when the free energy change is negative, i.e., when $\mu_{jo} < \mu_{ji}$. The spontaneous (i.e., passive) transport comes to a stop and the substance is in thermodynamic equilibrium when

$$\mu_{jo} = \mu_{ji}. \qquad (3.3)$$

The chemical potential of a nonelectrolyte in a dilute solution may be conveniently written as

$$\mu_j = \mu_{jo} + RT \ln c_j \qquad (3.4)$$

where μ_{jo} is the standard partial molal free energy of substance j and for a given solute and solvent it may be considered as a constant, independent of concentration; c_j is the molar concentration of substance j in the solution. From eq. (3.3) and (3.4) it then follows that in thermodynamic equilibrium

$$c_{ji} = c_{jo} \qquad (3.5)$$

i.e., the concentration of a nonelectrolyte is to be the same in the two solutions. We shall see later that with ions an additional electrical term has to be included in the expression for electrochemical potential and that when transport of the abundant species (the solvent) is described, the chemical potential cannot be expressed by its concentration and the hydrostatic pressure term cannot be neglected; slightly more complex conditions of equilibria are then obtained (see the chapters on equilibria of ions and transport of water).

Equation (3.2) is of assistance when the energy requirement for active transport processes is evaluated; when the transport proceeds "uphill", from a lower to a higher chemical potential, the Gibbs free energy of the system would increase at the rate

$$\frac{dG}{dt} = (\mu_{ji} - \mu_{jo})\frac{dn_j}{dt} \qquad (3.6)$$

or, per unit membrane area,

$$\frac{1}{A}\frac{dG}{dt} = (\mu_{ji} - \mu_{jo})J_j \tag{3.7}$$

where $\mu_{ji} > \mu_{jo}$ and

$$J_j = \frac{1}{A}\frac{dn_j}{dt}. \tag{3.8}$$

This is the flux of substance j, the amount in moles transferred per unit area in a unit of time. This increase in the free energy must be compensated for by an at least equal decrease in the free energy in the energy-providing chemical reactions in or at the membrane or in a coupled downhill transport of another substance. However, it is not a simple matter to evaluate experimentally the quantities appearing in eq. (3.8). When the nature of the metabolic energy-providing reacttions is known — most commonly it is the splitting of ATP — the energy available may be computed with a reasonable accuracy, especially in energy-depleted cells. It may be more difficult to evaluate the actual intracellular concentration and especially to estimate the net flux through the pumping mechanism J_j, not to be mistaken with the unidirectional flux of the substance measurable with tracers (*cf.* p. 203). Pumping mechanisms are, in general, reversible; thus the reversibility of the sodium pump is obvious not only from the possibility of ATP synthesis on its reversal (Glynn and Lew, 1970) but also from the existence of a critical energy barrier which it cannot overcome (Conway and Mullaney, 1961). For this reason there is always a back flux through the mechanism and each pump, in addition to the uphill transport, performs an exchange diffusion which does not require a supply of free energy.

3.2. THERMODYNAMICS OF THE STEADY STATE

The entropy of an isolated system which exchanges neither matter nor energy increases or remains constant. On the other hand, in systems which exchange energy or both energy and matter with their surroundings, the changes of entropy may be divided into the entropy flow $d_e S$ and entropy production $d_i S$

$$dS = d_e S + d_i S \tag{3.9}$$

The production of entropy in every macroscopic system is positive or nil

$$d_i S \geqq 0 \qquad (3.10)$$

whereas the entropy flow and hence also the entropy change dS can be either negative or positive. The production of entropy is a result of irreversible processes in the system — irreversible in the sense that their mathematical description is not invariant with respect to the sign of the time variable and that their results are never completely reversed. Even when the system itself is brought to the original state, there are necessarily changes left in other systems. Diffusion and heat conduction are typical examples of such irreversible processes.

The course of irreversible processes in the steady state is described by the phenomenological theory (i.e., a theory which does not take into account the inner structure of matter), called steady-state thermodynamics or thermodynamics of irreversible processes. Special forms of phenomenological relations of this thermodynamics have been known in the whole realm of physics for a long time: they describe a generalized flux J as being directly proportional to a generalized force X. Thus an electrical current is proportional to the gradient of electrical potential (Ohm's law), heat flow to the temperature gradient (Fourier's law) and diffusional flow to the concentration gradient (Fick's law). The principal aim of steady-state thermodynamics is to describe interactions between individual flows and to this end it makes a general assumption that any generalized flow J_i is proportional not only to its conjugate force X_i, but also in a higher or lesser degree to each of other generalized forces $X_j \, (j \neq i)$ operating in the system. Thus, e.g., temperature gradient gives rise not only to heat flow but also to electric current by the thermoelectric effect and to diffusional flow in the phenomenon of thermodiffusion. Hence the phenomenological relations may be written as

$$J_1 = L_{11}X_1 + L_{12}X_2 + \ldots + L_{1n}X_n$$
$$J_2 = L_{21}X_1 + L_{22}X_2 + \ldots + L_{2n}X_n \qquad (3.11)$$
$$\ldots\ldots\ldots\ldots\ldots\ldots\ldots\ldots\ldots\ldots\ldots$$
$$J_n = L_{n1}X_1 + L_{n2}X_2 + \ldots + L_{nn}X_n$$

where L_{ii} are straight coefficients relating the flows to their conjugate forces and L_{ij} $(i \neq j)$ are cross coefficients relating them to nonconjugate forces. Phenomena which cannot be at least approximately described by linear relations are not considered by steady-state thermodynamics. More conscisely the phenomenological relations may be written as

$$J_i = \sum_{j=1}^{n} L_{ij} X_j \qquad (i = 1, ..., n). \tag{3.12}$$

A proper choice of flows and forces requires the rate of entropy production in the system to be equal to the sum of flows multiplied by their conjugate forces

$$\frac{d_i S}{dt} = \sum_{i=1}^{n} J_i X_i > 0. \tag{3.13}$$

In isothermal system such a choice of flows and forces is preferable as results in the sum of products of flows with the conjugate forces giving the so-called dissipation function which is the rate of entropy production multiplied by absolute temperature

$$T \frac{d_i S}{dt} = \sum_{i=1}^{n} J_i X_i \tag{3.14}$$

where J'_is and X'_is now represent another set of flows.

When irreversible processes in a system are satisfactorily described by linear relations (3.11) and flows and forces are chosen in a way to satisfy eq. (3.13) or (3.14), Onsager's law

$$L_{ij} = L_{ji} \tag{3.15}$$

is valid for the cross coefficients of the linear relations. According to Onsager's law the matrix of phenomenological coefficients is symmetrical, being unchanged when its columns are replaced by its rows. The necessary conditions of the validity of Onsager's law are usually satisfied in the proximity of thermodynamic equilibrium. The practical value of Onsager's law is considerable; it allows to predict from the measurement of the dependence of flow J_i on force X_j the dependence of flow J_j on force X_i, which may be more difficult to measure. Onsager's law can be deduced mathematically from the principle of microscopic reversibility, stating that equilibrium on a molecular scale is not achieved by a cyclic process.

Following Denbigh (1951), the equivalence of Onsager's law with the principle of microscopic reversibility can be demonstrated for a simple system of three chemical reactions:

Let there be in a solution three tautomeric forms of some substance, denoted A, B and C, their concentrations being generally

a, b and c and in equilibrium $\bar{a}$, $\bar{b}$ and $\bar{c}$. The stability of the equilibrium concentrations $\bar{a}$, $\bar{b}$ and $\bar{c}$ could be imagined to result from a cyclic process

and no theorem of classical thermodynamics contradicts such scheme. The principle of microscopic reversibility, however, requires a detailed balancing of individual reactions

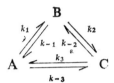

so that the following equations are valid

$$k_1 \bar{a} = k_{-1} \bar{b}, \tag{3.16a}$$

$$k_2 \bar{b} = k_{-2} \bar{c}, \tag{3.16b}$$

$$k_3 \bar{c} = k_{-3} \bar{a}. \tag{3.16c}$$

Expressing now the reaction rates by phenomenological equations, it will be obvious that the validity of equations of detailed balancing (3.16) is equivalent to the validity of Onsager's law (3.15).

The general kinetic equations of the system are the following

$$\frac{da}{dt} = -(k_1 + k_{-3})a + k_{-1}b + k_3 c, \tag{3.17a}$$

$$\frac{db}{dt} = k_1 a - (k_{-1} + k_2)b + k_{-2} c, \tag{3.17b}$$

$$\frac{dc}{dt} = k_{-3} a + k_2 b - (k_3 + k_{-2})c. \tag{3.17c}$$

Near equilibrium the actual concentrations deviate from the equilibrium ones only by small differences Δa, Δb and Δc, so that we can write

$$\Delta a = a - \bar{a} \tag{3.18}$$

and the rate of change of Δa is equal to that of the actual concentration a, i.e., to the reaction velocity v_A:

$$\frac{d\Delta a}{dt} = \frac{da}{dt} = v_A. \qquad (3.19)$$

Introducing a and da/dt from (3.18) and (3.19) into (3.17a) we obtain

$$v_A = -(k_1 + k_{-3})(\Delta a + \bar{a}) + k_{-1}(\Delta b + \bar{b}) + k_3(\Delta c + \bar{c}) =$$
$$= -(k_1 + k_{-3})\Delta a + k_{-1}\Delta b + k_3 \Delta c - (k_{-1} + k_{-3})\bar{a} + k_{-1}\bar{b} + k_3\bar{c}.$$

The sum of the last three terms, however, is equal to $d\bar{a}/dt$ and hence to zero; the equilibrium concentration does not change in time. Thus

$$v_A = -(k_1 + k_{-3})\Delta a + k_{-1}\Delta b + k_3 \Delta c. \qquad (3.20)$$

Let now $\bar{\mu}_A$ be the equilibrium chemical potential of substance A. Assuming the solution to be ideal and dilute, so that the chemical potential is proportional to the logarithm of concentration of substance A, we can write

$$\Delta\mu_A = \mu_A - \bar{\mu}_A = RT \ln\frac{a}{\bar{a}} = RT \ln\left(1 + \frac{\Delta a}{\bar{a}}\right) \qquad (3.21)$$

and when the deviation from the equilibrium concentration, Δa, is small

$$\Delta\mu_A = RT \frac{\Delta a}{\bar{a}} \qquad (3.22)$$

so that

$$\Delta a = \frac{\Delta\mu_A}{RT} \bar{a}. \qquad (3.23)$$

Introducing Δa from eq. (3.23) into eq. (3.20) and treating analogously the reaction velocities v_C and v_B we obtain

$$v_A = -\frac{(k_1 + k_{-3})\bar{a}}{RT}\Delta\mu_A + \frac{k_{-1}\bar{b}}{RT}\Delta\mu_B + \frac{k_3\bar{c}}{RT}\Delta\mu_C, \qquad (3.24a)$$

$$v_B = \frac{k_1\bar{a}}{RT}\Delta\mu_A - \frac{(k_{-1} + k_2)\bar{b}}{RT}\Delta\mu_B + \frac{k_{-2}\bar{c}}{RT}\Delta\mu_C, \qquad (3.24b)$$

$$v_C = \frac{k_{-3}\bar{a}}{RT}\Delta\mu_A + \frac{k_2\bar{b}}{RT}\Delta\mu_B - \frac{(k_3 + k_{-2})\bar{c}}{RT}\Delta\mu_C \qquad (3.24c)$$

which may be written

$$v_A = L_{AA}\Delta\mu_A + L_{AB}\Delta\mu_B + L_{AC}\Delta\mu_C, \quad (3.25a)$$

$$v_B = L_{BA}\Delta\mu_A + L_{BB}\Delta\mu_B + L_{BC}\Delta\mu_C, \quad (3.25b)$$

$$v_C = L_{CA}\Delta\mu_A + L_{CB}\Delta\mu_B + L_{CC}\Delta\mu_C \quad (3.25c)$$

where $L_{AB} = \dfrac{k_{-1}\bar{b}}{RT}$, $L_{BA} = \dfrac{k_1\bar{a}}{RT}$, and so on.

We can immediately see that when the equation of detailed balancing (3.16a)

$$k_1\bar{a} = k_{-1}\bar{b}$$

is satisfied, necessarily

$$L_{AB} = L_{BA}$$

so that Onsager's law follows from the principle of microscopic reversibility.

Methods of steady-state thermodynamics will be encountered in this book when discussing the phenomenological theory of transport of water as developed by Kedem and Katchalsky (1958).

Here at least the rigorous definition of active transport given by this type of thermodynamics (Kedem, 1961) may be mentioned, viz. that the transport is active if there is a demonstrable interaction between the transmembrane flow of the substance and a metabolic reaction at the membrane, which may be expressed by a suitable nonzero cross coefficient. As a result of this coupling the active transport can and commonly does (but need not) proceed uphill. When a transport is not coupled to a metabolic reaction but rather to the passive backflow of an actively transported substance it can also proceed uphill and may be conveniently denoted by Stein's (1967) term "secondary active transport".

3.3. NETWORK THERMODYNAMICS

In the following limited space only little can be presented about the intricate field of network thermodynamics beyond the reference to a conscise explanation of the basic ideas by Kedem (1972) and a thorough treatment of the subject by Oster and co-workers (1973). Unlike steady-state thermodynamics, the network thermodynamics is capable of treating theoretically also phenomena which are nonstationary and nonlinear. The basic approach of network thermodynamics is reticulation: each continuous system is mentally divided into homogeneous subsystems and each subsystem subdivided into reversible parts storing energy without dissipation and irreversible parts dissipating energy without storage. The whole system is then represented by a topological graph if the system is simple and by the so-called bond graph, if the system is more complex and involves energy transduction. The graph is equivalent to a set of differential equations describing the system and may be transformed into them using a suitable algorithm and, moreover, shows how the parts of the system are interconnected, i.e., reveals its topology.

The basic state variables of network thermodynamics can be divided into two classes: *effort* of force variables, also called "across" variables, since they can be estimated by a two-point measurement, and *flow* variables which can be estimated by one-point measurement and hence are also called "through" variables. Typical flow variables are electrical current, volume flow, molar flow and reaction rate; the conjugate effort variables being voltage, pressure difference, chemical potential and chemical affinity, respectively. An effort variable multiplied by the conjugate flow variable gives the energy rate or power.

Although network graphs are not to be confused with electrical analogues, the same topological constraints are applicable to both. Thus the flow variables are subject to Kirchhoff's current laws (KCL), being conserved at each node of the network, and the effort variables obey Kirchhoff's voltage law (KVL), having at each node a unique value. For this reason the flow variables are also called KVL variables and the effort variables KVL variables.

Physical properties of the system are expressed by "constitutive relations" between individual variables. The constitutive relation between a flow variable and the conjugate effort variable is given by the resistance of the system. Thus the dissipative flow of a substance across a membrane, J_m (in moles per second per unit area), is related to the difference of the chemical potential of the substance, $\Delta\mu$, by

$$J_m = \frac{\Delta\mu}{R_m} \qquad (3.26)$$

where R_m is the membrane resistance to flow. On the other hand, the constitutive relation between an integrated flow variable (electrical charge for electrical current, volume for volume flow, number of moles for molar flow and reaction advancement for reaction rate) and the effort variable is given by the capacitance of the subsystem. Thus, e.g., the differential or incremental capacitance of a membrane for a substance (all other variables being kept constant) is given by

$$C_m = \frac{dn_m}{d\mu_m} \quad (3.27)$$

where n_m is the number of moles of the substance per unit membrane area and μ_m the chemical potential of the substance in the membrane. The reversible flow of the substance, charging the membrane capacitance, J, is then given by

$$J = \frac{dn_m}{dt} = \frac{dn_m}{d\mu_m}\frac{d\mu_m}{dt} = C_m \frac{d\mu_m}{dt}. \quad (3.28)$$

If the membrane behaves as an ideal system, chemical potential may be expressed by $\mu_m = \text{const.} + RT \ln n_m$ and the capacitance is given by

$$C_m = \frac{n_m}{RT} = \frac{c_m V_m}{RT} \quad (3.29)$$

where c_m is the concentration of the substance in the membrane and V the volume of the membrane per unit area, i.e., membrane thickness.

In order to illustrate the representation of a simple system by topological graph we shall now consider the permeation of a nonelectrolyte across a homogeneous membrane by simple diffusion. In the topological graph in Fig. 3.1, drawn after Oster and co-workers (1973), J_m^1 and J_m^2 represent the dissipative flows of the nonelectrolyte into and out of the membrane, respectively. The constitutive relations involving these flows are

$$J_m^1 = \frac{\mu_1 - \mu_m}{R_1}, \quad (3.30)$$

$$J_m^2 = \frac{\mu_m - \mu_2}{R_2}$$

where

$$R_1 = R_2 = R_m/2 \quad (3.31)$$

the membrane being mentally subdivided into two dissipative elements and a capacitative element with capacitance C_m. There is a common reference potential μ_{ref} for the chemical potentials in the two solutions and in the membrane, the broken lines denoting the capacitative

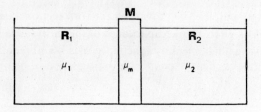

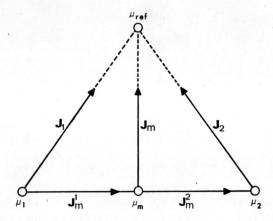

Fig. 3.1. Nonelectrolyte permeation across a homogeneous membrane represented by a topological graph (adapted from Oster et al., 1973). R_1, R_2 Reservoirs; M membrane.

elements and expressing the fact that there is no actual flow into the reference node.

J_1 is the "reservoir flow" from the first reservoir, $J_1(\mu_1 - \mu_{ref})$ being the power delivered by the reservoir. An analogous relation is valid for J_2. J_m, on the other hand, is the rate of nonelectrolyte accumulation in the membrane. All these are reversible flows.

Kirchhoff's current law written for the central node representing the membrane yields

$$J_m = J_m^1 - J_m^2. \tag{3.32}$$

Expressing the reversible flow J_m by eq. (3.28) and introducing for the two dissipative flows the explicit expressions (3.30) we obtain

$$C_m \frac{d\mu_m}{dt} = \frac{\mu_1 - \mu_m}{R_1} - \frac{\mu_m - \mu_2}{R_2} = \frac{\mu_1}{R_1} + \frac{\mu_2}{R_2} - \mu_m\left(\frac{1}{R_1} + \frac{1}{R_2}\right) \tag{3.33}$$

and since $R_1 = R_2 = R_m/2$ (eq. 3.31)

$$\frac{d\mu_m}{dt} = 2\frac{\mu_1 + \mu_2}{R_m C_m} - \frac{4}{R_m C_m}\mu_m. \quad (3.34)$$

On introducing the relaxation time τ_m of charging or discharging the membrane with the permeant,

$$\tau_m = \frac{R_m C_m}{4} \quad (3.35)$$

eq. (3.34) may be written

$$\tau_m \frac{d\mu_m}{dt} = \frac{\mu_1 + \mu_2}{2} - \mu_m. \quad (3.36)$$

This is easily integrated to describe the time course of charging or discharging the membrane with the permeant. Thus, when the membrane initially free of the permeant is exposed at time zero to two reservoirs with the values of the permeant chemical potential equal to μ_1 and μ_2, respectively, the integration yields

$$\mu_m = \frac{\mu_1 + \mu_2}{2}\left(1 - e^{-\frac{t}{\tau_m}}\right). \quad (3.37)$$

The relaxation time for a simple membrane is easily interpreted in terms of membrane thickness Δx and the diffusion coefficient of the permeant in the membrane, D_m. Introduction of an expression for membrane resistance

$$R_m = RT\,\Delta x/c_m D_m \quad (3.38)$$

and of an expression for membrane capacitance (cf. eq. 3.29)

$$C_m = c_m\,\Delta x/RT \quad (3.39)$$

(where Δx is equal to the membrane volume per unit of its area) into eq. (3.35) yields (Oster et al., 1973):

$$\tau_m = \frac{R_m C_m}{4} = \frac{RT\,\Delta x}{4c_m D_m} \cdot \frac{c_m\,\Delta x}{RT} = \frac{(\Delta x/2)^2}{D_m}. \quad (3.40)$$

SYNOPSIS

The phenomenological and hence general laws of thermodynamics are of manifold assistance in the classification and description of transport phenomena. Criteria of equilibrium of classical thermodynamics allow to discriminate between two types of stationary distribution of substances across membranes: equilibrium distribution and steady-state distribution resulting from active transport. Further, classical thermodynamics makes it possible to evaluate the rate of energy supply required by active transport transferring a substance to a higher free-energy level. Thermodynamics of the steady state ("irreversible thermodynamics") is suited for description of interactions between various steady flows and, by Onsager's law, it reduces the necessary number of experimental measurements. Finally, network thermodynamics is capable of treating nonstationary and nonlinear phenomena. In network thermodynamics each system is reticulated into homogeneous subsystems and reversible and irreversible parts and represented by a graph. The graph is equivalent to a set of differential equations describing the system and displays, at the same time, the topology of the system.

4. TRANSPORT OF NONELECTROLYTES

"And therefore I scruple not to propose the Principles of Motion above-mention'd, they being of very general Extent, and leave their Causes to be found out."

Isaac Newton, Opticks

Depending on their size, degree of hydrophobicity and specific structural features, molecules can cross biological membranes by three principally distinct mechanisms: (1) Nonspecific "diffusion" through the lipid and, in a limited degree, through the polar parts of the membrane; (2) specific "carrier" transport—this category includes, as a subgroup, systems where the transported substance is altered chemically; (3) mechanisms involving profound, even if transient, changes in membrane architecture, such as pinocytosis, penetration of biopolymers.

The first two categories will be dealt with in the present chapter, the third will be discussed separately later (chapter 7). Likewise, the movement of water, actually belonging to the first category, will be treated in the special chapter 6.

Although our primary concern here is with the movement across membranes, it will aid in understanding the more advanced kinetic treatment if we review briefly the features of diffusion in a continuous (homogeneous) phase.

4.1. PRINCIPLES OF DIFFUSION

Particles dissolved or suspended in liquids are subject to permanent thermal movement. In the absence of external fields, such as electrical field which influences considerably the movement of charged particles, or high gravitational fields obtainable in an ultracentrifuge, which raise the gravity of heavier particles to appreciable values, there is no preferential direction in particle movement. As a result of the laws of probability, more particles leave the volume element in which they are concentrated while a smaller number of particles flow during the same time into the element from its dilute surroundings. For this reason, concentration gradients in solutions tend to disappear with time and the process of equalizing of particle concentrations, called diffusion, is commonly observed.

The rate of diffusion follows an important phenomenological law, called Fick's first law. When diffusion proceeds or is observed in one direction only, it has the form

$$\frac{1}{A}\frac{dn}{dt} = -D\frac{\partial c}{\partial x}. \qquad (4.1)$$

The left-hand side represents the number of particles diffusing in a unit of time per unit area normal to the direction of the x-axis, along which the diffusion proceeds; in transport studies this quantity is commonly called flux and is usually denoted by J. According to Fick's first law it is directly proportional, by the diffusion coefficient D, to the concentration gradient $\partial c/\partial x$, prevailing at a given time at given x. The minus sign expresses the fact that when the concentration gradient is a positive number, i.e. when concentration increases in the direction of the positive x-axis, the diffusion flow proceeds in the opposite direction.

It will be seen later (p. 193) that Fick's first law may serve as a convenient approximation when diffusion across a thin layer separating two mixed reservoirs is to be described. However, when describing diffusion proceeding across greater distances in continuous systems, the same equation contains four variables even in the simplest, one-dimensional case. For this purpose Fick's first law is transformed into a partial differential equation, called simply the diffusion equation or Fick's second law. In the one-dimensional case the procedure gives

$$\frac{\partial c}{\partial t} = D\frac{\partial^2 c}{\partial x^2} \qquad (4.2)$$

so that the rate of concentration change is proportional to the second derivative of the concentration in space. Phenomenologically, Fick's first law may be considered as an approximation of a specific case of the general Teorell's (1953) formula for the flux of a substance:

$$\text{Flux} = \text{Mobility} \cdot \text{Concentration} \cdot \text{Total driving force} \qquad (4.3)$$

where the flux J is the amount of substance in moles which in unit time penetrates per unit area normal to the direction of the transport:

$$J = \frac{1}{A}\frac{dn}{dt}. \qquad (4.4)$$

In the absence of external fields the system tends to an equilibrium in which the chemical potential of solute is the same at each point of the available space (see the Thermodynamics of Transport, p. 169). Hence the suitable total driving force to be introduced into Teorell's equation to describe simple diffusion is the negative of the space gradient of the chemical potential; in the one-dimensional case equal to the partial derivate of the chemical potential μ with respect to x, $\partial \mu / \partial x$. Teorell's equation (4.3) then takes the form

$$\frac{1}{A}\frac{dn}{dt} = -Uc\frac{\partial \mu}{\partial x} \qquad (4.5)$$

where U is the mobility and c the concentration at x at time t. Using the expression for chemical potential of solute in a dilute solution of ideal behaviour

$$\mu = \mu_0 + RT \ln c \qquad (4.6)$$

eq. (4.5) may be written

$$\frac{1}{A}\frac{dn}{dt} = -UcRT\frac{\partial \ln c}{\partial x} \qquad (4.7)$$

and, since $d \ln y = dy/y$,

$$\frac{1}{A}\frac{dn}{dt} = -RTU\frac{\partial c}{\partial x}. \qquad (4.8)$$

Denoting the quantity RTU by D we obtain

$$\frac{1}{A}\frac{dn}{dt} = -D\frac{\partial c}{\partial x} \qquad (4.9)$$

which is Fick's first law, derived by its author as an analogy to Fourier's law, describing the flow of heat.

Fick's second law (the partial differential equation of diffusion) may be derived from Fick's first law as follows.

Let there be two infinitesimally distant parallel planes normal to the x-axis, one at x, another at $x + \mathrm{d}x$. The concentration gradient being $\partial c/\partial x$ at the former plane, it is equal to

$$\frac{\partial c}{\partial x} + \frac{\partial(\partial c/\partial x)}{\partial x}\mathrm{d}x = \frac{\partial c}{\partial x} + \frac{\partial^2 c}{\partial x^2}\mathrm{d}x$$

at the latter. Hence also the diffusional fluxes across the two planes differ, being

$$\frac{1}{A}\frac{\mathrm{d}n}{\mathrm{d}t} = -D\frac{\partial c}{\partial x}$$

and

$$\frac{1}{A}\frac{\mathrm{d}n'}{\mathrm{d}t} = -D\left(\frac{\partial c}{\partial x} + \frac{\partial^2 c}{\partial x^2}\mathrm{d}x\right)$$

respectively. Their difference is the rate of change of the number of moles of solute between the two planes per unit area:

$$\frac{1}{A}\frac{\mathrm{d}n - \mathrm{d}n'}{\mathrm{d}t} = D\frac{\partial^2 c}{\partial x^2}\mathrm{d}x \qquad (4.10)$$

and, since the volume included between the planes per unit area is numerically equal to $\mathrm{d}x$, the rate of concentration change between the planes is

$$\frac{\partial c}{\partial t} = \frac{1}{A\,\mathrm{d}x}\frac{\mathrm{d}n - \mathrm{d}n'}{\mathrm{d}t}. \qquad (4.11)$$

Combining equations (4.10) and (4.11) we obtain

$$\frac{\partial c}{\partial t} = D\frac{\partial^2 c}{\partial x^2}$$

which is Fick's second law (eq. 4.2). When observable diffusional flow is not limited to one dimension of space, a more general form of the diffusion equation must be used:

$$\frac{\partial c}{\partial t} = D\nabla^2 c \qquad (4.12)$$

where the symbol ∇^2 is the operator nabla squared (Laplacean) and takes a form according to the choice of space coordinates, which in their turn depend on the geometry of the system observed.

Solutions of the diffusion equation (4.2) or (4.12) are of interest for a biologist studying diffusion into or out of tissue samples of simple geometrical forms, provided that the process is not rate-limited by membranes (otherwise the problem of diffusion would be reduced to the problem of permeation, see p. 192). Problems of this kind may be encountered when diffusion in extracellular spaces is followed, or when the solute permeates across cell membranes extremely rapidly. Solution of eq. (4.2) for a plane sheet of thickness d, exposed at $t = 0$ on both sides to concentration c_0 of the solute of which the initial concentration inside the sheet is zero (so that $t = 0$, $0 \leq x \leq d$, $c = 0$) is (Hill, 1928):

$$c = c_0 \left[1 - \frac{4}{\pi} \left(e^{-D\pi^2 t/d^2} \sin \frac{\pi x}{d} + \frac{1}{3} e^{-9D\pi^2 t/d^2} \sin \frac{3\pi x}{d} + \right. \right.$$
$$\left. \left. + \frac{1}{5} e^{-25 D\pi^2 t/d^2} \sin \frac{5 x}{d} + \ldots \right) \right] \quad (4.13)$$

which represents a rapidly convergent series and gives the concentration c at any x (distance from one of the surface planes) inside the sheet, at any time t. Especially useful are formulas giving the fractional equilibration m_t/m_∞ of simple geometric shapes, where m_t is the amount of the solute present in the body at time t and m_∞ the amount present after full equilibrium has been reached. The fractional equilibration of the above plane sheet may be calculated as

$$\frac{m_t}{m_\infty} = \frac{\int_0^d c\,dx}{c_0 d}$$

and is equal to (Hill, 1928):

$$\frac{m_t}{m_\infty} = 1 - \frac{8}{\pi^2} \left(e^{-D\pi^2 t/d^2} + \frac{1}{9} e^{-9D\pi^2 t/d^2} + \frac{1}{25} e^{-25 D\pi^2 t/d^2} + \ldots \right) \quad (4.14)$$

Fractional equilibration of a sphere of radius r is given by

$$\frac{m_t}{m_\infty} = 1 - \frac{6}{\pi^2} \left(e^{-D\pi^2 t/r^2} + \frac{1}{4} e^{-4D\pi^2 t/r^2} + \frac{1}{9} e^{-9D\pi^2 t/r^2} + \ldots \right) \quad (4.15)$$

and fractional equilibration of a cylinder of radius r by

$$\frac{m_t}{m_\infty} = 1 - 4\left(\frac{1}{\mu_1^2} e^{-\mu_1^2 Dt/r^2} - \frac{1}{\mu_2^2} e^{-\mu_2^2 Dt/r^2} + \ldots\right) \quad (4.16)$$

where μ's are the zeros of the Bessel function J_0, $\mu_1 = 2.4048$, $\mu_2 = 5.5201$, $\mu_3 = 8.6537$, $\mu_4 = 11.7915$, *etc.*

The diffusion coefficients D of substances of low molecular weight are of the order of 10^{-5} cm^2 s^{-1}, ranging from $2.5 \cdot 10^{-5}$ for water to $0.5 \cdot 10^{-5}$ for sucrose. It should be added that, as shown by Hartley and Crank (1949), the rate of diffusion is determined by the intrinsic diffusion coefficient of an individual substance only in the so-called self-diffusion experiments, where no net concentration change occurs and the process of diffusion is followed with tracers. The reason for this is that only pure diffusion can be observed under such conditions, whereas a net transfer of a substance is always accompanied by the bulk flow of the solution. Since the intrinsic rates of pure diffusion of interdiffusing substances (solute and solvent) differ, there is a tendency to build up a hydrostatic pressure difference in the course of a net transfer, relieved by a bulk flow which enhances the flow of the less mobile solute and slows down the flow of the more mobile solvent. The rate of interdiffusion is then determined by a single mutual diffusion coefficient, depending on the conditions of the experiment. The effect of solvent diffusion, however, is likely to be small in the case of diffusion of minor components of the solution.

As shown by Einstein (1905) the diffusion equation (4.2) can be derived not only from the phenomenological Fick's first law but also directly from the very basic principles. In his famous derivation of the theory of Brownian movement, Einstein (1905) considers n particles in a liquid, each of them being displaced during a short time interval τ by some individual distance Δ in the direction of the x-axis. A probability law describes the extent of the displacement; the number of particles dn being displaced during the time interval τ by a distance between Δ and $\Delta + \mathrm{d}\Delta$ is given by

$$\mathrm{d}n = n\Phi(\Delta)\,\mathrm{d}\Delta \quad (4.17)$$

where

$$\int_{-\infty}^{\infty} \Phi(\Delta)\,\mathrm{d}\Delta = 1$$

since each particle is subject to some displacement between minus and plus infinity, zero displacements included. Moreover

$$\Phi(\Delta) = \Phi(-\Delta) \qquad (4.18)$$

movements of particles in positive and negative direction being equally probable. (Actually, the function Φ deviates from zero only for very small values of Δ). The concentration of particles is a function of the coordinate x and the time t, $c(x, t)$. The concentration distribution at time $t + \tau$ is equal to the distribution at time t with the x-coordinate of each particle changed by its proper displacement Δ:

$$c(x, t + \tau) = \int_{\Delta = -\infty}^{\Delta = \infty} c(x + \Delta, t)\,\Phi(\Delta)\,\mathrm{d}\Delta. \qquad (4.19)$$

Since τ is very small, the left-hand side of the equation may be expressed by

$$c(x, t + \tau) = c(x, t) + \tau\frac{\partial c}{\partial t} \qquad (4.20)$$

and the function $c(x + \Delta, t)$ under the integration sign may be expanded in a Taylor series:

$$c(x + \Delta, t) = c(x, t) + \Delta\frac{\partial c(x, t)}{\partial x} + \frac{\Delta^2}{2}\frac{\partial^2 c(x, t)}{\partial x^2}\ldots. \qquad (4.21)$$

Eq. (4.19) then has the form

$$c + \tau\frac{\partial c}{\partial t} = c\int_{-\infty}^{\infty}\Phi(\Delta)\,\mathrm{d}\Delta + \frac{\partial c}{\partial x}\int_{-\infty}^{\infty}\Delta\Phi(\Delta)\,\mathrm{d}\Delta +$$
$$+ \frac{\partial^2 c}{\partial x^2}\int_{-\infty}^{\infty}\frac{\Delta^2}{2}\Phi(\Delta)\,\mathrm{d}\Delta\ldots. \qquad (4.22)$$

As a result of (4.18) the second, fourth, etc., terms on the right-hand side vanish. The integral in the first right-hand side term is equal to 1 by eq. (4.17). The integral in the third term is a constant value which may be put equal to the constant time τ multiplied by another constant D:

$$\int_{-\infty}^{\infty}\frac{\Delta^2}{2}\Phi(\Delta)\,\mathrm{d}\Delta = \tau D. \qquad (4.23)$$

The other odd terms are negligibly small. Eq. (4.22) then can be written as

$$\frac{\partial c}{\partial t} = D\frac{\partial^2 c}{\partial x^2}$$

which is the partial differential diffusion equation (4.2).

Einstein (1905) then solves the diffusion equation to describe the time-dependent space distribution of particles which, at time zero, are concentrated in the immediate vicinity of the plane at $x = 0$. Thus, for all x different from zero, $c(x, t) = 0$ for $t = 0$ and

$$\int_{-\infty}^{\infty} c(x, t)\, dx = n \tag{4.24}$$

where n is the total number of particles, conserved during the diffusional process. With these conditions the solution is

$$c(x, t) = \frac{n}{\sqrt{4\pi D}} \frac{e^{-\frac{x^2}{4Dt}}}{\sqrt{t}}. \tag{4.25}$$

It may be easily shown, by taking the appropriate derivatives, that the function (4.25) satisfies the diffusion equation. Thus, since

$$\frac{d(u(t)/v(t))}{dt} = \frac{1}{v^2}\left(v\frac{du}{dt} - u\frac{dv}{dt}\right),$$

$$\frac{\partial c(x, t)}{\partial t} = \frac{n}{\sqrt{4\pi D}} \frac{\left(\sqrt{t}\,\dfrac{x^2}{4Dt^2} - \dfrac{1}{2\sqrt{t}}\right)e^{-\frac{x^2}{4Dt}}}{t},$$

whereas

$$\frac{\partial c(x, t)}{\partial x} = \frac{n}{\sqrt{4\pi D}} \frac{-\dfrac{x}{2Dt}e^{-\frac{x^2}{4Dt}}}{\sqrt{t}};$$

and since

$$\frac{d(u(x)\cdot v(x))}{dx} = u\frac{dv}{dx} + v\frac{du}{dx},$$

$$\frac{\partial^2 c(x, t)}{\partial x^2} = \frac{n}{\sqrt{4\pi D}} \frac{\left(\dfrac{x^2}{4D^2 t^2} - \dfrac{1}{2Dt}\right)e^{-\frac{x^2}{4Dt}}}{\sqrt{t}} \frac{\sqrt{t}}{\sqrt{t}} =$$

$$= \frac{n}{\sqrt{4\pi D}} \frac{1}{D} \frac{\left(\sqrt{t}\,\dfrac{x^2}{4Dt^2} - \dfrac{1}{2\sqrt{t}}\right)e^{-\frac{x^2}{4Dt}}}{t}$$

so that

$$\frac{\partial c(x, t)}{\partial t} = D\frac{\partial^2 c(x, t)}{\partial x^2}.$$

The concentration change at distance x and time t (equal for zero initial concentration to the function $c(x, t)$ itself) divided by the total number of particles is equal to the probability of an individual particle being displaced by a distance x during time t. Hence the arithmetic means of the squares of displacements may be evaluated as (Fürth, 1956)

$$\bar{x}^2 = \frac{1}{n} \int_{-\infty}^{\infty} c(x, t) x^2 \, dx = \frac{1}{\sqrt{4\pi Dt}} \int_{-\infty}^{\infty} e^{-\frac{x^2}{4Dt}} x^2 \, dx = 2Dt. \quad (4.26)$$

For those who might be interested in the procedure we describe this integration in details. First, the substitution $y = x^2/4Dt$ is done so that $x^2 = 4Dty$, $x = 2\sqrt{Dt}\sqrt{y}$ and $dx = (\sqrt{Dt}/\sqrt{y}) \, dy$. The integral (4.26) thus becomes

$$\bar{x}^2 = \frac{1}{\sqrt{4\pi Dt}} \int_{-\infty}^{\infty} e^{-y} \cdot 4Dty \frac{\sqrt{Dt}}{\sqrt{y}} \, dy =$$
$$= \frac{2Dt}{\sqrt{\pi}} \int_{-\infty}^{\infty} e^{-y} \sqrt{y} \, dy = \frac{4Dt}{\sqrt{\pi}} \int_{0}^{\infty} e^{-y} \sqrt{y} \, dy.$$

The integration can now be carried out per partes according to

$$\int u \, dv = uv - \int v \, du$$

with $u = \sqrt{y}$ and $dv = e^{-y} \, dy$, so that $du = dy/(2\sqrt{y})$ and $v = -e^{-y}$. The integral becomes

$$\bar{x}^2 = \frac{4Dt}{\sqrt{\pi}} \left\{ [-e^{-y}\sqrt{y}]_0^{\infty} + \frac{1}{2} \int_0^{\infty} \frac{e^{-y}}{\sqrt{y}} \, dy \right\}.$$

The first term on the right-hand side vanishes at both boundaries, the second is simplified by the substitution $y = u^2$, $\sqrt{y} = u$, and $dy = 2u \, du$, so that

$$\frac{1}{2} \int_0^{\infty} \frac{e^{-y}}{\sqrt{y}} \, dy = \int_0^{\infty} e^{-u^2} \, du.$$

The value of this integral has been known since Euler's time, being $\sqrt{\pi}/2$. Thus

$$\bar{x}^2 = 2Dt$$

as indicated by eq. (4.26).

This relationship, often written as $\sqrt{\bar{x}^2} = \sqrt{2Dt}$, gives a vivid picture of the rate at which diffusion proceeds. Since the diffusion coefficients of common small molecules are of the order of

10^{-5} cm^2 s^{-1}, the mean displacement of these molecules is approximately 4.4 µm s^{-1}, 34.6 µm min^{-1}, 0.27 cm h^{-1}, 1.31 cm d^{-1}, so that diffusion is justly described as a rapid process for short and slow one for longer distances.

The excellent book on Diffusion Processes by Jacobs (1967) is recommended to the reader interested in a clear and complete treatment of the theory of diffusion.

4.2. DIFFUSION ACROSS MEMBRANES

Unlike diffusion in space which is properly described by solutions of the partial differential equation (Fick's second law), permeation across a thin membrane by a diffusional process may be described to a good approximation by an ordinary differential equation, *viz.* Fick's first law as well as its modifications taking into account the saturability of mediated diffusion and/or volume changes of the compartment into which the diffusion across the membrane proceeds. This results from the fact that the equilibration process in the media at the two sides of the membrane is mostly much faster than the process of permeation across the membrane; even with rapidly permeating substances it is usually sufficient to introduce a correction for the diffusion in the unstirred layers adjacent to the membrane and behaving like high permeability membranes in series with the membrane proper.

The intricacies of permeation by saturable processes are the subject of all the following sections of this chapter and we shall see that in most cases it is not even necessary to integrate the modified laws of Fick to determine the characteristics of a saturable process. Estimation of the so-called initial velocities is based on the fact that measurable quantities of a substance crossing the membrane do not change the concentrations in the compartments in question appreciably and hence nonintegrated modified laws of diffusion in which differentials are replaced by small finite differences can be used to describe the process quantitatively.

The aim of the present section is to discuss the differential equations governing the permeation of a nonelectrolyte across a membrane by a nonsaturable process (simple diffusion) and to show how they can be integrated in special cases. It will be seen that analogous integrated equations describe the tracer exchange in steady state over a much wider range, from nonsaturable equilibrating transport of a nonelectrolyte to saturable active transport of ions.

Fick's first law relates the flux of a substance, J (amount of substance in moles crossing a unit surface perpendicular to the direction of the movement in unit time) to the concentration gradient of the substance

$$J = -D\frac{dc}{dx} \quad (4.27)$$

where D is the diffusion coefficient. If the flux is expressed in mol cm^{-2} s^{-1}, the length x in cm and concentration in mol cm^{-3}, the dimensions of D are cm^2 s^{-1}.* For a thin membrane, in a steady state of transport, the derivative of concentration may be replaced with the quotient of a finite concentration difference across the membrane thickness

$$J = -D\frac{\Delta c}{\Delta x} = -D\frac{c_{II} - c_I}{l} = P(c_I - c_{II}) \quad (4.28)$$

where $P(=D/l)$ is the permeability coefficient across the membrane of a substance permeating by simple diffusion according to a diffusion coefficient across the given membrane equal to D. The units of P are cm s^{-1}.

Equation (4.28) indicates that the rate of transport will depend on the concentration of the transported solute. If the initial rate is followed (or if the unidirectional flux inward is measured) the dependence is between J and c_I; if the net flux is examined, the dependence will be between J and Δc (Fig. 4.1).

The value of P includes an important constant for every given solute and every given membrane, *viz.* the membrane: water partition coefficient** defined as

$$K = c_{Im}/c_I = c_{IIm}/c_{II} \quad (4.29)$$

(the equation assumes identical solvent properties of both bulk solutions and the same lipid composition near the two membrane faces). The indexed values refer to concentrations just within the membrane where it holds that

$$J = P_m(c_{Im} - c_{IIm}). \quad (4.30)$$

---

* The dimension of D does not change if concentration is expressed in g cm^{-3} and flux in g cm^{-2}s^{-1}.

** This is experimentally approximated satisfactorily by using olive oil or olive oil plus oleic acid (for basic solutions) as membrane substitute.

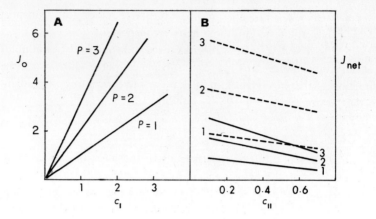

Fig. 4.1. Concentration dependence of diffusional flow across a membrane. In **A**, the initial (or unidirectional) flux is plotted against concentration at the starting (*cis*) side; in **B**, the net flow is plotted against concentration at the *trans* side for $c_I = 1$ (full lines) and $c_I = 2$ (dashed lines) at three values of P (1, 2, 3).

Hence

$$J = P_m K(c_I - c_{II}) = \frac{D_m}{l} K(c_I - c_{II}). \quad (4.31)$$

It should be appreciated that the transmembrane permeability is thus comprised of two factors, the distribution between water and membrane and the "diffusion" across the membrane.

The diffusion coefficient included in the fundamental eq. (4.27) is itself a function of the size and shape of the molecule. For spherical molecules, where the Stokes–Einstein equation* holds, it is seen (Davson and Danielli, 1952) that

$$DM^{1/3} = \text{constant}$$

where M is the molecular weight. In other words, the relationship between $\log D$ and $\log M$ should give a straight line of slope equal to -0.33. This is actually found for diffusion in water of large molecules of proteins. For small molecules, ranging from hydrogen to a trisaccharide, the empirical slope is steeper, corresponding to

$$DM^{1/2} = \text{constant}$$

---

* $D = RT/6N\pi\eta r$, N being the Avogadro constant, η the viscosity of the medium and r the particle radius.

apparently because the size of the molecules is such that the surrounding water medium is not "seen" by them as a continuum.

These relationships have also been applied with greater or lesser success to diffusion across membranes, the usual way of plotting being log $(PM^{1/2})$ vs. log K, when points lying within a broad straight band are obtained (cf. Collander, 1949; Stein, 1967) (Fig. 4.2).

For a long time, these data have been taken as indicative that molecules pass through the lipid parts of the membrane merely by

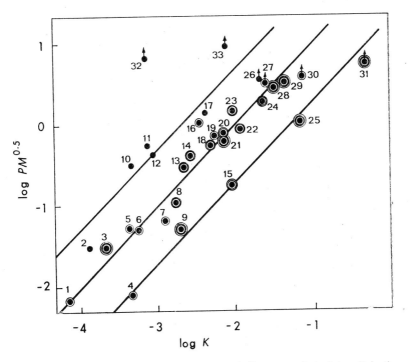

Fig. 4.2. The permeability of *Chara ceratophylla* to nonelectrolytes. P is the permeability constant, M the molecular weight, K the partition coefficient between olive oil and water. The size of the circles is related to the molar refraction, ranging from less than 15 for simple dots, to 15—22, 22—30 and more than 30 for the points with increasing number of concentric circles. 1 Glycerol, 2 urea, 3 hexamethylenetetramine, 4 dicyanodiamide, 5 methylurea, 6 lactamide, 7 thiourea, 8 ethylurea, 9 diethylmalonamide, 10 ethylene glycol, 11 formamide, 12 acetamide, 13 dimethylurea, 14 glycerol methyl ether, 15 monoacetin, 16 propionamide, 17 cyanamide, 18 succinimide, 19 propylene glycol, 20 glycerol ethyl ether, 21 diethylurea, 22 chlorhydrin, 23 butyramide, 24 valeramide, 25 diacetin, 26 ethanol, 27 urethylan, 28 antipyrin, 29 trimethyl citrate, 30 urethane, 31 triethyl citrate, 32 water, 33 methanol.

virtue of their lipid solubility (and, of course, the more slowly the greater the molecule). However, the deviating data on rather lipophilic molecules, on the one hand, and on very small hydrophilic molecules, such as water, on the other, have led to several developments of the model.

Thus, if compounds with a log K of 0 and higher are considered, the Collander graph of Fig. 4.2 ceases to be linear but actually follows a parabola as shown in Fig. 4.3. The behaviour has been explained

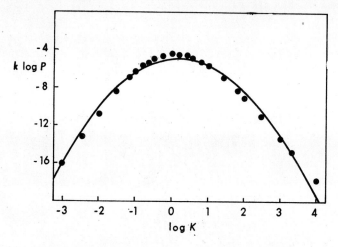

Fig. 4.3. Dependence of concentration in a distant compartment (approximated by permeability) on the partition coefficient of compounds between ether and water. The points are fitted by a parabola using the least-squares method. (According to Penniston et al., 1969).

in analogy with equations derived by Hansch and Fujita (1964) who showed that a standard pharmacological response relates to the hydrophobicity of a solute in a rather complicated way, implying that for highly hydrophobic substances (so as to be practically insoluble in water) the rate of transport across the membrane is diminished:

$$\log (1/c) = -k(\log K)^2 + k' \log K + k'' + \varrho\sigma \qquad (4.32)$$

where c is the concentration resulting in equal flux, k, k', and k'' are empirical constants and ϱ and σ are Hammett constants, defined as $\log (k_d/k_0) = \sigma\varrho$, k_d and k_0 being reaction rate constants of a derivative and of the parent compound, respectively; σ depends only on the character and position of a substituent, ϱ on the type of reaction involved. Thus, to elicit the same effect (or flux in an approximation) a compound must be present at a higher concentration if its partition coefficient is too high (as K increases, so does c, the concentration required to produce the same response).

If c is replaced with the permeability constant to which it may be proportional, relationships of the type of eq. (4.32) are seen to fit experimental data very well indeed (cf. Penniston et al., 1969).

The deviations concerning small hydrophilic molecules have been treated recently by Lieb and Stein (1969). The concept views the membrane as a polymer sheet which undergoes thermal movement whereby "holes" are transiently opened so as to permit the passage of molecules across the membrane. This is conceptually related to the still later theory of "kinks" moving through the membrane with a "diffusion coefficient" of 10^{-5} cm^2 s^{-1}. The kinks are mobile structural defects in the hydrocarbon phase of the membrane (Träuble, 1971).

In this development, the D_m is defined as kM_r^{-s} where M_r is the relative molecular weight referred to methanol as unity, k is a constant and s is the so-called differential mass selectivity coefficient, its value being $0.3-0.5$ in water but as high as 3.5 in a biological membrane. This results in a steeper differentiation according to size than expected on the basis of diffusion in a continuous aqueous medium. Thus, the diffusion coefficients in the molecular weight range from 45 (formamide) to 122 (erythritol) lie between $1.6 \cdot 10^{-5}$ and $8 \cdot 10^{-6}$ cm^2 s^{-1} in water (a factor of 2) but between $1.4 \cdot 10^{-8}$ and $2 \cdot 10^{-10}$ cm^2 s^{-1} in *Chara ceratophylla* (a factor of 70).

Another, conceptually different, extension of the sieving-plus-lipophilicity hypothesis, was the introduction of the pore concept, envisaging small permanent hydrophilic pores that would permit the passage of small water-soluble molecules across the membrane. This concept evolved from measuring the so-called reflection coefficient σ (discussed on p. 312). Using the development by Kedem and Katchalsky (1958) and Kedem (1961) one can show that

$$P_m = (1 - \sigma) L_p RT / K \bar{V}_s \qquad (4.33)$$

where L_p is the hydraulic permeability coefficient (cf. eq. 6.41) and $\bar{V}_s$ is the partial molal volume of the solute. Thus, the reflection coefficient σ is related to the permeability coefficient in a straightforward manner.

The use of σ for computing the equivalent pore radius has been found to be somewhat more objectionable, the principal flaw being in the fact that even in a membrane with no pores, $\sigma < 1$ for a permeant solute which contributes to the volume flow (see p. 314). Still, reproducible values have been obtained for various cells, using the relationship

$$1 - \sigma = \frac{[2(1 - a/r)^2 - (1 - a/r)^4] \times {} \atop {} \times [1 - 2.104a/r + 2.09(a/r)^3 - 0.95(a/r)^5]}{[2(1 - a_W/r)^2 - (1 - a_W/r)^4] \times {} \atop {} \times [1 - 2.104a_W/r + 2.09(a_W/r)^3 - 0.95(a_W/r)^5]} \quad (4.34)$$

where r is the equivalent pore radius, a is the radius of the permeating molecule and a_W the radius of a water molecule (Renkin, 1954; Goldstein and Solomon, 1960). The equivalent pore radii thus assessed range from 0.42 nm in human erythrocytes to 0.56 nm in *Necturus* kidney to 2.3 nm in a Visking dialysis tubing.

Although the existence of permanent hydrophilic pores in the membrane is still contested their function at least in bulk flow of solutions appears to be generally accepted.

The magnitude of the permeability constant estimated from the flux and bulk concentrations of solute according to eq. (4.28) is usually in error because of the fact that layers of solution adjacent to the membrane surface are not stirred vigorously enough, with the consequence that the solute concentration in this unstirred layer may differ substantially from that in the bulk phase. The unstirred layers (also called Nernst diffusion layers) can be both morphological (such as the cell wall of bacteria or yeasts or the glycocalyx of epithelial cells) as well as (as layers of slow laminar flow) located directly in the solution itself, the thickness of the former being perhaps 1 µm, that of the latter ranging from 20 to 500 µm, depending on the efficiency of mechanical stirring (Dainty, 1963).

Due to the presence of the unstirred layer, the concentration difference across the membrane proper is less than that between the bulks of the solutions. The situation is depicted schematically in Fig. 4.4, in which the so-called effective thickness of the unstirred layer at each side of the membrane is shown. Layers of this thickness would exert the observed effects if the concentration gradients in them were constant. In fact, the concentration profiles are not linear but have the form shown approximately by the dot-and-dash lines in the figure.

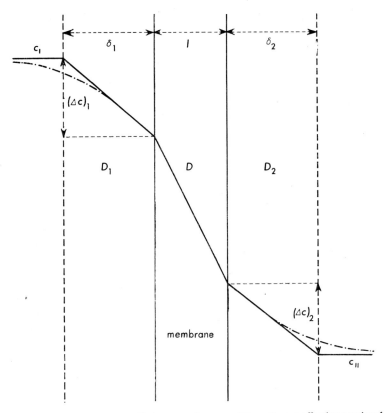

Fig. 4.4. Concentration profile in a membrane and the adjacent effective unstirred layers. The diffusion coefficients are shown in capital D's.

The true permeability properties are expressed by the permeability coefficient

$$P_{\text{true}} = \frac{J}{c_\text{I} - c_\text{II} - (\Delta c)_1 - (\Delta c)_2} \tag{4.35}$$

rather than by eq. (4.28). Since

$$(\Delta c)_1 = \frac{J}{\frac{D_1}{\delta_1}} \quad \text{and} \quad (\Delta c)_2 = \frac{J}{\frac{D_2}{\delta_2}}$$

it may be shown that

$$P_{\text{true}} = \frac{1}{\frac{1}{P_{\text{measured}}} - \frac{\delta_1}{D_1} - \frac{\delta_2}{D_2}}. \tag{4.36}$$

Most diffusion coefficients are of the order of magnitude of 10^{-5} cm^2 s^{-1} and since the thickness of the unstirred layers is of the order of 100 μm (10^{-2} cm), the "permeability" of an unstirred layer D/δ is of the order of 10^{-3} cm s^{-1}. Hence measured permeability coefficients as high as 10^{-3} cm s^{-1} show that the process is controlled by the diffusion in the unstirred layers and give little information about the properties of the membrane. Those of the order of 10^{-4} cm s^{-1} require a considerable correction.

The permeability constants are markedly influenced by temperature, something that rules out diffusion through large aqueous pores but does not exclude the possibility of passage through narrow pores where the hydration molecules would have to be stripped before the permeation. Values of activation energy of 40–80 kJ mol^{-1} are not uncommon, the main contribution being represented by the translocation across the membrane.

Fick's law for simple diffusion across a thin membrane (eq. 4.28) may be used to express concentration changes in a membrane-surrounded compartment of surface A and volume V, which are brought about by the permeation of the diffusing substance. The total inflow of the substance per unit time being $J \cdot A$, the rate of the concentration change due to permeation is

$$\frac{dc_{II}}{dt} = \frac{PA}{V}(c_I - c_{II}). \qquad (4.37)$$

It should be observed that c_{II} can change even in the absence of any permeation process, due to changes in compartment volume. If we assume that there is no permeation of the substance in question across the membrane the total amount of substance in compartment II will be constant

$$c_{II}V = \text{constant}. \qquad (4.38)$$

By taking derivatives with respect to time we obtain

$$V\frac{dc_{II}}{dt} + c_{II}\frac{dV}{dt} = 0$$

or

$$\frac{dc_{II}}{dt} = -\frac{c_{II}}{V}\frac{dV}{dt} \qquad (4.39)$$

which is the rate of concentration change due to changing volume of the compartment (e.g., a cell). Superposition of the concentration changes due to the permeation process (4.37) and to the changes of compartment volume yields

$$\frac{dc_{\text{II}}}{dt} = \frac{PA}{V}(c_{\text{I}} - c_{\text{II}}) - \frac{c_{\text{II}}}{V}\frac{dV}{dt}. \qquad (4.40)$$

Although volume changes of the compartment may be caused by various factors, those which interest us most are brought about by osmotic effects of the permeating substance itself. As far as the behavior of the compartment may be approximated by that of an ideal osmometer, the rate of volume change is proportional to the difference between the internal (II) and the external (I) osmolarity:

$$\frac{dV}{dt} = k\left(c_{\text{II}} + \frac{m}{V} - c_{\text{I}} - \frac{m_0}{V_0}\right) \qquad (4.41)$$

where m is the amount of other osmotically active substances present in the osmometer and m_0/V_0 is the initial osmolarity in the osmometer, presumably equal to that of the outer medium before addition of the permeating substance. Equations (4.40) and (4.41) are equivalent to those given by Jacobs (1952) and are not easy to solve. One possibility is to calculate (under reasonable assumptions about the value of m, such as $m = m_0$, the permeation of other substances being negligible during the experiment) small changes of c_{II} and V in turns and, from the small finite differences obtained, to construct curves approximating the time course of concentration and volume (Sigler and Janáček, 1971). Another way is to determine the course of volume changes experimentally. It should be added that equations (4.40) and (4.41) may represent a good approximation only in the case of reflection coefficient σ for the solute practically equal to 1 (i.e., when there is in the membrane no appreciable interaction between the flow of a slowly permeating solute and the flow of water), otherwise equations (4.40) and (4.41) have to be corrected in accordance with equations (6.34) and (6.35); see p. 111. Equations so modified were developed and successfully used by Johnson and Wilson (1967).

It is of great advantage if the volume changes of the compartment (cell) during permeability measurements are negligibly small, as can be always achieved (unless the saturability of the process is tested) by using a minute concentration of radioactively labelled permeant, or when the volume changes are minimized by the presence of a rigid cell wall. The same objective is accomplished by applying a small amount of tracer to the system which is in a steady state with respect to the unlabelled substance. Then the rate of the concentration change of the permeant is given by eq. (4.37), where P is a constant, in the case of simple diffusion across the membrane. In laboratory experiments it is often feasible to employ a large outer compartment, a reservoir, so that c_I is constant to a good approximation. Eq. (4.37) can then be easily integrated to obtain the time course of concentration:

$$\int_{c_{II}=0}^{c_{II}} \frac{dc_{II}}{c_I - c_{II}} = \frac{PA}{V} \int_{t=0}^{t} dt \qquad (4.42)$$

yielding

$$\ln \frac{c_I - c_{II}}{c_I} = -\frac{PA}{V} t \qquad (4.43)$$

or

$$c_{II} = c_I \left(1 - e^{-\frac{PA}{V} t}\right) \qquad (4.44)$$

which describes the exponential equilibration of concentrations. When the outflow of a substance from a preloaded compartment into a reservoir with practically zero concentration is measured the situation is even simpler, the decay of the internal concentration being described by

$$c_{II(t)} = c_{II(0)} e^{-\frac{PA}{V} t} \qquad (4.45)$$

which, in the logarithmic, linear form can be conveniently used to determine the coefficient PA/V from the slope of a straight line (cf. Fig. 4.5)

$$\ln \frac{c_{II(t)}}{c_{II(0)}} = 2.303 \log \frac{c_{II(t)}}{c_{II(0)}} = -\frac{PA}{V} t. \qquad (4.46)$$

An expression frequently used in this connection is the half-time of equilibration defined as

$$\ln 0.5 = -\ln 2 = -\frac{PA}{V} t_{0.5} \quad \text{or} \quad t_{0.5} = 0.693 V/PA. \qquad (4.46')$$

When the outer volume V_I is of a size comparable to that of the inner compartment, V_{II}, the concentration c_I in eq. (4.37) cannot be any more considered as a constant, but must be expressed from the condition

$$c_I V_I + c_{II} V_{II} = c_{I(0)} V_I \qquad (4.47)$$

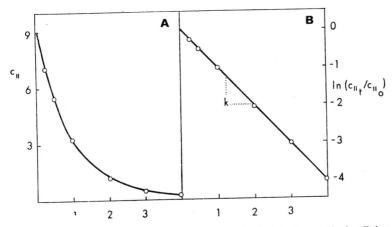

Fig. 4.5. Diffusional outflow of solute from cells. In **A**, the residual cellular concentration is plotted against time; in **B**, the logarithm of the ratio of initial and actual concentrations is plotted against time. The rate constant k is given by the slope of the straight line.

and the equilibration of compartment II is described by

$$c_{II} = c_{I(0)} \frac{V_I}{V_I + V_{II}} \left(1 - e^{-\frac{V_I+V_{II}}{V_I V_{II}} PAt}\right). \quad (4.48)$$

In the case of mediated diffusion the above equations do not apply, since P is then not a constant, but rather a function of concentration; in the case of an active transport, moreover, the stationary concentration in the two compartments is not the same. But even there the equilibration of a tracer, resulting in the same specific activity of the labelled substance in the two compartments, is described by analogous equations provided that the tracer is added in a small amount to the system in a steady state. The appropriate equations are solutions of the differential equation

$$\frac{dc_{II}^*}{dt} = \frac{J_s}{V_{II}} A \frac{c_I^*}{c_I} - \frac{J_s}{V_{II}} A \frac{c_{II}^*}{c_{II}} \quad (4.49)$$

where c^*'s are concentrations of the tracer and J_s is the steady-state unidirectional flux, i.e., a constant under the given experimental conditions.

The equation for tracer equilibration, equivalent to eq. (4.44) then will be

$$c_{II}^* = c_I^* \frac{c_{II}}{c_I} \left(1 - e^{-\frac{A}{V_{II}} \frac{J_s}{c_{II}} t}\right) \quad (4.50)$$

that equivalent to eq. (4.45) is

$$c_{II}^* = c_{II(0)}^* \, e^{-\frac{A}{V_{II}} \frac{J_s}{c_{II}} t} \tag{4.51}$$

and finally that equivalent to (4.48)

$$c_{II}^* = c_{I(0)}^* \frac{c_{II} V_I}{c_I V_I + c_{II} V_{II}} \left(1 - e^{-\frac{(c_I V_I + c_{II} V_{II}) J_s}{c_I V_I c_{II} V_{II}} At}\right) \tag{4.52}$$

the validity of which in a steady state is general and which enable us to estimate the value of the steady-state unidirectional flux of substance from the data on tracer exchange.

When there are several compartments present in a parallel or a series arrangement the situation is more complex, but still it may be described by a sum of several exponential terms and the experimental curves may be analyzed graphically (e.g., Kotyk and Janáček, 1975) or fitted by a sum of exponential functions using a computer (e.g., Berman et al., 1962).

Compounds transported across biological membranes by some kind of "simple", non-carrier diffusion are of many types. Besides the smallest hydrophilic molecules, such as water, dimethylsulfoxide formamide, etc., there is a vast variety of lipid-soluble compounds, ranging from methanol to complicated drugs which apparently cross membranes via the lipid domains; of naturally produced compounds of chemotherapeutic importance, one may name the penicillins, macrolides, rifamycins, actinomycin D, puromycin and chloramphenicol (cf. Franklin, 1974). Streptomycin, bacitracin and several other antibiotics penetrate across the cell membrane only after disrupting the external architecture of the cell.

On the other hand, even relatively small polar substances, such as glycols, monosaccharides, amino acids, cannot cross the membranes of most cells to any considerable degree by simple diffusion. There are usually specific carriers for such substances present in the cell membrane.

Apparent diffusion of a number of compounds can be induced by the application of various membrane-active substances, typically the polyene antibiotics, e.g., filipin, nystatin, etruscomycin, pimaricin, N-acetylcandidin, amphotericin B, and azalomycin (e.g., van Zupthen et al., 1971).

Filipin

[Chemical structure of Filipin]

Amphotericin B

[Chemical structure of Amphotericin B]

The mechanism of action these antibiotics appears to consist in binding to the sterol component of cells, preferring ergosterol to cholesterol, and thus forming aqueous "pores" which are in a continuous process of forming and breaking up. The holes thus formed range in size from a diameter of some 0.4 nm after nystatin to 12.5 nm after filipin, depending also on the concentration of the antibiotic. Table 4.1 shows some examples of the permeability changes involved.

TABLE 4.1. Effect of amphotericin B on the permeability properties of artificial sterol membranes (adapted from Dennis et al., 1970)

Molecule	P cm s^{-1} . 10^6	
	control	with 10^{-7} M amphotericin B
Water	630	1310
Urea	< 5	900
Glycerol	< 5	220
Ribose	< 5	67

Another group of antibiotics known to cause the leakage (= diffusion) of substances across the membrane are the tyrocidines, colistins and polymyxins which probably bind to cardiolipin, phosphatidyl choline and phosphatidyl serine through an electrostatic interaction. Cyclic peptides known to make nonselective holes in membranes are the surfactin from *Bacillus subtilis* and monamycin from *Streptomyces* species.

Polymyxin B$_1$

L-leu – DAB – DAB – L-thr – DAB – DAB – L-thr – DAB – MOA

D-phe – DAB

(MOA = 6-methyloctanoic acid, DAB = diaminobutyric acid)

Tyrocidine A

L-pro – L-phe – D-phe – L-asn – L-gln

D-phe – L-leu – L-orn – L-val – L-tyr

Surfactin

L-glu – L-leu – D-leu – L-val – L-asp

COCH$_2$CHO – L-leu – D-leu

(CH$_2$)$_9$

CH

(CH$_3$)$_2$

There are a number of various other naturally produced compounds which decrease the resistance to permeation of biological as well as artificial membranes. However, their effects being specific for various ions (and usually not associated with actual pore formation), they will be dealt with in the chapter on the transport of ions.

4.3. KINETICS OF MEDIATED TRANSPORT

The plasma membranes, mitochondrial and chloroplast membranes and possibly (but not certainly) other cell membranes contain mechanisms which allow for the specific transport of many substances, both of uncharged and ionic character, from one side of the membrane to the other. All the systems of this type are characterized besides their specificity by the fact that the rate of transport through their mediation rises with concentration only to a certain limiting value: they are saturable. The most straightforward (and probably justified in most cases) explanation for this phenomenon is that the transported solute binds transiently to a specific receptor on the membrane surface and only after this binding can it be translocated to the other side. It is of no consequence whether the receptor itself with the bound solute moves across the membrane or whether it transmits it to another membrane molecule which moves from one side to the other or, finally, whether it releases the solute into an internal "pool" from which it is attached by an analogous receptor at the other membrane face. The only property that such saturable transports must possess is that a given bidning site for substrate must never be exposed simultaneously to both external media. If this should happen, some of the typical properties of saturable transport, such as countertransport, would be lost.

4.3.1. Mediated or facilitated diffusion

The simplest of all the mechanisms so far proposed (but not necessarily the one universally valid) is the concept of a mobile carrier, the kinetics of which will be developed first.

If the carrier binds a single substrate we may depict the situation as follows:

$$\begin{array}{ccc} \text{I} & & \text{II} \\ CS_I & \underset{k_4}{\overset{k_3}{\rightleftarrows}} & CS_{II} \\ k_1s_I \updownarrow k_2 & & k_6s_{II} \updownarrow k_5 \\ C_I & \underset{k_7}{\overset{k_8}{\rightleftarrows}} & C_{II} \end{array}$$

4.3.1.1. Steady-state approach

If no assumption is made as to the rate limitation of the system the expression for the rate of flow J_s of substrate S is derived either rom the set of steady-state equations

$$J_s = k_1 c_I s_I - k_2 c s_I = k_3 c s_I - k_4 c s_{II} =$$
$$= k_5 c s_{II} - k_6 c_{II} s_{II} = k_7 c_{II} - k_8 c_I \quad (4.53)$$

and the carrier conservation equation

$$2c_t = c_I + c s_I + c_{II} + c s_{II} \quad (4.54)$$

or by one of the abbreviated techniques described by King and Altman (1956), Wong and Hanes (1962) or Fromm (1970) (*cf.* Kotyk, 1974)

$$J_s = c_t \frac{k_1 k_3 k_5 k_7 s_I - k_2 k_4 k_6 k_8 s_{II}}{\begin{array}{l}(k_7 + k_8)(k_2 k_4 + k_2 k_5 + k_3 k_5) + \\ + k_1[k_5(k_3 + k_7) + k_7(k_3 + k_4)] s_I + \\ + k_6[k_4(k_2 + k_8) + k_8(k_2 + k_3)] s_{II} + \\ + k_1 k_6 (k_3 + k_4) s_I s_{II}\end{array}} \quad (4.55)$$

If the system has no energy input we may speak of mediated or facilitated diffusion which is characterized by the fact that net flow ceases ($J_s = 0$) when $s_I = s_{II}$. From this it follows that $k_1 k_3 k_5 k_7 = k_2 k_4 k_6 k_8$ so that the numerator of eq. (4.55) may be written as $c_t k_1 k_3 k_5 k_7 (s_I - s_{II})$. (In systems of active transport to be dealt with later, $k_1 k_3 k_5 k_7 \neq k_2 k_4 k_6 k_8$.)

The initial rate of uptake (when $s_{II} = 0$) is defined by a much simpler expression, *viz.*

$$J_{s(0)} = c_t \frac{k_1 k_3 k_5 k_7 s_I}{\begin{array}{l}(k_7 + k_8)(k_2 k_4 + k_2 k_5 + k_3 k_5) + \\ + k_1[k_5(k_3 + k_7) + k_7(k_3 + k_4)] s_I\end{array}} \quad (4.56)$$

which, like all expressions relating to transport by a univalent carrier, is formally identical with the Michaelis–Menten equation of enzyme kinetics $v = Vs/(K_m + s)$ or $J_s = J_{max} s/(K_T + s)$. Here

$$J_{max} = c_t \frac{k_3 k_5 k_7}{k_5(k_3 + k_7) + k_7(k_3 + k_4)}$$

and

$$K_T = \frac{(k_7 + k_8)(k_2 k_4 + k_2 k_5 + k_3 k_5)}{k_1[k_5(k_3 + k_7) + k_7(k_3 + k_4)]}.$$

If we work with very low concentrations of substrate (well below the apparent K_T) eq. (4.56) reduces to a diffusion-like process

$$J_{s(0)} = c_t \frac{k_1 k_3 k_5 k_7 s_I}{(k_7 + k_8)(k_2 k_4 + k_2 k_5 + k_3 k_5)} \quad (4.57a)$$

while at very high concentrations (well above the apparent K_T) a zero-order process is obtained so that

$$J_{s(0)} = c_t \frac{k_3 k_5 k_7}{k_3 k_5 + k_5 k_7 + k_3 k_7 + k_4 k_7}. \quad (4.57b)$$

Equation (4.55) does not show the unidirectional fluxes. These can be derived either by a somewhat lengthy iterative procedure (cf. Britton, 1966) or by the King and Altman or Fromm method proceeding from the following model:

$$\begin{array}{c}
\text{I} \qquad\qquad\qquad\quad \text{II} \\
\mathbf{CS_I} \underset{k_4}{\overset{k_3}{\rightleftarrows}} \mathbf{CS_{II}} \\
k_1 s_I \updownarrow \,\, k_2 \qquad k_6 s_{II} \updownarrow \,\, k_5 \\
\mathbf{C_I} \underset{k_7}{\overset{k_8}{\rightleftarrows}} \mathbf{C_{II}} \\
k_2 \updownarrow \,\, k_1 s_I^* \qquad k_5 \updownarrow \,\, k_6 s_{II}^* \\
\mathbf{CS_I^*} \underset{k_4}{\overset{k_3}{\rightleftarrows}} \mathbf{CS_{II}^*}
\end{array}$$

where S* is he labelled form of S.

Because of the chemical identity of S and S* both the dissociation constants and the translocation constants are the same for S and S*. The expression for the flux from left to right is then

$$\overrightarrow{J_{s^*}} = c_t \frac{k_1 k_3 k_5 s_I^* [k_7 + k_2 k_4 k_6 s_{II}/(k_2 k_4 + k_2 k_5 + k_3 k_5)]}{\text{denominator}} \quad (4.58a)$$

the denominator being as in eq. (4.55). The flux from right to left is given by

$$\overleftarrow{J_{s^*}} = c_t \frac{k_2 k_4 k_6 s_{II}^* [k_8 + k_1 k_3 k_5 s_I/(k_2 k_4 + k_2 k_5 + k_3 k_5)]}{\text{denominator}} \quad (4.58b)$$

Both fluxes are thus seen to depend not only on the *cis* but also on the *trans* concentrations of the substrate ($\overrightarrow{J_{s^*}}$ on s_{II} and $\overleftarrow{J_{s^*}}$ on s_I). It is a question of the magnitude of the translocation constants (k_3, k_4, k_7, k_8) whether the effect will result in stimulation or inhibition.

If the system is intrinsically symmetrical, *i.e.* the translocation constants in both directions are equal ($k_3 = k_4$; $k_7 = k_8$) and the rate constants of association and dissociation, respectively, are equal at the two sides ($k_1 = k_6$; $k_2 = k_5$), eq. (4.55) for the overall rate of flow becomes

$$J_s = c_t \frac{k_1 k_3 k_5 k_7 (s_I - s_{II})}{2k_5 k_7 (k_5 + 2k_3) + k_1 (k_3 k_5 + k_5 k_7 + 2k_3 k_7)(s_I + s_{II}) + 2k_1^2 k_3 s_I s_{II}} \quad (4.59)$$

and the unidirectional fluxes will be

$$\overrightarrow{J_{s^*}} = c_t \frac{k_1 k_3 k_5 s_I^*[k_7 + k_1 k_3 s_{II}/(2k_3 + k_5)]}{\text{denominator}} \quad (4.60a)$$

and

$$\overleftarrow{J_{s^*}} = c_t \frac{k_1 k_3 k_5 s_{II}^*[k_7 + k_1 k_3 s_I/(2k_3 + k_5)]}{\text{denominator}} \quad (4.60b)$$

the denominator being the same as in eq. (4.59). In this case, the transport rate expression may be written as

$$J_s = \frac{J_{\max} K_T (s_I - s_{II})}{(K_T + s_I)(K_T + s_{II}) - x^2 K_T^2} \quad (4.61)$$

with $K_T = \alpha k_5/k_7$, $J_{\max} = c_t k_7/2\alpha$, $x = (1 - \alpha)/\alpha$, where $\alpha = (k_3 k_5 + k_5 k_7 + 2k_3 k_7)/2k_3 k_5$ (*cf.* Lieb and Stein, 1971).

For very low concentrations of S, the initial rate derived from eq. (4.59) reduces to

$$J_{s(0)} = c_t/2 \frac{k_1 k_3 s_I}{2k_3 + k_5} \quad (4.62a)$$

while for very high concentrations to

$$J_{s(0)} = c_t \frac{k_3 k_5 k_7}{k_3 k_5 + k_5 k_7 + 2k_3 k_7} \quad (4.62b)$$

Macroscopically, this system cannot be distinguished from one where no two constants are identical but where $K_{CS_I} k_8/k_3 = K_{CS_{II}} k_7/k_4$, K_{CS_I} being k_2/k_1, $K_{CS_{II}}$ being k_5/k_6 (*cf.* Geck, 1971).

An expression which is identical with eq. (4.55) but which allows a simple calculation of flows when $s_I = s_{II}$ was derived by Regen and Morgan (1964)

$$J_s = c_t \left[\frac{As_\text{I}}{1 + Bs_\text{I} + \dfrac{(s_\text{I} - s_\text{II})C}{1 + s_\text{II}/D}} - \frac{As_\text{II}}{1 + Bs_\text{II} + \dfrac{(s_\text{II} - s_\text{I})/C'}{1 + s_\text{I}/D}} \right] \quad (4.63)$$

where $A = k_1 k_3 k_5 k_7 (= k_2 k_4 k_6 k_8)/\alpha\beta$;
$B = k_1 k_7 \gamma / k_2 k_4 \alpha = k_6 k_8 \gamma / k_3 k_5 \alpha$;
$C = k_1 k_3 k_5 [k_2 k_4 - k_7 (k_2 + k_3 + k_4)]/\alpha\beta k_2 k_4$;
$C' = k_2 k_4 k_6 [k_3 k_5 - k_8 (k_3 + k_4 + k_5)]/\alpha\beta k_3 k_5$;
$D = k_7 \beta / k_2 k_4 k_6$; and
$\alpha = k_7 + k_8$;
$\beta = k_2 k_4 + k_2 k_5 + k_3 k_5$;
$\gamma = k_3 + k_4$.

4.3.1.2. Equilibrium approach

While the above expressions are more universal than those that will follow, they are bulky and in many cases a rather substantial simplification is justified. One makes use of the assumption that the translocation constants are much smaller than the dissociation constants at the membrane surface, whereupon*

$$J_s = 2k_c k_{cs} c_t \frac{s'_\text{I} - s'_\text{II}}{2(k_c + k_{cs} s'_\text{I} s'_\text{II}) + (k_c + k_{cs})(s'_\text{I} + s'_\text{II})} \quad (4.64)$$

where $k_c = k_7 = k_8$, $k_{cs} = k_3 = k_4$; $s'_\text{I} = s_\text{I}/K_\text{CS} = s_\text{I} k_1/k_2$; $s_\text{II} = s_\text{II}/K_\text{CS} = s_\text{II} k_6/k_5$.

The initial rate ($s_\text{II} = 0$) is then

$$J_{s(0)} = 2k_c k_{cs} c_t \frac{s'_\text{I}}{2k_c + s'_\text{I}(k_c + k_{cs})} \quad (4.65)$$

so that $J_\text{max} = 2k_c k_{cs} c_t / (k_c + k_{cs})$ and $K_T = 2k_c K_\text{CS}/(k_c + k_{cs})$. Like with eq. (4.60), one can derive simple formulae for uptake at very low or very high concentrations of substrate.

The sum of unidirectional fluxes is

$$J_{s^*} = 2k_{cs} c_t \frac{s^{'*}_\text{I}(k_c + k_{cs} s'_\text{II}) - s^{'*}_\text{II}(k_c + k_{cs} s'_\text{I})}{2(k_c + k_{cs} s'_\text{I} s'_\text{II}) + k_c(s'_\text{I} + s'_\text{II})} . \quad (4.66)$$

---

* The formula can be derived either by introducing the simplifying assumptions into eq. (4.55) or from the assumption of equilibria at the surfaces and setting up carrier flux equations (for J_C and J_{CS}), the sum of which must be equal to zero (cf. Kotyk and Janáček, 1975).

The ratio of k_{cs}/k_c ($= \varrho$) can be determined experimentally by comparing the initial rate of efflux of S into an equilibrium concentration of S ($J_=$) with that into a substrate-free medium (J_0).

$$J_0 = -2c_t k_c k_{cs} s_{II}^* / [2k_c K_{CS} + s_{II}(k_c + k_{cs})] \quad (4.67a)$$

and

$$J_= = -c_t k_{cs} s_{II}^* / (K_{CS} + s) \quad (4.67b)$$

(here $s = s_I = s_{II}$).

By dividing these two equations we obtain

$$\varrho = \frac{2K_{CS}(1-x) + s(2-x)}{sx} \quad (4.68a)$$

where $x = J_0/J_=$.

For very high values of s, moreover,

$$\varrho = (2-x)/x. \quad (4.68b)$$

In this equilibrium model, like in the previous steady-state one, if $k_{cs} > k_c$, one will observe the phenomenon of trans acceleration or the preloading effect, i.e., that the initial rate of uptake of substrate will be increased by preincubation with the same or a related substrate.

We may proceed further in simplifying the rate expression by assuming ϱ to be equal to one ($k_c = k_{cs} = k$) whereupon

$$J_s = kc_t/2 \left(\frac{s_I}{s_I + K_{CS}} - \frac{s_{II}}{s_{II} + K_{CS}} \right) \quad (4.69)$$

this being applicable probably only in rare cases (cf. Regen and Morgan, 1964). For the initial rate of uptake then

$$J_{s(0)} = kc_t/2 \frac{s_I}{s_I + K_{CS}} \quad (4.70)$$

so that $J_{max} = kc_t/2$ and $K_T = K_{CS}$.

The unidirectional fluxes here are determined by the individual terms in brackets, thus

$$\overrightarrow{J_{s*}} = kc_t \frac{s_I^*}{s_I^* + K_{CS}} \quad (4.71a)$$

and

$$\overleftarrow{J_{s*}} = kc_t \frac{s_{II}^*}{s_{II}^* + K_{CS}}. \quad (4.71b)$$

Thus, in this case, the unidirectional fluxes are independent of each other and of the trans concentration.

If two substrates are present which share the same carrier, the simplest model yields the following equation:

$$J_{s(r)} = J_{max}\left(\frac{s_I}{s_I + K_{CS}(1 + r_I/K_{CR})} - \frac{s_{II}}{s_{II} + K_{CS}(1 + r_{II}/K_{CR})}\right) \quad (4.72)$$

the expression (for s_{II}, $r_{II} = 0$) being identical with that for competitive inhibition in enzyme kinetics.

The three cases discussed so far (eq. 4.56, 4.65 and 4.70) are thus seen to be indistinguishable if only the apparent J_{max} and K_T are evaluated. All of them, moreover, will obey first-order kinetics at very low concentrations of substrate when

$$J_s = J_{max}(s_I - s_{II}) \quad (4.73)$$

This presents a danger when studying systems with a high K_T (when the values of s used are relatively small), when one is inclined to define the situation as simple, rather than mediated, diffusion. Other properties should always be examined concurrently, such as inhibition by heavy metal ions (UO_2^{2+}, Th^{4+}, La^{3+}) or competition with a high-affinity analogue, etc.

At high values of s, all the systems studied yield

$$J_s = J_{max}K'(1/s_{II} - 1/s_I) \quad (4.74)$$

where K' is a composite constant, depending on the degree of complexity of the system (it is equal to K_{CS} according to eq. 4.69).

At higher levels of substrate, the actual J_s is thus very substantially dependent on the magnitude of s_{II}, particularly at its low levels (beginning of uptake experiments).

The similarity between the three expressions mentioned above extends further. One can derive various parameters related to J_{max} and K_T by the application of tracers (usually the radioactively labeled form of substrate examined). The movement of a tracer in a steady state of the system (when $s_I = s_{II}$ or, more generally, when $J_s = 0$) always obeys the laws of diffusion (cf. the cases of various compartments on p. 203).

Taking the case described by eq. (4.71) we see for $s_I = s_{II} = s$ that

$$\vec{J}_{s*} \equiv ds_{II}^*/dt = J_{max}(s_I^* - s_{II}^*)/(s + K_{CS}). \quad (4.75)$$

Integration from 0 to s_{II} yields

$$J_{max}t = (s + K_{CS}) \ln [s_I^*/(s_I^* - s_{II}^*)] \qquad (4.76)$$

the half-time of tracer equilibration (when $s_{II} = 0.5s_I$) being

$$t_{0.5} = 0.693(s + K_{CS})/J_{max}. \qquad (4.77)$$

Plotting $t_{0.5}$ against different concentrations of s yields a straight line, intersecting the abscissa at $-K_{CS}$ ($= -K_T$) and having a slope of $0.693/J_{max}$ ($= 1.386/c_t k$).

In the medium-complex case of eq. (4.66) an analogous plot will intersect the abscissa at $-K_{CS}$ (here, however, $K_T = 2k_c K_{CS}/(k_c+k_{cs})$!) and the slope will be $0.693/k_{cs}c_t$ (J_{max} is equal to $2k_c k_{cs} c_t/(k_c + k_{cs})$).

Unfortunately, in the steady-state case of eq. (4.58a, b) an analogous plot will yield a complex J_{max} and a constant related to the K_T in a complicated manner.

An interesting feature of carrier-mediated transports is the difference in the rate of equilibration of a tracer in the presence of a larger amount of unlabeled substance and that of the analytical substance itself.

Taking the case when cells do not change volume on taking up solute and when the size of the external medium is much greater than that of cells, the half-time of equilibration of a tracer in simple diffusion is equal to the half-time of equilibration of an analytical substance, viz. $t_{0.5} = 0.693/k$ (cf. eq. 4.43 and 4.46'). On the other hand, in carrier-mediated diffusion, the tracer half-equilibration time (taking the simplest model) is $t_{0.5} = 0.693(s + K_{CS})/J_{max}$ (cf. eq. 4.77) while the half-equilibration of the total concentration is given by

$$t_{0.5} = (s + K_{CS})(0.693 + 0.193s/K_{CS})/J_{max}. \qquad (4.78)$$

Hence, for a high-affinity process, where $K_{CS} = 10^{-4} M$ and for $s = 0.1 M$ we have for the tracer equilibration time $0.0694 \, M/J_{max}$, while for the analytical half-equilibration time $19.39 \, M/J_{max}$. (For the situation where the cells respond to osmotic changes in the medium and when their amount is comparable with that of the medium, a much more complicated case, consult LeFevre and McGinnis (1960).)

The above consideration is directly related to the well-known Ussing (1949) flux ratio which states that only in simple diffusion will the ratio of fluxes $\vec{J_s}/\overleftarrow{J_s}$ be equal to the ratio of concentrations s_I/s_{II}. In carrier-mediated transport (taking again the simplest case) we have, indeed, $\vec{J_s}/\overleftarrow{J_s} = (s_I/s_{II})(K_{CS} + s_{II})/(K_{CS} + s_I) < 1$, unless, of course, $s_I = s_{II}$.

The ratio shows the interesting property of carrier fluxes that at high concentrations (when $s > K_{CS}$) the ratio will be practically equal to unity even if s_I is very different from s_{II}.

It may be of importance to decide whether the translocation across the membrane is rate-limiting (then the "equilibrium" equations 4.64, etc., apply) or whether none of the rate constants is pronouncedly smaller or, finally, whether the surface dissociation is much slower than the translocation. While the last-named situation is very unlikely to occur (uphill transport most of countertransport and could not be observed), distinction between the first two cases deserves attention.

The simplest and oldest approach is due to Wilbrandt (1954) who suggested plotting J_s against s_I for various s_{II} (Fig. 4.6).

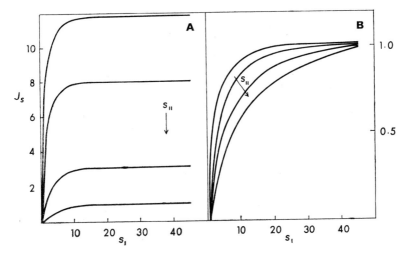

Fig. 4.6. Differentiation between a carrier transport with rate limitation by the translocation across the membrane (**A**) and with no pronounced rate limitation (**B**). The values of s_{II} in this particular example (taken from Wilbrandt, 1954) range from 0.08 to 0.13, 0.3 and 0.8.

The other approach was developed by Hoare and is founded on comparing the maximum rate of influx into preequilibrated cells with the maximum rate of influx into empty cells. The latter is derived from eq. (4.56) as

$$\vec{J}_{max} = \frac{c_t k_3 k_5 k_7}{k_7(k_3 + k_4) + k_5(k_3 + k_7)} \quad (4.79a)$$

while the former (from eq. 4.63) is

$$\overleftrightarrow{J}_{max} = \frac{c_t k_2 k_3 k_4 k_5}{(k_3 + k_4)[k_5(k_2 + k_3) + k_2 k_4]}. \tag{4.79b}$$

If the translocation is rate-limiting ($k_1, k_2, k_5, k_6 \gg k_3, k_4, k_7, k_8$)

$$\overrightarrow{J}_{max}/\overleftrightarrow{J}_{max} = k_7(k_3 + k_4)/k_4(k_3 + k_7) \tag{4.80a}$$

which can have values both smaller and greater than unity, depending on the relative mobilities k_4 and k_7. If the translocation is not rate-limiting (all the constants are of the same order of magnitude) we have

$$\overrightarrow{J}_{max}/\overleftrightarrow{J}_{max} = \frac{k_7(k_3 + k_4)[k_5(k_2 + k_3) + k_2 k_4]}{k_2 k_4 [k_7(k_3 + k_4) + k_5(k_3 + k_7)]} \tag{4.80b}$$

which is always greater than unity, particularly if the surface reaction between substrate and carrier should become rate-limiting. Thus, whenever the above ratio is found to be less than one we can postulate a translocation-limited mechanism.

All the above carrier mechanisms (as well as any other saturable mechanism with spatially or functionally separated events at the two membrane surfaces) display the phenomenon of countertransport which consists in an uphill movement of one carrier substrate at the expense of the downhill movement of another substrate, no metabolic energy being required.

There are two basic ways of demonstrating countertransport (it will be shown on the simple case of eq. 4.72).

1. If cells are preloaded with substrate R so that $r_I = r_{II}$ and then S is added (at that moment $s_{II} = 0$) the movement of R will be described by

$$J_r = J_{max}\left(\frac{r'}{1 + s' + r'} - \frac{r'}{1 + r'}\right) \tag{4.81}$$

where $r' = r/K_{CR}$; $s' = s/K_{CS}$). This is less than zero indicating that R will move out of cells and hence against its concentration gradient (initially $r_I = r_{II}$!).

2. The rate of uptake of a minute concentration of labelled substrate into empty cells is compared with that into cells preloaded with the substrate (Fig. 4.7) when an overshoot is observed. This demonstration of countertransport (due to Miller, 1965) is superior

to the first case because it involves no osmotic effects. In the first case, a countertransport-like movement of substrate out of cells might be produced by osmotic pressure increase and hence cell shrinkage due to adding the high concentration of substrate S.

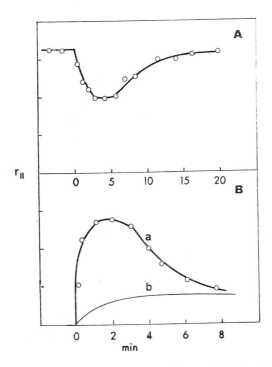

Fig. 4.7. Countertransport according to Rosenberg and Wilbrandt (1957) using the example of D-xylose in baker's yeast (**A**), and according to Miller (1965) using the example of galactose in human erythrocytes. (**B**) Curve a refers to cells preincubated with unlabeled galactose, curve b to a control without preincubation.

Unstirred layers (cf. p. 198) will affect the apparent K_T of a saturable transport process (Winne, 1973). The transport rate of S across the unstirred layer (in mol s^{-1} cm^{-2}) may be written as

$$J_s = (D/\delta)(s_B - s_M) \qquad (4.82)$$

where D is the diffusion coefficient in the unstirred layer (cm^2 s^{-1}), δ the layer thickness (cm), and s_B and s_M are the molarities of substrate in the mixed bulk phase and at the membrane surface, respectively. Combining eq. (4.82) with the simple eq. (4.70) ($s_M \equiv s_I$) we can solve for s_M and introduce the suitable root into eq. (4.82). The bulky expression

$$J_s =$$
$$= (D/\delta)[0.5(K_T + s_B + J_{max}\delta/D) \pm \sqrt{0.25(K_T - s_B + J_{max}\delta/D)^2 + s_B K_T}] \quad (4.83)$$

simplifies for $s \ll K_T$ to

$$J_s = s_B/(\delta/D + K_T/J_{max}) \quad (4.84a)$$

and for $s \gg K_T$ to

$$J_s = J_{max} \quad (4.84b)$$

and permits to show that for $J_s = 0.5 J_{max}$ (when s_B is the concentration giving rise to half maximum transport) we have

$$s_B = K_T + 0.5 J_{max}\delta/D \quad (4.85)$$

The value of K_T obtained experimentally is thus greater than the true one (cf. Fig. 4.8). The presence of interfering unstirred layers is

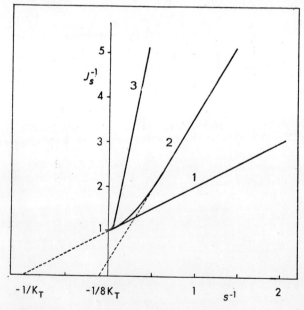

Fig. 4.8. Influence of unstirred layer on a Lineweaver—Burk plot of carrier-mediated transport. Curve **1**, no unstirred layer; curve **2**, $K_T = 0.5 J_{max}\delta/D$; curve **3**, $5 K_T = 0.5 J_{max}\delta/D$. Values used: $J_{max} = 1$ nmol s^{-1} cm^{-2}, $K_T = 1$ mM, $D = 5 \cdot 10^{-6}$ cm^2 s^{-1}, $\delta = 0.1$ mm (curve 2) or 0.5 mm (curve 3). Even with the lower thickness of the unstirred layer the error in the estimation of K_T amounts to eight-fold. (Adapted from Winne, 1973.)

indicated if, on increasing the rate of stirring, the experimental value of K_T decreases.

If one sets $D/\delta = P$ and rearranges eq. (4.82) one obtains for the unstirred layer at side I

$$s_I = s_{B,I} - J_s/P_1 \qquad (4.86a)$$

and similarly for side II

$$s_{II} = s_{B,II} + J_s/P_2. \qquad (4.86b)$$

Manipulation of these equations in combination with eq. (4.55) (for an extremely lucid treatment see Lieb and Stein, 1974) permits to determine the effective permeability of the unstirred layer by plotting $1/J_s$ against $s_B/(J_\infty - J_s)$ (Fig. 4.9) where J_s is the net flow rate from

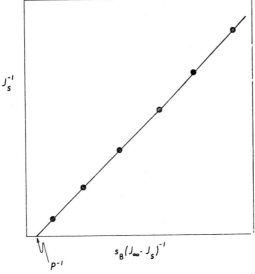

Fig. 4.9. Determination of unstirred layer permeabilities. (According to Lieb and Stein, 1974.)

an extremely high cis-concentration against a concentration s_B, J_∞ is the net flow rate from the same high cis-concentration when $s_B = 0$. Starting from a very high $s_{B,I}$ against different values of $s_{B,II}$, the intercept of the line in Fig. 4.9 denotes $1/P_2$, i.e., the unstirred layer permeability inside the cell in the usual arrangement.

Underestimation of the role of unstirred layers on transport kinetics has in fact led to attempts to supplant the carrier model with more sophisticated versions, e.g., that by Lieb and Stein (1970) where tetrameric proteins were assumed to span the membrane, possessing high- and low-affinity subunits equally distributed on both sides. Another model was that of introvertive hemiports lining the membrane faces (LeFevre, 1973) and another still that by Naftalin (1970) who postulated the existence of transmembrane pores lined with binding sites, the substrate having restricted chances of transfer from one site to the other.

However, a recent report by Regen and Tarpley (1974) demonstrates rather convincingly that a carrier of the type described by equation (4.55) and unstirred layers at both sides of the membrane provide an explanation for all the existing experimental findings. Thus, although the polymeric models would provide greater versatility particularly if cooperative interactions were allowed, there is at present no need to replace the conceptually simpler mobile carrier model. An extremely clear review of the situation can be found in Le Fevre (1975).

4.3.1.3. Two-site carrier

In principle, multi-site carriers can exist and at least a few cases have been described in active or coupled transports (adenosinetriphosphatase transport of Na^+ and K^+: Repke and Schön, 1973; symport of two sodium ions with amino acid in pigeon erythrocytes: Vidaver and Shepherd, 1968). They may be kinetically indistinguishable from single-site carriers if the sites are mutually independent or they may show some kind of cooperativity. If it is negative (corresponding to a Hill coefficient of enzyme kinetics less than one), high concentrations of substrate would decrease the actual rate of uptake (a type of excess-substrate inhibition). If it is positive (corresponding to a Hill coefficient greater than one) it may be due either to a greater affinity of any subsequent site for its substrate, or to a greater translocation constant of complexes with more than one molecule of bound substrate.

A rate equation proceeding from the equilibrium assumption (translocation constants being rate-limiting) for two competing substrates S and R and two binding sites* has the form

$$J_{s(r)} = 2c_t \frac{x_{II} y_I - x_I y_{II}}{x_{II} w_I + x_I w_{II}} \tag{4.87}$$

* Any greater number of cooperating binding sites will produce qualitatively the same effects.

where $x = s'(k_{cs} + k_{css}s'' + k_{csr}r''') +$
$+ r'(k_{cr} + k_{crr}r'' + k_{crs}s''') + k_c$
$y = s'(k_{cs} + k_{csr}r'' + 2k_{css}s'') + k_{crs}s'''r'$
$w = 1 + s'(1 + s'' + r''') + r'(1 + r'' + s''')$.

Here $s' = s/K_{CS}$, $s'' = s/K_{CSS}$, $s''' = s/K_{CRS}$ and analogously for r', r'' and r'''.

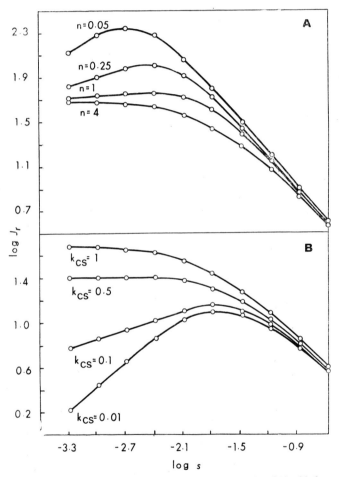

Fig. 4.10. Dependence of the rate of uptake of labelled solute (J_r) added to various concentrations of unlabelled solute S on the concentration of the latter (s), in a two-site carrier system with positive cooperativity. In **A**, n denotes the ratio of K_{CSS}/K_{CS}; in **B**, the translocation constant of the singly charged carrier is varied. (Adapted from Wilbrandt, W. and Kotyk, A. (1964). *Naunyn—Schmiedebergs Arch. exp. Path. u. Pharmak.* **249**, 279.)

The equation predicts several interesting phenomena.
1. The dependence of initial rate on concentration will be sigmoid.
2. If cells are prequilibrated with R and the S is added and its rate of uptake plotted against r (preferably in logarithmic coordinates), the curve will pass through a maximum rather than have a zero slope at low and a slope of -1 at high concentrations (Fig. 4.10).
3. In an experiment with countertransport of 1st type (p. 216) using two forms of the same substrate, the movement of preequilibrated R at the moment of adding S will be given by

$$J_r = sr(1 - r's'' - r'r'')a/D$$

where a is a complex positive constant and D the denominator containing only positive terms. Thus, for relatively high concentrations of S and R, $J_r < 0$ (*countertransport*) while for very low concentrations of S and R, $J_r > 0$ (the so-called *cotransport*). Here the addition of the new substrate will actually cause more of the originally equilibrated substrate to move into cells.

It should be noted that the same phenomena are to be observed in a system with no pronounced rate limitation by transmembrane translocation although the appropriate equations are prohibitively complicated.

4.3.2. Primary active transport

4.3.2.1. Kinetics

The only difference between the kinetics of mediated diffusion and that of active carrier transport is in the fact that for active transport energy is "fed" into one of the steps in such a way that for J_s to become zero we have $s_{II}/s_I \neq 1$. In the case described by eq. (4.55) the zero net flow condition is satisfied for $s_{II}/s_I = k_1 k_3 k_5 k_7 / k_2 k_4 k_6 k_8$. Thus, an increase (by metabolic energy) of any of the odd-numbered constants or a decrease of any of the even-numbered ones will result in accumulation of substrate S.

Using a stratagem of enzyme kinetics (*cf.* Cleland, 1963) one can convert the unwieldy groups of rate constants of eq. (4.55) into more meaningful kinetic constants. The equation can be rewritten as

$$J_s = \frac{N_1 s_I - N_2 s_{II}}{\text{const.} + D_1 s_I + D_2 s_{II} + D_{12} s_I s_{II}}$$

or

$$J_s/J_{max}^{\rightarrow} = \frac{s_I - s_{II}/K_{eq}}{K_{T_1}(1 + s_{II}/K_{T_2}) + s_I(1 + s_{II}/K_i)}$$

where $K_{eq} = N_1/N_2 = K_{T_2}J_{max}^{\rightarrow}/K_{T_1}J_{max}^{\leftarrow}$; $K_{T_1} = \text{const.}/D_1$; $K_{T_2} = \text{const.}/D_2$; $K_i = D_1/D_{12}$; $J_{max}^{\rightarrow} = N_1/D_1$. On setting $\Delta s = s_I - s_{II}/K_{eq}$ and inverting the last equation we obtain

$$1/J_s = \left[\frac{K_{T_1}}{J_{max}^{\rightarrow}}(1 + s_{II}/K_{T_2}) + \frac{s_{II}}{J_{max}^{\rightarrow}K_{eq}}(1 + s_{II}/K_i)\right](1/\Delta s) + \frac{1}{J_{max}^{\rightarrow}}(1 + s_{II}/K_i)$$

On plotting $1/J_s$ against $1/\Delta s$ for different values of s_{II} we arrive at a set of $J_{max_{app}}$'s which are related to true $J_{max}^{\rightarrow}$ by

$$1/J_{max_{app}} = s_{II}/J_{max}^{\rightarrow}K_t + 1/J_{max}^{\rightarrow}$$

A plot of $1/J_{max_{app}}$ against s_{II} will yield $1/J_{max}^{\rightarrow}$ as the intercept with the ordinate and $-K_i$ as the intercept with the abscissa. Similarly, the intercepts with the abscissa in the $1/J_s$ vs. $1/\Delta s$ plot will yield a set of $K_{T_{app}}$'s which are related to the true K_T's by

$$K_{T_{app}} = K_{T_1}\frac{1 + s_{II}/K_{T_2}}{1 + s_{II}/K_i} + s_{II}/K_{eq}$$

Since K_{T_1} is known from measuring the initial velocity of uptake at different concentrations of S_I and K_i was determined in the previous replot and since K_{eq} can be determined readily from the steady-state accumulation ratio (for mediated diffusion it is equal to 1) we can calculate the value of K_{T_2} and hence $J_{max}^{\leftarrow}$ can be derived from the definition of K_{eq} (cf. Cuppoletti and Segel, 1975; Betz et al., 1975).

The above mechanism has a serious drawback in that it predicts constant s_{II}/s_I ratios irrespective of concentration. It is generally observed, however, that these ratios decrease with increasing concentrations, either tending toward unity or decreasing further, the value of s_{II} being then seemingly constant (Fig. 4.11).

Hence somewhat more elaborate models have to be developed, *viz.* for the first case a model where the carrier exists in two forms with different affinities at each of the membrane sides, the conversion between them being relatively slow; for the second case a model where the movement across the membrane of the association-

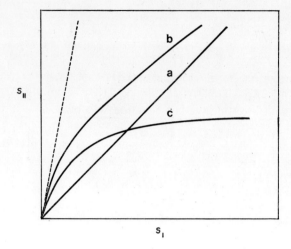

Fig. 4.11. Dependence of steady-state intracellular concentration (s_{II}) on the external concentration (s_I). Curve a (e.g. D-xylose in *Saccharomyces cerevisiae*) reflects a mediated diffusion; curve b (e.g. D-xylose in *Rhodotorula glutinis*) reflects an active transport which resembles mediated diffusion at high s_I (then $s_{II}/s_I = 1$); curve c (e.g. D-xylose in *Candida beverwijkii*) reflects an active transport which is strictly dependent on a source of energy which becomes less available at high values of s_I (then s_{II} rises very slowly). The dashed line depicts the case of a constant ratio over the entire concentration range as predicted by the simple models.

dissociation reaction is obligatorily dependent on the presence of a source of energy which is only present in a limited amount.

The first case may be depicted as follows:

$$\begin{array}{ccc}
\text{I} & & \text{II} \\
\text{CS}_\text{I} & \underset{k_4}{\overset{k_3}{\rightleftarrows}} & \text{CS}_{\text{II}} \\
k_1 s_\text{I} \updownarrow k_2 & & k_6 s_{\text{II}} \updownarrow k_5 \\
\text{C}_\text{I} & \underset{k_{18}}{\overset{k_{17}}{\rightleftarrows}} & \text{C}_{\text{II}} \\
k_{15} \updownarrow k_{16} & & k_8 \updownarrow k_7 \\
\text{Z}_\text{I} & \underset{k_{20}}{\overset{k_{19}}{\rightleftarrows}} & \text{Z}_{\text{II}} \\
k_{13} \updownarrow k_{14} s_\text{I} & & k_{10} \updownarrow k_9 s_{\text{II}} \\
\text{ZS}_\text{I} & \underset{k_{11}}{\overset{k_{12}}{\rightleftarrows}} & \text{ZS}_{\text{II}}
\end{array}$$

C represents the carrier with high affinity for substrate, Z the carrier with low affinity. Metabolic energy may in this case stimulate either k_7 or k_{15}. Even if the association-dissociation reactions between C, Z and S are considered to be in equilibrium and symmetrical at the two faces and if $k_3 = k_4 = k_{11} = k_{12} = k_{cs}$, $k_{17} = k_{18} = k_{19} = k_{20} = k_c$, the rate equation is quite bulky (here $s' = s/K_{CS}$ and $s'' = s/K_{ZS}$)

$$J_s = J_{CS} + J_{ZS} = 2c_t k_{cs} \frac{X_{II}(s'_I + s''_I \alpha\beta) - X_I(s'_{II} + s''_{II}\beta\gamma)}{A_I X_{II} + A_{II} X_I} \quad (4.88)$$

with $A_I = 1 + s'_I + \alpha\beta + s''_I\alpha\beta$
$A_{II} = 1 + s'_{II} + \beta\gamma + s''_{II}\beta\gamma$
$X_I = k_c(1 + \alpha\beta) + k_{cs}(s'_I + s''_I\alpha\beta)$
$X_{II} = k_c(1 + \beta\gamma) + k_{cs}(s'_{II} + s''_{II}\beta\gamma)$

where $\alpha = (k_c + k_{cs}s'_I)/(k_c + k_7 + k_{cs}s'_{II})$
$\beta = k_7/k_{15}$
$\gamma = (k_c + k_{15} + k_{cs}s''_I)/(k_c + k_{cs}s''_{II})$.

The accumulation ratio s_{II}/s_I may be shown to be equal to

$$\frac{[K_{ZS}k_{15}(k_c + k_7) + K_{CS}k_c k_7]}{[K_{CS}k_7(k_c + k_{15}) + K_{CS}k_c k_{15}]} \quad (4.89)$$

for low concentrations of substrate (i.e., a constant) but equal to one for high concentrations of substrate.

To account for the values of s_{II}/s_I decreasing below unity, the most expedient model appears to be

$$\begin{array}{ccc}
\text{I} & & \text{II} \\
\text{CS} & \underset{d}{\overset{cq}{\rightleftarrows}} & \text{CS}' \\
as_I \updownarrow b & & fs_{II} \updownarrow e \\
\text{C} & \underset{g}{\overset{h}{\rightleftarrows}} & \text{C}'
\end{array}$$

where Q is the source of energy required for the movement of S from side I to side II.

The steady-state concentration ratio of substrate S is

$$s_{II}/s_I = acegq/bdfh. \quad (4.90)$$

However, since $q_t = q + cs'$, q will decrease with increasing substrate concentration (limited, obviously, by the amount of c_t present). Since the relative concentration of CS' can be computed to be equal to

$$(acfqs_I s_{II} + acgqs_I + cfhqs_{II} + bfhs_{II})/c_t$$

a cubic equation for s_{II} can be set up:

$$b^2 d^2 f^3 h^2 (as_I + h) s_{II}^3 +$$
$$+ bdf^2 h \{bdeh(g + h) + as_I [bde(g + h) + dgh(b + e) +$$
$$+ egh(b + cc_t - cq_t)] + a^2 egs_I^2 (d + cc_t - cq_t)\} s_{II}^2 +$$
$$+ aefgs_I \{bdh(bd + be - ceq_t)(g + h) + as_I [bcdgh(c_t - q_t) +$$
$$+ bdgh(d + e) + bceghc_t - ceq_t(bdh + dgh + bgh + bdg)] -$$
$$- s_I^2 a^2 cdegq_t\} s_{II} - a^2 ce^2 g^2 s_I^2 q_t (d + e)(ags_I + bg + bh) = 0 \quad (4.91)$$

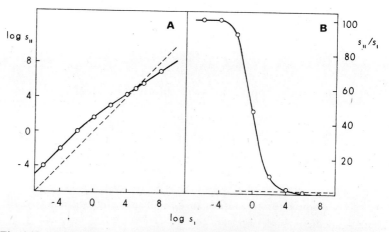

Fig. 4.12. Dependence of steady-state intracellular concentration s_{II} (**A**) and of the s_{II}/s_I ratio (**B**) on s_I, in a system with limited energy supply and with only the energized form of carrier mobile (see scheme on p. 224). The dashed lines denote distribution according to diffusion equilibrium.

An example of a numerical solution is shown in Fig. 4.12. Other models can explain this type of observation, their common feature being that the source of energy is of limited availability and that only the "energized" form can cross the membrane.

4.3.2.2. Combined systems

The case may arise that a compound is transported both by a saturable process and by simple diffusion as described on p. 192.

The situation is readily recognized by several characteristic features. 1. The inhibitory effect of a competing substrate is only partial even at extreme concentrations of inhibitor. 2. The effect of energy poisons (in the case of active transport) is only partial. 3. A plot of $J_{s(0)}$ against s or of $J_{s(0)}^{-1}$ against s^{-1} has the shape shown in Fig. 4.13.

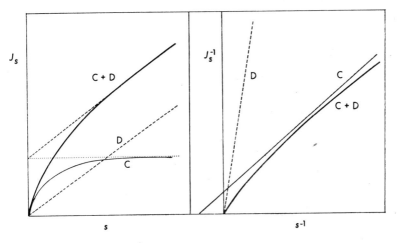

Fig. 4.13. A transport system combining diffusion with carrier transport. The flow-versus-concentration plot as well as the Lineweaver-Burk plot using reciprocal values is broken down to the diffusion (D) and carrier (C) components. (Adapted freely from Neame, K. D. and Richards, T. G. (1972). *Elementary Kinetics of Membrane Carrier Transport*. Blackwell Scientific Publications, Oxford.)

An analysis of such an event is fairly simple and follows from the figure.

A slightly more complicated case is the rather frequent transport of one substrate by two carrier systems. A plot of $J_{s(0)}^{-1}$ against s^{-1} is amenable to extracting from it all the pertinent data according to Neal (1972); cf. Kotyk and Janáček (1975).

A system of active transport operating simultaneously with a simple diffusion (a pump-and-leak model) will plot just as shown in Fig. 4.13 but the steady-state accumulation ratio will be generally given (from eq. 4.55 combined with 4.28) by

$$s_{II}/s_I = \frac{P(\alpha + \beta s_I + \gamma s_{II} + \delta s_I s_{II}) + k_1 k_3 k_5 k_7}{P(\alpha + \beta s_I + \gamma s_{II} + \delta s_I s_{II}) + k_2 k_4 k_6 k_8} \quad (4.92)$$

α, β, γ, and δ being positive constants. Thus the ratio is nearly constant for very low concentrations and tends toward unity at high ones.

A qualitatively identical result is obtained from any of the simplified versions of the uptake equation.

4.3.2.3. Energetics

Calculation of the energy (in $J\,s^{-1}$) required to maintain a given concentration ratio proceeds from the modified Nernst formula, *viz.*

$$\text{Energy/time} = J_s RT \ln (s_{II}/s_I) \qquad (4.93)$$

J_s being as usual the unidirectional flux inward. While the ratio of s_{II}/s_I is determined with greater or lesser accuracy without much difficulty, an exact expression for the flux is uncommonly difficult and can be generally obtained only for a system in surface equilibrium (*cf.* eq. 4.88). The most expedient way is to split eq. (4.93) into expressions relating to the fluxes of the individual carrier species, these being of first order, e.g.,

$$\text{Energy/time} = k_c[(c_I - c_{II})\,RT\ln(c_I/c_{II}) +$$
$$+ (z_I - z_{II})\,RT \ln (z_I/z_{II})] + k_{cs}[(cs_I - cs_{II})\,RT \ln (cs_I/cs_{II}) +$$
$$+ (zs_I - zs_{II})\,RT \ln (zs_I/zs_{II})]. \qquad (4.94)$$

This equation predicts among other things that for low values of s the energy consumption will be greater than for high values of s, this being a consequence of the s_{II}/s_I approaching unity at very high concentrations (when $\ln (s_{II}/s_I) = 0$). However, a straightforward calculation is hardly ever possible, there being systems requiring energy only for translocation but not for maintaining the gradient (*cf.* Kotyk and Říhová, 1972), as well as others where s_{II} falls below s_I at very high concentrations and still energy is required for the transport inward (e.g., Deák and Kotyk, 1968). On the other hand, if a leak is present in parallel with an active transport, the energy required for maintaining a concentration ratio may be more than calculated from the above formula.

Moreover, as shown by Rosenberg and Wilbrandt (1963), the economy of the system may differ, depending on the ratio of affinities at the two membrane sides as well as on the efficiency with which one of the carrier forms is changed to the other.

4.3.3. Coupled transport

Among the numerous uphill-transporting systems now known to operate in cell membranes, the secondary mechanisms have come into clear prominence so that some authors will even doubt the existence of true primary active transports of nonelectrolytes. It is indeed of a distinct advantage to the cell if it can use as a source of energy the gradient of an ion that it expels into the medium perhaps for other reasons; thus H^+ must be removed from cells to maintain the intracellular pH at a reasonable level and Na^+ is undesirable in cells because it counteracts the activating effect of K^+ on many enzymes. Still, some overzealous supporters of ubiquitous ion-nonelectrolyte coupling may overlook the fact that even if, say sodium ions will drive sugars, amino acids, as well as other ions into cells, each of the substrate types requires its specific carrier or at least recognition protein so that the number of components of a coupled transport system is probably the same as that of a primary active transport.

Kinetically speaking, a ternary complex of carrier-nonelectrolyte-ion is the form commonly assumed to mediate this type of transport. The ternary complex may be formed in one of two ways:
(1) carrier + ion + substrate (C + A + S), or
(2) carrier + substrate + ion (C + S + A) as shown schematically below:

$$
\begin{array}{c}
\text{I} \qquad\qquad\quad k_{as} \qquad\quad \text{II} \\
\quad\quad CAS_I \;\rightleftharpoons\; CAS_{II} \\
K_2 \;\updownarrow\quad\; k_a \;\;\updownarrow \\
\quad\quad CA_I \;\rightleftharpoons\; CA_{II} \\
K_1 \;\updownarrow\quad\; k_c \;\;\updownarrow \qquad\qquad \text{Path I}\\
\quad\quad C_I \;\rightleftharpoons\; C_{II} \\
K_3 \; b\updownarrow a \quad k_s \; b\updownarrow a \qquad\qquad \text{Path II}\\
\quad\quad CS_I \;\rightleftharpoons\; CS_{II} \\
K_4 \; d\updownarrow c \quad k_{sa} \; d\updownarrow c \\
\quad\quad CSA_I \;\rightleftharpoons\; CSA_{II}
\end{array}
$$

If the system is in surface equilibrium so that dissociation constants $K_1 - K_4$ may be used to describe the partial steps, the overall expression for Path I is

$$J_s = \frac{2c_t k_{as}}{\alpha_I \beta_{II} + \alpha_{II} \beta_I} (\beta_{II} s_I a_I / K_1 K_2 - \beta_I s_{II} a_{II} / K_1 K_2) \qquad (4.95)$$

with $\alpha = 1 + a/K_1 + s \cdot a/K_1K_2$;
$\beta = k_c + k_a a/K_1 + k_{as} s \cdot a/K_1K_2$.

The initial rate $J_{s(0)}$ (for $s_{II} = 0$) is given by Michaelis–Menten type expressions with

$$J_{max} = \frac{2c_t(k_cK_1 + k_a a_{II})k_{as}}{K_1(k_{as} + k_c) + a_{II}(k_{as} + k_a)} \quad (4.96a)$$

and

$$K_T = \frac{K_2[(K_1 + a_{II})(k_cK_1 + k_a a_I) + (K_1 + a_I)(k_cK_1 + k_a a_{II})]}{a_I[K_1(k_{as} + k_c) + a_{II}(k_{as} + k_a)]}. \quad (4.96b)$$

The steady-state accumulation ratio (for $J_s = 0$) is given by

$$s_{II}/s_I = \frac{a_I(k_cK_1 + k_a a_{II})}{a_{II}(k_cK_1 + k_a a_I)}. \quad (4.96c)$$

If the ion-carrier complex is immobile ($k_a = 0$) we have

$$J_{max} = \frac{2c_t k_c k_{as} K_1}{K_1(k_{as} + k_c) + a_{II}k_{as}}, \quad (4.97a)$$

$$K_T = \frac{k_c K_1 K_2 (2K_1 + a_I + a_{II})}{a_I[K_1(k_{as} + k_c) + k_{as}a_{II}]} \quad (4.97b)$$

and

$$s_{II}/s_I = a_I/a_{II}. \quad (4.97c)$$

If the formation of the ternary complex proceeds via path II the corresponding rate is expression is given by

$$J_s = \frac{2c_t}{\alpha_I \beta_{II} + \alpha_{II}\beta_I}\left[\beta_{II}\left(\frac{k_s s_I}{K_3} + \frac{k_{sa}s_I a_I}{K_3 K_4}\right) - \beta_I\left(\frac{k_s s_{II}}{K_3} + \frac{k_{sa}s_{II} a_{II}}{K_3 K_4}\right)\right] \quad (4.98)$$

where $\alpha = 1 + s/K_3 + s \cdot a/K_3 K_4$; $\beta = k_c + k_s s/K_3 + k_{sa} s \cdot a/K_3 K_4$. The parameters of the initial rate expression are

$$J_{max} = \frac{2c_t k_c(k_s K_4 + k_{sa} a_I)}{K_4(k_s + k_c) + a_I(k_{sa} + k_c)} \quad (4.99a)$$

and

$$K_T = \frac{2k_c K_3 K_4}{K_4(k_s + k_c) + a_I(k_{sa} + k_c)} \quad (4.99b)$$

while

$$s_{II}/s_I = \frac{k_s K_4 + k_{sa} a_I}{k_s K_4 + k_{sa} a_{II}}.$$ (4.99c)

If the activation is essential ($k_s = 0$) we have

$$J_{max} = \frac{2c_t k_c k_{sa} a_I}{k_c K_4 + a_I(k_{sa} + k_c)},$$ (4.100a)

$$K_T = \frac{2k_c K_3 K_4}{k_c K_4 + a_I(k_{sa} + k_c)}$$ (4.100b)

and

$$s_{II}/s_I = a_I/a_{II}.$$ (4.100c)

The above expressions predict several possibilities of influencing the various parameters by the presence of A at the *cis* or at the *trans* side (Table 4.2). They simplify greatly if one assumes equal translocation constants for all the carrier forms.

TABLE 4.2. Effect of activating ion on transport parameters in a coupled transport in surface equilibrium

Mechanism	Ion at *cis* side	Ion at *trans* side
Path I		
for $ks = 0$	K_T	K_T, J_{max}
for $ks = 0$	K_T	K_T, J_{max}
and K_1 large	K_T	K_T
Path II		
for $k_s = 0$	K_T, J_{max}	—
for $k_s = 0$	K_T, J_{max}	—
and K_4 large	J_{max}	—
Random formation of CSA complex	K_T, J_{max}	K_T, J_{max}

A special case arises when the system contains C, CA, CS, CAS and when the dissociation constant of C−S is the same as that of CA−S. Then no accumulation is predicted and $s_{II}/s_I = 1$.

The membrane potential obviously plays a role in the accumulation of a nonelectrolyte by a coupled process. Since it is usually oriented with its negative side inward it will result in an attraction for the positively charged carrier complex toward the inside face of the membrane. Whatever the charge of the free carrier, the binding of a cation will increase it by one positive unit; in path II which is simpler to treat kinetically, the ternary complex CSA will have a positive charge with respect to C and CS. If this fact is reflected in $\overrightarrow{k_{sa}} > \overleftarrow{k_{sa}}$ ($\equiv k' > k$) the accumulation ratio for the path II mechanism is

$$s_{II}/s_I = \frac{k_s K_4 + k' a_I}{k_s K_4 + k a_{II}} \qquad (4.101a)$$

and, for the case of immobile CS,

$$s_{II}/s_I = k' a_I / k a_{II}. \qquad (4.101b)$$

This expression has the important consequence that even for $a_I/a_{II} < 1$ there may be a concentrative transport of S as long as $k'/k > a_I/a_{II}$.

Expressed quantitatively, the ratio of ternary complex fluxes may be written, in analogy to Ussing's flux ratio (1949)

$$\frac{J_{CSAI}}{J_{CSAII}} \equiv \frac{k' c s a_I}{k c s a_{II}} = \frac{c s a_I}{c s a_{II}} e^{-nF\varphi/RT} \text{ and } \frac{k'}{k} = e^{-z} \qquad (4.102)$$

where φ is the membrane potential in mV, F is the Faraday (96 500 coulombs/equivalent), n is the charge of the ion. Thus a ratio of fluxes equal to 100 is to be expected for a potential of 120 mV (negative inside) even if $a_I = a_{II}$.

Like in primary active transport, the accumulation ratios appear to be constant over the whole range of concentrations unless further assumptions are introduced. If, say, the values of a_I and a_{II} are very small or if K_4 is very large, or if $k_s \gg k_{sa}$, the accumulation ratio will tend toward unity. Likewise, in a steady-state approach (pertaining to the symmetrical path II of scheme on p. 229) the accumulation ratio is given by

$$s_{II}/s_I = \frac{k_s(d + 2k_{sa}) + c k_{sa} a_I}{k_s(d + 2k_{sa}) + c k_{sa} a_{II}} \qquad (4.103)$$

and hence for $k_s \gg k_{sa}$ or for d/c very large, s_{II}/s_I will approach unity.

However, to express the fact that the ratio approaches unity only at high concentrations, and to account for values of the accumulation ratio less than one, one has to assume a "dissipation" of the ion gradient by the carrier-substrate flux just as derived at eq. (4.91). For a compulsory coupling ($k_{cs} = 0$) it will always hold that $s_{II}/s_I = a_I/a_{II}$ but a_I will be indirectly proportional to the amount of CS_I and hence S_I.

Somewhat more complex models of coupled transport have been designed to accommodate the simultaneous effects of two cations (e.g., Na^+ and H^+) by Curran and associates (1970) and by Frizzell and Schultz (1970).

4.4. CHEMICAL NATURE OF NONELECTROLYTE TRANSPORT SYSTEMS

Kinetic evidence has provided invaluable information about the properties of various transport systems but, like in the corresponding stage of enzymology, the time has come to isolate and define the chemical infividuals participating in transport. In contrast with enzymology, however, the work was hampered by the lack of any specific reaction of the transport components except binding their substrate. This led to the necessity of working with rather large quantities of material but helped in developing refined techniques of measuring ligand binding to macromolecules.

4.4.1. Group-translocation systems

In the field of nonelectrolytes the greatest amount of detailed information is available on a system that may be designated as group translocation rather than carrier-mediated transport, *viz.* the phosphotransferase system. Kundig and co-workers reported its existence in 1964 in *Escherichia coli* and since then it was discovered in *Salmonella typhimurium, Bacillus subtilis, Aerobacter aerogenes, Lactobacillus plantarum, Streptococcus lactis, Staphylococcus aureus, Rhodospirillum rubrum* and several *Mycoplasma* species. It is thus widely distributed among bacteria but it is not present in eukaryotic cells.

The function of the system is to phosphorylate sugars while translocating them from the medium into the cell. In Gram-negative species, the sugars thus phosphorylated are D-glucose, D-fructose D-mannose, mannitol, sorbitol, D-glucosamine, *N*-acetyl-D-glucosamine, 2-deoxy-D-glucose and β-glucosides. In Gram-positive species, the list probably includes also D-galactose, various pentoses, lactose, sucrose, trehalose, melibiose, maltose, melezitose, and glycerol. As far as is known, the position phosphorylated in monosaccharides is usually at the terminal carbon (5 or 6) but fructose is phosphorylated at carbon 1 and so is lactose.

The overall process common to all species is the following:

$$\text{phosphoenolpyruvate} + \text{enzyme I} \xrightleftharpoons{Mg^{2+}} \text{P-enzyme I} + \text{pyruvate}$$

$$\text{P-enzyme I} + \text{HPr} \rightleftharpoons \text{HPr-P} + \text{enzyme I}$$

$$\text{HPr-P} + \text{sugar} \xrightleftharpoons{\text{E II, factor III, } Mg^{2+}} \text{sugar-P} + \text{HPr}$$

The third reaction has been split up in the case of lactose in *Staphylococcus aureus* into

$$3 \text{ HPr-P} + \text{factor III}^{lac} \xrightleftharpoons{\text{EIIA}} \text{factor III}^{lac}\text{-P}_3 + 3 \text{ HPr}$$

$$\text{factor III}^{lac}\text{-P}_3 + 3 \text{ lactose} \xrightleftharpoons{\text{EIIB}^{lac}} 3 \text{ lactose-P} + \text{factor III}^{lac}$$

Hence one high-energy phosphate is required for the transport of one sugar molecule. Of the protein components involved, HPr (the heat-stable protein of molecular weight of about 9 000), enzyme I (molecular weight about 80 000) and factor III (molecular weight about 35 000) are soluble while enzyme II (molecular weight about 40 000) is membrane bound. There is a pronounced lipid requirement (for phospatidyl glycerol) of the phosphotransferase system in *Escherichia coli*.

The HPr protein appears to have an identical primary structure in the Gram-negative species and to be phosphorylated at a histidine residue.

There are at least three other group-translocating systems assumed to operate in nonelectrolyte transport.

1. The γ-glutamyltranspeptidase described in greatest detail in kidney cortex cells (Bodnaryk, 1972; Meister, 1973) uses intracellular glutathione as one substrate and extracellular amino acid (all natural amino acids except proline) as the second substrate, giving rise to γ-glutamylamino acid and cysteinylglycine. The γ-glutamyl derivative is then hydrolyzed to the originally transported amino acid and 5-oxoproline which is then used for the resynthesis of glutathione. The system operates rather inefficiently, requiring three molecules of ATP for the resynthesis of glutathione and hence for the transport of a single amino acid.

2. In *Escherichia coli*, adenine is transported with concomitant phosphoribosylation to adenylic acid (Hochstadt-Ozer and Stadtman, 1971).

3. In the brush-border membranes of the intestine, sucrose is transported only after previous splitting to its constituent monosaccharides. The system has been highly purified and incorporated in a functional state into an artificial phospholipid membrane (Storelli et al., 1972). It remains to be explored whether any of the numerous disaccharidases present in the intestine are capable of a similar feat.

4.2.2. Oxidoreductive systems

Substantial evidence has accumulated on the intimate coupling between the oxidation of D-lactate or the artificial substrate ascorbate-phenazine methosulfate and the transport of sugars, amino acids and some ions in specially prepared membrane vesicles of *Escherichia coli, Salmonella typhimurium, Pseudomonas putida, Proteus mirabilis, Bacillus megaterium, Bacillus subtilis, Micrococcus denitrificans, Mycobacterium phlei* and *Staphylococcus aureus* (see, e.g., Kaback, 1972; Konings et al., 1971).

Other substrates that may be used with different degrees of efficiency are α-glycerol phosphate and much less L-lactate, DL-α-hydroxybutyrate, and even formate.

The sugars thus involved are β-galactosides, galactose, arabinose, glucose-6-phosphate, gluconate and glucuronate; amino acids include all natural ones except glutamine (and asparagine ?), arginine, methionine and ornithine; the cations thus transported are possibly K^+ and Rb^+ (after treatment of the vesicles with valinomycin).

The postulated mechanism is depicted in an abridged form as follows:

$$\begin{array}{c} \text{lac} \quad \text{pyr} \\ \searrow \nearrow \\ C_{ox}S_I \longrightarrow C_{red}S_{II} \\ S_I \uparrow \downarrow \quad \text{lac} \quad \text{pyr} \quad \uparrow \searrow S_{II} \\ C_{ox_I} \longleftarrow \searrow \nearrow \longrightarrow C_{red_{II}} \\ \swarrow \nwarrow \\ cyt^{2+} \quad cyt^{3+} \end{array}$$

The theory of oxidoreductive transport is not without problems, particularly in that it requires different oxidative pathways for different transported solutes and in that it has an even more attractive alternative for most of the arguments rallied in its support. This other possibility is affiliated with the proton-motive force generated by a proton-extruding ATPase or a proton-generating oxidative system

TABLE 4.3. Binding proteins for nonelectrolytes isolated from various cell plasma membranes

Transported substrate	Source	Molecular weight	K_D in vitro (M)	K_T in vivo (M)	Reference	Note
L-Leucine	Escherichia coli	36 000	$7 \cdot 10^{-7}$	10^{-6}	Furlong and Weiner (1970)	Binds also isoleucine and valine
L-Leucine	Escherichia coli	36 000	10^{-6}	10^{-6}	Anraku (1968a, b, c)	The J protein
L-Histidine	Salmonella typhimurium	~26 000	$1.4 \cdot 10^{-7}$	$3 \cdot 10^{-8}$	Ames and Lever (1970, 1972)	
L-Histidine	Salmonella typhimurium	25 000	$1.5 \cdot 10^{-6}$	$3 \cdot 10^{-8}$	Rosen and Vasington (1971)	
L-Lysine	Escherichia coli	28 000	$1.5 \cdot 10^{-6}$	$5 \cdot 10^{-7}$	Rosen (1971)	Binds also arginine and ornithine
L-Arginine	Escherichia coli	27 700	$3 \cdot 10^{-8}$	$4 \cdot 10^{-7}$	Rosen (1973)	Specific
L-Arginine	Saccharomyces cerevisiae	5 000	$4 \cdot 10^{-4}$	10^{-5}	Opekarová et al. (1975)	
L-Glutamine	Escherichia coli	26 000	$3 \cdot 10^{-7}$	$8 \cdot 10^{-8}$	Weiner et al. (1970)	
L-Glutamate	Escherichia coli	30 000	10^{-6}	10^{-6} (Na+) 10^{-5} (no Na+)	Willis and Furlong (1974)	Contains an exposed S—S bond
L-Cystine	Escherichia coli	27 000	10^{-7}	10^{-8}	Berger and Heppel (1972)	
L-Phenylalanine	Comamonas	26 000	10^{-8}—10^{-7}	$2 \cdot 10^{-5}$	Kuzuya et al. (1971)	
L-Tryptophan	Neurospora crassa	200 000	$8 \cdot 10^{-5}$	$5 \cdot 10^{-5}$	Wiley (1970)	Binds also phenylalanine and leucine
D-Galactose	Escherichia coli	35 000	10^{-7} and 10^{-4}	$4 \cdot 10^{-7}$	Boos et al. (1972)	Exists in two conformations
D-Glucose	Saccharomyces cerevisiae	40 000	$1.1 \cdot 10^{-3}$	$7 \cdot 10^{-3}$	Horák and Kotyk (1973)	Lipoprotein
D-Glucose	Rat kidney cortex	60 000	$6.7 \cdot 10^{-7}$	$7 \cdot 10^{-5}$	Thomas (1973)	Binds also NEM and phlorizin

Transported substrate	Source	Molecular weight	K_D in vitro (M)	K_T in vivo (M)	Reference	Note
D-Glucose	Rabbit jejunum	2×120 000	10^{-3}	10^{-3}	Storelli et al. (1972)	Sucrase-isomaltase
D-Glucose	Aspergillus nidulans	29 200	$5.3 \cdot 10^{-4}$	10^{-4}	Desai and Modi (1975)	
L-Arabinose	Escherichia coli	35 000	$2.2 \cdot 10^{-6}$	10^{-6}	Hogg and Englesberg (1969); Schleif (1969)	High-affinity system
D-Ribose	Salmonella typhimurium	31 000	$3.3 \cdot 10^{-7}$	—	Aksamit and Koshland (1972)	
D-Ribose	Escherichia coli	—	$2 \cdot 10^{-6}$	—	Hazelbauer and Adler (1971)	Involved in chemotaxis
β-Galactosides	Escherichia coli	30 000	$7 \cdot 10^{-5}$	—	Jones and Kennedy (1968)	The M protein
Maltose	Escherichia coli	40 000	$1.5 \cdot 10^{-6}$ and 10^{-5}	—	Kellermann and Szmelcman (1974)	Possibly in two forms
Glucose-1-phosphate	Agrobacterium tumefaciens	35 000 (I) 43 000 (II)	$8 \cdot 10^{-7}$ $1.3 \cdot 10^{-6}$	$4.5 \cdot 10^{-6}$	Fukui and Isobe (1973)	
D-Glucose and L-histidine	Intestinal brush border	55 000	—	—	Faust and Shearin (1974)	Na$^+$-activated; separate binding of glucose and histidine
Thiamine	Escherichia coli	42 000	$3 \cdot 10^{-7}$	—	Griffith et al. (1971)	
Thiamine	Escherichia coli	36 000	$2 \cdot 10^{-8}$	—	Nishimune and Hayashi (1973)	
Riboflavin	Escherichia coli	48 000	$3 \cdot 10^{-5}$	—	Griffith et al. (1971)	
Cyanocobalamine	Escherichia coli	22 000 or 200 000	$5 \cdot 10^{-7}$	10^{-8}	Taylor et al. (1972)	

and the accompanying membrane potential. One can easily visualize that the oxidation of various substrates simply gives rise to either a pH gradient across the membrane or to a membrane potential (negative inside) which can serve as driving forces for the translocation of nonelectrolytes.

4.4.3. Binding proteins

Several years ago the most promising way to extracting "molecular-level" information on the transport of solutes across membranes appeared to be the solubilization of membrane components that would bind the substrate without altering it and the extraction of which would reduce the transport capacity of the cell *in vivo*. Particularly due to the elegant method of osmotic shock (*cf.* p. 57) it has been possible to solubilize and purify a number of such binding proteins from bacteria. Chemical methods of extraction had to be used for obtaining such binding proteins from eukaryotic cells and tissues. The present state of affairs is summarized in Table 4.3.

The role of the binding proteins is by no means unequivocally accepted. It is very likely that most of them represent indeed essential components of the corresponding transport systems — in many cases there is evidence from gene mutations that this is so. In a most remarkable review of the subject, Slayman (1973) lists no less than 199 mutations affecting transport in microorganisms plus several score in multicellular organisms. Of these, at least seven involve the structural genes for the various binding proteins.

However, what the binding proteins do in transport remains unclear. It is possible, although not widely believed, that some of them might actually act as carriers (particularly those of large molecular weight from nonbacterial sources) but it is rather likely that the function is to recognize the substrate and transmit it further to one of the "effector" molecules which may include: (*1*) chemotactic proteins bringing about movement toward a suitable substrate concentration (in motile bacteria) (Hazelbauer and Adler, 1971); (*2*) carrier proteins which actually translocate the substrate across the membrane (with or without requirement for energy), (*3*) effectors of pinocytotic invaginations in protozoan as well as other cells.

More information on binding proteins can be found on p. 294 and in Kotyk and Janáček (1975).

4.5. DISTRIBUTION AND ROLE OF NONELECTROLYTE TRANSPORTS

There are two main groups of compounds that have been studied in considerable detail, *viz.* the sugars and the amino acids. In addition to these, some data are available on the transport of purines and pyrimidines, polyols, vitamins, organic acids and small organic molecules (*cf.* Kotyk and Janáček, 1975).

All the specific, saturable, systems appear to aid in providing nutrition for the cells while the low-molecular waste products probably diffuse from cells by nonsaturable processes.

In all prokaryotic cells, the major groups are transported by energy-requiring systems against gradients of concentration of up to 10 000 : 1, depending on species and substrate. In the case of sugars, the energy may come from ATP but it does so usually indirectly through coupling with a gradient of H^+ ions which drive the solute inward. It may also derive from an oxidoreductive system, either by direct participation of the system in the transport of the nonelectrolyte or by the formation of a suitable membrane potential. Finally, a phosphotransferase system can be of importance.

With amino acids, the situation is not as varied but not any clearer for that. Oxidative energy is involved with some amino acids but ATP is required for others. Moreover, some amino acid transports even in prokaryotes require Na^+ as a symport cation, the specificity of the requirement differing from species to species.

While monosaccharides use carriers with a relatively broad specificity, some disaccharides and particularly amino acids are transported by systems with specificities that may be restricted to a single or to two chemical individuals. Still, there is a great deal of overlapping and one amino acid may be transported by as many as four different systems in a given species.

In lower eukaryotes (yeasts and fungi) the specificity of uptake is rather similar to that in bacteria but several major differences may be discerned in the involvement of energy. Thus, monosaccharides are transported in some species only by mediated diffusion while in others an active transport is involved. The energy required for this transport may be derived from H^+ gradients but possibly a high-molecular weight polyphosphate may have this function is some fungi (Kulaev, 1975).

In yeasts, amino acids are transported by multiple carriers against high intracellular concentrations but they cannot return to the outside even in exchange for an amino acid present in the medium. Moreover, a pronounced trans inhibition (like in *Streptomyces*) is found both in yeasts and fungi, resulting in a decreased uptake rate when high intracellular concentrations are reached.

There is relatively little information on nonelectrolyte transport in plants with the exception of algae where proton symports have been described for monosaccharides, energy being derived either from ATP or from the photosynthetic cycle. In higher plants, transmembrane movement appears to be usually by mediated diffusion but active transports are also known.

Animal tissues can be divided into several groups in this context. In epithelia, such as intestinal or kidney brush border, most monosaccharides and amino acids are transported in symport with Na^+, although probably primary active transport may also exist there. In muscle, Ehrlich ascites cells and erythrocytes, monosaccharides use mediated diffusion systems while amino acids are generally transported in conjunction with one or more sodium ions. In the extreme case of mammalian anucleate erythrocytes, both monosaccharides and amino acids are translocated by a mediated diffusion process.

The time is hardly ripe to make generalizations for lack of comparative data but it appears justified to conclude that active transports (either primary or secondary) have developed (or survived, as the case may be) in cells that depend on variable external environment to keep alive, or in cells that specialize in such transport within a higher-order framework of a differentiated organism. On the contrary, cells or organs that are not being exposed to substantial changes in the external concentration of nutrients (cultivated yeasts as well as sarcosomes) have lost this ability.

SYNOPSIS

Uncharged molecules are transported across biological membranes either by simple diffusion or by various mediated processes.

Simple diffusion, obeying Fick's laws of diffusion, proceeds dither across the lipid parts of membranes (most organic molecules, drugs) or, rarely, across water-filled channels, possibly only of statistical nature (primarily water).

Mediated processes require the presence of a specific membrane component which recognizes the substrate and either translocates it itself or transmits it to another membrane component, the carrier proper, for transmembrane movement. These processes are characterized by their saturability at high substrate concentrations and usually serve in the transport of sugars, amino acids and other metabolites.

The mediated processes either proceed without participation of metabolic energy (facilitated diffusion) or in coupling with a chemical reaction, such as splitting of ATP (primary active transport) or in coupling with downhill flow of a driving ion, usually H^+ in microorganisms and some plants, or Na^+ in animal tissues and some plants (secondary active transport).

There may be multiple binding sites on the recognition or carrier proteins.

A special type of transport is that by group translocation wherein the chemical nature of the substrate is altered during the transport process (typically the sugar phosphorylation by bacterial phosphotransferases).

5. TRANSPORT OF IONS

"...and perhaps electrical Attraction may reach to such small distances, even without being excited by Friction."
   Isaac Newton, *Opticks*

5.1. EQUILIBRIA OF IONS

5.1.1. A simple membrane equilibrium and membrane potentials

We can begin the discussion of ion equilibria by considering a particularly simple practical instance. Let there be two solutions, say 10 mM and 100 mM potassium chloride, separated by a cation-permeable membrane. The membrane may be made, e.g., of a cation-exchange resin and contain pores lined with fixed negative charges. If the pores are sufficiently narrow and the density of fixed charges high, only potassium ions can move through the membrane from one negatively charged site to another, whereas chloride anions are prevented by electrostatic repulsion from entering the pores. In the absence of an external circuit, connecting the two solutions via electrodes and metallic conductors, the system will achieve an equilibrium with respect to ions. Water will not be in equilibrium, being osmotically driven from the more dilute toward the more concentrated solution; its equilibrium, however, can be easily achieved by applying a suitable hydrostatic pressure to the concentrated solution or adjusting the osmolarity of the dilute one with a nonpermeating nonelectrolyte.

Equilibria of the two ion species, however, will be of widely different types. The equilibrium of chloride ions, a result of the membrane impermeability, is of a mechanic-electrostatic type and without a drastic change in the membrane structure there is no net transport of chloride ions across the membrane, as well as no exchange observable with tracers. The equilibrium of potassium cations, on the other hand, is a thermodynamical one, being the result of a mutual balance of active tendencies of the system. A very small amount of potassium ions has actually crossed the membrane during the equilibration process charging the capacity of the system to an electrical potential, the gradient of which is exactly equivalent to the concentration gradient of the potassium ions and thus prevents their further diffusional transfer. Net transfer of potassium ions from one side of the membrane to the other is thus not a spontaneous process and no work can be drawn from such a transfer. Hence the partial molal free energy, the so-called electrochemical potential of potassium ions (the maximum amount of work which would be obtained by transferring one mole of potassium ions to some chosen standard state) must be the same in the two solutions. The electrochemical potential may be expressed by

$$\tilde{\mu} = \mu_0 + RT \ln a + zF\varphi \tag{5.1}$$

where μ_0 depends neither on the concentration of the ion nor on the electrical potential, but rather on the nature of the solvent; R is the gas constant, equal to 8.314 J mol^{-1} K^{-1} (or voltcoulomb mol^{-1} K^{-1}, joule being equal to voltcoulomb) and T the absolute temperature in degrees of Kelvin. The activity a may be considered as the molar concentration of the ion, corrected for mutual interactions of all ions; $a = fc$, where f is the activity coefficient, in very dilute solutions equal to 1, in ordinary physiological salines having the value of about 0.76. The number of elementary charges per ion or its valency is denoted by z and is a signed quantity (+1 for K$^+$, −1 for Cl$^-$, +2 for Ca^{2+}, etc.), and F is the Faraday number, approximately 96 500 coulomb per mole of univalent ion or per gramequivalent of a polyvalent ion. Finally, φ is the electrical potential; its difference between two points corresponds to the amount of electrical work performed when a unit charge is transferred between the two points. If the charge is 1 coulomb and the potential difference 1 volt, the electrical work done is one voltcoulomb, or joule. Denoting the values corresponding to the concentrated solution by subscript i and those in the dilute solution by o, the equilibrium of an ion of valency z is described by

$$\mu_{0o} + RT \ln a_o + zF\varphi_o = \mu_{0i} + RT \ln a_i + zF\varphi_i. \quad (5.2)$$

Water being the solvent in both solutions, $\mu_{0i} = \mu_{0o}$ and

$$\varphi_i - \varphi_o = \frac{RT}{zF} \ln \frac{a_0}{a_i} = 2.303 \frac{RT}{zF} \log \frac{a_0}{a_i} \quad (5.3)$$

or, if we neglect the difference between the activity coefficients and cancel them out from the ratio of the activities,

$$\varphi_i - \varphi_o = 2.303 \frac{RT}{zF} \log \frac{c_o}{c_i}. \quad (5.4)$$

The values of the coefficient $2.303RT/F$ for various temperatures are given in Table 5.1. In the particular case discussed above, $c_o = 10$ mM, $c_i = 100$ mM, $\log(c_o/c_i) = -1$ and $z = +1$; hence the electrical potential difference across the membrane, $\varphi_i - \varphi_o$, will be -58.2 mV at 20 °C, -61.6 mV at 37 °C, etc.

TABLE 5.1. Values of $2.3029RT/F$ in mV at different temperatures. Calculated for $R = 8.3147 \cdot 10^3$ mV . C . mol^{-1} . K^{-1}, $F = 96\,490$ C . mol^{-1} and $T = (t + 273.2)$ °C

t	mV	t	mV
0	54.2	25	59.2
5	55.2	30	60.2
10	56.2	35	61.2
15	57.2	37	61.6
20	58.2		

The equilibrium electrical potential difference (the Nernst–Donnan potential) between two solutions can be measured by connecting a suitable millivoltmeter to the two solutions by electrodes. The electrical potential difference between the electrode and the solution is, in general, a function of the solution composition. In Fig. 5.1**A**, **B**, two limiting cases are shown, the electrodes reacting reversibly either with the ion which is in thermodynamic equilibrium or with the impermeant ion. To measure directly the electrical potential difference between two solutions one must use a device which does not contribute by new potential differences to the circuit. This is usually satisfied by reference electrodes (any electrodes immersed in a medium of constant composition, calomel electrodes being most popular), the solutions of which are connected with the measured

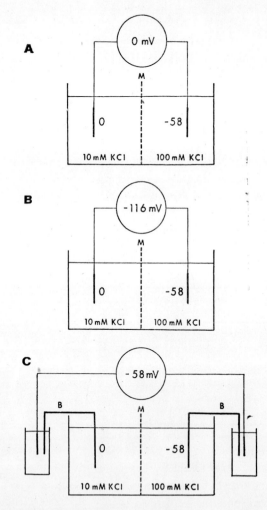

Fig. 5.1. Electrical potential difference measured in a circuit with a cation-permeable membrane M and (**A**) cation-sensitive electrodes (potassium amalgam or potassium-sensitive glass, etc.), (**B**) anion-sensitive electrodes (silver-silver chloride), (**C**) bridges and reference electrodes.

solutions via suitable bridges. The bridges contain a concentrated salt solution with oppositely charged ions of the same mobility (e.g., potassium chloride), the solution being suitably immobilized by, say, an agar gel. The liquid junctions between the bridges and solutions thus develop only small junction potentials and in symmetrical arrangement their difference is likely to be still smaller; hence the possibility to measure the true difference of the electrical potential across a membrane (the so-called membrane potential) as shown in Fig. 5.1C.

The last principle mentioned is of particular importance in biology: the so-called microelectrodes, introduced by Ling and Gerard (1949) are, in fact, microbridges pulled from borosilicate glass and filled usually with $3\,M$ KCl. Using suitable micromanipulators they can be introduced into living cells to measure electrical potential differences across cell membranes.

It is of some interest to calculate the amount of electrical charge (a departure from electroneutrality) responsible for a membrane potential. Cell membrane potentials are of the order of 100 mV or 10^{-1} V; the membrane capacity being about 1 microfarad per cm^2 or 10^{-6} F cm^{-2}, the charge per cm^2 membrane may be calculated from the well-known condenser formula charge = capacity · voltage, as 10^{-7} coulombs per cm^{-2}, which corresponds by the Faraday number (about 96 500 coulomb per mol) to some 10^{-12} mol of an univalent ion. This amount of ions is certainly not detectable by analytical techniques. Presently we will encounter the problem of how such charges are distributed in space (in connection with the distribution of the electrical potential in the so-called diffuse double layers).

The example of the membrane equilibrium discussed above is an especially simple instance of the important Gibbs−Donnan equilibria (with only one impermeant anion and one impermeant cation present); we shall now derive formulae governing Gibbs−Donnan equilibria in a more general case.

5.1.2. Gibbs—Donnan equilibrium

Fig. 5.2 shows a system in which a slightly more complex Gibbs−Donnan equilibrium is established than that discussed in the previous section. The system consists of two compartments; the right one is enclosed by rigid elastic walls (so that a hydrostatic pressure in excess of the atmospheric pressure may be attained in it) and contains nondiffusible anions A^{n-} in the equivalent concentration $[A^-] = n[A^{n-}]$, where $[A^{n-}]$ is the concentration in moles per unit volume. It is immaterial how the nondiffusible anions are restrained from leaving the compartment; they may represent a part of a rigid meshwork or the membrane between the two compartments may be impermeable to them. The solution in the right-hand compartment may be called the Donnan phase. The solution in the left-hand compartment could, of course, also contain nondiffusible anions or cations, without making the derivation of equilibrium relations much more complicated. In our case it contains only diffusible ions, the potassium cation and the chloride anion.

Equilibrium relations of the system can be derived using the conditions of equality of electrochemical potential in the two phases (see eq. 5.2):

$$\mu_{0K^+} + RT \ln a_{K^+_o} + F\varphi_o = \mu_{0K^+} + RT \ln a_{K^+_i} + F\varphi_i,$$
$$\mu_{0Cl^-} + RT \ln a_{Cl^-_o} - F\varphi_o = \mu_{0Cl^-_i} + RT \ln a_{Cl^-_i} - F\varphi_i \quad (5.5)$$

and the condition of electroneutrality in each of the two phases:

$$a_{K^+_o} - a_{Cl^-_o} = 0,$$
$$a_{K^+_i} - a_{Cl^-_i} - [A^-] = 0. \quad (5.6)$$

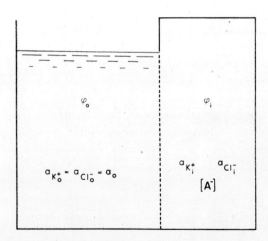

Fig. 5.2. A system in Gibbs—Donnan equilibrium. $[A^-]$, equivalent concentration of a nondiffusible anion; a's, activities of individual ions, φ's, electrical potentials in the two solutions.

From conditions (5.5) it may be immediately derived that

$$\varphi_i - \varphi_o = \frac{RT}{F} \ln \frac{a_{K^+o}}{a_{K^+i}} = \frac{RT}{F} \ln \frac{a_{Cl^-i}}{a_{Cl^-o}} \tag{5.7}$$

or

$$\frac{a_{K^+i}}{a_{K^+o}} = \frac{a_{Cl^-o}}{a_{Cl^-i}} = r \tag{5.8}$$

where r is called the Donnan ratio.

Combining eq. (5.6) and (5.8) and solving the quadratic equations obtained the concentrations of diffusible ions in the Donnan phase may be expressed as functions of the concentration of the nondiffusible anion and of the concentration in the external medium:

$$a_{K^+i} = \frac{1}{2}(\sqrt{[A^-]^2 + 4a_0^2} + [A^-]),$$

$$a_{Cl^-i} = \frac{1}{2}(\sqrt{[A^-]^2 + 4a_0^2} - [A^-]) \tag{5.9}$$

where

$$a_0 = a_{K^+o} = a_{Cl^-o}.$$

The osmotic pressure difference between the two phases can be expressed by using van't Hoff's formula $\pi = RTa$ as

$$\Delta\pi = RT\left(\frac{1}{n}[A^-] + a_{K^+i} + a_{Cl^-i} - 2a_0\right) \tag{5.10}$$

and combination of eq. (5.10) with equations (5.9) yields

$$\Delta\pi = RT\left(\frac{1}{n}[A^-] + \sqrt{[A^-]^2 + 4a_0^2} - 2a_0\right). \tag{5.11}$$

An equally large increment of hydrostatic pressure develops in the Donnan phase to achieve water equilibrium in the system.

Let us now assume that, in addition to potassium and chloride ions, diffusible calcium ions will be present in the system, so that an additional condition of equality of the electrochemical potential will apply:

$$\mu_{0Ca^{2+}} + RT \ln a_{Ca^{2+}o} + 2F\varphi_o =$$
$$= \mu_{0Ca^{2+}} + RT \ln a_{Ca^{2+}i} + 2F\varphi_i \tag{5.12}$$

from which it may be deduced that the Donnan ratio r, defined by eq. (5.8) may be also expressed by

$$r = \frac{\sqrt{a_{Ca^{2+}o}}}{\sqrt{a_{Ca^{2+}i}}} \tag{5.13}$$

or

$$\frac{a_{Ca^{2+}o}}{a_{Ca^{2+}i}} = r^2. \tag{5.14}$$

It may be seen, that the equilibrium ratio of the activities of a bivalent ion is the square of that for a univalent one.

A number of instructive numerical examples of Donnan distributions may be found in the monograph "Electrolytes and Plant Cells" by Briggs and co-workers (1961).

5.1.3. Diffuse electrical double layer

The electrical charge responsible for the electrical potential of a salt solution is represented by a deviation from electroneutrality, by a certain, generally very small, difference between the total amount of cationic and anionic charges in the solution. Unlike the hypothetical electrical charges of electrostatics, ions are subject not only to electrical forces but their electrical attraction to an oppositely charged surface is counteracted by their thermal motion. Interplay of the two factors results in a spatial extension of the charge in solution, the double layer of charges being of a diffuse character, as shown schematically in Fig. 5.3. As a result of this the electrical potential, too, is a function of the space coordinate; its value in the bulk of the solution is not attained by an abrupt step at the membrane surface but rather it decays slowly from the value at the surface which we may call φ_0 to the constant value in the distant bulk, which we may put arbitrarily equal to zero. The aim of the present section is to give an approximate description of this variation of the electrical potential in the diffuse or Gouy–Chapman double layer. Corrections for finite dimensions of ions and specific adsorption of ions at the surface which form the basis of Stern's theory of the double layer are omitted here, the present theory of cell membrane potentials being still too crude for these refinements. The electrical potential difference across the mobile part of the double layer is of importance in description of the electrokinetic phenomena and is commonly called the ζ-potential. A comprehensive and penetrating treatment of the double layer theory may be found in Overbeek (1952).

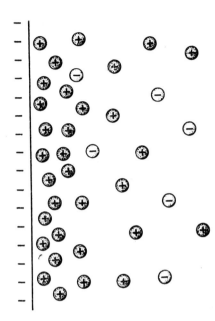

Fig. 5.3. Distribution of ions at a charged surface.

The interplay of electrical forces and random thermal movements of ions is mathematically expressed by a combination of Poisson's electrostatic equation with Boltzmann's statistical law. Poisson's equation is an expression of coulombic interactions and relates the divergence of the electrical potential to charge density. In the unidimensional case we are considering (the surface is presumably flat and all variables change in the direction of the x-axis only) the divergence is expressed by a second derivative and Poisson's equation may be written as

$$\frac{d^2\varphi}{dx^2} = -\frac{\varrho}{\varepsilon} \qquad (5.15)$$

where ϱ is the charge density and ε the absolute dielectric constant. The system of units used is the rationalized SI system (meter−kilogram−second−ampere; rationalized since factors like 4π have been excluded from equations where they were not expected to occur from the geometry). The absolute dielectric constant, ε, is equal to the relative dielectric constant, ε_r (equal to 1 for a vacuum and about 80 for water), multiplied by ε_0, the electric constant, depending only on the choice of units:

$$\varepsilon = \varepsilon_r \varepsilon_0. \tag{5.16}$$

Boltzmann's statistical law relates the concentration (rigorously activity) of ion j at the point where the electrical potential is φ to its concentration in the distant bulk of the solution, where the electrical potential is set equal to zero:

$$c_j = c_{j\infty} e^{-zF\varphi/RT}. \tag{5.17}$$

For simplicity, we will assume that only potassium chloride is present in the solution so that the concentration of potassium ions in the bulk is equal to the concentration of chloride ion

$$c_{K^+\infty} = c_{Cl^-\infty} = c. \tag{5.18}$$

Hence, Boltzmann's law written for potassium and chloride ions yields

$$c_{K^+} = c\, e^{-F\varphi/RT},$$
$$c_{Cl^-} = c\, e^{F\varphi/RT}. \tag{5.19}$$

The local charge density is equal to the difference between the local concentration of cations and anions multiplied by the Faraday number F

$$\varrho = F(c_{K^+} - c_{Cl^-}) \tag{5.20}$$

and may be introduced into Poisson's equation (5.15), giving

$$\frac{d^2\varphi}{dx^2} = -\frac{Fc}{\varepsilon}\left(e^{-F\varphi/RT} - e^{F\varphi/RT}\right). \tag{5.21}$$

Equation (5.21) can be solved exactly (as shown by Overbeek, 1952); for the present purpose of a rough estimate of the extent of the diffuse double layer, it is sufficient to solve it for the special case of small potentials φ; hence the approximative character of the solution when higher potential differences are involved. For small exponents the exponential functions may be approximated by the first two terms of its development in series according to Maclaurin's theorem

$$e^y \approx 1 + y.$$

Such an approximation of the two exponential terms in eq. (5.21) results in

$$\frac{d^2\varphi}{dx^2} = \frac{2F^2c}{RT\varepsilon}\varphi = \varkappa^2\varphi. \tag{5.22}$$

As may be easily verified by taking derivatives twice, the function

$$\varphi = \varphi_0 e^{-\varkappa x} \tag{5.23}$$

satisfies eq. (5.22) as well as the boundary conditions, giving $\varphi = \varphi_0$ for $x = 0$ and $\varphi = 0$ for infinite x. The thickness of the double layer can be arrived at by means of δ, the Debye thickness, which is defined as

$$\delta = \frac{1}{\varkappa} \tag{5.24}$$

i.e.

$$\delta = \sqrt{\frac{RT\varepsilon}{2F^2c}} = \sqrt{\frac{RT\varepsilon_r\varepsilon_0}{2F^2c}}. \tag{5.25}$$

As is obvious from eq. (5.23), δ characterizes the exponential decay of the electrical potential with distance x; for $x = \delta$, potential φ_0 decreases to $1/e$, i.e., to 36.8% of its value. Moreover, it can be shown that the capacity of the diffuse double layer is equal to the capacity of a plate condenser with the distance of plates equal to δ. According to the exact solution of eq. (5.21) the decay of the electrical potential in the diffuse layer is only roughly exponential (Overbeek 1952), but it may be still characterized by the quantity δ, commonly called the thickness of the diffuse double layer. The thickness δ may be calculated from the eq. (5.25), using $R = 8.31 \cdot 10^3$ kg m^2 s^{-2} K^{-1} kmol^{-1}; $F = 9.65 \cdot 10^7$ A s kmol^{-1}; $\varepsilon_0 = 8.85 \cdot 10^{-12}$ A^2 s^4 m^{-3} kg^{-1}. For 20 °C (= 293 K) this gives

$$\delta = 3.4 \cdot 10^{-11} \sqrt{\frac{\varepsilon_r}{c}}. \tag{5.26}$$

In the SI system the concentration emerges in kmol m^{-3}, which is numerically equal to concentrations in mol l^{-1} or molarity. Using ε_r for water equal to 80, eq. (5.26) gives the values of δ summarized for various concentrations of the uni-univalent electrolyte (KCl in the present example)

Concentration (kmol m^{-3})	Thickness δ (nm)
10^{-5}	96.0
10^{-4}	30.4
10^{-3}	9.6
10^{-2}	3.04
10^{-1}	0.96

When various ions of molarity c_j and valency z_j are present in the solution, a more general form of formula (5.25) can be used to calculate the thickness of the diffuse double layer

$$\delta = \sqrt{\frac{RT\varepsilon}{F^2 \Sigma z_j^2 c_j}} \qquad (5.27)$$

whence it follows that the higher the valency of the ions present in the solution, the more compressed the diffuse double layer will be.

Although the above considerations were carried out for a phase boundary between a solid (e.g., a metal) and an ion-containing solution they apply equally well to the dependence of the electrical potential on the space coordinate near the contact of two ionic systems, where double layers are formed. Double diffuse double layers necessarily exist at the boundaries between a membrane and solutions, as expressed by Agin (1967): "The existence of a diffuse layer outside the membrane is at least implicitly accepted by most, but for anything except a metal, a diffuse layer must exist internally as well". Hence the situation encountered at the membrane boundaries is likely to be analogous to that at the interface between two inmiscible liquids, treated theoretically, e.g., by Verwey and Overbeek (1948; see also Overbeek, 1952). If the boundary potential drop due to oriented dipoles is neglected, the course of the electrical potential at the interface may resemble that shown in Fig. 5.4. When the double diffuse double layer is as a whole electrically neutral

$$\int_{-\infty}^{0} \varrho_1(x)\,dx + \int_{0}^{\infty} \varrho_2(x)\,dx = 0 \qquad (5.28)$$

from which it may be shown, using Poisson's equation (5.15) that

$$\varepsilon_1 \left(\frac{d\varphi_1}{dx}\right)_{x=0} = \varepsilon_2 \left(\frac{d\varphi_2}{dx}\right)_{x=0} \qquad (5.29)$$

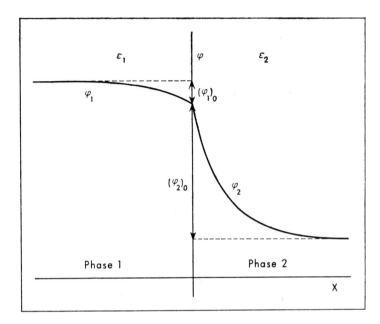

Fig. 5.4. Distribution of the electrical potential at the interface of two liquid phases. (Adapted schematically from Verwey and Overbeek, 1948.)

i.e., the discontinuity in the derivative of the potential at the interface is in this case entirely determined by the different dielectric constants (Verwey and Overbeek, 1948).

To account for the net fixed charge, resulting from the dissociation of ionizable polar head groups of phospholipids, Ciani and co-workers (1973) assume a uniform distribution of the charge at the membrane surface with density σ. Equation (5.29) then becomes

$$\varepsilon_1 \left(\frac{d\varphi_1}{dx}\right)_{x=0} - \varepsilon_2 \left(\frac{d\varphi_2}{dx}\right)_{x=0} = \sigma. \tag{5.30}$$

They also take into account the possible existence of a discontinuous change of electrical potential

$$\varphi_{1(x=0)} - \varphi_{2(x=0)} = a \tag{5.31}$$

resulting from the uniform distribution of a sheet of dipoles on the membrane surface; a is assumed to be a constant, independent of the applied electrical field as well as of the composition of the solutions.

Estimation of the thickness of the internal diffuse double layer within the membrane would involve some uncertainty, the dielectric constant and the concentrations in the regions or channels through which ions permeate not being known.

5.2. ELECTRODIFFUSION AND MEMBRANE POTENTIALS

5.2.1. Introduction

After the equilibrium membrane potentials treated before we shall now deal with potential differences on membranes across which a net ion transport takes place. Unlike the equilibrium potentials, the nonequilibrium ones will be seen to be functions not only of ion concentrations, but also of ion mobilities and partition coefficients between the medium and the membrane. By assuming zero mobilities of all but one ion the formulae describing nonequilibrium potentials must be, of course, reduced to the correct equilibrium relation for the given ion.

For the sake of simplicity it will be assumed in the following that steady-state conditions prevail, i.e., that fluxes of ions are constants independent of time. Those interested more deeply in the kinetic phenomena especially on excitable membranes are referred to the scholarly treatments of this subject by Cole (1965) and Agin (1972). The steady state required by the following derivations will be the result of both the stable properties of the membrane and the stable concentrations at the membrane. It will be immaterial for the derivations how the stability of the concentrations at the membrane will be ensured, three possibilities of achieving steady concentrations of sodium and potassium ions at the two sides of a membrane being shown in Fig. 5.5. In the first case (**a**), the concentrations are steady due to the vastness of the two reservoirs; such a situation is most simply obtained in experiments with artificial membranes. In the second case (**b**), which may be typical of a number of nonpolar cells, the steady concentrations are preserved by the operation of a one-to-one coupled electroneutral sodium-potassium exchange pump. Note that although the net flow of the two ions across the membrane is zero, the passive flows of ions, representing a transport of electrical charge, are not zero. In the third case (**c**), there is a large reservoir at one side of the membrane, whereas the composition of the compartment at the other side is kept constant by a pumping mechanism at the opposite pole of the cell. A situation of this kind may be found in ion-transporting epithelial layers.

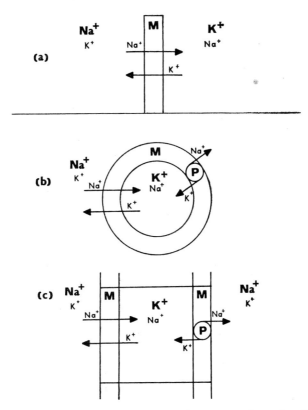

Fig. 5.5. Various mechanisms ensuring steady concentrations of sodium and potassium ions at a membrane. **M** Membrane, **P** pump. The size of the concentration symbols corresponds roughly to the concentration levels. Further explanations in the text.

To derive formulae describing the differences of the electrical potential between the solutions at the two sides of the membrane we will consider first the electrodiffusion in the membrane proper and neglect the steps of electrical potential at its two boundaries. The membrane (or rather the regions of the membrane through which the ions permeate) will be treated as a continuous medium. Such an approach, when applied to the thin cell membrane, is certainly not free of theoretical objections; if the penetration of an ion across a layer corresponds to only a small number of transitions across energy barriers, a discontinuous approach based on the absolute rate theory rather than a continuous one appears to be warranted. Although there are already some promising attempts in this direction (e.g., Ciani, 1965), the practical applicability of the theory may be still rather remote. Following Planck, Henderson, and Goldman we shall calculate the difference of the electrical potential between two planes inside a continuous layer as a function of ion concentrations at the two planes and of the ion mobilities in the layer. In this way we shall arrive at an approximate value of the potential difference across the membrane proper, provided that various simplifying assumptions be introduced and the concentrations of ions at the surface, just inside the membrane, be known (see Fig. 5.6).

Later we shall consider the difference of electrical potential across the two boundaries, between the membrane and the adjacent solu-

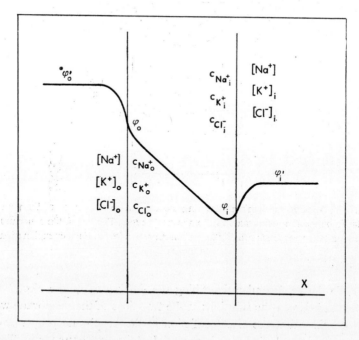

Fig. 5.6. Electrical potential differences across a membrane and its boundaries, the concentration symbols shown being as used in the appropriate equations in the text. The potential profile is shown highly schematically.

tions, derive formulae for the potential difference across the whole membrane and inquire how this picture is changed by the presence of an electrogenic pump, i.e., of an active mechanism translocating a net electrical charge across the membrane.

5.2.2. The electrodiffusion equation— general considerations

When discussing the equilibrium membrane potentials we observed that the condition of thermodynamic equilibrium of an ion is that its electrochemical potential be the same at any point of the space available for its movement. From this observation one is led to the conclusion that the driving force of an ion flux is the gradient of its electrochemical potential in space. Even this assumption may represent only an approximation for the cell membrane, for it apparently requires both the condition of local electroneutrality (for the electrochemical potential to be defined at a point) and the condition of thermal equilibrium (for Einstein's interpretation of the diffusion coefficient $D = RTU$) to be satisfied if it is to be valid strictly (see Agin, 1967, for discussion and references).

When the electrochemical potential changes in space in one direction only, say, in the direction of the x-axis, it will be a function solely of time t and of x, and its gradient will be equal to the partial derivative with respect to x:

$$\text{grad } \tilde{\mu}(t, x) = \frac{\partial \tilde{\mu}}{\partial x}. \tag{5.32}$$

Thus the general transport equation due to Teorell (1953)

FLUX = MOBILITY . CONCENTRATION . DRIVING FORCE

will have for the transport of an ion j the form*

$$J_j = -c_j U_j \frac{\partial \tilde{\mu}_j}{\partial x}. \tag{5.33}$$

The electrochemical potential may be expressed by (see p. 244)

$$\tilde{\mu}_j = \mu_{j0} + RT \ln a_j + z_j F \varphi \tag{5.34}$$

---

* The minus sign stems from the fact that when the derivative is a positive quantity, $\tilde{\mu}$ increases in the direction of the positive x-axis and the flux proceeds in the opposite direction.

where $a_j = fc_j$ is the activity of the ion, its molar concentration multiplied by the activity coefficient. Introducing this expression into eq. (5.33) we obtain

$$J_j = -c_j U_j \left(RT \frac{\partial \ln a_j}{\partial x} + z_j F \frac{\partial \varphi}{\partial x} \right). \quad (5.35)$$

Now, the well-known identity from differential calculus $d \ln y = dy/y$ can be used, so that

$$d \ln a = \frac{da}{a} = \frac{d(fc)}{fc} = \frac{f}{f} \frac{dc}{c} = \frac{dc}{c}$$

and eq. (5.35) is changed into

$$J_j = -RTU_j \frac{\partial c_j}{\partial x} - z_j F U_j c_j \frac{\partial \varphi}{\partial x} \quad (5.36)$$

which is the familiar Nernst–Planck equation. The first term on the right-hand side describes pure diffusion, the second term is the expression for migration in an electrical field or electric drift. The Nernst–Planck differential equation of electrodiffusion may thus be considered as a superposition of Fick's and Ohm's laws. It may be noted that not the activity coefficient but only its dependence on x was neglected in the above derivation.

A rigorous derivation of a more general differential equation describing the ionic flux was carried out by methods of steady-state thermodynamics by Schlögl (1964). Schlögl's derivation is described in detail in section 5.2.3. The Schlögl equation may be written

$$J_j = c_j v - RTU_j \frac{\partial c_j}{\partial x} - z_j F U_j c_j \frac{\partial \varphi}{\partial x} - RTU_j \frac{\partial \ln f_j}{\partial x} -$$

$$- U_j M_j c_j \left(\frac{\bar{V}_j}{M_j} - \frac{\bar{V}_n}{M_n} \right) \frac{\partial p}{\partial x} \quad (5.37)$$

where $\bar{V}_j$ and $\bar{V}_n$ are the partial molal volumes and M_j and M_n the molecular weights of the ion and of the solvent, respectively, and v is the volume flow. Comparing the Nernst–Planck equation (5.36) with Schlögl's equation (5.37) it can be seen that the former neglects (apart from the mutual interactions between ions, omitted explicitly even in Schlögl's derivation) (1) flow of ion due to volume flow, (2) dependence of activity coefficient on the x-coordinate and (3) pressure diffusion described by the last term of the Schlögl equation. It may be seen that in pressure diffusion, ions with a specific volume greater than that of solvent molecules move in the direction of the negative pressure gradient, whereas those less voluminous move in the opposite direction.

Expressing the electrochemical potential or each of its components in J mol^{-1}, flux in mol cm^{-2} s^{-1} and concentration in mol cm^{-3}, mobility U is seen to have the dimension of cm^2 s^{-1} J^{-1} mol. In electrochemistry, mobility u is often used with the dimension of cm^2 s^{-1} V^{-1}. The relation between the two mobilities is

$$u_j = |z_j| FU_j.$$

The equation of electrodiffusion (5.36) relates the value of the flux to the values of differential quotients, which are not amenable to a direct experimental measurement; only finite differences of concentrations or of the electrical potential over finite distances can be measured. In order to establish the relation between the flux and the finite differences, electrodiffusion equation has to be integrated over a finite distance. There are, however, two unknown functions of x in the equation and an additional condition is required to carry out the integration. Two assumptions concerning the nature of these functions have been widely used. According to the first, originally applied by Henderson (1907, 1908), the concentration of each individual ion is a linear function of the distance; according to the other, due to Goldman (1943), such a linear function is represented by the electrical potential. (The first condition is satisfied in a situation for which it was formerly intended, i.e. in mixed boundaries, and hence it is not likely to be applicable to unmixed systems like membranes. The second condition will be seen to be satisfied in locally electroneutral systems when the ionic strength is a constant independent of the space coordinate and will be used as a simplifying assumption in derivations of Goldman's formulae, which can be used as approximations to describe cell membrane potentials in steady state.)

When, however, the aim of the calculation is to supply only a formula for the electrical potential difference across a continuous layer as a function of ionic concentrations at its boundaries, rather than expressions for fluxes of individual ions, there is a different, although less simple, approach. The mathematical procedure is due to Planck (1890) and involves only the assumption of local, or microscopic, electroneutrality. We will see later that the condition of local electroneutrality may not be satisfied in cell membranes (for a discussion see Agin, 1967). Still, it will be attempted in the following to describe Planck's derivation in a great detail both for its intrinsic merit (Cole, 1965, speaks about its beauty and power mostly lost in the numerous rederivations, which can be fully appreciated in the original) and to show under which conditions Planck's transcendental equation for the electrical potential difference (Planck, 1890) reduces to the explicit formula of Goldman (1943).

Thus the following three sections will be concerned, respectively, with the derivation of Schlögl's differential equation of electrodiffusion, of Henderson's formula and of Planck's formula for the difference of electrical potential in continuous ionic systems. Those interested more in derivations of formulae directly applicable to cell membranes may skip the following three sections.

5.2.3. Schlögl's (1964) derivation of the general differential equation of electrodiffusion

The derivation, based on the principles of steady-state thermodynamics of continuous systems, aims at calculating the entropy production in the liquid phase inside membrane pores. The liquid phase contains n components, the relative abundance of each component being expressed by its partial density $\varrho_j = m_j/V$, the "mean velocity" of each component being denoted by $\mathbf{v}_j$. The product $\varrho_j \mathbf{v}_j$ then corresponds to the mass flow of component j. "Barycentric velocity" (the velocity of the centre of mass) is defined as

$$\mathbf{v} \equiv \sum_{j=1}^{n} \varrho_j \mathbf{v}_j / \varrho \qquad (5.38)$$

where $\varrho = \Sigma \varrho_j$ and the relative movement of component j with respect to the movement of the common centre of mass is characterized by vector $\mathbf{M}_j$

$$\mathbf{M}_j = \varrho_j(\mathbf{v}_j - \mathbf{v}). \qquad (5.39)$$

From (5.38) and (5.39) it follows that

$$\sum_{j=1}^{n} \mathbf{M}_j = \sum_{j=1}^{n} \varrho_j \mathbf{v}_j - \sum_{j=1}^{n} \varrho_j \cdot \sum_{j=1}^{n} \varrho_j \mathbf{v}_j / \varrho = \sum_{j=1}^{n} \varrho_j \mathbf{v}_j - \varrho \sum_{j=1}^{n} \varrho_j \mathbf{v}_j / \varrho = 0. \qquad (5.40)$$

Assuming that no chemical reactions take place, the following equations of continuity hold

$$\frac{\partial \varrho_j}{\partial t} = -\operatorname{div}(\varrho_j \mathbf{v}_j) \qquad (5.41)$$

$$\frac{\partial \varrho}{\partial t} = -\operatorname{div}(\varrho \mathbf{v}) \qquad (5.42)$$

expressing the fact that no substance is lost – spatial change of flow is equal to temporal change of density. Furthermore, since ϱ (unlike ϱ_j) in condensed systems depends only little on composition, equation

$$\operatorname{div} \mathbf{v} \approx 0 \qquad (5.43)$$

is assumed to be approximately valid.

Assuming the system to be in thermal equilibrium, entropy production per unit volume of the system, σ, multiplied by absolute temperature (the so-called dissipation function) is given by

$$T\sigma = T\sigma_{\text{visc}} - \sum_{j=1}^{n} \mathbf{M}_j \operatorname{grad} \frac{\tilde{\mu}_j}{M_j} \qquad (5.44)$$

where σ_{visc} is the contribution to entropy production due to the dissipation of convective energy by friction, $\tilde{\mu}_j$ the electrochemical potential of component j, M_j its molecular weight and hence $\tilde{\mu}_j/M_j$ the electrochemical potential per unit of mass of component j. Using eq. (5.39), the second right-hand term of eq. (5.44) may be rewritten as

$$-\sum_{j=1}^{n} \mathbf{M}_j \operatorname{grad} \frac{\tilde{\mu}_j}{M_j} = -\sum_{j=1}^{n} \mathbf{v}_j \varrho_j \operatorname{grad} \frac{\tilde{\mu}_j}{M_j} + \mathbf{v} \sum_{j=1}^{n} \varrho_j \operatorname{grad} \frac{\tilde{\mu}_j}{M_j}. \qquad (5.45)$$

The sum in the second right-hand term of eq. (5.45) may be transformed using the following thermodynamical considerations:

The first law of thermodynamics for an open system which performs only volume work can be written as

$$dU = TdS - pdV + \sum_{j=1}^{n} \mu_j dn_j \qquad (5.46)$$

where μ_j is the chemical potential of component j. From the definition of Gibbs free energy

$$G \equiv U - TS + pV$$

a mathematical identity follows by differentiation

$$dG \equiv dU - S\,dT - T\,dS + V\,dp + p\,dV. \qquad (5.47)$$

Adding identity (5.47) to the first law (5.46) we obtain an expression for the first law in the form

$$dG = -S\,dT + V\,dp + \sum_{j=1}^{n} \mu_j\,dn_j. \qquad (5.48)$$

The Gibbs free energy is an extensive variable of state; if, at constant temperature and pressure, the number of moles of each component is increased by a certain factor, the Gibbs free energy of the system is increased by the same factor. Hence the Gibbs free energy is a homogeneous function of the first degree of the number of moles n_j and, according to Euler's theorem, may be written as

$$G = \sum_{j=1}^{n} n_j \left(\frac{\partial G}{\partial n_j}\right)_{p,T} = \sum_{j=1}^{n} n_j \mu_j. \qquad (5.49)$$

By differentiating eq. (5.49) we obtain

$$dG = \sum_{j=1}^{n} \mu_j\,dn_j + \sum_{j=1}^{n} n_j\,d\mu_j \qquad (5.50)$$

and, on subtracting eq. (5.48) from this, we have

$$\sum_{j=1}^{n} n_j\,d\mu_j = -S\,dT + V\,dp \qquad (5.51)$$

which is the well-known Gibbs–Duhem equation. Dividing this equation by V and observing that

$$\sum_{j=1}^{n} \frac{n_j}{V}\,d\mu_j = \sum_{j=1}^{n} \frac{n_j M_j}{V} \frac{d\mu_j}{M_j} = \sum_{j=1}^{n} \frac{m_j}{V} \frac{d\mu_j}{M_j} = \sum_{j=1}^{n} \varrho_j \frac{d\mu_j}{M_j}$$

we obtain

$$\sum_{j=1}^{n} \varrho_j \frac{d\mu_j}{M_j} = -\frac{S}{V}\,dT + dp. \qquad (5.52)$$

If the particles of component j are charged the gradient of their chemical potential balances in equilibrium the force exerted on them by the electrical potential gradient; in other words, their electrochemical potential $\tilde{\mu}_j = \mu_j + z_j F\varphi$ has the same value everywhere in the system and (5.52) may be written as

$$\sum_{j=1}^{n} \varrho_j \frac{d\tilde{\mu}_j}{M_j} = -\frac{S}{V} dT + dp + \sum_{j=1}^{n} \frac{\varrho_j z_j F}{M_j} d\varphi. \quad (5.53)$$

Defining $\varrho_{el} = \Sigma \varrho_j z_j F/M_j$ (the electrical space charge per unit volume) and proceeding from differentials to gradients, we obtain

$$\sum_{j=1}^{n} \varrho_j \operatorname{grad} \frac{\tilde{\mu}_j}{M_j} = -\frac{S}{V} \operatorname{grad} T + \operatorname{grad} p + \varrho_{el} \operatorname{grad} \varphi \quad (5.54)$$

where the first term on the right-hand side vanishes in thermal equilibrium. Under this condition, eq. (5.45) may be written as

$$-\sum_{j=1}^{n} \mathbf{M}_j \operatorname{grad} \frac{\tilde{\mu}_j}{M_j} = -\sum_{j=1}^{n} \mathbf{v}_j \varrho_j \operatorname{grad} \frac{\tilde{\mu}_j}{M_j} + \mathbf{v} \operatorname{grad} p + \mathbf{v}\varrho_{el} \operatorname{grad} \varphi. \quad (5.55)$$

This equation is equivalent with

$$-\sum_{j=1}^{n} \mathbf{M}_j \operatorname{grad} \frac{\tilde{\mu}_j}{M_j} = \operatorname{div}(\mathbf{v}p) + \varrho_{el} \operatorname{div}(\mathbf{v}\varphi) -$$

$$- \sum_{j=1}^{n} \operatorname{div}\left(\varrho_j \mathbf{v}_j \frac{\tilde{\mu}_j}{M_j}\right) - \sum_{j=1}^{n} \frac{\tilde{\mu}_j}{M_j} \frac{\partial \varrho_j}{\partial t} \quad (5.56)$$

as shown by the following consideration: $\operatorname{div}(\mathbf{v}p) = p \operatorname{div} \mathbf{v} + \mathbf{v} \operatorname{grad} p$ where $p \operatorname{div} \mathbf{v} = 0$ (from eq. 5.43) and, similarly, $\varrho_{el} \operatorname{div}(\mathbf{v}\varphi) = \varrho_{el}\varphi \operatorname{div} \mathbf{v} + \varrho_{el}\mathbf{v} \operatorname{grad} \varphi$ where again $\varrho_{el}\varphi \operatorname{div} \mathbf{v} = 0$, and, finally,

$$-\sum_{j=1}^{n} \operatorname{div}\left(\varrho_j \mathbf{v}_j \frac{\tilde{\mu}_j}{M_j}\right) = -\sum_{j=1}^{n} \varrho_j \mathbf{v}_j \operatorname{grad} \frac{\tilde{\mu}_j}{M_j} - \sum_{j=1}^{n} \frac{\tilde{\mu}_j}{M_j} \operatorname{div}(\varrho_j \mathbf{v}_j)$$

where the first term on the right-hand side appears in eq. (5.55), whereas the second is (by eq. 5.41) equal to $+\sum \frac{\tilde{\mu}_j}{M_j} \frac{\partial \varrho_j}{\partial t}$. (Hence it cancels out with the last term of eq. 5.56.)

Eq. (5.56) is now introduced into the dissipation function (5.44); integration over the volume between two cross-sections of the membrane at x and $x + \Delta x$ is carried out. The integration may be carried out over the whole membrane since outside the pores, in the matrix of the membrane, the integrands vanish. When the divergence theorem

$$\int_V \operatorname{div} \mathbf{v} \, dV = \int_S \mathbf{v} \, d\mathbf{S}$$

is used, integration yields

$$T\int_V \sigma \, dV = T\int_V \sigma_{\text{visc}} \, dV - \sum_{j=1}^{n} \int_V \frac{\tilde{\mu}_j}{M_j} \frac{\partial \varrho_j}{\partial t} \, dV +$$

$$+ \int_S p\mathbf{v} \, d\mathbf{S} + \varrho_{\text{el}} \int_S \varphi \mathbf{v} \, d\mathbf{S} - \sum_{j=1}^{n} \int_S \varrho_j \mathbf{v}_j \frac{\tilde{\mu}_j}{M_j} \, d\mathbf{S}. \qquad (5.57)$$

In a steady state, the second term on the right-hand side is considered to be negligibly small (representing only rather small temporal fluctuations). The surface integrals are replaced by integrals over the two cross-sections of the membrane (the edge area of thickness Δx being small) which, in their turn, are assumed to be equivalent to integrals over the corresponding temporally and spatially fluctuating equipotential surfaces. Hence the quantities p, φ and $\tilde{\mu}$, which correspond to the medium in the pores of the membrane and are functions of x only, can be factored out before the integration symbols. The remaining surface integrals have the values

$$\int_S \mathbf{v} \, d\mathbf{S} = \pm A\mathbf{v}$$

$$\int_S \varrho_j \mathbf{v}_j \, d\mathbf{S} = AJ_j M_j$$

where A is the membrane surface, $\mathbf{v}$ the barycentric velocity of the fluid in pores calculated over the whole membrane cross-section, J_j the molar flow of component j again expressed across the whole cross-section and M_j its molecular weight. The surface vectors $d\mathbf{S}$ being directed outwards from the volume of integration, their direction is opposite at x and at $x + \Delta x$. Hence the minus sign corresponds to distance x and the plus sign to $x + \Delta x$. Introducing the values of the surface integrals into eq. (5.57) and replacing the resulting differences $p_{x+\Delta x} - p_x$ with $\frac{dp}{dx} \Delta x$, etc., we obtain after division by the volume of the considered membrane layer, $A \Delta x$, the dissipation function per unit volume of the membrane

$$T\sigma = T\sigma_{\text{visc}} + \mathbf{v}\left(\frac{dp}{dx} + \varrho_{\text{el}} \frac{d\varphi}{dx}\right) - \sum_{j=1}^{n} J_j \frac{d\tilde{\mu}_j}{dx}. \qquad (5.58)$$

However, it may be shown, by "Gedanken" experiments using devices consisting of a closed (hence $\Sigma dU = 0$) isothermal ($T = \text{const.}$) system with one isobaric ($p = \text{const.}$) and one isochoric ($V = \text{const.}$) subsystem, that the whole rate of entropy production when ions are transported across membrane is accounted for by the dissipation function

$$T\sigma = -\sum_{j=1}^{n} J_j \frac{d\tilde{\mu}_j}{dx} \qquad (5.59)$$

(see Schlögl, 1964). Comparing eq. (5.58) and (5.59) we have

$$T\sigma_{\text{visc}} = v\left(-\frac{dp}{dx} - \varrho_{\text{el}}\frac{d\varphi}{dx}\right). \qquad (5.60)$$

At this stage, Schlögl defines the mean mass flow of component j as

$$M_j = \frac{1}{A}\int_S \varrho_j(\mathbf{v}_j - \mathbf{v})\,d\mathbf{S} = J_j M_j - \varrho_j \mathbf{v} \qquad (5.61)$$

or,

$$J_j = \frac{M_j}{M_j} + \frac{\varrho_j \mathbf{v}}{M_j}. \qquad (5.62)$$

Introducing (5.60) and (5.62) into (5.58) we get

$$T\sigma = v\left(-\frac{dp}{dx} - \varrho_{\text{el}}\frac{d\varphi}{dx}\right) + v\left(\frac{dp}{dx} + \varrho_{\text{el}}\frac{d\varphi}{dx}\right) -$$
$$- \sum_{j=1}^{n} \frac{M_j}{M_j}\frac{d\tilde{\mu}_j}{dx} - \sum_{j=1}^{n} \frac{\varrho_j \mathbf{v}}{M_j}\frac{d\tilde{\mu}_j}{dx} \qquad (5.63)$$

and using eq. (5.53) for an isothermal system

$$\sum_{j=1}^{n} \varrho_j \frac{d\tilde{\mu}_j}{M_j} = dp + \varrho_{\text{el}}\,d\varphi \qquad (5.64)$$

the second and fourth term on the right-hand side of eq. (5.63) are seen to cancel out and the dissipation function takes the form

$$T\sigma = v\left(-\frac{dp}{dx} - \varrho_{\text{el}}\frac{d\varphi}{dx}\right) - \sum_{j=1}^{n} \frac{M_j}{M_j}\frac{d\tilde{\mu}_j}{dx}. \qquad (5.65)$$

The expression may be seen to represent the sum of the products of fluxes with their conjugate local (x-dependent) driving forces. We can now write the linear relations of steady-state thermodynamics, according to which each flow is linearly dependent not only on its conjugate force, but also on forces conjugate to flows with which the former flow interacts. Attention is called here by Schlögl to the Curie principle, according to which processes of unlike tensorial order do not interact in isotropic systems. The production of entropy by inner frinction σ_{visc} may be represented as a sum of two second-order tensors (even if their tensorial character is obscured by an averaging integration procedure in the present derivation) whereas gradients of the electrochemical potentials are obviously vectors, *i.e.* first-order tensors. Hence the barycentric velocity expressed across the whole cross-section of the membrane (the volume flow) will be given by the simple hydrodynamic relation

$$\mathbf{v} = k\left(-\frac{\mathrm{d}p}{\mathrm{d}x} - \varrho_{\text{el}}\frac{\mathrm{d}\varphi}{\mathrm{d}x}\right) \tag{5.66}$$

whereas the flows of individual substances j will be related only to gradients of the electrochemical potentials per unit mass

$$\mathsf{M}_j = -\sum_{j=1}^{n} L_{jk}\frac{1}{M_k}\frac{\mathrm{d}\tilde{\mu}_k}{\mathrm{d}x} \quad (j=1,\ldots,n). \tag{5.67}$$

According to eq. (5.40)

$$\sum_{j=1}^{n} \mathsf{M}_j = 0 \tag{5.68}$$

and hence, as follows on introducing (5.67) into (5.68)

$$\sum_{j=1}^{n} L_{jk} = 0 \quad (k=1,\ldots,n). \tag{5.69}$$

In view of Onsager's reciprocity relations, $L_{jk} = L_{kj}$, also the condition,

$$\sum_{k=1}^{n} L_{jk} = 0 \quad (j=1,\ldots,n) \tag{5.70}$$

is valid, or

$$\sum_{k=1}^{n-1} L_{jk} = -L_{jn} \quad (j=1,\ldots,n) \tag{5.71}$$

where $j = 1, \ldots, n-1$ denote solutes and $j = n$ the solvent. Hence eq. (5.67) may be written as

$$\mathsf{M}_j = -\sum_{k=1}^{n} L_{jk}\left(\frac{1}{M_k}\frac{\mathrm{d}\tilde{\mu}_k}{\mathrm{d}x} - \frac{1}{M_n}\frac{\mathrm{d}\tilde{\mu}_n}{\mathrm{d}x}\right) \quad (j=1,\ldots,n-1). \tag{5.72}$$

Neglecting now any possible interactions between individual solutes,

Schlögl sets $L_{jk} = 0$ with the exception of L_{jj} (and L_{jn} which, however, does not appear in eq. (5.72) explicitly). Hence

$$M_j = -L_{jj}\left(\frac{1}{M_j}\frac{d\tilde{\mu}_j}{dx} - \frac{1}{M_n}\frac{d\tilde{\mu}_n}{dx}\right) \quad (5.73)$$

and, using eq. (5.62), the molar flow J_j of the ionic solute may be expressed as

$$J_j = c_j v - \frac{L_{jj}}{M_j}\left(\frac{1}{M_j}\frac{d\tilde{\mu}_j}{dx} - \frac{1}{M_n}\frac{d\tilde{\mu}_n}{dx}\right) \quad (5.74)$$

where $c_j = \varrho_j/M_j$ is the molar concentration.

Writing now the electrochemical potential as

$$\tilde{\mu}_j = \mu_{0j} + \bar{V}_j p + z_j F\varphi + RT \ln a_j$$

where $\bar{V}_j$ is the partial molal volume and a_j, the activity of an ion, may be expressed as $f_j c_j$ (activity coefficient multiplied by molar concentration), and considering a_n (activity of abundant solvent) as approximately constant, we get (for $z_n = 0$)

$$d\tilde{\mu}_j - d\tilde{\mu}_n = (\bar{V}_j - \bar{V}_n)\,dp + z_j F\,d\varphi +$$
$$+ RT\,d\ln c_j + RT\,d\ln f_j.$$

Finally, on introducing the last expression into eq. (5.74), we obtain the Schlögl equation

$$J_j = c_j v - RTU_j\frac{dc_j}{dx} - z_j FU_j c_j\frac{d\varphi}{dx} -$$
$$- RTU_j c_j \frac{d\ln f_j}{dx} - U_j M_j c_j\left(\frac{V_j}{M_j} - \frac{V_n}{M_n}\right)\frac{dp}{dx} \quad (j = 1, ..., n). \quad (5.75)$$

5.2.4. Henderson's equation—potential difference across a continuous layer with constant concentration gradients of individual ions

Henderson's condition of linear dependence of concentration on the distance for each ionic species, valid more or less strictly in mixed boundaries, may be written as

$$c_j = c_{jo} + \frac{c_{ji} - c_{jo}}{d}x \quad (5.76)$$

so that the concentration gradient for each ionic species is

$$\frac{dc}{dx} = \frac{c_{ji} - c_{jo}}{d}. \tag{5.77}$$

Introducing these two expressions into the Nernst–Planck equation (5.36), written for the steady-state conditions in ordinary derivatives, we have

$$J_j = -RTU_j \frac{dc}{dx} - z_j F U_j c_j \frac{d\varphi}{dx} =$$

$$= -RTU_j \frac{c_{ji} - c_{jo}}{d} - z_j F U_j \left(c_{jo} + \frac{c_{ji} - c_{jo}}{d} x \right) \frac{d\varphi}{dx}. \tag{5.78}$$

Now, in the absence of an external circuit, no current flows across the membrane in the steady state

$$\sum_{j=1}^{n} z_j J_j = 0 \tag{5.79}$$

i.e.

$$RT \frac{1}{d} \left(\sum_{j=1}^{n} z_j U_j c_{ji} - \sum_{j=1}^{n} z_j U_j c_{jo} \right) +$$

$$+ \frac{d\varphi}{dx} F \left[\sum_{j=1}^{n} z_j^2 U_j c_{jo} + \frac{x}{d} \left(\sum_{j=1}^{n} z_j^2 U_j c_{ji} - \sum_{j=1}^{n} z_j^2 U_j c_{jo} \right) \right] = 0. \tag{5.80}$$

After separation of variables, integration can be carried out easily

$$\int_{\varphi_o}^{\varphi_i} d\varphi = -\frac{RT}{F} \frac{1}{d} \int_0^d \left(\sum_{j=1}^{n} z_j U_j c_{ji} - \sum_{j=1}^{n} z_j U_j c_{jo} \right) \times$$

$$\times \frac{dx}{\left[\sum_{j=1}^{n} z_j^2 U_j c_{jo} + \frac{x}{d} \left(\sum_{j=1}^{n} z_j^2 U_j c_{ji} - \sum_{j=1}^{n} z_j^2 U_j c_{jo} \right) \right]} \tag{5.81}$$

yielding

$$\varphi_i - \varphi_o = -\frac{RT \sum_{j=1}^{n} z_j U_j c_{ji} - \sum_{j=1}^{n} z_j U_j c_{jo}}{F \sum_{j=1}^{n} z_j^2 U_j c_{ji} - \sum_{j=1}^{n} z_j^2 U_j c_{jo}} \ln \frac{\sum_{j=1}^{n} z_j^2 U_j c_{ji}}{\sum_{j=1}^{n} z_j^2 U_j c_{jo}} \tag{5.82}$$

which is Henderson's equation for the difference of electrical potential across a layer in which the concentration gradients of individual ionic species may be considered to be constant.

5.2.5. Planck's procedure—potential difference across a microscopically electroneutral continuous layer

Apart from the validity of the equation of electrodiffusion (5.36) Planck's procedure assumes the validity of the condition of local electroneutrality and further that all ions present are of the same valency (say, all univalent). The condition of local or microscopic electroneutrality is expressed by

$$\sum_{j=1}^{n} c_{j^+} = \sum_{j=1}^{n} c_{j^-} = c \qquad (5.83)$$

where c_{j^+} is the local concentration of a cation, c_{j^-} the local concentration of an anion and c the total concentration of cations or anions at any point. The total concentration of cations being equal to the total concentration of anions, the same must be true about their derivatives with respect to time

$$\sum_{j=1}^{n} \frac{\partial c_{j^+}}{\partial t} = \sum_{j=1}^{n} \frac{\partial c_{j^-}}{\partial t}. \qquad (5.84)$$

Planck (1890) relates the time derivatives of concentrations at first to fluxes and then to gradients of concentrations and of the electrical potential. The flux of an ion across a plane of unit area, perpendicular to the x-axis is J at the coordinate x and $J + (\partial J/\partial x)\,\Delta x$ at the coordinate $(x + \Delta x)$. Accumulation of the ion between the two planes during the time Δt is thus

$$-\frac{\partial J}{\partial x} \Delta x \, \Delta t$$

per unit area which, in the volume of Δx cm^3 (between the two unit areas), amounts to a change of concentration

$$\Delta c = -\frac{\partial J}{\partial x} \Delta t$$

and in the limit

$$\frac{\partial c}{\partial t} = -\frac{\partial J}{\partial x}. \qquad (5.85)$$

The flux J, however, is given by the electrodiffusion equation (5.36) and hence we can write for cations ($z = +1$)

$$\frac{\partial c_{j^+}}{\partial t} = RTU_{j^+}\frac{\partial^2 c_{j^+}}{\partial x^2} + FU_{j^+}\frac{\partial}{\partial x_t}\left(c_{j^+}\frac{\partial \varphi}{\partial x}\right) \qquad (5.86)$$

and for anions

$$\frac{\partial c_{j-}}{\partial t} = RT U_{j-} \frac{\partial^2 c_{j-}}{\partial x^2} - FU_{j-} \frac{\partial}{\partial x}\left(c_{j-} \frac{\partial \varphi}{\partial x}\right). \quad (5.87)$$

Introducing equations (5.86) and (5.87) into (5.84) we have

$$RT \sum_{j=1}^{n} U_{j+} \frac{\partial^2 c_{j+}}{\partial x^2} + F \sum_{j=1}^{n} U_{j+} \frac{\partial}{\partial x}\left(c_{j+} \frac{\partial \varphi}{\partial x}\right) =$$
$$= RT \sum_{j=1}^{n} U_{j-} \frac{\partial^2 c_{j-}}{\partial x^2} - F \sum_{j=1}^{n} U_{j-} \frac{\partial}{\partial x}\left(c_{j-} \frac{\partial \varphi}{\partial x}\right) \quad (5.88)$$

which can be integrated term by term, giving

$$RT \sum_{j=1}^{n} U_{j+} \frac{\partial c_{j+}}{\partial x} + F \sum_{j=1}^{n} U_{j+} c_{j+} \frac{\partial \varphi}{\partial x} =$$
$$= RT \sum_{j=1}^{n} U_{j-} \frac{\partial c_{j-}}{\partial x} - F \sum_{j=1}^{n} U_{j-} c_{j-} \frac{\partial \varphi}{\partial x} \quad (5.89)$$

or

$$RT \frac{\partial \mathbf{U}}{\partial x} + F\mathbf{U} \frac{\partial \varphi}{\partial x} = RT \frac{\partial \mathbf{V}}{\partial x} - F\mathbf{V} \frac{\partial \varphi}{\partial x} \quad (5.90)$$

where

$$\mathbf{U} = \sum_{j=1}^{n} U_{j+} c_{j+} \quad \text{and} \quad \mathbf{V} = \sum_{j=1}^{n} U_{j-} c_{j-}. \quad (5.91)$$

Eq. (5.90) may be solved for the local gradient of electrical potential, giving

$$\frac{\partial \varphi}{\partial x} = -\frac{RT}{F} \frac{\dfrac{\partial (\mathbf{U} - \mathbf{V})}{\partial x}}{\mathbf{U} + \mathbf{V}} \quad (5.92)$$

This equation cannot be integrated, since the dependence of ionic concentrations (and hence also of $\mathbf{U}$ and $\mathbf{V}$) on x is not known. It is for this reason that further derivation is restricted to steady-state conditions, eq. (5.92) being preserved to be used later. In a steady state, the time derivatives of concentrations vanish, so that (5.86) and (5.87) may be written as

$$RT \frac{d^2 c_{j+}}{dx^2} + F \frac{d}{dx}\left(c_{j+} \frac{d\varphi}{dx}\right) = 0, \quad (5.93)$$

$$RT \frac{d^2 c_{j-}}{dx^2} - F \frac{d}{dx}\left(c_{j-} \frac{d\varphi}{dx}\right) = 0. \quad (5.94)$$

The concentrations and the electrical potential in a steady state being functions of x only, ordinary derivatives instead of partial derivatives are used. Equations (5.93) and (5.94) may now be integrated, giving

$$RT\frac{dc_{j+}}{dx} + Fc_{j+}\frac{d\varphi}{dx} = A_j, \tag{5.95}$$

$$RT\frac{dc_{j-}}{dx} - Fc_{j-}\frac{d\varphi}{dx} = B_j. \tag{5.96}$$

Summation over all ionic species yields

$$RT\frac{dc}{dx} + Fc\frac{d\varphi}{dx} = \mathbf{A}, \tag{5.97}$$

$$RT\frac{dc}{dx} - Fc\frac{d\varphi}{dx} = \mathbf{B} \tag{5.98}$$

where

$$\mathbf{A} = \sum_{n=1}^{n} A_j, \quad \mathbf{B} = \sum_{n=1}^{n} B_j \tag{5.99}$$

and c is the total concentrations of cations or anions, defined by eq. (5.83). Adding the eq. (5.97) and (5.98) we obtain

$$2RT\frac{dc}{dx} = \mathbf{A} + \mathbf{B} \tag{5.100}$$

which can be integrated readily, yielding

$$2RTc = (\mathbf{A} + \mathbf{B})x + \text{const.} \tag{5.101}$$

It is thus seen that from the condition of local electroneutrality and from the condition of steady state it follows that the total concentration of ions is a linear function of the distance. If d is the thickness of the continuous layer, c_0 the total concentration of cations or anions at $x = 0$ and c_i the concentration at $x = d$, the total local concentration at any x between zero and d may be expressed by

$$c = c_o + \frac{c_i - c_o}{d}x. \tag{5.102}$$

Subtracting, on the other hand, eq. (5.98) from eq. (5.97) we have

$$2Fc\frac{d\varphi}{dx} = \mathbf{A} - \mathbf{B}. \tag{5.103}$$

Combining equations (5.102) and (5.103) we obtain

$$\frac{d\varphi}{dx} = \frac{(A - B)d}{2F[(c_i - c_o)x + c_o d]} \qquad (5.104)$$

which integrates from $x = 0$ to $x = d$ to

$$\varphi_i - \varphi_o = \frac{(A - B)d}{2F(c_i - c_o)} \ln \frac{c_i}{c_o} \qquad (5.105)$$

which may be written

$$\varphi_i - \varphi_o = \frac{RT}{F} \ln \xi \qquad (5.106)$$

where

$$\xi = \left(\frac{c_i}{c_o}\right)^{\frac{(A-B)d}{2(c_i-c_o)RT}}. \qquad (5.107)$$

It is, however, necessary to exclude the constants **A** and **B** from this expression so that the dependence of ξ (and hence, also of $\varphi_i - \varphi_o$) only on concentrations and mobilities of ions is obtained. To this end, Planck multiplies each of the equations (5.95) and (5.96) with the appropriate mobility, summing afterwards over all cation and anion species and using symbols defined by (5.91):

$$\sum_{j=1}^{n} U_{j^+} A_j = RT \frac{dU}{dx} + FU \frac{d\varphi}{dx} \qquad (5.108)$$

$$\sum_{j=1}^{n} U_{j^-} B_j = RT \frac{dV}{dx} - FV \frac{d\varphi}{dx}. \qquad (5.109)$$

Using eq. (5.105), it is easy to prove that the right-hand sides of eq. (5.108) and (5.109) are equal. Hence the same must be true about their left-hand sides:

$$\sum_{j=1}^{n} U_{j^+} A_j = \sum_{j=1}^{n} U_{j^-} B_j = C. \qquad (5.110)$$

Now, the eq. (5.104) and (5.110) are introduced into each of eq. (5.108) and (5.109) yielding two linear first-order equations:

$$\frac{dU}{dx} + U \frac{(A - B)d}{2RT[(c_i - c_o)x + c_o d]} = \frac{C}{RT}, \qquad (5.111)$$

$$\frac{dV}{dx} - V \frac{(A - B)d}{2RT[(c_i - c_o)x + c_o d]} = \frac{C}{RT}. \qquad (5.112)$$

The form of these equations is

$$\frac{dy}{dx} + f(x) y = g(x)$$

and hence the form of their solution is

$$y = e^{-\int f(x)dx}\left[\int e^{\int f(x)dx} g(x)\, dx + \text{const.}\right]$$

i.e.

$$U = \frac{2C[(c_i - c_o)x + c_o d]}{2(c_i - c_o)RT - (A - B)d} +$$

$$+ [(c_i - c_o)x + c_o d]^{-\frac{(A-B)d}{2(c_i-c_o)RT}} \cdot \text{const.} \quad (5.113)$$

$$V = \frac{2C[(c_i - c_o)x + c_o d]}{2(c_i - c_o)RT + (A - B)d} +$$

$$+ [(c_i - c_o)x + c_o d]^{\frac{(A-B)d}{2(c_i-c_o)RT}} \cdot \text{const.} \quad (5.114)$$

At $x = 0$, $U = U_0$ and $V = V_0$; at $x = d$, $U = U_i$ and $V = V_i$; equations (5.113) and (5.114) are written in pairs (for each of the two boundary conditions), the constants of the last integration are eliminated and ξ defined by eq. (5.107) is introduced

$$\xi U_i - U_o = \frac{2Cd(\xi c_i - c_o)}{2(c_i - c_o)RT + (A - B)d}, \quad (5.115)$$

$$V_i - \xi V_o = \frac{2Cd(c_i - \xi c_o)}{2(c_i - c_o)RT - (A - B)d}. \quad (5.116)$$

C is eliminated by division of the two equations

$$\frac{\xi U_i - U_o}{V_i - \xi V_o} = \frac{2(c_i - c_o)RT - (A - B)d}{2(c_i - c_o)RT + (A - B)d} \cdot \frac{\xi c_i - c_o}{c_i - \xi c_o} \quad (5.117)$$

and $(A - B)$ is eliminated by substitution from eq. (5.107)

$$\frac{\xi U_i - U_o}{V_i - \xi V_o} = \frac{\ln\frac{c_i}{c_o} - \ln \xi}{\ln\frac{c_i}{c_o} + \ln \xi} \cdot \frac{\xi c_i - c_o}{c_i - \xi c_o}. \quad (5.118)$$

This is the final equation of Planck; introducing in to it the total concentrations defined by eq. (5.83) and the quantities **U** and **V** defined by equations (5.91) the quantity ξ is determined from which the potential difference across a continuous layer may be calculated using the simple eq. (5.106). It is not easy to calculate ξ from eq. (5.118); due to its transcendent form, ξ occurs both inside and outside the argument of the logarithmic function. The easiest solution is perhaps a graphical one (MacInnes, 1961): both the left-hand side and the right-hand side of eq. (5.118) is plotted in the same graph as a function of various assumed values of ξ. The point of the intersection of the two curves then corresponds to the value of which solves the equation (5.118) for given values of U_o, V_o, U_i, V_i, c_o and c_i.

In order to learn when the condition of local electroneutrality is strictly satisfied in steady-state ionic systems, the Poisson equation may be used, a general expression for electrostatic interactions which, for a unidimensional case, may be written as

$$\frac{d^2\varphi}{dx^2} = -\frac{\varrho}{\varepsilon} \qquad (5.119)$$

where φ is the electrical potential, ϱ the charge density and ε the absolute dielectric constant. When local electroneutrality prevails, the charge density is zero:

$$F \sum_{j=1}^{n} z_j c_j = \varrho = 0 \qquad (5.120)$$

(where F is again the Faraday number, z_j the valency of the j-th ionic species and c_j its concentration) so that the Poisson equation turns into the Laplace equation

$$\frac{d^2\varphi}{dx^2} = 0 \qquad (5.121)$$

which by integration gives

$$\frac{d\varphi}{dx} = \text{const.} \qquad (5.122)$$

As stressed by Agin (1967) any attempt to calculate potential profiles in a regime using condition (5.120) but not the condition (5.122) is physically inadmissible. Now, following Polissar (1954), we multiply the Nernst–Planck differential equation of electrodiffusion

(5.81) which can be written for steady-state conditions (c and φ being functions of x only) as

$$J_j = -RTU_j \frac{dc_j}{dx} - z_j F U_j c_j \frac{d\varphi}{dx} \quad (5.123)$$

by z_j/FU_j and carry out sumation over all ionic species present. After rearrangement we obtain

$$\frac{d\varphi}{dx} \sum_{j=1}^{n} z_j^2 c_j = -\frac{RT}{F} \frac{d}{dx} \sum_{j=1}^{n} z_j c_j - \frac{1}{F} \sum_{j=1}^{n} \frac{z_j J_j}{U_j}. \quad (5.124)$$

The first term on the right-hand side vanishes when local electroneutrality prevails, the second does so only either when there is no electrodiffusional flow J_j (equilibrium), or when all ions have the same mobility (mobility then can be taken out before the summation sign and the sum of electric currents $z_j J_j$ is zero uder open-circuit conditions). Now. if local electroneutrality prevails, at least some of the ionic species present have different mobilities and the system is in a steady state rather than in equilibrium (so that J_j = const. $\neq 0$) and also mobilities are constants independent of x, eq. (5.124) than shows

$$\frac{d\varphi}{dx} = \frac{k}{\sum_{j=1}^{n} z_j^2 c_j}. \quad (5.125)$$

This equation is compatible with the Poisson equation for zero charge density (5.122) only when $\sum_{j=2}^{n} z_j^2 c_j$, the local ionic strength, is a constant independent of x, and hence only under this condition is the condition of local electroneutrality strictly satisfied. When only univalent ions are present in significant concentrations, eq. (5.125) becomes equivalent to eq. (5.103) of Planck and the requirement for a strict total electroneutrality is that the total concentration of ions is a constant independent of x. In a locally electroneutral steady-state system, however, the total concentration is always a linear function of x, as shown by eq. (5.101) and hence using eq. (5.102) we can see that the total concentration of ions is a constant independent of x when it is the same at two different points, i.e., when $c_i = c_o$.

The condition of equality of total concentrations ($c_i = c_o$) not only makes the condition of local electroneutrality valid and hence also the condition of constant field exactly satisfied but, at the same time, it greatly simplifies the transcendental equation of Planck

(5.118) by converting it into the equation of Goldman. On introducing the condition $c_i \approx c_o$ into eq. (5.118) its right-hand side becomes equal to one, so that

$$\xi = \frac{U_o - V_i}{U_i + V_o} \qquad (5.126)$$

and, using eq. (5.106) and equations (5.91) we have

$$\varphi_i - \varphi_o = \frac{RT}{F} \ln \frac{U_o + V_i}{U_i + V_o} = \frac{RT}{F} \ln \frac{\sum_{j=1}^{n} U_{j^+} c_{j^+o} + \sum_{j=1}^{n} U_{j^-} c_{j^-i}}{\sum_{j=1}^{n} U_{j^+} c_{j^+i} + \sum_{j=1}^{n} U_{j^-} c_{j^-o}} \qquad (5.127)$$

which indeed is the equation for electrical potential difference by Goldman (1943).

However, the above derivations which were carried out for a layer of a continuous and uniform medium, can probably be applied to the interior of a membrane or of ion-permeable regions of a membrane only when the membrane is so thick that the diffuse double layers extending inside it from the phase boundaries may be neglected — in the double layers the condition of local electroneutrality is violated. As we have seen already (p. 253) the effective thickness of an electrical diffuse double layer (the Debye length) is directly proportional to the square root of the relative dielectric constant and inversely proportional to the square of the concentration beyond the extent of the double layer. The relative dielectric constant in the ion-permeable regions of cell membranes may be quite low (between 1 for a vacuum and about 80 for water) but if the concentration of ions in these regions is low, too, the effective thickness of the double layer may be greater than the thickness of the cell membrane. Under this condition, the assumption of constant field again represents a useful approximation (Cole, 1965), which is the better the smaller the ratio of membrane thickness to the effective thickness of the double layer. Thus we can recognize two cases where the constant-field condition for the valitidy of the Goldman equation (5.127) is approximately satisfied: (*1*) in thick membranes when the local ionic strength is approximately constant; (*2*) in thin membranes when the concentration of ions in the membrane or in the ion-permeable membrane regions is low.

The Goldman equation (5.127) can be derived in a much simpler way if the constant-field assumption is introduced at the very beginning. This derivation is shown in the next section.

5.2.6. Goldman's procedure—potential difference across a continuous layer with constant field

The constant-field condition may be expressed by

$$\frac{\partial \varphi}{\partial x} = \frac{\varphi_i - \varphi_o}{d} \qquad (5.128)$$

where $\varphi_i - \varphi_o$ is the electrical potential difference across a continuous layer (say, across the membrane proper) and d is the thickness of the layer. We can introduce this condition into the differential equation of electrodiffusion (5.36), assuming, moreover, steady-state conditions, so that J_j will be a constant and c_j a function of x only (hence ordinary derivatives):

$$J_j = -RTU_j \frac{dc_j}{dx} - z_j F U_j c_j \frac{\varphi_i - \varphi_o}{d} \qquad (5.129)$$

we separate the variables and integrate across the layer (from $x = 0$, $c_j = c_{jo}$, to $x = d$, $c_j = c_j$):

$$\int_0^d dx = -RTU_j \int_{c_{jo}}^{c_{ji}} \frac{dc_j}{J_j + z_j F U_j \frac{\varphi_i - \varphi_o}{d} c_j}$$

which gives

$$d = \frac{RTd}{z_j F(\varphi_i - \varphi_o)} \ln \frac{J_j + z_j F U_j \frac{\varphi_i - \varphi_o}{d} c_{ji}}{J_j + z_j F U_j \frac{\varphi_i - \varphi_o}{d} c_{jo}}.$$

The last equation may be solved for J_j, yielding

$$J_j = z_j F U_j \frac{\varphi_i - \varphi_o}{d} \cdot \frac{c_{ji} - c_{jo} e^{-z_j F(\varphi_o - \varphi_i)/RT}}{e^{-z_j F(\varphi_o - \varphi_i)/RT} - 1}. \qquad (5.130)$$

This is the familiar constant-field expression for the steady-state flux of an individual ionic species, applicable in this form to the inside of a continuous layer. It is of interest to see how this equation is reduced, when the absence of either the concentration gradient or of the electrical potential gradient is assumed.

In the first case, when $c_{ji} = c_{jo} = c_j$, it yields

$$J_j = -z_j F U_j c_j \frac{\varphi_i - \varphi_o}{d} \qquad (5.131)$$

which describes the steady-state migration of an ion in the electrical field.

In the other case, however, when $\varphi_i - \varphi_o$ is zero, eq. (5.130) is an indeterminate quantity 0/0. L'Hôpital's rule has to be used, according to which

$$\lim \frac{f(x)}{g(x)} = \lim \frac{f'(x)}{g'(x)},$$

the limit of the ratio of two functions is equal to the limit of the ratio of their derivatives. We consider the right-hand side of eq. (5.130) as a quotient of two functions of the variable $\varphi_i - \varphi_o$, we calculate the derivatives of the numerator and of the denominator and in the quotient of the two derivatives we let the variable $\varphi_i - \varphi_o$ vanish. In this way we obtain the correct steady-state formula

$$J_j = RTU_j \frac{c_{jo} - c_{ji}}{d} \tag{5.132}$$

describing the diffusion of an ion in the absence of the electrical field.

We may observe a certain asymmetry of the formula (5.130); the external concentration of the ion c_{jo} is multiplied by an exponential factor, whereas its internal concentration, c_{ji}, is not. This is a natural result of our choice of the external potential φ_o, as a reference; the exponential factor multiplying the internal concentration thus contains $\varphi_o - \varphi_o = 0$ in the exponent and hence reduces to 1.

To calculate the value of the electrical potential difference across a constant-field layer, use can be made of the condition that when there is no external electric circuit, the sum of the individual ionic currents across the layer must be zero,

$$\sum_{j=1}^{n} z_j F J_j = 0. \tag{5.133}$$

Simple formulae for the electrical potential difference are obtained when all ions present in significant concentrations are of the same valency (say, univalent), for then the denominators in the flux formulae of the type of (5.130) can be brought to the same form and therefore may be cancelled out from equations of the type of (5.133). Thus the current carried across the layer by an univalent cation will be given by

$$FJ_{j^+} = F^2 U_{j^+} \frac{\varphi_i - \varphi_o}{d} \frac{c_{j^+i} - c_{j^+o} e^{-F(\varphi_i - \varphi_o)/RT}}{e^{-F(\varphi_i - \varphi_o)/RT} - 1} \tag{5.134}$$

and that carried by an univalent anion by

$$FJ_{j^-} = -F^2 U_{j^-} \frac{\varphi_i - \varphi_o}{d} \frac{c_{j-i} - c_{j-o} e^{F(\varphi_i - \varphi_o)/RT}}{e^{F(\varphi_i - \varphi_o)/RT} - 1} =$$

$$= F^2 U_{j^-} \frac{\varphi_i - \varphi_o}{d} \frac{c_{j-i} e^{-F(\varphi_i - \varphi_o)/RT} - c_{j-o}}{e^{-F(\varphi_i - \varphi_o)/RT} - 1}. \quad (5.135)$$

From the condition (5.133) that under open-circuit conditions the sum of individual currents is zero we have (when only univalent ions are present)

$$\sum_{j=1}^{n} FJ_{j^+} + \sum_{j=1}^{n} FJ_{j^-} = 0. \quad (5.136)$$

It follows that

$$\sum_{j=1}^{n} U_{j^+} c_{j^+ i} - \sum_{j=1}^{n} U_{j^+} c_{j^+ o} e^{-F(\varphi_i - \varphi_o)/RT} =$$

$$= \sum_{j=1}^{n} U_{j^-} c_{j^- i} e^{-F(\varphi_i - \varphi_o)/RT} - \sum_{j=1}^{n} U_{j^-} c_{j^- o} \quad (5.137)$$

or

$$e^{-F(\varphi_i - \varphi_o)/RT} = \frac{\sum_{j=1}^{n} U_{j^+} c_{j^+ i} + \sum_{j=1}^{n} U_{j^-} c_{j^- o}}{\sum_{j=1}^{n} U_{j^+} c_{j^+ o} + \sum_{j=1}^{n} U_{j^-} c_{j^- i}} \quad (5.138)$$

i.e.

$$\varphi_i - \varphi_o = \frac{RT}{F} \ln \frac{\sum_{j=1}^{n} U_{j^+} c_{j^+ o} + \sum_{j=1}^{n} U_{j^-} c_{j^- i}}{\sum_{j=1}^{n} U_{j^+} c_{j^+ i} + \sum_{j=1}^{n} U_{j^-} c_{j^- o}}.$$

For example, if only potassium, sodium and chloride ions are present in significant concentrations, the electrical potential difference across a constant-field layer will then be given by

$$\varphi_i - \varphi_o = \frac{RT}{F} \ln \frac{U_{K^+} c_{K^+ o} + U_{Na^+} c_{Na^+ o} + U_{Cl^-} c_{Cl^- i}}{U_{K^+} c_{K^+ i} + U_{Na^+} c_{Na^+ i} + U_{Cl^-} c_{Cl^- o}}. \quad (5.139)$$

An important (although rather obvious) observation is now in order: ions which are in thermodynamic equilibrium do not enter into eq. (5.138); their fluxes are identical in both directions and the current carried by them is therefore zero and need not be taken into account (refer to the condition of zero net total electric current across an open-circuited membrane, eq. 5.136). Thus, e.g., when in the above system with sodium, potassium and chloride ions the chloride ions are in thermodynamic equilibrium everywhere inside the continuous layer,

$$\varphi_i - \varphi_o = \frac{RT}{F} \ln \frac{U_{K^+} c_{K^+o} + U_{Na^+} c_{Na^+o}}{U_{K^+} c_{K^+i} - U_{Na^+} c_{Na^+i}} = \frac{RT}{F} \ln \frac{c_{Cl^-i}}{c_{Cl^-o}}. \quad (5.140)$$

5.2.7. Constant-field equation for potential difference across the whole membrane

When accepting formula (5.139) or (5.140) as an expression for the electrical potential difference across the membrane proper we are faced with the task of relating the concentrations of ions at the boundaries inside the membrane to those in the adjacent media as well as of accounting for the steps of the electrical potential between the media and the membrane (see once more Fig. 5.6).

The assumption of equilibrium between the membrane and the adjacent media simplifies the derivation, but even under such an assumption it would be incorrect to relate the concentration of an ion just inside the membrane to its concentration in the adjacent medium by a constant partition coefficient. Ions are charged particles and their equilibrium distribution between two phases is thus a function not only of the chemical nature of the two media but also of the electrical potential difference between them. Such electrical potential difference between phases of different chemical composition can never be measured experimentally for the very reason that other forces beside the potential gradient are operative in transitions of really existing electrical charges (ions) between the two phases. The difference between the pair of such phase potentials, however, does contribute to the overall potential difference across the membrane and this is why these individual steps must be considered (Polissar, 1954).

The equilibrium of, say, the potassium ion at the outer surface of the membrane is expressed by equality of its electrochemical potentials and may be written (referring again to Fig. 5.6 and neglecting activity coefficients) as

$$\mu'_{0K^+} + RT \ln [K^+]_o + F\varphi'_o = \mu_{0K^+} + RT \ln c_{K^+o} + F\varphi_o \quad (5.141)$$

where μ'_{0K^+} and μ_{0K^+} are standard potentials (the part of partial concentration and of the electrical potential) for the medium and the membrane, respectively. From eq. (5.141) it follows that

$$\ln c_{K^+_o} = \frac{\mu'_{0K^+} - \mu_{0K^+}}{RT} + \ln [K^+]_o + \frac{F(\varphi'_o - \varphi_o)}{RT} \quad (5.142)$$

or

$$c_{K^+_o} = e^{\frac{\mu'_{0K^+} - \mu_{0K^+}}{RT}} [K^+]_o e^{\frac{F(\varphi'_o - \varphi_o)}{RT}} = k_{K^+}[K^+]_o e^{-FE_o/RT} \quad (5.143)$$

where $k_{K^+} = e^{\frac{\mu'_{0K^+} - \mu_{0K^+}}{RT}}$ is the partition (or distribution) coefficient of potassium ions and $E_o = \varphi_o - \varphi'_o$. Analogously, for concentrations of all ions considered in the membrane it may be written

$$c_{Na^+_o} = k_{Na^+}[Na^+]_o e^{-FE_o/RT}$$
$$c_{Na^+_i} = k_{Na^+}[Na^+]_i e^{FE_i/RT}$$
$$c_{K^+_o} = k_{K^+}[K^+]_o e^{-FE_o/RT}$$
$$c_{K^+_i} = k_{K^+}[K^+]_i e^{FE_i/RT}$$
$$c_{Cl^-_o} = k_{Cl^-}[Cl^-]_o e^{FE_o/RT}$$
$$c_{Cl^-_i} = k_{Cl^-}[Cl^-]_i e^{-FE_i/RT}. \quad (5.144)$$

Introducing relations (5.144) into eq. (5.139) for the potential difference across the membrane proper, $\varphi_i - \varphi_o$, and calculating the overall membrane potential $E = \varphi'_i - \varphi'_o = E_o + \varphi_i - \varphi_o + E_i$, we make use of the identity

$$E_o + E_i \equiv \frac{RT}{F} \ln \left(e^{FE_o/RT} \cdot e^{FE_i/RT} \right) \quad (5.145)$$

to obtain

$$E = \frac{RT}{F} \ln \frac{U_{K^+}k_{K^+}[K^+]_o + U_{Na^+}k_{Na^+}[Na^+]_o + U_{Cl^-}k_{Cl^-}[Cl^-]_i e^{F(E_o - E_i)/RT}}{U_{K^+}k_{K^+}[K^+]_i + U_{Na^+}k_{Na^+}[Na^+]_i + U_{Cl^-}k_{Cl^-}[Cl^-]_o e^{F(E_o - E_i)/RT}} \quad (5.146)$$

Finally, we may introduce permeability coefficients

$$P_{K^+} = RTU_{K^+}k_{K^+}/d$$
$$P_{Na^+} = RTU_{Na^+}k_{Na^+}/d \quad (5.147)$$
$$P_{Cl^-} = RTU_{Cl^-}k_{Cl^-}/d$$

with which eq. (5.146) takes the form

$$E = \frac{RT}{F} \ln \frac{P_{K^+}[K^+]_o + P_{Na^+}[Na^+]_o + P_{Cl^-}[Cl^-]_i e^{F(E_o - E_i)/RT}}{P_{K^+}[K^+]_i + P_{Na^+}[Na^+]_i + P_{Cl^-}[Cl^-]_o e^{F(E_o - E_i)/RT}} \quad (5.148)$$

or, when chloride ions are in a passive-flux equilibrium or when in the membrane their permeability is negligibly low

$$E = \frac{RT}{F} \ln \frac{P_{K^+}[K^+]_o + P_{Na^+}[Na^+]_o}{P_{K^+}[K^+]_i + P_{Na^+}[Na^+]_i}. \quad (5.149)$$

If P_{K^+} is much greater than the other permeability coefficients, formulae (5.148) or (5.149) reduce to that for the equilibrium potential of potassium ions, for other terms in the nominator and the denominator may be neglected and the potassium permeability coefficients cancel out. This state of affairs apparently corresponds to many instances of resting potentials of muscle and nerve membranes. If, on the other hand, the sodium permeability of such membranes increases, the membrane potential decreases and even changes its sign, the distribution of sodium ions across the membranes being opposite to that to that of potassium ions and becoming more and more important in the generation of the membrane potential. Such a change of membrane permeability forms the basis of the membrane potential change, called the action potential, according to the well-known sodium theory of Hodgkin and Huxley (see, e.g., Katz, 1966).

5.2.8. Constant-field equation for steady-state membrane potential in the presence of an electrogenic sodium pump

Formula (5.149) has been derived under the assumption that active sodium transport, if present, is electroneutral, i.e., that the sodium pump performs obligatory coupled exchange of one sodium ion for one potassium ion. However, it has been shown by Thomas (1972) that, in a steady state, eq. (5.149) includes even the case of an electrogenic sodium pump. The following assumptions must be made. (1) The net passive flux of sodium and potassium ion is described by eq. (5.130), or, more generally, by any equation of the form

$$J = U(c_i - c_o\, e^{-F(\varphi_i - \varphi_o)/RT}) f(\varphi_i - \varphi_o) \quad (5.150)$$

where $f(\varphi_i - \varphi_o)$ is an arbitrary function of the potential difference across the membrane proper, $\varphi_i - \varphi_o$.

(2) Due to the steady-state condition, the ion concentrations are steady and hence the sum of net passive flux J and net active flux J_{act} is zero for each of the two ions:

$$J_{K^+} + J_{K^+ act} = 0, \qquad (5.151)$$

$$J_{Na^+} + J_{Na^+ act} = 0. \qquad (5.152)$$

(3) Finally, there is a given coupling ratio r giving the number of sodium ions pumped from the cell for each of the potassium ions taken actively in:

$$rJ_{K^+ act} + J_{Na^+ act} = 0. \qquad (5.153)$$

From (5.151), (5.152) and (5.153) it follows that also

$$rJ_{K^+} + J_{Na^+} = 0. \qquad (5.154)$$

Now, introducing the explicit expressions for net passive fluxes (5.150) we obtain

$$rU_{K^+}(c_{K^+ i} - c_{K^+ o}\, e^{-F(\varphi_i - \varphi_o)/RT})f(\varphi_i - \varphi_o) +$$
$$+ U_{Na^+}(c_{Na^+ i} - c_{Na^+ o}\, e^{-F(\varphi_i - \varphi_o)/RT})f(\varphi_i - \varphi_o) = 0 \qquad (5.155)$$

where $f(\varphi_i - \varphi_n)$ may be cancelled and the electrical potential difference across the membrane proper, $\varphi_i - \varphi_o$, can be expressed as follows:

$$\varphi_i - \varphi_o = \frac{RT}{F} \ln \frac{rU_{K^+}c_{K^+ o} + U_{Na^+}c_{Na^+ o}}{rU_{K^+}c_{K^+ i} + U_{Na^+}c_{Na^+ i}}. \qquad (5.156)$$

Finally, the same procedure of accounting for the electrical potential steps at the membrane boundaries and relating the ion concentrations inside the membrane to those in the adjacent media as used in the ast section yields

$$E = \frac{RT}{F} \ln \frac{rP_{K^+}[K^+]_o + P_{Na^+}[Na^+]_o}{rP_{K^+}[K^+]_i + P_{Na^+}[Na^+]_i} \qquad (5.157)$$

where r is the number of sodium ions actively extruded per one potassium ion actively transported in the opposite direction. When the pump is electroneutral (coupled one-to-one) $r = 1$ and (5.157) reduces to (5.149). When, on the other hand, the pump functions in a purely electrogenic manner, r tends to infinity, the sodium terms become negligible and the expression reduces to that for equilibrium potassium potential. An important conclusion may be drawn from the last derivation: in a steady state, the electrogenic or electroneutral character of the pump cannot be inferred from the knowledge of the membrane potential and of equilibrium potentials of individual ions alone. As is obvious from eq. (5.157) the steady-state value of the membrane within the limits given by the equilibrium potentials. Variations of the membrane potential within these limits may result equally well from variations of passive permeabilities and of the coupling ratio. Hence the character of the pump must be established by other means, by a direct search for coupling interdependence of individual active fluxes or for nonsteady-state potentials outside the range of equilibrium potentials.

5.2.9. The Hodgkin-Horowicz equation

There is an alternative approach to the description of membrane potentials, based on equivalent circuits or electrical analogues of the membrane and discussed, e.g., by Dainty (1960) and Finkelstein and Mauro (1963). Although known for years, this approach has not been fully appreciated until a recent penetrating study by Jaffe (1974). He showed the relations of the type of eq. (5.159) to be

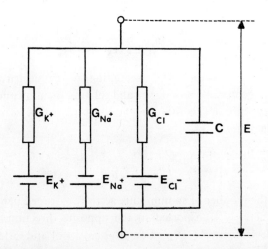

Fig. 5.7. Equivalent circuit for a cell membrane. C Membrane capacitance, G's conductances for individual ion species, E_j's electromotive forces, equal to equilibrium potentials of the individual species (e.g., $E_{K^+} = (RT/F) \ln ([K^+]_o/[K^+]_i)$.

better suited for describing potential differences across membranes of many, perhaps all, cells than are Goldman-type equations; in many voltage-concentration relations found experimentally and described in the literature the conductances in eq. (5.159) behave as constants whereas the permeabilities in Goldman-type equations do not. Thus, different ion species appear to use relatively separate pathways across cell membranes. Jaffe calls such relations the Hodgkin−Horowicz equations, since Hodgkin and Horowicz (1959) used them explicitly.

The equations may be derived as follows. Let us consider an equivalent circuit for the membrane system with potassium, sodium and chloride ions, the steady state of which was described above using the Goldman equation (5.148) or (5.149). The circuit is shown in Fig. 5.7. The membrane potential E is seen to be generated by electromotive forces (equal to equilibrium potentials of individual ionic species) which are connected in series with corresponding integral conductances. The current through each of the circuit branches is given by

$$I_j = G_j(E_j - E) \tag{5.158}$$

and from the condition that the sum of the currents be zero under open-circuit conditions when the capacity charge is constant (steady state) the membrane potential may be expressed as

$$E = \frac{G_{K^+}E_{K^+} + G_{Na^+}E_{Na^+} + G_{Cl^-}E_{Cl^-}}{G_{K^+} + G_{Na^+} + G_{Cl^-}}. \tag{5.159}$$

It may be seen that if the conductance of a certain ionic species exceeds considerably the other conductances, the membrane potential will be close to the equilibrium potential of that species. When an electrogenic pumping mechanism is present another branch with electromotive force of the pump and the pump conductivity has to be included in the circuit.

Theoretical description of cellular ion transport is likely to profit considerably in the near future from the comprehensive physical theory of ion transport across modified artificial phospholipid membranes, developed by Markin and Chizmadjev (1974). A very interesting approach to the theoretical description of ion transport across the cell membrane is represented by the stochastic method of Györgyi and Sugar (1974), used by the authors to describe the alkali cation transport in erythrocyte membranes.

5.3. CHEMICAL NATURE OF ION-TRANSLOCATING SYSTEMS

Most translocations of the major cations are accomplished at the expense of ATP in a reaction catalyzed by one or another membrane-bound adenosinetriphosphatase.

5.3.1. Na, K-Adenosinetriphosphatase

The enzyme that has aroused greatest interest and about which perhaps more is known than about all the other transport agencies taken together is the sodium-potassium stimulated, magnesium-activated adenosinetriphosphatase; for the most recent cross section of views see the Annals of the New York Academy of Sciences, **242**, 1−741 (1974).

The Na, K-ATPase is a widely distributed plasma membrane enzyme, occurring in all animal organs so far tested, as well as in protozoans, higher plants and some algae (Bonting, 1970), with the notable exception of some erythrocytes (dog) and of yeasts. In bacteria it plays a minor role (if any) in Na^+ and K^+ transport. All the Na, K-ATPase systems and preparations have a number of properties in common. Thus, they all require Mg^{2+} for activity; the apparent half-saturation constant for Na^+ is of the order of 10 mM, that for K^+ of the order of 1 mM, there being probably two sets of sites, the i-sites preferring sodium, the o-sites preferring potassium (cf. Skou, 1972); the pH optimum is at about 7.5; they are all inhibited by various glycosides, typically ouabain (it causes 50 % inhibition at concentrations ranging from 10^{-7} to 10^{-4} M); they are all slightly stimulated by 10^{-9} to 10^{-8} M ouabain*; the ratio of cations transported under optimal conditions per ATP molecule split is 2.6 ±0.19, reflecting the frequently observed exchange of 3 Na^+ for 2 K^+ per 1 ATP.

ATP can be partly replaced with other nucleotides, the relative efficiencies being as follows. ATP : dATP : CTP : ITP : GTP : UTP =
= 100 : 49 : 2.3 : 24 : 0.6 : 0.6.

The requirement for both Na^+ and K^+ is shown in Fig. 5.8, the different sodium : potassium affinity ratios of the two sets of sites being

---

* There appear to be two binding sites for ouabain in the eel electric organ but only one in cat brain.

reflected in the asymmetry of the curve. The coupling ratio is, however, rather flexible and under suitable conditions one can find an electroneutral exchange of Na^+ for K^+. Moreover, there exists a $Na^+ : Na^+$ exchange as well as a $K^+ : K^+$ exchange, the various possibilities depending in a complicated way on the ratios of Na^+ to K^+ outside and inside cells but most distinctly also on intracellular concentrations of ATP, ADP and inorganic phosphate. (e.g., Garrahan and Glynn, 1967). Actually, an activity series for the "potassium" site has been established as $Tl^+ > K^+ > Rb^+ > NH_4^+ > Cs^+ > Li^+$.

Magnesium can be replaced with less efficiency by Mn^{2+} or Co^{2+} but inhibition ensues in the presence of Fe^{2+}, Ca^{2+}, Sr^{2+}, Ba^{2+}, Be^{2+}, Zn^{2+} or Cu^{2+}.

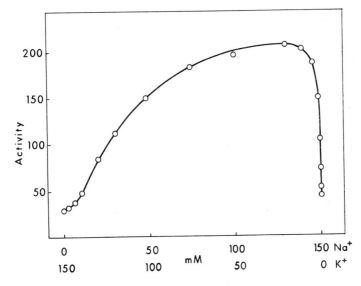

Fig. 5.8. The effect of sodium and potassium ions on the Na, K-adenosinetriphosphatase from ox brain microsomes. The experiment was carried out with 6 mM Mg^{2+}, 3 mM ATP, at pH 7.4 and 37 °C. The ATPase activity is expressed in μmol liberated P per hour per mg protein. (Redrawn from Skou, 1972.)

As new observations are reported, the existing models of ATPase action are being amended and enlarged so that up to the present about a dozen serious suggestions have been advanced in this direction. The general consensus is that there are several partial reactions involved:

$$E_1 + ATP \xrightleftharpoons{Mg_2^+, Na^+} E_1 \sim P + ADP$$

$$E_1 \sim P \xrightleftharpoons{Mg_2^+} E_2 - P$$

$$E_2 - P + K^+ \rightleftharpoons K - E_2 - P$$

$$K - E_2 - P + H_2O \rightleftharpoons K - E_2 + \text{phosphate}$$

$$K - E_2 \rightleftharpoons E_2 + K^+$$

$$E_2 \xrightleftharpoons{Na^+} E_1$$

The Na^+-dependent phosphorylase activity thus includes the formation of a phosphorylated enzyme intermediate, formerly assumed to involve the phosphate bond at the γ-carboxyl of glutamic acid, now known to be at the β-carboxyl group of L-aspartic acid (Post and Orcutt, 1973). All the steps are seen to be reversible and, indeed, under suitable conditions, ATP can be generated by the Na^+ pump running in reverse. There are apparently at least two binding sites for Mg^{2+}, one with a high affinity for the cation which is required for the ATP-ADP exchange reaction, another with a much lower affinity, which is required for the actual hydrolysis giving rise to inorganic phosphate.

The K^+-dependent phosphatase activity has been viewed by some to represent an independent enzyme entity.

The property that is germane to the transport function is the vectorial orientation of the enzyme complex in the membrane. In the above set of reactions, E_1 is thought to be exposed to the inside of the cell where it binds Na^+, then moves over to the other side (this being apparently an exergonic step) where it binds K^+ to transport it inward. However, there is evidence that sites reactive with sodium as well as potassium are present at both membrane faces (Hoffman and Tosteson, 1971).

It is rather attractive to view the ATPase molecule as an oligomeric enzyme, this being supported by the purification attempts as well as by cooperative kinetics of interactions between sodium (low) and potassium (high) (e.g., Robinson, 1970). A membrane-spanning tetramer with an internal cavity (involving half-site reactivity in analogy with enzyme kinetics) was suggested in this connection by Stein and co-workers (1973) and a flip-flop mechanism of action of an oligomer (in analogy to the "tight" and "relaxed" state of monomers of allosteric enzyme kinetics) by Repke and Schön (1973). The last-named model has been developed to a high level of sophistication

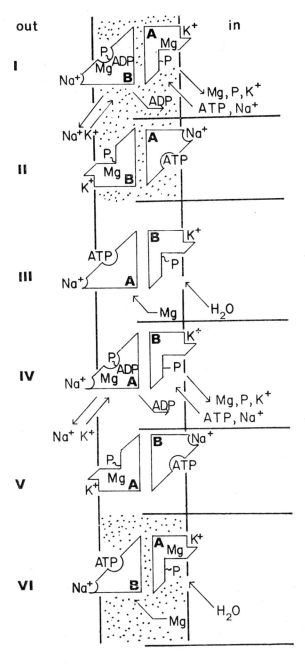

Fig. 5.9. The sequence of steps in the flip-flop model of Na, K-ATPase as suggested by Repke and Schön (1973), based on two subunits A and B.

and shown to be thermodynamically least demanding. Due to the tight coupling between the two states which flip from one membrane face to the other in unison with a symmetrical set of monomers, the changes in free energy involved in the individual steps are cut down to values roughly one-fourth those without the flip-flop coupling. The model is shown schematically in Fig. 5.9.

Ouabain is believed primarily to bind to the E_2 form but possibly allotopically with respect to Na^+ or K^+. There are peculiar aspects of this inhibition, such as its being time-dependent; this is apparently due to ouabain stabilizing (and inactivating) one of the enzyme conformations which is thus gradually removed from the membrane transport cycle.

A number of Na, K-ATPases have been isolated from membranes and purified to a considerable degree, the extent of optimal purification being rather difficult to determine as enzyme activity is frequently lost as lipids are removed or the quaternary structure of the molecule is disrupted. The apparent molecular weight of the whole preparation ranges from 190 000 to 560 000, depending on the source and method of determination (e.g., Kepner and Macey, 1968; Atkinson et al., 1971; Uesugi et al., 1971; Kyte, 1971). However, sodium dodecyl sulfate electrophoresis of the preparations shows it to be composed of probably two types of polypeptide chains, the "α" one having a molecular weight of over 100 000, the "β" one about 50 000. The whole molecule may thus be an $\alpha_2\beta_2$ tetramer or a $\alpha_2\beta_4$ hexamer or possibly have a more complicated quaternary structure. The latest work of Hokin (1974) proceeding from the shark rectal gland and the eel electric organ, shows the light chain to be a glycoprotein and the oligomer to be perhaps of the $\alpha_4\beta_2$ type. It is probably the heavy chain which is phosphorylated during ATPase operation (Avruch and Fairbanks, 1972; Nakao et al., 1974).

The isolated preparations show an affinity for Mg^{2+}, Na^+, K^+ and ouabain in varying degrees and always one for ATP ($K_D =$ $= 0.2$ mM), ADP ($K_D = 0.5-2$ mM) but practically none for CTP, GTP, ITP and UTP ($K_D = 0.1-1$ M). The binding of ATP is affected by univalent as well as bivalent cations.

Preparations of Na,K-ATPase have been combined with artificial phospholipid membranes and marked effects on conductance for Na^+ and K^+ have been observed (e.g., Jain et al., 1972). Shamoo (1974) shows that two α and one β polypeptide chains must be combined in an artificial system to produce the ionophoretic effect.

5.3.2. Ca-Adenosinetriphosphatase

The transport of calcium ions has been studied with special emphasis on the intestine and on sarcoplasmic reticulum, the implication of an ATPase in the transport being clear in the case of muscle reticulum. The mechanism resembles that of Na,K-ATPase:

$$
\begin{aligned}
E_1 + ATP &\rightleftharpoons E_1\text{—}ATP \\
E_1\text{—}ATP + 2Ca^{2+} &\rightleftharpoons E_1\text{—}ATP\text{—}Ca_2 \\
E_1\text{—}ATP\text{—}Ca_2 &\rightleftharpoons E_1{\sim}P\text{—}Ca_2 + ADP \\
E_1{\sim}P\text{—}Ca_2 &\rightarrow E_2\text{—}P\text{—}Ca_2 \\
E_2{\sim}P\text{—}Ca_2 &\xrightleftharpoons{Mg^{2+}} E_2 + \text{phosphate} + 2\,Ca^{2+} \\
E_2 &\rightleftharpoons E_1
\end{aligned}
$$

The E_1 form is assumed to occur on the surface of the microsomes while E_2 exists at the inside face of the microsomal membrane.

Like the Na,K-ATPase, it has a requirement for phospholipids and for Mg^{2+} (although here it aids primarily in the phosphatase-type reaction).

The most acceptable explanation of the various kinetic data is one of a half-site reactivity enzyme capable of existing in two conformations.

A calcium pump has now been reconstituted in vesicles of sarcoplasmic reticulum by addition of isolated ATPase and dioleyllecithin (Warren *et al.*, 1974).

Calcium-binding proteins have been isolated from the intestine (Fullmer and Wasserman, 1973) and from the sarcoplasmic reticulum (Ostwald and MacLennan, 1974). The Ca^{2+}-binding intestinal protein has been obtained from various animals and has a molecular weight ranging from 12 000 in the cow and the rat to 28 000 in the chick. It is produced in response to vitamin D_3, has a dissociation constant with Ca^{2+} of $10^{-6}\,M$ and binds probably four calcium ions per molecule. It is not altogether clear whether it might be a part of a Ca^{2+}-dependent ATPase as suggested by Melancon and DeLuca (1970).

The six Ca^{2+}-binding proteins from sarcoplasmic reticulum (including the two components of calsequestrin) are distinct from the Ca^{2+}-ATPase, range in molecular weight from 20 000 to 55 000 and in the K_D with Ca^{2+} from $5 \cdot 10^{-3}\,M$ to $2.5 \cdot 10^{-6}\,M$. However, their implication in calcium transport does not seem likely.

5.3.3. Other adenosinetriphosphatases

There is some evidence that even Mg^{2+}, Mn^{2+} and other bivalent cations may use the energy of ATP for their transport but the chemical nature of the process is not yet fully understood.

It is rather likely that Mg^{2+}-activated ATPases are active particularly in microorganisms where they translocate H^+ ions outward (probably 2 H^+/ATP). A number of Mg^{2+}-activated ATPases have been isolated and purified and it is surmised that some may be the active H^+ translocators, the best example being provided by the work on *Streptococcus faecalis* (Schnebli and Abrams, 1970; Harold et al., 1969; Abrams and Baron, 1967) where the Mg^{2+}-ATPase was shown to have a molecular weight of 385 000 (apparently an $\alpha_6\beta_6$ oligomer) and to resemble in its inhibitor sensitivity the cation (particularly H^+) transport in the intact cells.

Various ATPases were found to be implicated in the transport of anions, the best understood being the translocation of bicarbonate in gastric oxyntic cells (e.g., Wiebelhaus et al., 1971). The enzyme is also activated by Mg^{2+}.

Transport of chloride anions, in some animal and plant cells, often correlates with an ATPase activity although little is known about the mechanism.

5.3.4. Ion-binding proteins

Besides the Ca^{2+} binding, vitamin D-dependent protein isolated from the intestine of various animals, the Ca^{2+}-binding proteins and the Mg^{2+}-activated proteins from sarcoplasmic reticulum (Meissner, 1973), there are two membrane proteins known from bacteria to be involved in ion transport. One is the sulfate-binding protein from *Salmonella typhimurium* (Pardee, 1967) of molecular weight of 32 000, with a K_D *in vitro* of $3 \cdot 10^{-5}$ M. The other is a protein from *Escherichia coli*, probably a component of the high-affinity system for phosphate transport (Medweczky and Rosenberg, 1970). It has a molecular weight of 42 000 and a K_D *in vitro* of $7 \cdot 10^{-7}$ M.

5.3.5. Transport of ferric ions

Several systems of membrane transport of iron are known in some detail.
The one in *Mycobacterium smegmatis* (Ratledge and Marshall, 1972) involves the iron-chelating compound mycobactin

which transports ferric ions inward where they are reduced to the ferrous form by NADH. Another system is of relatively wide distribution in bacteria and includes one or another of the chelating agents sideramines which act as membrane carriers (Franklin, 1974)

In *Escherichia coli*, Fe^{3+} ions are transported in a chelate with enterobactin (enterochelin) which is a cyclic form of 2,3-dihydroxy-

N-benzoylserine, or in a chelate with citrate. The former system also occurs in *Salmonella typhimurium* (O'Brien and Gibson, 1970; Young et al., 1967; Pollack and Neilands, 1970).

In *Bacillus megaterium*, Fe^{3+} ions enter cells in a coordination complex with schizokinen (Mullis et al., 1971):

$$\begin{array}{c}\text{schizokinen structure}\end{array}$$

In animal organs, particularly in the gut, the transmembrane movement of iron (mainly in the ferrous form) is apparently a function of the intracellular concentration of apoferritin (molecular weight 450 000) which contains as much as 23% Fe by weight, and of the plasma concentration of transferrin (or siderophilin), a Fe^{2+}-binding globulin of molecular weight of 90 000, with 5.5% content of carbohydrate. Whether the proteins are involved in the transmembrane movement itself remains unclear.

5.3.6. Ionophores

Although their role in natural ion transport is disputable, there exist a number of compounds, mostly of bacterial or fungal origin, which permit the passage across membranes of various ions, either very selectively or with a broad spectrum. They are generally designated as ionophores.

The membranes sensitive to their attack include mitochondria, chloroplasts, Gram-positive and some Gram-negative bacteria, erythrocytes, as well as artificial phospholipid membranes. However, it cannot be excluded that other membranes may be susceptible to the ionophoretic action of some of these agents.

The chemical nature of ionophores is rather heterogeneous. They may be divided into peptides, depsipeptides, macrotetrolides and cyclic polyethers.

The peptides can be either linear (gramicidin A) or cyclic (gramicidin S, tyrocidin A, PV peptide, antamanide, alamethicin).

Gramicidin A

HCO—L-val—gly—L-ala—D-leu—L-ala—D-val—D-val—
—D-val—L-tyr—D-leu—L-tyr—D-leu—L-tyr—D-leu—
—L-tyr—NHCH$_2$CH$_2$OH

Gramicidin S

L-val—L-orn—L-leu—D-phe—L-pro
| |
L-pro—L-glu—L-asp—D-phe—L-phe

Tyrocidin A

L-val—L-orn—L-leu—D-phe—L-pro
| |
L-tyr—L-glu—L-asp—D-phe—L-phe

PV peptide (artificially synthesized)

L-val—D-pro—D-val—L-pro—L-val—D-pro
| |
L-pro—D-val—D-pro—L-val—L-pro—D-val

Antamanide

L-ala—L-phe—L-phe—L-pro—L-pro
| |
L-pro—L-pro—L-val—L-phe—L-phe

Alamethicin

AIB—Glu—Pro—AIB—Ala—AIB—Ala—Gln—AIB*
| |
     Gln
AIB—Val—Pro—AIB—Leu—Gly—AIB—Val

The depsipeptides are probably the most important group of ionophores and include valinomycin, various enniatins and beauvericin.

Valinomycin

[L-lactate—L-val—HIV—D-val]$_3$

Enniatin A

[HIV—L-Me-ile]$_3$

* AIB, α-aminoisobutyric acid; HIV, D-hydroxyisovaleric acid; Me, methyl.

Enniatin B

$$[\text{HIV}-\text{L-Me-val}]_3$$

Enniatin C

$$[\text{HIV}-\text{L-Me-leu}]_3$$

Beauvericin

$$[\text{HIV}-\text{L-Me-phe}]_3$$

The macrotetrolides or actins are derivatives of nonactinic acid with four lactonated moieties in the molecule:

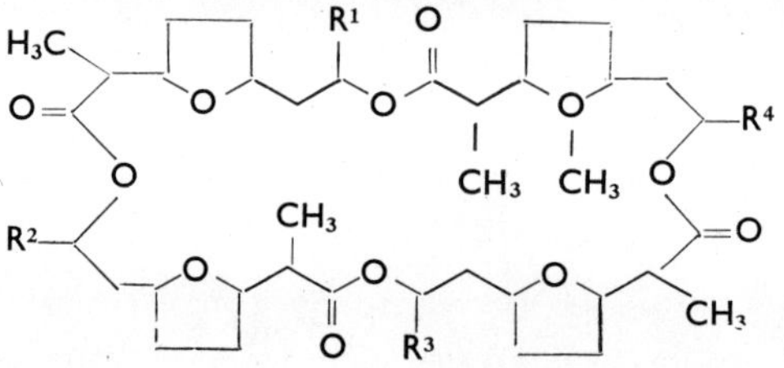

When $R^1 = R^2 = R^3 = R^4 = CH_3$, it is nonactin; when $R^1 = R^2 = R^3 = CH$ and $R^4 = C_2H_5$, it is monactin; when $R^1 = R^3 = CH_3$ and $R^2 = R^4 = C_2H_5$, it is dinactin; when $R^1 = CH_3$ and $R^2 = R^3 = R^4 = C_2H_5$, it is trinactin.

The cyclic polyethers are all artificial compounds of the basic formula

$$[CH_2CH_2O]_n$$

with various substituents at the methylene groupings. They are often called crowns, thus dibenzo-18-crown-6 is

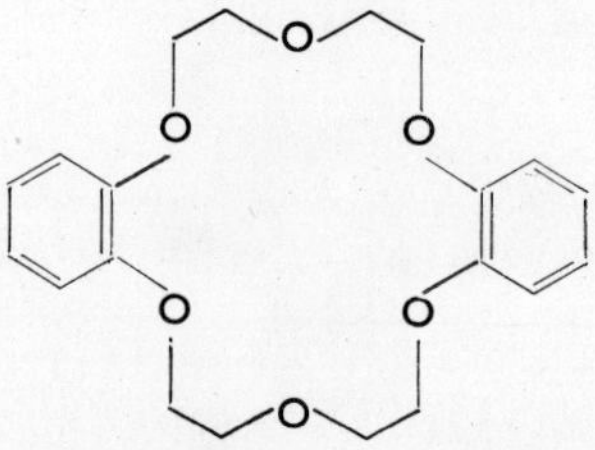

There are crown-type ionophores which are highly selective toward a particular ionic species. Thus

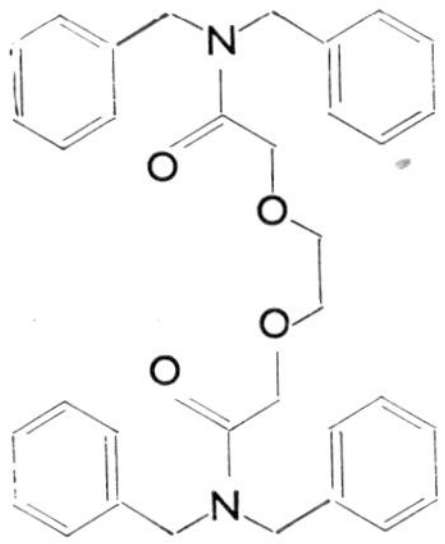

for Na⁺

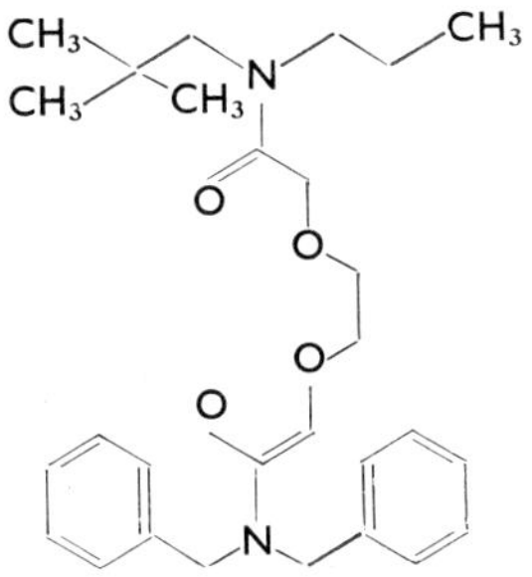

for Ca²⁺,

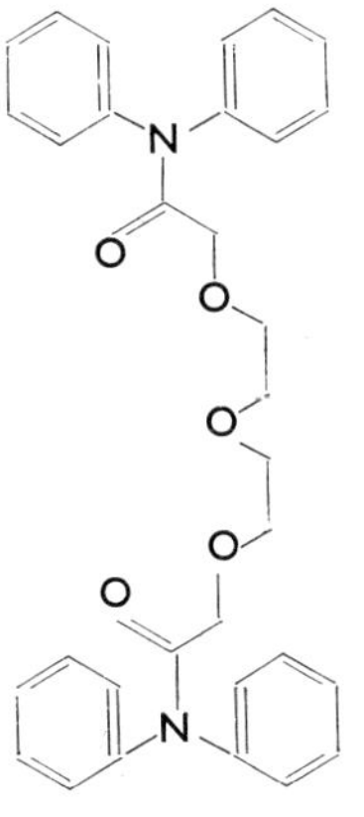

for Ba²⁺

Natural analogues to the cyclic polyethers become cyclic only on formation of a metal-binding complex (monensin or nigericin).

Monensin

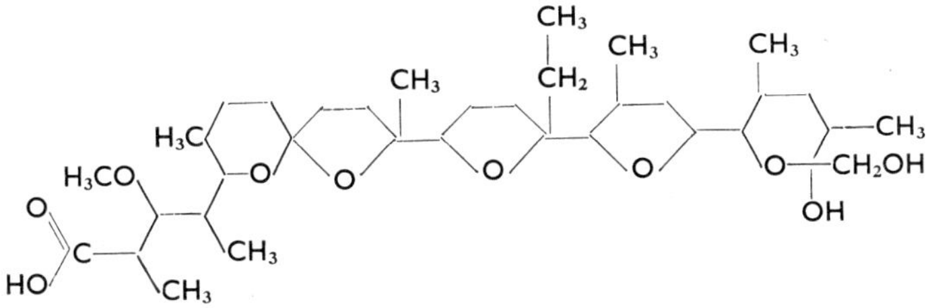

Nigericin

A special ionophore has been prepared that binds preferentially bivalent cations and was designated as X-537A (*cf.* Pressman and de Guzman, 1974). The relative transport capacity mediated by it for various ions is as follows: $Ca^{2+} : Cs^+ : Ba^{2+} : Sr^{2+} : Rb^+ : Na^+ = 100 : 99 : 75 : 45 : 38 : 33$.

X-537A

It is a characteristic of all ionophores that they are attached to the plasma or organelle membrane without any particular requirement for the lipid composition.*

Most ionophores, in particular valinomycin and the enniatins, actins and crowns, are known to form complexes with alkaline metal ions, some with remarkable selectivity especially for K^+. Thus the stability constant of the valinomycin-K^+ complex (in ethanol) is $1.9 \cdot 10^6\ M^{-1}$, while that of the valinomycin-Na^+ complex (in methanol) only $12\ M^{-1}$. Others, such as the enniatins and actins, display a difference of one order of magnitude in favor of K^+ while

---

* It is perhaps no coincidence that cyclic peptides of a completely different field of interest are similarly attached to plasma membranes—the phallotoxins and amatoxins found in the deadly toadstools of the genus *Amanita* which are now known to disrupt membranes causing leakage of Ca^{2+} and K^+ from cells.

monensin actually prefers Na$^+$ over K$^+$. The binding of ion in a coordination complex is accompanied by a conformational change of the molecule (*cf.* Ivanov *et al.*, 1969; Dobler *et al.*, 1969; Kilbourn *et al.*, 1967).

The resulting effect of the above ionophores is a facilitation of transport of one or more cations which, under some conditions, may result in an electrogenic building up of a membrane potential. Several examples of action of valinomycin on bacteria or bacterial vesicles, resulting in a "short-circuiting" for K$^+$ ions, K$^+$ efflux and generation of a transient membrane potential are cited in support of the membrane electrical potential difference acting as a driving force in the translocation of various nonelectrolytes.

Nigericin and monensin, having the same lipid solubility in the protonated and nonprotonated forms, facilitate the exchange (antiport) of K$^+$ for H$^+$ and of Na$^+$ for H$^+$, respectively. Unlike with valinomycin which in complex with a cation bears a positive charge, both monensin and nigericin are electrically neutral when carrying a cation.

In contrast with all the above-mentioned agents, gramicidin A apparently forms channels across the lipid membranes but these are not as large as to permit hydrodynamic flow (in distinction to amphotericin B, filipin and others; *cf.* p. 205). In fact, the channels can be somewhat cation-specific, the specificity sequence being:

$$H^+ > NH_4^+ \geqq Cs^+ > Rb^+ \geqq K^+ > Na^+ > Li^+$$

(*cf.* Hladky and Haydon, 1970). Unlike the previous ionophores, whose action ceases as membranes freeze (assume the crystalline state), the action of gramicidin A is not markedly affected even at the transition temperature.

For a most scholarly review of the subject of ionophores the reader is referred to Koryta (1973).

5.4. DISTRIBUTION AND ROLE OF ION TRANSPORT

As suggested in the introduction, transport of ions by cellular membranes is intimately connected with other life processes and serves them in many ways. On the one hand, the transport regulates the electrolyte composition of the cell interior and the free energy stored in the form of ion gradients across the cell membrane can be used in several ways; on the other hand, in more highly developed organisms the transcellular transport across specialized epithelial cells serves the homeostasis of the organism as a whole.

In the animal kingdom a prominent role is played by the active transport of sodium and potassium ions, mediated by membrane Na,K-ATPase (see p. 288), commonly called the sodium pump. Two important properties of the sodium pump, which are not often stressed, may be mentioned here: the pump is a reversible mechanism and the sodium-potassium coupling ratio seems to be a function of the opposing gradients. The reversibility of the pump is obvious from the existence of a critical energy barrier across which no further net extrusion of sodium from the cell can proceed; as found by Conway (Conway and Mullaney, 1961), pumping of sodium from muscle fibers is prevented by a sufficiently high gradient of sodium electrochemical potential. Moreover, in erythrocytes, ATP synthesis was shown to result from a backflow of sodium ions through the pump (Glynn and Lew, 1970). Thus, even if the passive permeability of the cell membrane to sodium ions (the sodium leak) were nil, the sodium pump would create only a definite sodium gradient in accordance with the Le Chatelier principle—a reversible mechanism is opposed and finally balanced by the gradient which it creates. An interesting positive feedback, a change in the sodium-to-potassium coupling ratio, seems to oppose the Le Chatelier effect. In snail neurones, the pump becomes more coupled and hence less electrogenic when it operates against a greater gradient (Kostyuk et al., 1972). Similarly, in epithelial cells of the anuran urinary bladder, the sodium pump appears to be electrogenic when it extrudes sodium ions downhill, into a choline or potassium solution (Essig and Leaf, 1963; Frazier and Leaf, 1963) but obligatorily tightly coupled when it extrudes the same ions against a gradient into a sodium solution (Janáček et al., 1972). Coupling diminishes the opposing effect of electrical forces and thus, at the same energy consumption rate, reduces the effort (see p. 178).

Perhaps the most spectacular exploitation of the sodium electrochemical potential gradient across the cell membrane is displayed in the generation of the action potential in nerve and muscle fibres. As explained by the well-known sodium theory (see p. 284) a sudden pronounced increase in membrane permeability to sodium which takes place in the membrane depolarized by local currents, belonging to a propagated electrical impulse, brings about a repolarization of the membrane. The polarity change, called the action potential, amplifies the signal and promotes its further propagation along cable-like structures.

The sodium leak, governed by the sodium gradient and membrane sodium permeability, apparently plays in many cases a more important role than that of a primitive overflow feedback dissipating the metabolic energy, being coupled to the inflow of nutrients and thus permitting their accumulation in cells by a secondary active transport (see p. 299). Thus the flow of sugars (Crane, 1965) and of amino acids (Schultz and Curran, 1970) across the mucosal membranes of intestinal cells as well as the accumulation of amino acids by ascites tumor cells (Heinz, 1972) are, according to considerable experimental evidence, governed by the sodium electrochemical potential gradient across the membrane.

The great importance of transcellular ion transport is best displayed in specialized organs like kidney, the remarkable function of which is assisted and the study of which is rendered difficult by the great morphological complexity of the organ. Simpler tissues, notably skins and urinary bladders of anurans, frogs and toads, have been extensively used in studies of transcellular ion transport. These organs not only survive for many hours in simple salt solutions at room temperature; they may be used as partitions in split chambers with identical solutions at the two sides and the electrical potential difference across the tissue reduced to zero by passing a definite electrical current across it from an external battery. This procedure forms the basis of the ingenious short-circuit current technique introduced by Ussing and Zerahn (1951). The current required to bring the spontaneous potential across the tissue to zero is an exact and convenient measure of active ion flows across the preparation, for there is no gradient of any electrochemical potential across the tissue under the above conditions and hence net passive ion flows come to a stop. With this technique it was demonstrated that in most anuran skins and in anuran bladders, sodium ions are almost exclusively subject to an *in vitro* active transcellular transport, since potassium ions pumped in the opposite direction return passively into their original compartment. An exception is represented by the skin of the South American frog, *Leptodactyllus ocellatus*, performing *in vitro* a considerable transcellular active chloride transport (Zadunaisky et al. 1963). Pharmacological actions of drugs and stimulatory effects of hormones (neurohypophysial hormones, aldosterone, insulin, etc.) were conveniently studied in this manner and metabolic relations of the active sodium transport were established. As shown by Vieira and co-workers (1972) each frog skin has its own stoichiometric ratio of sodium ions actively transported per molecule of oxygen consumed for the transport, which is changed neither by transport stimulation with antidiuretic hormone, nor by transport inhibition with ouabain (in individual skins ratios from 7.1 to 30.9 were found). On the other hand, it was demonstrated by Sarkadi and Schubert (1972) that the ratio varies with temperature, medium pH and sodium osmolarity, being lowest under physiological conditions, where the efficiency of energy conversion exhibited the highest value.

An interesting feature of ion transport in plant cells, as compared with the above mechanisms found in animals, is the relative importance of ATP-independent anionic pumps. Thus in the alga *Nitella translucens*, together with an active accumulation of potassium ions dependent on ATP synthesized by photophosphorylation an active accumulation of chloride anions directly coupled to the second photosynthetic system was demonstrated by MacRobbie (1965).

In microorganisms, including bacteria and yeasts, alkaline cations are translocated by systems not coupled to the Na,K-ATPase, the general observation being that of an $H^+ - K^+$ exchange. It appears that a proton-generating ATP-ase may be instrumental in the process. The transport of bivalent cations utilizes specific carrier systems, energized mostly by ATP. Anions, such as chloride, bromide, iodide, nitrate, are distributed probably passively across the membrane according to the existing membrane potential; on the other hand, phosphate and sulfate appear to require specific binding proteins as well as metabolic energy for transport.

SYNOPSIS

Thermodynamic characterization of membrane equilibria of ions in terms of measurable quantities, concentrations and electrical potential differences, permits to recognize stationary distributions of ions resulting from active transport. Combination of electrostatic and statistical laws gives an estimate of double-layer thickness and, thereby, a vivid picture of electrical relations and ion distributions in the proximity of membrane boundaries.

Steady-state membrane potentials across cellular membranes may be described by approximate solutions of the differential equation of electrodiffusion, accounting both for the steps of the electrical potential at the membrane boundaries and for electrogenic active ion transport. An alternative description of membrane potentials can be approached using electrical analogues.

The mechanism of active ion transport is linked to the biochemistry and comparative physiology of membrane ATPases. Mediated transport of ions, both natural and induced, is the result of chemical properties of a heterogeneous group of chelating agents, often called ionophores.

Active ion transport, distributed throughout the living world, creates ionic gradients serving as free energy reservoirs available for various physiological functions and contributes to the homeostasis of living systems. In animal cells the ATPase-mediated sodium and potassium transport appears to be of prevailing importance, in plant cells anion transport directly coupled to photosynthesis may be equally important, in microorganisms an exchange of K^+ for H^+ driven by various energy sources is almost ubiquitous.

6. TRANSPORT OF WATER

"*All Birds, Beasts and Fishes, Insects, Trees, and other Vegetables, with their several Parts, grow out of Water and watry Tinctures and Salts, and by Putrefaction return again into watry Substances.*"

Isaac Newton, Opticks

6.1. STEADY-STATE THERMODYNAMICS OF WATER PERMEATION

The whole field of water transport in cells and tissues was recently reviewed in an extensive and scholarly manner by House (1974). The content of the present chapter is much more modest being mostly limited to the phenomenological description of water transport across membranes by the methods of steady-state thermodynamics (see p. 172), as developed by Kedem and Katchalsky (1958).

The two most common and principal causes of water movement across a membrane from one compartment into another are the following: (*1*) An excess of hydrostatic pressure in the first compartment; (*2*) an excess of concentration of a solute to which the membrane is less permeable than to water, in the second compartment. The second type of water movement, known as osmosis, involves a number of interesting theoretical problems. The theory of osmotic water flow is relatively simple when the membrane behaves as ideally semipermeable, i.e., when it is permeable to water and completely impermeable to the solute. In general, however, the membrane is permeable, although not in the same degree, to both water and the solute. As may be easily envisaged, when water and the solute follow inside such a leaky membrane a common pathway (say, water-filled pores) the flows of the two interact frictionally and the magnitude of each depends on the degree of interaction.

As discussed previously (p. 173), interactions of this type between steady flows are satisfactorily treated by the thermodynamics of the steady state, which assumes each generalized flow in the system to be proportional to all generalized forces in the system. This approach is not free of theoretical criticism when applied to thin cell membranes which, according to Schlögl (1969), cannot be considered as a continuous phase. Still, for the time being, the treatment by Kedem and Katchalsky (1958) appears to provide the best, even if approximative, quantitative description of water flows across biological cell membranes. The simple case of ideally semipermeable membranes will be seen to be a limiting case of this description.

For simplicity, let there be only one solute present in the system, so that there are two flows across the membrane: flux of water, J_w, and flux of solute, J_s. Both of them are proportional to the gradient of chemical potential of water and the gradient of chemical potential of the solute:

$$J_w = L_{ww}\left(-\frac{d\mu_w}{dx}\right) + L_{ws}\left(-\frac{d\mu_s}{dx}\right), \quad (6.1)$$

$$J_s = L_{ss}\left(-\frac{d\mu_s}{dx}\right) + L_{sw}\left(-\frac{d\mu_w}{dx}\right). \quad (6.2)$$

As shown by Katchalsky (1961) the equations may be integrated, by the method of Kirkwood (1954), for steady-state conditions, when the fluxes are independent of the x-coordinate, giving

$$J_w = L'_{ww}\Delta\mu_w + L'_{ws}\Delta\mu_s, \quad (6.3)$$

$$J_s = L'_{ss}\Delta\mu_s + L'_{sw}\Delta\mu_w \quad (6.4)$$

where $\Delta\mu$'s are the differences of chemical potential of water and solute across the membrane, Onsager's law (p. 174) is satisfied by the new phenomenological cross coefficients

$$L'_{ws} = L'_{sw}. \quad (6.5)$$

It is now important to develop equations (6.3) and (6.4) in terms of directly measurable quantities, differences of solute concentration, Δc_s, and of the hydrostatic pressure, Δp, across the membrane. Assuming that equilibrium prevails at the membrane surfaces, the differences of chemical potentials between the two surfaces will be the same as those between the two solutions. As long as the solutions are considered ideal, the chemical potentials may be expressed as proportional to the logarithm of the mole fraction of the substance and also, by its partial molal volume, to hydrostatic pressure. The differences of chemical potential across the membrane thus can be written as

$$\mu_w = \bar{V}_w \Delta p + RT \Delta \ln x_w, \qquad (6.6)$$

$$\mu_s = \bar{V}_s \Delta p + RT \Delta \ln x_s. \qquad (6.7)$$

Since

$$x_w = \frac{c_w}{c_w + c_s} \quad \text{and} \quad x_s = \frac{c_s}{c_w + c_s} \qquad (6.8)$$

we may write

$$\Delta \ln x_w = \Delta \ln \frac{c_w}{c_w + c_s} = -\Delta \ln \frac{c_w + c_s}{c_w} =$$
$$= -\Delta \ln \left(1 + \frac{c_s}{c_w}\right) \approx -\Delta \frac{c_s}{c_w} = -\frac{\Delta c_s}{\tilde{c}_w} \qquad (6.9)$$

where $\tilde{c}_w$ is the mean concentration of water in the two solutions. The approximation $\ln(1 + x) \approx x$, valid for small x, was used here. Similarly,

$$\ln x_s = \Delta \ln \frac{c_s}{c_w + c_s} \approx \ln \frac{c_s}{c_w} =$$
$$= \Delta (\ln c_s - \ln c_w) \approx \Delta \ln c_s = \frac{\Delta c_s}{\tilde{c}_s} \qquad (6.10)$$

where $\tilde{c}_s$ is the mean concentration of the solute in the two solutions, approximations $c_w + c_s \approx c_w$ (since $c_w \gg c_s$) and $\Delta \ln c_w \approx 0$ having been used. Hence equations (6.6) and (6.7) may be written as

$$\Delta \mu_w = \bar{V}_w \Delta p - \frac{RT \Delta c_s}{\tilde{c}_w} \qquad (6.11)$$

$$\Delta \mu_s = \bar{V}_s \Delta p + \frac{RT \Delta c_s}{\tilde{c}_s}. \qquad (6.12)$$

The dissipation function (see p. 174) which represents the sum of products of individual fluxes and their conjugate forces may then be written as

$$\frac{d_i S}{dt} T = J_w \Delta\mu_w + J_s \Delta\mu_s = J_w \left(\overline{V}_w \Delta p - \frac{RT \Delta c_s}{\tilde{c}_w} \right) +$$

$$+ J_s \left(\overline{V}_s \Delta p + \frac{RT \Delta c_s}{\tilde{c}_s} \right) = (J_w \overline{V}_w + J_s \overline{V}_s) \Delta p +$$

$$+ \left(\frac{J_s}{\tilde{c}_s} - \frac{J_w}{\tilde{c}_w} \right) RT \Delta c_s = J_V \Delta p + J_D RT \Delta c_s. \tag{6.13}$$

Two new fluxes were thus defined; volume flow J_V

$$J_V = J_w \overline{V}_w + J_s \overline{V}_s \tag{6.14}$$

conjugate to the force Δp, and exchange flow, J_D, which corresponds to the velocity of solute relative to water in the membrane

$$J_D = \frac{J_s}{\tilde{c}_s} - \frac{J_w}{\tilde{c}_w} \tag{6.15}$$

conjugate to the force $RT \Delta c_s$, which is called osmotic pressure. The phenomenological equations may now be written in terms of new flows and forces as

$$J_V = L_p \Delta p + L_{pD} RT \Delta c_s, \tag{6.16}$$

$$J_D = L_D RT \Delta c_s + L_{pD} \Delta p \tag{6.17}$$

since Onsager's relation $L_{pD} = L_{Dp}$ is again valid.

The cross coefficient L_{pD} shows how large a volume flow is generated by osmotic pressure $RT \Delta c_s$ on a given membrane and how large an ultrafiltration is brought about on the same membrane by the difference in hydrostatic pressure Δp. With very coarse membranes, such as those of sintered glass, both osmosis and ultrafiltration are negligible, and hence $L_{pD} = 0$ for such membranes. On the other hand, with ideally semipermeable membranes, volume flow J_V is given solely by the volume flow of water

$$J_V = J_w \overline{V}_w \tag{6.18}$$

and

$$J_D = -\frac{J_w}{c_w} \tag{6.19}$$

since J_s in eq. (6.15) is zero for ideally semipermeable membranes. Assuming that the solutions are dilute, $1/\bar{c}_w \approx \bar{V}_w$ ($\approx 18 \text{ cm}^3 \text{ mol}^{-1}$), and from equations (6.18) and (6.19) it follows that

$$J_V + J_D = 0 \qquad (6.20)$$

for ideally semipermeable membranes. Combining eq. (6.20) with equations (6.16) and (6.17) we obtain

$$(L_p + L_{pD})\Delta p + (L_D + L_{pD})RT\Delta c_s = 0. \qquad (6.21)$$

This equation is satisfied for an ideally semipermeable membrane for all values of hydrostatic and osmotic pressure differences, so that, for such membranes, obviously

$$L_p = -L_{pD} = L_D. \qquad (6.22)$$

Thus, eq. (6.16) for an ideally semipermeable membrane becomes

$$J_V = L_p(\Delta p - RT\Delta c_s). \qquad (6.23)$$

This is an interesting equation since it shows that in an ideally semipermeable membrane the volume flow is proportional to the difference in hydrostatic pressure by the same coefficient as to the difference in osmotic pressure. This fact implies that hydrostatic and osmotic pressures generate volume flow by the same mechanism. Thus in a porous membrane, where hydrostatic pressure produces a mass flow through pores, osmotic flow must proceed by the same mechanism, rather than each water molecule diffusing separately in the concentration gradient of water and encountering on its way considerably higher friction. The origin of the osmotic mass flow was explained by Dainty and Meares (Dainty, 1963). No solute can penetrate into the pores of an ideally semipermeable membrane and hence there is a juxtaposition of a layer of solution and of a layer of pure water at the pore opening. The concentration of water being higher in the latter layer, the jumps of water molecules from it into the solution are more frequent than in the opposite direction and leave behind more vacancies. Hence the density and, as a result of this, the hydrostatic pressure, is less in this part of the pore than in the surroundings. The difference in the hydrostatic pressure is relieved by mass flow from the other end of the pore, rather than from the juxtaposed solution, where the movement of water is opposed by its concentration gradient. The osmotic pressure is thus seen to produce the mass flow *via* the hydrostatic pressure. The hydrostatic pressure is, however, localized only in the proximity of pore openings, does not distort the membrane and makes much higher pressures practicable in osmotic experiments than in hydrostatic experiments.

Coefficient L_p is called the hydraulic conductivity of the membrane and characterizes completely the permeability properties of an ideally semipermeable membrane. The general membrane, permeable also to the solute, requires for its description the knowledge of two other coefficients, L_D and L_{pD}, or better still, of two related coefficients, ω and σ, which are more convenient in the interpretation of experimental data. Their meaning and relation to the former coefficients may be derived as follows:

The flow of solute, J_s, is usually measured experimentally rather than exchange flow, J_D. In dilute solutions the flow of solute can be expressed using eq. (6.14), in which $J_s \bar{V}_s$ is very small as compared with $J_w \bar{V}_w$ and may be neglected, and eq. (6.15) as

$$J_s = (J_V + J_D)\tilde{c}_s. \tag{6.24}$$

The flow of solute, J_s, is often measured with tracers in the absence of volume flow, J_V. For $J_V = 0$ it follows from eq. (6.16) that

$$\Delta p = -\frac{L_{pD}}{L_p} RT \Delta c_s. \tag{6.25}$$

Substituting from eq. (6.25) and (6.17) into (6.24) for $J_V = 0$ we obtain

$$J_s = \frac{L_p L_D - L_{pD}^2}{L_p} \tilde{c}_s RT \Delta c_s. \tag{6.26}$$

Coefficient ω was introduced by Kedem and Katchalsky (1958) as

$$\omega = \frac{L_p L_D - L_{pD}^2}{L_p} \tilde{c}_s \tag{6.27}$$

so that eq. (6.26) becomes

$$J_s = RT\omega \, \Delta c_s \tag{6.28}$$

where $RT\omega$ is the permeability coefficient of a nonelectrolyte. Staverman's reflection coefficient σ was defined by Kedem and Katchalsky as

$$\sigma = -\frac{L_{pD}}{L_p}. \tag{6.29}$$

With this definition eq. (6.25) becomes

$$\Delta p = \sigma RT \Delta c_s. \qquad (6.30)$$

According to eq. (6.30) the volume flow across a membrane is zero when Δp, the observed osmotic pressure, is equal to the theoretical osmotic pressure $RT \Delta c_s$ multiplied by the reflection coefficient. For an ideally semipermeable membrane $\sigma = 1$, since then $L_p = L_{pD}$ (eq. 6.22). In leaky membranes $0 < \sigma < 1$ and the observed osmotic pressure, measured as hydrostatic pressure preventing volume flow, is less than the theoretical osmotic pressure. Phenomenological equations may now be rewritten using the new coefficients. From equations (6.16) and (6.29) the volume flow is

$$J_V = L_p(\Delta p - \sigma RT \Delta c_s) \qquad (6.31)$$

and the flow of solute can be expressed by introducing eq. (6.29) into (6.27), so that

$$\omega = (L_D - L_p \sigma^2) \tilde{c}_s \qquad (6.32)$$

and by combining equations (6.24), (6.31), (6.17) with (6.29) and (6.32), yielding

$$J_s = \tilde{c}_s L_p (1 - \sigma) \Delta p + \left[\omega - \tilde{c}_s L_p (1 - \sigma) \sigma \right] RT \Delta c_s. \qquad (6.33)$$

Finally, equations (6.33) and (6.31) combined give

$$J_s = RT\omega \Delta c_s + \tilde{c}_s (1 - \sigma) J_V. \qquad (6.34)$$

Equations (5.31) and (6.34) describe completely the simple transport of solute and water across a membrane.

Eq. (6.31) may be generalized to account for the presence of the difference of an impermeant solute concentration across the membrane, Δc_i:

$$J_V = L_p(\Delta p - RT \Delta c_i - \sigma RT \Delta c_s). \qquad (6.35)$$

In the absence of a hydrostatic pressure difference the volume flow vanishes when

$$RT \Delta c_i = \sigma RT \Delta c_s \qquad (6.36)$$

so that the reflection coefficient can be estimated as

$$\sigma = -\frac{\Delta c_i}{\Delta c_s}. \qquad (6.37)$$

Two more points deserve our attention. Especially with highly permeable solutes the effect of unstirred layers (see p. 198) becomes very important and the difference of solute concentration, Δc_s, between the solutions at the membrane surfaces can be considerably less than that between the bulks of the two solutions, so that appropriate corrections have to be applied (see p. 199).

Further it is of importance to realize, that although frictional interactions in the common pathway of water and the solute is the principal cause reducing the value of the reflection coefficient, the reflection coefficient of a permeating solute is always slightly less than one, even when the solute pathway and water pathway are distinct. Let us assume that the pathways in some membrane are indeed distinct. Then the volume flow across the solute-impermeable pathway is given solely by the volume flow of water and, moreover, the reflection coefficient is equal to unity for this pathway. Eq. (6.31) for this pathway becomes

$$J_w \bar{V}_w = L_p(\Delta p - RT \Delta c_s). \tag{6.38}$$

The volume flow of solute across the water-impermeable pathway under the condition of zero total volume flow (under which the reflection coefficient is measured) is, according to eq. (6.34), equal to

$$J_s \bar{V}_s = \bar{V}_s RT \omega \, \Delta c_s. \tag{6.39}$$

Since the sum of the two flows is zero (being equal to J_V), it follows that

$$\frac{p}{RT \Delta c_s} = 1 - \frac{\bar{V}_s \omega}{L_p} \tag{6.40}$$

which according to eq. (6.30), is equal to the measured value of the reflection coefficient. Hence the conclusion about a common pathway (pores) for solute and water is justified only when

$$\sigma < 1 - \frac{\bar{V}_s \omega}{L_p}. \tag{6.41}$$

6.2. THE STATE OF WATER IN CELLS

The accessibility of intracellular water to solutes penetrating across the cell membrane is of importance in considerations of equilibrium distribution between the external medium and the cell interior. The history of experimental approaches to the problem together with much useful information on the properties of water was recently reviewed by House (1974). The present chapter restricts itself to the description of the most recent development in two principal experimental approaches to the elucidation of the state of water in cells, viz., comparison of properties of intracellular and extracellular water by nuclear magnetic resonance, and direct measurement of the distribution of various solutes between the medium and the cell water.

Considerable progress in NMR studies of the state of intracellular water was achieved by Civan and Shporer (1972) who used $H_2{}^{17}O$ as the NMR probe to examine the state of water in the frog striated muscle. $H_2{}^{17}O$ not only resembles the properties of ordinary water better than ^2H_2O used previously but also, because of the large quadrupolar interaction of the ^{17}O nucleus in $H_2{}^{17}O$, it is particularly sensitive to changes in molecular movement. The authors found the total intensity of the ^{17}O signal of $H_2{}^{17}O$ in frog striated muscle to represent only about three-fourths of the maximum anticipated intensity. As shown by the authors, however, an immobilization of approximately one-fourth of the intracellular water by adsorption or binding to a solid phase represents only one (and apparently not the most probable) of the possible explanations of the reduction of the NMR signal in the muscle. As shown by the same authors (Shporer and Civan, 1972) already for NMR spectra of ^{23}Na, alternative interpretations are possible, viz., tumbling of a part of intracellular water molecules in anisotropic regions or a rapid exchange of free water molecules with a bound water fraction which may be very small.

These possibilities appear to be in better agreement with new data on nonelectrolyte distribution in muscle. As observed by Miller (1974), the distribution ratio between intracellular water (the content of which was determined by drying overnight at 98 °C) and external medium in mouse diaphragm is frequently equal to unity (for methanol, ethylene glycol, glycerol, 2-deoxyerythritol, 2-deoxy-D-ribose and, at 34 °C in the presence of insulin, also for D-xylose). The finding indicates that there is no appreciable non-solvent water in this tissue. Distribution ratios for other substances which are less than one appear to be adequately explained in terms of intracellular compartmentation, with membranes surrounding the individual compartments which are either impermeable for the substance in question or which extrude it actively.

SYNOPSIS

Whereas the water flow across an ideally semipermeable membrane can be adequately described as being proportional to the difference of hydrostatic and osmotic pressure across the membrane, in membranes which are leaky for solutes this simple description fails mainly because it cannot account for interactions between water flow and solute flow in the membrane. The best contemporary approach to the phenomenological description of water flow across membranes permeable to solutes is offered by the formalism of steady-state thermodynamics. The reflection coefficient introduced by this theory is a quantitative expression of the fact that the osmotic pressure of a permeating substance is less than that of a nonpermeating one at the same concentration, especially if there is a common path for water and solute in the membrane and the two flows can interact.

Concerning the state of water in cells, the modern interpretation of nuclear magnetic resonance studies does not appear to contradict the new data on the distribution of various nonelectrolytes between the cell and its surroundings: practically all cell water can act as a solvent and its reduced accessibility for solutes may be explained by intracellular compartmentation and active transport.

7. TRANSPORT BY SPECIAL MECHANISMS

"At least, I see nothing of Contradiction in all this."
  Isaac Newton, *Opticks*

There is considerable evidence showing that, along with carrier-mediated processes of transport of organic substrates or ions there are rather special mechanisms which permit the uptake or release of substances both in the low and in the high-molecular weight range.

7.1. OLIGOPEPTIDE PERMEASES

The first to be dealt with is actually a carrier-type mechanism which, however, can handle molecules of up to a molecular weight of nearly one thousand. Two such systems, oligopeptide permeases, are known from *Escherichia coli* (for a review see Payne and Gilvarg, 1971) and from *Salmonella typhimurium* (Ames *et al.*, 1973).

The operability of the system is somewhat surprising in view of findings with other compounds (in particular sugars) which cannot ride on a carrier once they exceed a certain size although they are bound to it (e.g., maltose binding but no transport in human erythrocytes; Beneš and Kotyk, 1976). On the other hand, the ability to transport oligopeptides may have some bearing on the rather unexpected finding of a freely reversible uptake of histone by *Escherichia coli* (Pavlasová and Stejskalová, 1972).

7.2. PINOCYTOSIS

There is another way of carrying large molecules across the cell membranes which is by nature indiscriminate but which, nonetheless, may function only after a certain stimulus is received. The mechanism is called pinocytosis and may be described morphologically as the transport of vesicles enclosing liquid drops, either from the cytoplasm outward (exocytosis; particularly in cells excreting protein products,

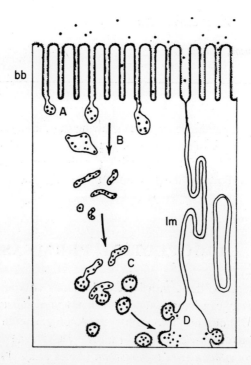

Fig. 7.1. Diagram of antibody transport across the brush-border membranes (bb) of the small intestine of a newborn rat. At A, antibodies are bound selectively to specialized pits in the membrane (*cf.* Fig. 2.35**D**); at B, the antibodies are absorbed by pinocytosis; at C, the pinocytotic vesicles turn into spherical coated vesicles; at D, the vesicles discharge the antibodies at the lateral membrane (lm) by reverse pinocytosis. (Adapted from Rodewald, R. (1973). *J. Cell Biol.* **58**, 189.)

such as the pancreatic acinar cells) or from the medium inward (endocytosis; most pronouncedly in protozoans but apparently also in various animal cells, notably in some tissue culture cells exhibiting an avidity for certain components of the surrounding medium, and

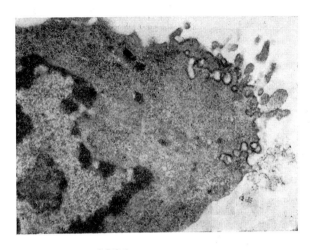

— 1000 nm

Fig. 7.2. An activated spleen lymphocyte from newborn pig showing abundant endocytosis of extraneous material and formation of pinocytotic vesicles. (Courtesy of Dr. I. Trebichavský, Institute of Microbiology, ČSAV, Prague.)

in the cells lining the blood capillaries). The sequence of events of pinocytosis, both inward and outward, is shown schematically in Fig. 7.1.

Inward pinocytosis obviously serves mainly the nutrition of cells and absorption by epithelial lining and sets in only after a certain food stimulus is received (apparently the outer envelope receptors are involved). Although this stimulus may be quite specific the cell engulfs with the drop of extracellular fluid inadvertently all that is contained in it (Fig. 7.2). This fact can be made use of for bringing into unwanted (e.g., cancer) cells displaying an increased endocytosis such compounds as will destroy them.

Outward pinocytosis (Fig. 7.3) is a process typical of all hormone-producing cells (cf. Fig. 2.54) and has been observed as the sole transport process for both acetylcholine and epinephrine from the nerve end plate to the muscle receptors across the synaptic

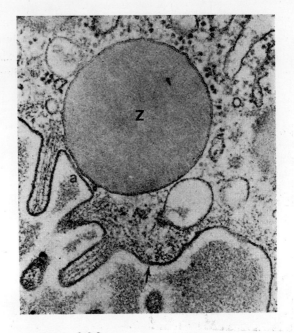

Fig. 7.3. Discharge of a zymogen granule (Z) from a pancreatic exocrine cell into the lumen of the duct. The point of confluence of the granule with the plasma membrane is shown at **a**. (Taken with permission from Jamieson, 1972, as in Fig. 2.54**B**.)

cleft. This fact brings out an important, previously unsuspected, ability of pinocytosis, *viz.* to proceed at a rapid rate, comparable with that of carrier transport so that a nerve impulse may be transmitted (Fig. 7.4).

7.3. UPTAKE OF NUCLEIC ACIDS AND SPECIAL PROTEINS

There is now little doubt that in a number of instances, both nucleic acids and proteins enter into cells after having traversed the membrane as such. A case in point is the transfer of genetic information (encoded in deoxyribonucleic acid) into a recipient competent bacterial cell.*

* Analogous processes probably take place in some animal cells.

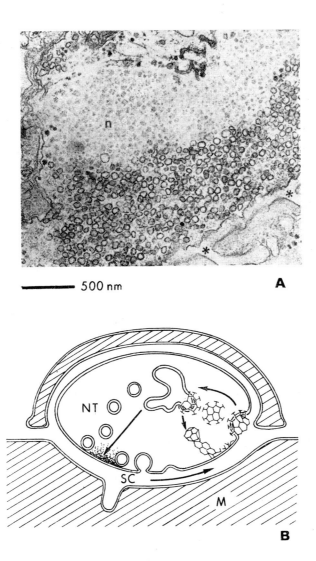

Fig. 7.4. **A**. A portion of the neuromuscular junction and muscle fiber of the pectoral muscle of a frog. Numerous synaptic vesicles are present in the nerve terminal (n) about to discharge into the synaptic cleft (marked with asterisks). (Reproduced from Ceccarelli, B., Hurlblut, W. P., and Mauro, A. (1973). *J. Cell Biol.* **57** 499.) **B**. Schematic drawing of the path of a sinaptic vesicle from the nerve terminal (NT) through the sympatic cleft (SC) and back. The content of the vesicle (acetylcholine or noradrenaline) is discharged in the cleft and trapped by the muscle (M) surface receptors. (Adapted from Heuser, J. E. and Reese, T. S. (1973). *J. Cell Biol* **57**, 315.)

If we disregard processes such as conjugation (which involves direct contact of heterosexual cells), transduction (which requires the penetration of the cell by a phage particle) there are transformation and transfection wherein a free DNA molecule of another bacterium or a phage, respectively, penetrates the competent cell.

The state of competence itself requires the transmission of an endogenously produced and plasma membrane-bound protein called the activator or competence factor, to a noncompetent cell. The competence factor has a molecular weight of about 10 000 and is rather species-specific. At the bacterial surface it is bound to a receptor of about 50 000 mol. wt. and in an interaction with it "sets the stage" for the admission of the transformation DNA. Protein and RNA synthesis are required for the preparatory process. An agglutinin is formed at this stage which modifies the surface features of the cytoplasmic membrane.

After arriving at the cytoplasmic membrane (apparently in regions of cell envelope growth; *cf.* the mesosome) the DNA is adsorbed to the surface similarly to a virus particle — an irreversible energy-independent process, whereupon it is drawn in by an energy-requiring process and, if homologous, is incorporated into the bacterial chromosome. The process is completed in a matter of seconds or at most minutes. A molecular description of the penetration of DNA through the membrane is not possible at present (for more information see, e.g., Tomasz, 1973).

Likewise, the uptake of proteins, be it colicins or interferon, obviously requires specific receptor sites at the membrane surface which then (possibly by a cooperative process) alter the local permeability in such a way that even the large molecules can be drawn in. It is of interest that some of the receptors involved in nucleic acid uptake are actually nonresponsive to binding of polynucleotides of molecular weights lower than 10^5.

Analogies between the state of membrane proteins (e.g., Na,K-adenosinetriphosphatase) and the uptake activity toward large molecules or even viruses stress the necessity of regarding the membrane as an interacting continuum (*cf.* Koch and Fehér, 1973).

One is tempted to visualize a part of a mosaic fluid membrane to be dislodged upon binding the large molecule in such a way as to permit its penetration which may be accomplished either by a coordinated binding-and-releasing process or simply by the local physical forces, such as surface tension or tendency toward maximum hydrophobic-hydrophobic interactions in the membrane. Once the protein is bound to the receptor which is mobile in the membrane, one should not be surprised by very fast reorientation or rotation of the whole complex—relaxation times of 10^1-10^2 µs have been calculated for molecules of mol. wt. about 50 000 in lipid membranes (Sackmann *et al.*, 1973).

Although this is obviously in the realm of speculations one should keep in mind the dynamic character of the membrane and the possibility of fast-spreading (cooperatively transmitted) signals along its surface which can, within fractions of a second, alter the permeability properties either generally or at a site responsible for a particular process. But there is much to be learned yet and we should strive with zest and zeal to learn it.

SYNOPSIS

Large molecules, such as proteins and nucleic acids, generally do not use membrane carriers but cross the cell membrane either by pinocytosis or by specific, enzyme-driven processes. Pinocytosis, however, may serve the transport even of small molecules, such as neuro-transmitters at the synaptic cleft.

REFERENCES

Abercrombie, M. and Heaysman, J. E. M. (1953). *Exptl. Cell Res.* **5**, 111.
Abragam, A. (1961). *The Principles of Nuclear Magnetism.* Oxford University Press, London.
Abrahamsson, S., Pascher, I., Larsson, K., and Karlsson, K. A. (1972). *Chem. Phys. Lipids* **8**, 152.
Abrams, A. and Baron, C. (1967). *Biochemistry* **6**, 225.
Agin, D. (1967). *Proc. Natl. Acad. Sci. USA* **57**, 1232.
Agin, D. (1972). In: *Foundations of Mathematical Biology* (ed. by Rosen, R.), vol. I, p. 253. Academic Press, New York—London.
Aksamit, R. and Koshland, D. E. Jr. (1972). *Biochem. Biophys. Res. Commun.* **48**, 1348.
Ames, G. F. and Lever, J. (1970). *Proc. Nat. Acad. Sci.* **66**, 1096.
Ames, G. F. and Lever, J. (1972). *J. Biol. Chem.* **247**, 4309.
Ames, B. N., Ames, G. F.—L., Young, J. D., Tsuchiya, D., and Lecocq, J. (1973). *Proc. Natl. Acad. Sci. USA* **70**, 456.
Anraku, Y. (1968*a*). *J. Biol. Chem.* **243**, 3116.
Anraku, Y. (1968*b*). *J. Biol. Chem.* **243**, 3123.
Anraku, Y. (1968*c*). *J. Biol. Chem.* **243**, 3128.
Atkinson, A., Gatenby, A. D., and Lowe, A. C. (1971). *Nature* **233**, 145.
Avruch, J. and Fairbanks, G. (1972). *Proc. Natl. Acad. Sci. USA* **69**, 1216.
Baddiley, J., Blumson, N. L., and Douglas, L. J. (1968). *Biochem. J.* **110**, 565.
Bangham, A. D. (1968). In: *Progress in Biophysics and Molecular Biology* (ed. by Butler, J. A. V. and Noble, D.). Pergamon Press, New York.
Bangham, A. D., Hill, M. W. and Miller, N. G. A. (1974). In: *Methods in Membrane Biology* (ed. by Korn, E. D.), vol. **1**, p. 1. Plenum Press, New York—London.
Barash, H. and Halpern, Y. S. (1971). *Biochem. Biophys. Res. Commun.* **45**, 681.
Baraud, J., Mawrice, A., and Napias, C. (1970). *Bull. Soc. Chim. Biol.* **52**, 421.
Beneš, I. and Kotyk, A. (1976). *Can. J. Biochem.*, **54**, 99.
Benson, A. A. (1968). In: *Membrane Models and the Formation of Biological Membranes* (ed. by Bolis, L. and Pethica, D. A.). North Holland, Amsterdam.

Berger, E. A. and Heppel, L. A. (1972). *J. Biol. Chem.* **247**, 7684.
Berman, M., Shahn, E., and Weiss, M. F. (1962). *Biophys. J.* **2**, 275.
Bervaes, J. C. A. M. and Kuiper, P. J. C. (1975). Personal communication.
Betz, A. L., Gilboe, D. D. and Drewes, L. R. (1975). *Biochim. Biophys. Acta* **401**, 416.
Beychok, S. (1967). In: *Poly-α-amino Acids—Protein Models for Conformation Studies* (ed. by Fasman, G. D.). Marcel Dekker, New York.
Blasie, J. K. and Worthington, C. R. (1969). *J. Mol. Biol.* **39**, 417.
Bloj, B., Morero, R. O., and Farias, R. W. (1973). *FEBS Lett.* **38**, 101.
Bodnaryk, R. P. (1972). *Canad. J. Biochem.* **50**, 524.
Bonting, S. L. (1970). In: *Membranes and Ion Transport* (ed. by Bittar, E. E.), vol. I, p. 257. Willey Interscience, London.
Boos, W., Gordon, A. S., Hall, R. E., and Price, H. D. (1972). *J. Biol. Chem.* **247**, 917.
Bretscher, M. S. (1973). *Science* **181**, 622.
Briggs, G. E., Hope, A. B., and Robertson, R. N. (1961). *Electrolytes and Plant Cells*. Blackwell Scientific Publications, Oxford.
Britton, H. G. (1966). *J. Theoret. Biol.* **10**, 28.
Bruckdorfer, K. R., Demel, R. A., de Gier, J., and van Deenen, L. L. M. (1969). *Biochim. Biophys. Acta* **181**, 334.
Caccam, J. F., Jackson, J. J., and Eylar, E. N. (1969). *Biochem. Biophys. Res. Commun.* **35**, 505.
Cantrell, A. C. (1973). *Biochim. Biophys. Acta* **311**, 381.
Caspar, P. L. D. and Kirschner, D. A. (1971). *Nature New Biol.* **231**, 46.
Chapman, D. (1969). *Lipids* **4**, 251.
Chapman, D., Kamat, V. B., de Gier, J., and Penkett, S. A. (1968). *J. Mol. Biol.* **31**, 101.
Chapman, D. and Wallach, D. F. H. (1968). In: *Biological Membranes* (ed. by Chapman, D.) p. 125. Academic Press, London—New York.
Chapman, D. and Dodd, G. H. (1971). In: *Structure and Function of Biological Membranes* (ed. by Rothfield, L. I.), p. 13. Academic Press, New York—London.
Chapman, D. and Urbina, J. (1971). *FEBS Lett.* **12**, 169.
Cho, K. Y. and Salton, M. R. J. (1966). *Biochim. Biophys. Acta* **116**, 73.
Ciani, S. (1965). *Biophysik* **2**, 368.
Ciani, S. M., Eisenman, G., Laprade, R., Szabo, G. (1973). In: *Membranes* (ed. by Eisenman, G.), Vol. **2**, p. 61. Marcel Dekker, New York.
Civan, M. M. and Shporer, M. (1972). *Biophys. J.* **12**, 404.
Clarke, M. (1971). *Biochem. Biophys. Res. Commun.* **45**, 1063.
Cleland, W. W. (1963). *Biochim. Biophys. Acta* **67**, 104.
Colacicco, G. (1972). *Ann. N. Y. Acad. Sci.* **195**, 224.
Cole, K. S. (1965). *Physiol. Rev.* **45**, 340.
Coleman, R. (1973). *Biochim. Biophys. Acta* **300**, 1.
Collander, R. (1949). *Physiol. Plant.* **2**, 300.
Conway, E. J. and Mullaney, M. (1961). In: *Membrane Transport and Metabolism* (ed. by Kleinzeller, A. and Kotyk, A.), p. 117. Academic Press, New York.
Cotman, C., Brown, D. H., Harrell, B. W., and Anderson, N. G. (1970). *Arch. Biochem. Biophys.* **136**, 436.

Crane, R. K. (1965). *Fed. Proc.* **24**, 1000.
Cronan, J. E. Jr., Birge, C. H., and Vagelos, P. R. (1969). *J. Bacter.* **100**, 601.
Cronan, J. E. Jr., Roy, T. K., and Vagelos, R. P. (1970). *Proc. Natl. Acad. Sci. USA* **65**, 737.
Cuppoletti, J. and Segel, I. H. (1975). *J. Theoret. Biol.* **53**, 125.
Curran, P., Hajjar, J. J., and Glynn, I. M. (1970). *J. Gen. Physiol.* **55**, 297.
Curtis, A. S. G. (1967). *The Cell Surface.* Logos Press and Academic Press, London.
Dainty, J. (1960). *Symp. Soc. Exptl. Biol.* **14**, 140.
Dainty, J. (1963). *Adv. Bot. Res.* **1**, 279.
Davson, H. and Danielli, J. F. (1952). *The Permeability of Natural Membranes.* Cambridge Univ. Press, Cambridge.
Dawson, R. M. C., Herrington, N., and Lindsay, D. B. (1960). *Biochem. J.* **77**, 226.
Deák, T. and Kotyk, A. (1968). *Folia Microbiol.* **13**, 212.
Dehlinger, P. J. and Schimke, R. T. (1971). *J. Biol. Chem.* **246**, 2574.
de Kruyff, B., van Dijck, P. W. M., Goldbach, R. W., Demel, R. A., and van Deenen, L. L. M. (1973). *Biochim. Biophys. Acta*, **330**, 269.
Denbigh, K. G. (1951). *The Thermodynamics of the Steady State.* Methuen & Co., Ltd. London.
Dennis, V. W., Stead, N. W., and Andreoli, T. E. (1970). *J. Gen. Physiol.* **55**, 375.
De Robertis, E. (1971). *Science*, **171**, 963.
Desai, J. D. and Modi, V. V. (1975). *Experientia* **31**, 160.
Dewey, M. M. and Barr, L. (1970). In: *Current Topics in Membranes and Transport* (ed. by Bronner, F. and Kleinzeller, A.), vol. I, p. 1. Academic Press, New York—London.
Diesterhaft, M. and Freese, E. (1974). *Fed. Proc.* **33**, Abstr. 964.
Djerassi, C. (1960). *Optical Rotatory Dispersion.* Mc Graw—Hill, New York.
Dobler, M., Dunitz, J. D., and Krajewski, J. (1969). *J. Mol. Biol.* **42**, 603.
Dreyer, W. J., Papermaster, D. S., and Kühn, H. (1972). *Ann. N. Y. Acad. Sci.* **195**, 61.
Edsall, J. T. and Wyman, J. (1958). *Biophysical Chemistry.* Academic Press, New York.
Einstein, A. (1905). *Ann. Physik* **17**, 549.
Engelman, D. M. and Morowitz, H. J. (1968). *Biochim. Biophys. Acta* **150**, 385.
Ernster, L. and Kuylenstierna, B. (1970). In: *Membranes of Mitochondria and Chloroplasts* (ed. by Racker, E.), p. 172. Van Nostrand—Reinhold, New York.
Essig, A. and Leaf, A. (1963). *J. Gen. Physiol.* **46**, 505.
Faust, R. G. and Shearin, S. J. (1974). *Nature* **248**, 60.
Fettiplace, R., Andrews, D. M. and Haydon, D. A. (1971). *J. Membr. Biol.* **5**, 277.
Finean, J. B. (1958). *Exptl. Cell Res.* Suppl. 5, 18.
Finean, J. B. (1973). In: *Form and Function of Phospholipids* (ed. by Ansell, G. B., Hawthorne, J. N., and Dawson, R. M. C.), p. 171. Elsevier, Amsterdam—London—New York.
Finkelstein, A. and Mauro, A. (1963). *Biophys. J.* **3**, 215.
Folch-Pi, J. and Stoffyn, P. J. (1972). *Ann. N. Y. Acad. Sci.* **195**, 86.

Fox, C. F., Law, J. H., Tsukagoshi, N., and Wilson, G. (1970). *Proc. Natl. Acad. Sci. USA* **67**, 598.
Franklin, T. J. (1974). In: *Industrial Aspects of Biochemistry* (ed. by Spencer, B.), p. 549. North Holland/American Elsevier, Amsterdam—London—New York.
Frazier, H. S. and Leaf, A. (1963). *J. Gen. Physiol.* **46**, 491.
Freysz, L., Bieth, R., and Mandel, P. (1969). *J. Neurochem.* **16**, 1417.
Frizzel, R. A. and Schultz, S. G. (1970). *J. Gen. Physiol.* **56**, 462.
Fromm, H. J. (1970). *Biochem. Biophys. Res. Commun.* **40**, 692.
Frommhertz, P. (1970). *FEBS Lett.* **11**, 205.
Frye, L. D. and Edidin, M. (1970). *J. Cell. Sci.* **7**, 391.
Fukui, S. and Isobe, K. (1973). *Biochim. Biophys. Acta*, **328**, 114.
Fullmer, C. S. and Wasserman, R. H. (1973). *Biochem. Biophys. Acta* **317**, 172.
Furlong, C. E. and Weiner, J. H. (1970). *Biochem. Biophys. Res. Commun.* **38**, 1076.
Fürth, R. (1956). In: A. Einstein, *Investigations on the Theory of the Brownian Movements* (ed. with notes by Fürth, R.), p. 101. Dover Publications, New York.
Galliard, T. (1973). In: *Form and Function of Phospholipids* (ed. by Ansell, G. B., Hawthorne, J. N., and Dawson, R. M. C.), p. 253. Elsevier, Amsterdam—London—New York.
Garrahan, P. J. and Glynn, I. M. (1967). *J. Physiol.* **192**, 237.
Geck, P. (1971). *Biochim. Biophys. Acta*, **241**, 462.
Getz, G. S. (1972). In: *Membrane Molecular Biology* (ed. by Fox, C. F. and Keith, A.), p. 386. Sinauer Ass., Stamford, Conn.
Ghuysen, J. M. and Shockman, G. D. (1973). In: *Bacterial Membranes and Walls* (ed. by Leive, L.), p. 37. Marcel Dekker, New York.
Glynn, I. M. and Lew, V. L. (1970). *J. Physiol.* **207**, 393.
Goldman, D. E. (1943). *J. Gen. Physiol.* **27**, 37.
Goldstein, D. A. and Solomon, A. K. (1960). *J. Gen. Physiol.* **44**, 11.
Goodwin, T. W. (1973). In: *Lipids and Biomembranes of Eukaryotic Microorganisms* (ed. by Erwin, J. A.), p. 1. Academic Press, New York—London.
Gorter, E. and Grendel, F. (1925). *J. Exptl. Med.* **41**, 439.
Green, D. E. (1974). *Biochim. Biophys. Acta*, **346**, 27.
Green, D. E. and Perdue, J. F. (1966). *Proc. Natl. Acad. Sci. USA* **55**, 1295.
Green, D. E., Haard, N. F., Lenaz, G., and Silman, H. I. (1968). *Proc. Natl. Acad. Sci. USA* **60**, 277.
Green, D. E. and Ji, S. (1972). *Bioenergetics* **3**, 159.
Green, D. E., Ji, S., and Brucker, R. F. (1972). *Bioenergetics* **4**, 527.
Greenfield, N., Davidson, B., and Fasman, G. (1967). *Biochemistry* **6**, 1630.
Greville, G. D. (1969). In: *Current Topics in Bioenergetics* (ed. by Sanadi, D. R.), vol. 3, p. 1. Academic Press, New York.
Griffith, T. W., Carraway, C., and Leach, F. R. (1971). *Fed. Proc.* **30**, 1115 Abs.
Grisham, C. M. and Barnett, R. E. (1973). *Biochemistry* **12**, 2635.
Gulik-Krzywicki, T., Shechter, E., Luzzati, V., and Faure, M. (1969). *Nature* **223**, 1116.
Gunstone, F. D. (ed.) (1970). *Topics in Lipid Chemistry*. Logos Press, London.
Gutfreund, H. (1972). *Enzymes: Physical Principles*. Wiley — Interscience, London.

Györgyi, S. and Sugár, I. (1974). *Abstr. 9th FEBS Meeting* Budapest, p. 245.
Hardman, K. D. and Ainsworth, C. F. (1973). *Biochemistry* **12**, 4442.
Hansch, C. and Fujita, T. (1964). *J. Am. Chem. Soc.* **86**, 1616.
Harold, F. M., Baarda, J. R., Baron, C., and Abrams, A. (1969). *J. Biol. Chem.* **244**, 2261.
Hartley, G. S. and Crank, J. (1949). *Trans. Faraday Soc.* **45**, 801.
Helmkamp, G. M. Jr., Harvey, M. S., Wirtz, K. W. A. and van Deenen, L. L. M. (1974). 9th FEBS Meeting, Budapest, Abstracts of Communications, p. 366.
Haurowitz, F. (1963). *The Chemistry and Function of Proteins*. Academic Press, New York.
Hazelbauer, G. L. and Adler, J. (1971). *Nature New Biol.* **230**, 101.
Hechemy, K. and Goldfine, H. (1971). *Biochem. Biophys. Res. Commun.* **42**, 245.
Hechter, O. (1965). *Fed. Proc.* **24** (2), S91.
Heinz, E. (ed.) (1972). *Na-Linked Transport of Organic Solutes*, p. 1. Springer Verlag, Berlin.
Heller, J. and Lawrence, M. A. (1970). *Biochemistry* **9**, 864.
Henderson, P. (1907). *Z. physik. Chem.* **59**, 118.
Henderson, P. (1908). *Z. physik. Chem.* **63**, 325.
Henning, U. and Schwartz, U. (1973). In: *Bacterial Membranes and Walls* (ed. by Leive, L.), p. 413. Marcel Dekker, New York.
Heppel, L. A. (1971). In: *Structure and Function of Biological Membranes* (ed. by Rothfield, L. I.), p. 223. Academic Press, New York—London.
Hill, A. V. (1928). *Proc. Roy. Soc. B.* **104**, 39.
Himmelspach, K., Westphal, O., and Teichman, B. (1971). *Eur. J. Immunol.* **1**, 106.
Hinman, N. D. and Phillips, A. H. (1970). *Science* **170**, 1222.
Hladky, S. B. and Haydon, D. A. (1970). *Nature*, **225**, 451.
Hochstadt-Ozer, J. and Stadtman, E. R. (1971). *J. Biol. Chem.* **246**, 5304.
Hodgkin, A. L. and Horowicz, P. (1959). *J. Physiol.* **148**, 127.
Hoffman, P. G. and Tosteson, D. C. (1971). *J. Gen. Physiol.* **58**, 438.
Hogg, R. and Englesberg, E. (1969). *J. Bacteriol.* **100**, 423.
Hokin, L. E. (1974). *Ann. N. Y. Acad. Sci.* **242**, 12.
Holden, J. T. (1968). *J. Theoret. Biol.* **21**, 97.
Holzwarth, G. (1972). In: *Membrane Molecular Biology* (ed. by Fox, C. F. and Keith, A.), p. 228. Sinauer Ass., Stamford, Conn.
Horák, J. and Kotyk, A. (1973). *Eur. J. Biochem.* **32**, 36.
House, C. R. (1974). *Water Transport in Cells and Tissues*. Edward Arnold (Publishers) Ltd., London.
Hsu, C. C. and Fox, C. F. (1970). *J. Bacter.* **103**, 410.
Hubbard, A. L. H. and Cohn, Z. C. (1972). *J. Cell. Biol.* **55**, 390.
Hubbell, W. L. and McConnell, H. M. (1971). *J. Am. Chem. Soc.* **93**, 314.
Hybl, A. and Dorset, D. (1970). *Biophys. Soc. Abstr.* **49**a.
Israelachvili, J. N. and Mitchell, D. J. (1975). *Biochim. Biophys. Acta* **389**, 13.
Itzkowitz, M. S. (1967), *J. Chem. Phys.* **46**, 3048.
Ivanov, V. T., Laine, I. A. Abdulaev, N. D., Senyavina, L. B., Popov, E. M., Ovchinnikov, Yu. A. and Shemyakin, M. M. (1969). *Biochem. Biophys. Res. Commun.* **34**, 803.
Jacobs, M. H. (1952). In: *Modern Trends in Physiology and Biochemistry* (ed. by E. S. G. Barron), p. 149. Academic Press, New York.

Jacobs, M. H. (1967). *Diffusion Processes*. Springer Verlag, Berlin—Heidelberg—New York (reprint from Ergebnisse der Biologie, Zwölfter Band, 1935).
Jacobs, M. H., Glassman, H. N. and Parpart, A. K. (1950). *J. Exptl. Zool.* **113**, 277.
Jaffe, L. F. (1974). *J. Theoret. Biol.* **48**, 11.
Jain, M. K., Strickholm, A., and Cordes, E. H. (1969). *Nature* **222**, 871.
Jain, M. K., White, F. P., Strickholm, A., Williams, E., and Cordes, E. H. (1972). *J. Membrane Biol.* **8**, 363.
Jakoby, W. D. (ed.) (1971). *Methods in Enzymology*, vol. XXII. Academic Press, New York—London.
Jakovcic, S., Getz, G. S., Rabinowitz, M., Jakob, H., and Swift, H. (1971). *J. Cell. Biol.* **48**, 490.
Janáček, K., Rybová, R., and Slavíková, M. (1972). *Biochim. Biophys. Acta* **288**, 221.
Ji, T. H. and Urry, D. W. (1969). *Biochem. Biophys. Res. Commun.* **34**, 404.
Johnson, A. R. and Davenport, J. B. (1971). *Biochemistry and Methodology of Lipids*. Wiley and Sons, New York.
Johnson, J. A. and Wilson, T. A. (1967). *J. Theoret. Biol.* **17**, 304.
Johnson, W. C., Silhavy, T. J. and Boos, W. (1975). *Appl. Miccobiol.* **29**, 405.
Jones, T. H. D. and Kennedy, E. P. (1968). *Fed. Proc.* **27**, 644.
Jost, J. P. and Rickenberg, H. V. (1971). *Annu. Rev. Biochem.* **40**, 741.
Jost, P., Wagoner, A. S., and Griffith, O. H. (1971). In: *Structure and Function of Biological Membranes* (ed. by Rothfield, L. I.), p. 83. Academic Press, New York—London.
Kaback, H. R. (1972). *Biochim. Biophys. Acta* **265**, 367.
Kagawa, Y. and Racker, E. (1971). *J. Biol. Chem.* **246**, 5477.
Kanemasa, Y., Akamatsu, Y., and Nojima, S. (1967). *Biochim. Biophys. Acta* **144**, 382.
Katchalsky, A. (1961). In: *Membrane Transport and Metabolism* (ed. by Kleinzeller, A. and Kotyk, A.), p. 69. Academic Press, New York.
Katchalsky, A. and Curran, P. F. (1965). *Nonequilibrium Thermodynamics in Biophysics*. Harvard University Press, Cambridge, Massachusetts.
Katz, B. (1966). *Nerve, Muscle, and Synapse*. Mc Graw-Hill, New York.
Kauzmann, N. (1959). *Adv. Prot. Chem.* **14**, 1.
Kedem, O. (1961). In: *Membrane Transport and Metabolism* (ed. by Kleinzeller, A. and Kotyk, A.), p. 87. Academic Press, New York.
Kedem, O. (1972). *J. Membrane Biol.* **10**, 213.
Kedem, O. and Katchalsky, A. (1958). *Biochim. Biophys. Acta* **27**, 229.
Kellermann, O. and Szmelcman, S. (1974). *Eur. J. Biochem.* **47**, 139.
Kepner, G. R. and Macey, R. I. (1968). *Biochim. Biophys. Acta* **163**, 188.
Kilbourn, B. T., Dunitz, J. D., Piodo, L. A. R., and Simon, W. (1967). *J. Mol. Biol.* **30**, 559.
King, E. L. and Altman, C. (1956). *J. Phys. Chem.* **60**, 1375.
Kirkwood, J. G. (1954). In: *Ion Transport across Membranes* (ed. by Clark, H. T.), p. 119. Academic Press, New York.
Klenk, H.D. (1973). In: *Biological Membranes* (ed. by Chapman, D. and Wallach, D. F. H.), vol. **2**, p. 145. Academic Press, London—New York.
Koch, A. S. and Fehér, G. (1973). *J. Gen. Vir.* **18**, 319.

Konings, W. N., Barnes, E. M. Jr., and Kaback, H. R. (1971). *J. Biol. Chem.* **246**, 5857.
Korn, E. D. (1968). *J. Gen. Physiol.* **52**, Suppl. 257.
Kornfeld, R. and Kornfeld, S. (1970). *J. Biol. Chem.* **245**, 2536.
Koryta, J. (1973). *Chem. listy* **67**, 897.
Kostyuk, P. G., Krishtal, O. A. and Pidoplichko, V. I. (1972). *J. Physiol.* **226**, 373.
Kotyk, A. (1973). *Biochim. Biophys. Acta* **300**, 183.
Kotyk, A. (1974). In: *Biophysics of Membrane Transport* (ed. by Miękisz, S. and Gomułkiewicz, J.), vol. **1**, p. 49. Inst. Plant Biol. Biophys. Agric. Academy, Wrocław.
Kotyk, A. and Řihová, L. (1972). *Biochim. Biophys. Acta* **288**, 380.
Kotyk, A. and Janáček, K. (1975). *Cell Membrane Transport: Principles and Techniques*. Plenum Press, New York.
Kroger, A. and Klingenberg, M. (1970). *Vitamins Hormones* **28**, 533.
Kubišta, V. (1974). *Vesmír* **53**, 230.
Kulaev, I. S. (1975). *Rev. Physiol. Biochem. Pharmacol.* **73**, 131.
Kundig, W., Ghosh, S., and Roseman, S. (1964). *Proc. Natl. Acad. Sci. USA* **52**, 1067.
Kundig, W. and Roseman, S. (1971). *J. Biol. Chem.* **246**, 1407.
Kuzuya, H., Bromwell, K. and Guroff, G. (1971). *J. Biol. Chem.* **246**, 6371.
Kyte, J. (1971). *J. Biol. Chem.* **246**, 4157.
Ladbrooke, B. D., Jenkinson, T. J., Kamat, V. B., and Chapman, D. (1968). *Biochim. Biophys. Acta* **164**, 101.
Laico, M. T., Ruoslahti, E. I., Papermaster, D. S., and Dreyer, W. J. (1970). *Proc. Natl. Acad. Sci. USA* **67**, 120.
Langmuir, I. (1933). *J. Chem. Phys.* **1**, 756.
Lapetina, E. G. and Michell, R. H. (1973). *FEBS Lett.* **31**, 1.
Law, J. H. and Snyder, W. R. (1972). In: *Membrane Molecular Biology* (ed. by Fox, C. L. and Keith, A. D.), p. 3. Sinauer Ass., Stamford, Conn.
LeFevre, P. G. (1973). *J. Membrane Biol.* **11**, 1.
LeFevre, P. G. (1975). *Ann. N. Y. Acad. Sci.* **264**, 398.
LeFevre, P. G. and McGinnis, G. F. (1960). *J. Gen. Physiol.* **44**, 87.
Lester, R. L., Smith, S. W., Wells, G. B., Rees, D. C. and Angus, W. W. (1974). *J. Biol. Chem.* **249**, 3388.
Lieb, W. R. and Stein, W. D. (1969). *Nature* **224**, 240.
Lieb, W. R. and Stein, W. D. (1970). *Biophys. J.* **10**, 585.
Lieb, W. R. and Stein, W. D. (1971). *J. Theoret. Biol.* **30**, 219.
Lieb, W. R. and Stein, W. D. (1974). *Biochim. Biophys. Acta* **373**, 178.
Linden, C. D., Wright, K. L., McConnell, H. M., and Fox, C. F. (1973). *Proc. Natl. Acad. Sci. USA* **70**, 2271.
Ling, G. N. and Gerard, R. W. (1949). *J. Cell. Comp. Physiol.* **34**, 382.
London, Y., Demel, R. A., van Kessel, W. S. M. G., Zahler, P., and van Deenen, L. L. M. (1974). *Biochim. Biophys. Acta* **332**, 69.
Longley, R. P., Rose, A. H., and Knights, B. A. (1968). *Biochem. J.* **108**, 401.
Lowenstein, J. M. (ed.) (1969). *Methods in Enzymology*, vol. XIV. Academic Press, New York—London.
Lucy, J. A. and Glauert, A. (1964). *J. Mol. Biol.* **8**, 727.
Lutz, M. and Breton, J. (1973). *Biochem. Biophys. Res. Commun.* **53**, 413.

Luzzati, V. (1968). In *Biological Membranes* (ed. by Chapman, D.), p. 71. Academic Press, New York.
Machtiger, N. A. and Fox, C. F. (1973). *Annu. Rev. Biochem.* **42,** 575.
MacInnes, D. A. (1961). *The Principles of Electrochemistry.* Dover Publications, New York.
MacRobbie, E. A. C. (1965). *Biochim. Biophys. Acta* **94,** 64.
Markham, R., Frey, S., and Hills, G. J. (1963). *Virology* **20,** 88.
Markin, V. S. and Chizmadjev, Yu. A. (1974). *Indutsirovannyi ionnyi transport.* Izd. Nauka, Moscow.
Martonosi, A. (1968). *J. Biol. Chem.* **243,** 71.
Matile, P. and Wiemken, A. (1967). *Arch. Mikrobiol.* **56,** 148.
McElhaney, R. and Tourtellotte, M. E. (1969). *Science* **164,** 433.
McElhaney, R. N., de Gier, J., and van Deenen, L. L. M. (1970). *Biochim. Biophys. Acta* **219,** 245.
Medweczky, N. and Rosenberg, H. (1969). *Biochim. Biophys. Acta* **192,** 369.
Medweczky, N. and Rosenberg, H. (1970). *Biochim. Biophys. Acta* **211,** 158.
Meissner, G. (1973). *Biochim. Biophys. Acta* **298,** 906.
Meister, A. (1973). *Science* **180,** 33.
Melancon, M. J. and DeLuca, H. F. (1970). *Biochemistry* **9,** 1658.
Miller, C. (1974). *Biochim. Biophys. Acta* **339,** 71.
Miller, D. M. (1965). *Biophys. J.* **5,** 417.
Mitchell, P. (1966). *Biol. Rev.* **41,** 445.
Mitchell, P. (1973). *Bioenergetics* **4,** 265.
Moffitt, W. and Moscowitz, A. (1959). *J. Chem. Phys.* **30,** 648.
Montal, M. and Mueller, P. (1972). *Proc. Natl. Acad. Sci. USA* **69,** 3561.
Morowitz, H. J. and Terry, T. (1969). *Biochim. Biophys. Acta* **183,** 276.
Mueller, P., Rudin, D. O., Ti Tien, H., and Wescott, W. C. (1962). *Nature* **194,** 979.
Mueller, P., Rudin, D. O., Ti Tien, H., and Wescott, W. C. (1964). In *Recent Progress in Surface Science* (ed. by Daneilli, J. F., Pankhurst, K. G. A., and Riddiford, A. C.), vol. **1,** p. 379. Academic Press, New York—London.
Mueller, P. and Rudin, D. O. (1968). *J. Theoret. Biol.* **18,** 222.
Mueller, P. and Rudin, D. O. (1969). *Curr. Topics Bioenergetics* **3,** 157.
Mullis, K. B., Pollack, J. R., and Neilands, J. B. (1971). *Biochemistry* **10,** 4894.
Naftalin, J. (1970). *Biochim. Biophys. Acta* **211,** 65.
Nägeli, K. W. and Cramer, K. (1855). *Pflanzenphysiologische Untersuchungen.* F. Schultess, Zürich.
Nakamura, K., Ostrovsky, D. N., Miyazawa, T., and Mizushima, S. (1974). *Biochim. Biophys. Acta* **332,** 329.
Nakao, M., Nakao, T., Hara, Y., Nagai, F., Yagasaki, S., Koi, M., Makagawa, A. and Kawai, K. (1974). *Ann. N. Y. Acad. Sci.* **242,** 24.
Neal, J. L. (1972). *J. Theoret. Biol.* **35,** 113.
Nikaido, H. (1973). In: *Bacterial Membranes and Walls* (ed. by Leive, L.), p. 131. M. Dekker, New York.
Nishimune, T. and Hayashi, R. (1973). *Biochim. Biophys. Acta* **328,** 124.
Njus, D., Sulzman, F. M., and Hastings, J. W. (1974). *Nature* **248,** 116.
Nossal, N. G. and Heppel, L. (1966). *J. Biol. Chem.* **241,** 3055.
Nurminen, T. and Suomalainen, H. (1971). *Biochem. J.* **125,** 963.
O'Brien, J. S. (1967). *J. Theoret. Biol.* **15,** 307.

O'Brien, I. G. and Gibson, F. (1970). *Biochim. Biophys. Acta* **215**, 393.
Ohta, H., Matsumoto, J., Kagano, K., Fujita, M., and Nakao, M. (1971). *Biochem. Biophys. Res. Commun.* **42**, 1127.
Okaya, Y. (1964). *Acta Crystallogr.* **17**, 1276.
Opekarová, M., Kotyk, A., Horák, J. and Kholodenko, V. (1975). *Eur. J. Biochim.* **59**, 373.
Osborn, M. J. (1969). *Annu. Rev. Biochem.* **38**, 501.
Oster, G. F., Perelson, A. S., and Katchalsky, A. (1973). *Quart. Rev. Biophys.* **6**, 1.
Ostwald, T. J. and MacLennan, D. H. (1974). *J. Biol. Chem.* **249**, 974.
Overbeek, J. T. G. (1952). In: *Colloid Science* (ed. by Kruyt, H. R.), vol. I, p. 115. Elsevier Publ. Co., Amsterdam.
Overath, P., Schairer, H.U., Hill, F. F., and Lamnek-Hirsch, I. (1971a). In: *The Dynamic Structure of Cell Membranes* (ed. by Wallach, D. F. H. and Fischer, H.), p. 149. Springer-Verlag, New York.
Overath, P., Hill, F. F., and Lamnek-Hirsch, I. (1971b). *Nature New Biol.* **234**, 264.
Overath, P. and Träuble, H. (1973). *Biochemistry* **12**, 2625.
Pardee, A. B. (1967). *Science* **156**, 1627.
Patterson, P. H. and Lennarz, W. J. (1970). *Biochem. Biophys. Res. Commun.* **40**, 408.
Pavlasová, E. and Stejskalová, E. (1972). *Folia Microbiol.* **17**, 471.
Payne, J. W. and Gilvarg, C. (1971). *Adv. Enzym.* **35**, 187.
Penniston, J. T., Beckett, L., Bentley, D. L. and Hansch, C. (1969). *Mol. Pharmacol.* **5**, 333.
Phillips, D. R. and Morrison, M. (1971). *FEBS Lett.* **18**, 95.
Phillips, S. K. and Cramer, W. A. (1973). *Biochemistry* **12**, 1170.
Pitot, H. C., Sladek, N., Raglaud, W., Murray, R. K., Moyer, G., Soling, H. D., and Jost, J. P. (1969). In: *Microsomes and Drug Oxidation* (ed. by Gillette, J. R.), p. 59. Academic Press, New York.
Planck, M. (1890). *Ann. Physik* **40**, 561.
Polissar, M. J. (1954). In: *The Kinetic Basis of Molecular Biology* (by Johnson, F. H., Eyring, H., and Polissar, M. J.), p. 515. John Wiley and Sons, New York.
Pollack, J. R. and Neilands, J. B. (1970). *Biochem. Biophys. Res. Commun.* **38**, 989.
Post, R. L. and Orcutt, B. (1973). In: *Organization of Energy-transducing Membranes* (ed. by Nakao, M. and Pocker, L.), p. 25. Tokyo University Press, Tokyo.
Pressman, B. C. (1970). In: *Membranes of Mitochondria and Chloroplasts* (ed. by Racker, E.), p. 213. Van Nostrrand-Reinhold, New York.
Pressman, B. C. and de Guzman, N. T. (1974). *Ann. N. Y. Acad. Sci.* **227**, 380.
Racker, E. (1970). In: *Membranes of Mitochondria and Chloroplasts* (ed. by Racker, E.), p. 127. Van Nostrand-Reinhold, New York.
Racker, E., Horstman, L. L., Kling, D., and Fessenden-Raden, J. M. (1969). *J. Biol. Chem.* **244**, 6668. Ratledge, C. and Marshall, B. J. (1972). *Biochim. Biophys. Acta* **279**, 58.
Ratledge, C. and Marshall, B. J. (1972). *Biochim. Biophys. Acta* **279**, 58.
Razin, S. (1972). *Biochim. Biophys. Acta* **265**, 241.
Redwood. W. R., Müldner, H., and Thompson, T. E. (1969). *Proc. Natl. Acad. Sci. USA* **64**, 989.

Regen, D. M. and Morgan, H. E. (1964). *Biochim. Biophys. Acta* **79**, 151.
Regen, D. M. and Tarpley, H. L. (1974). *Biochim. Biophys. Acta* **339**, 218.
Renkin, E. M. (1954). *J. Gen. Physiol.* **38**, 225.
Repke, K. R. H. and Schön, R. (1973). *Acta Biol. Med. Germ.* **31**, 19 K.
Reusch, V. M. Jr. and Burger, M. M. (1973). *Biochim. Biophys. Acta* **300**, 79.
Richardson, S. H., Hultin, H. O., and Green, D. E. (1963). *Proc. Natl. Acad. Sci. USA* **50**, 821.
Robertson, J. D. (1959), *Biochem. Soc. Symp.* **16**, 3.
Robinson, J. D. (1970). *Arch. Biochem. Biophys.* **139**, 17.
Romeo, D., Girard, A., and Rothfield, L. (1970a). *J. Mol. Biol.* **53**, 475.
Romeo, D., Girard, A., and Rothfield, L. (1970b). *J. Mol. Biol.* **53**, 491.
Rorive, G., Nielsen, R., and Kleinzeller, A. (1972). *Biochim. Biophys. Acta* **266**, 376.
Rosen, B. P. (1971). *J. Biol. Chem.* **246**, 3653.
Rosen, B. P. (1973). *J. Biol. Chem.* **248**, 1211.
Rosen, B. P. and Vasington, F. D. (1971). *J. Biol. Chem.* **246**, 5351.
Rosenberg, T. and Wilbrandt, W. (1957). *J. Gen. Physiol.* **41**, 289.
Rosenberg, T. and Wilbrandt, W. (1963). *J. Theoret. Biol.* **5**, 288.
Rosenheck, K. and Doty, P. (1961). *Proc. Natl. Acad. Sci. USA* **47**, 1775.
Rotman, B. and Papermaster, B. W. (1966). *Proc. Natl. Acad. Sci. USA* **55**, 134.
Rouser, G., Nelson, G. J., Fleischer, S., and Simon, G. (1968). In: *Biological Membranes* (ed. by Chapman, D.), p. 5. Academic Press, London—New York.
Sackmann, E. and Träuble, H. (1972). *J. Am. Chem. Soc.* **94**, 4482.
Sackmann, E., Träuble, H., Galla, K.J., and Overath, P. (1973). *Biochemistry* **12**, 5360.
Sarkadi, B. and Schubert, A. (**1972**). *Acta Biochim. Biophys. Acad. Sci. Hung.* **7**, 367.
Schleif, R. (1969). *J. Mol. Biol.* **46**, 185.
Schlögl, R. (1964). *Stofftransport durch Membranen.* Dr. D. Steinkopff-Verlag, Darmstadt.
Schlögl, R. (1969). *Quart. Rev. Biophys.* **2**, 305.
Schnaitman, C. A. (1969). *Proc. Natl. Acad. Sci. USA* **63**, 412.
Schnaitman, C. A. (1970). *J. Bacter.* **104**, 890.
Schnebli, H. P. and Abrams, A. (1970). *J. Biol. Chem.* **245**, 1115.
Schultz, S. G. and Curran, P. F. (1970). *Physiol. Rev.* **50**, 637.
Schwencke, J. S., Farias, G. and Rojas, M. (1971). *Eur. J. Biochem.* **21**, 137.
Seelig, J. (1970). *J. Am. Chem. Soc.* **92**, 3881.
Seelig, J., Axel, F. and Limacher, H. (1973). *Ann. N. Y. Acad. Sci.* **222**, 588.
Shamoo, A. E. (1974). *Ann. N. Y. Acad. Sci.* **242**, 389.
Shporer, M. and Civan, M. M. (1972). *Biophys. J.* **12**, 114.
Sigler, K. and Janáček, K. (1971). *Biochim. Biophys. Acta* **241**, 528.
Singer, S. J. and Nicolson, G. L. (1972). *Science* **175**, 720.
Siñeriz, F., Farías, R. N., and Trucco, R. E. (1973). *FEBS Lett.* **32**, 30.
Sjöstrand, F. S. (1963). *J. Ultrastruct. Res.* **9**, 561.
Sjöstrand, F. S. (1968). In: *Regulatory Functions of Biological Membranes* (ed. by Järnefelt, J.), p. 11. Elsevier Publ. Co., Amsterdam.
Skou, J. C. (1972). *Bioenergetics* **4**, 203.

Slater, E. C. (1958). *Rev. Pure Appl. Chem.* **8**, 221.
Slayman, C. (1973). In: *Current Topics in Membranes and Transport* (ed. by Bronner, F. and Kleinzeller, A.), vol. **4**, p. 1. Academic Press, New York—London.
Sonnenberg, M. (1971). *Proc. Natl. Acad. Sci. USA* **68**, 1051.
Spanner, D. C. (1964). *Introduction to Thermodynamics*. Academic Press, London—New York.
Steck, T. L. (1974). *J. Cell Biol.* **62**, 11.
Steck, T. L. and Wallach, D. F. H. (1970). *Methods Cancer Res.* **5**, 93.
Steck, T. L. and Fox, C. F. (1972). In: *Membrane Molecular Biology* (ed. by Fox, C. F. and Keith, A. D.), p. 27. Sinauer Ass., Stamford, Conn.
Stein, W. D. (1967). *The Movement of Molecules across Cell Membranes*. Academic Press, New York—London.
Stein, W. D. and Danielli, J. F. (1956). *Disc. Faraday Soc.* **21**, 238.
Stein, W. D., Lieb, W. R., Karlish, S. J. D., and Eilam, Y. (1973). *Proc. Natl. Acad. Sci. USA* **70**, 275.
Shipley, G. G., Leslie, R. B., and Chapman, D. (1969). *Nature* **222**, 561.
Storelli, C., Vögeli, H., and Semenza, G. (1972). *FEBS Lett.* **24**, 287.
Suomalainen, H., Nurminen, T., and Oura, E. (1973). In: *Progress in Industrial Microbiology* (ed. by Hockenhull, B. J. D.), vol. **13**, p. 109. Churchill Livingstone, Edinburgh—London.
Szubinska, B. (1971). *J. Cell Biol.* **49**, 747.
Takacs, B. J. and Holt, S. C. (1971). *Biochim. Biophys. Acta* **233**, 278.
Taylor, R. T., Norrell, S. A., and Hanna, M. L. (1972). *Arch. Biochem. Biophys.* **148**, 366.
Teorell, T. (1953). In: *Progress in Biophysics and Biophysical Chemistry* (ed. by Butler, J. A. V. and Randall, J. T.), vol. **3**, p. 305. Academic Press, New York.
Thomas, L. (1973). *Biochim. Biophys. Acta* **291**, 454.
Thomas, R. C. (1972). *Physiol. Rev.* **52**, 563.
Tien, H. T., Carbone, S., and Dawidowicz, E. A. (1966). *Nature* **212**, 718.
Tomasz, A. (1973). In: *Bacterial Membranes and Walls* (ed. by Leive, L.), p. 321. M. Dekker, New York.
Tourtellotte, M. E. (1972). In: *Membrane Molecular Biology* (ed. by Fox, C. F. and Keith, A. D.), p. 439. Sinauer Ass., Stamford, Conn.
Träuble, H. (1971). *J. Membrane Biol.* **4**, 193.
Träuble, H. and Haynes, D. H. (1971). *Chem. Phys. Lipids* **7**, 324.
Träuble, H. and Overath, P. (1973). *Biochim. Biophys. Acta* **307**, 491.
Träuble, H. and Eibl, H. (1974). *Proc. Natl. Acad. Sci. USA* **71**, 214.
Tsukagoshi, N., Fielding, P., and Fox, C. F. (1971). *Biochem. Biophys. Res. Commun.* **44**, 497.
Tsukagoshi, N. and Fox, C. F. (1973*a*), *Biochemistry* **12**, 2816.
Tsukagoshi, N. and Fox, C. F. (1973*b*). *Biochemistry* **12**, 2822.
Tsukagoshi, N., Tamura, G., and Arima, K. (1970*a*). *Biochim. Biophys. Acta* **196**, 204.
Tsukagoshi, N., Tamura, G., and Arima, K. (1970*b*). *Biochim. Biophys. Acta* **196**, 211.
Uesugi, S., Dulak, N., Dixon, J. F., Hexum, T. D., Dahl, J. L., Perdue, J. F., and and Hokin, L. E. (1971). *J. Biol. Chem.* **246**, 531.

Ussing, H. H. (1949). *Acta Physiol. Scand.* **19**, 43.
Ussing, H. H. (1960). In: *The Alkali Metal Ions in Biology*. Handbuch der experimentellen Pharmakologie, vol. **13**, p. 10. Springer-Verlag, Berlin—Göttingen—Heidelberg.
Ussing, H. H. and Zerahn, K. (1951). *Acta Physiol. Scand.* **23**, 110.
Vandenheuvel, F. A. (1963). *J. Am. Oil Chem. Soc.* **40**, 455.
Vanderkooi, G. and Capaldi, R. A. (1972). *Ann. N. Y. Acad. Sci.*, **195**, 135.
Vanderkooi, G. and Green, D. E. (1970). *Proc. Natl. Acad. Sci. USA* **66**, 615.
Vanderkooi, G. and Sundaralingam, M. (1970). *Proc. Natl. Acad. Sci. USA* **67**, 233.
van Zupthen, H., Demel, R. A., Norman, A. W., and van Deenen, L. L. M. (1971). *Biochim. Biophys. Acta* **241**, 310.
Verkeleeij, A. J., Zwaal, R. F. A., Roelofsen, B., Comfurius, P., Kastelijn, D., and van Deenen, L. L. M. (1973). *Biochim. Biophys. Acta* **323**, 178.
Verwey, E. J. W. and Overbeek, J. T. C. (1948). *Theory of the Stability of Lyophobic Colloids*. Elsevier Publ. Co., New York.
Vidaver, G. A. and Shepherd, S. L. (1968). *J. Biol. Chem.* **243**, 6140.
Vieira, F. L., Caplan, S. R., and Essig, A. (1972). *J. Gen. Physiol.* **59**, 60.
Vincent, J. M. and Skoulios, A. E. (1966). *Acta crystallogr.* **20**, 432.
von Mohl, H. (1851). *Grundzüge der Anatomie und Physiologie der vegetabilischen Zelle*. Braunschweig.
von Szyszkowski, B. (1908). *Z. physik. Chem.* **64**, 385.
Wallach, D. F. H. (1967). In: *The Specificity of Cell Surfaces* (ed. by Davis, B. D. and Warren, L.), p. 129. Prentice-Hall, Englewood Cliffs, N. J.
Wallach, D. F. H. (1969). *J. Gen. Physiol.* **54**, 3s.
Wallach, D. F. H. (1971). In: *The Dynamic Structure of Cell Membranes* (ed. by Wallach, D. F. H. and Fischer, H.), p. 181. Springer-Verlag, New York.
Wallach, D. F. H. (1972). *The Plasma Membrane*. Springer-Verlag, New York—Heidelberg—Berlin.
Wallach, D. F. H. and Gordon, A. S. (1968). In: *Regulatory Functions of Biological Membranes* (ed. by Järnefelt, J.), p. 11. Elsevier Publ. Co., Amsterdam.
Wallach, D. F. H. and Lin, P. S. (1973). *Biochim. Biophys. Acta* **300**, 211.
Warren, G. B., Toon, P. A., Birdsall, N. J. M., Lee, A. G. and Metcalfe, J. C. (1974). *Proc. Natl. Acad. Sci. USA* **71**, 622.
Weiner, J. H., Berger, E. A., Hamilton, M. N., and Heppel, L. A. (1970). *Fed. Proc.* **29**, 341.
Weiner, J. H. and Heppel, L. A. (1971). *J. Biol. Chem.* **246**, 6933.
Wiebelhaus, V. D., Sung, C. P., Helander, H. F., Shah, G., Blum, A. L. and Sachs, G. (1971). *Biochim. Biophys. Acta* **241**, 49.
Wiemken, A. and Nurse, P. (1973). *Proc. IIIrd Internat. Special. Symp. Yeasts*, Part II, p. 331. Print OY, Helsinki.
Wilbrandt, W. (1954). *Symp. Soc. Expt. Biol.* **8**, 136.
Wiley, W. R. (1970). *J. Bacter.* **103**, 656.
Willis, R. C. and Furlong, C. E. (1974). *Fed. Proc.* **33**, Abstr. no. 963.
Wilson, G. and Fox, C. F. (1971a). *J. Mol. Biol.* **55**, 49.
Wilson, G. and Fox, C. F. (1971b). *Biochem. Biophys. Res. Commun.* **44**, 503.
Winne, D. (1973). *Biochim. Biophys. Acta* **298**, 27.
Winzler, R. J. (1969). In: *Red Cell Membrane Structure and Function* (ed. by Jamieson, G. A. and Greenwalt, T. J.), p. 157. Lippincott, Philadelphia.

Wirtz, K. W. A. and Zilversmit, D. B. (1968). *J. Biol. Chem.* **243,** 3596.
Wirtz, K. W. A., Kamp, H. H. and van Deenen, L. L. M. (1972). *Biochim. Biophys. Acta* **274,** 606.
Wolfe, L. S. (1964). *Can. J. Biochem.* **42,** 971.
Wong, J. T. F. and Hanes, J. (1962). *Can. J. Biochem. Physiol.* **40,** 763.
Yamashita, S. and Racker, E. (1969). *J. Biol. Chem.* **244,** 1220.
Yang, J. T. (1967). In: *Poly-α-amino Acids* (ed. by Fasman, G.), p. 239. M. Dekker, New York.
Yariv, J. J., Kalb, Katchalski, E., Goldman, R., and Thomas, E. W. (1969). *FEBS Lett.* **5,** 173.
Young, I. G., Cox, C. B. and Gibson, F. (1967). *Biochim. Biophys. Acta* **141,** 319.
Zadunaisky, J. A., Candia, O. A., and Chiardini, D. J. (1963). *J. Gen. Physiol.* **47,** 393.
Zahler, P. (1969). In: *Modern Problems of Blood Preservation*, p. 1. G. Fischer-Verlag, Stuttgart.

SUBJECT INDEX

Absorbance, 65, 67
Absorption spectra, 65
Absorptivity, 65
Accumulation ratio, 223, 225, 230, 232
4-Acetamido'-4-isothiocyanostilbene-2,2'-disulfonic acid, 82, 124
Acetate kinase, 133
Acetylcholine, 110, 146
Acetylcholinesterase, 133
Acholeplasma, 37, 48, 84, 88
Acholeplasma laidlawii, 52, 55, 56, 91, 92
Acinar cell, 102, 157, 319
Aconitate hydratase, 150
Actinomycin D, 204
Actins, 300
Action potential, 284, 302
Active transport, 170, 177, 222, 239—241
Active transport, secondary, 170, 177, 241, 303
Activity coefficient, 260
Acyl-carrier protein, 110
Acyl-CoA dehydrogenase, 108
Acyl-CoA synthetase, 109, 150
Adenosinetriphosphatase (ATPase), 58, 107, 110, 133, 235, 294
Adenosinetriphosphatase, Ca, 109, 293*ff*
Adenosinetriphosphatase, Mg, 92, 93, 133
Adenosinetriphosphatase, Na,K, 92, 99, 106, 109, 133, 220, 288*ff*
Adenylate cyclase, 106, 109, 146, 147
Adenylate kinase, 150
Adrenocorticotropin, 147
Aerobacter aerogenes, 52, 233
Aerococcus, 136, 138
Agglutinins, 131, 147, 322
Agglutinogen BA, 141
Agrobacterium tumefaciens, 237
Alamethicin, 296—297
Aldolase, 100

Aldosterone, 304
Alkenyl-glycerophosphinicocholine hydrolase, 109
Aluminum, 146
Amanita, 300
Amatoxins, 300
Amine oxidase (flavin-containing), 107, 125, 150
Amino-acid oxidase, 125
Aminopeptidase, 163
Ammonium ion, 289
Amoeba, 37, 145
Amphipathic lipids, 28
Amphotericin B, 204—205, 301
1-Anilinonaphthalene-8-sulfonate, 80, 82
Anionic pumps, 304
Antamanide, 296—297
Antibody transport, 318
Antigen CA, 141
Antigenicity, 130
Antigens, 60
Antitoxins, 147
Apyrase, 125
Arabinogalactan, 143
Arabinose, 235, 237
Arboviruses, 61
Arginine, 235, 236
Arrhenius plot, 91, 92, 93
Arsanilinochloromethoxyacridine, 82
Arthrobacter crystallopoietes, 136
Artificial membranes, 95*ff*, 287
Arylsulfatase, 125
Ascites tumor cells (see also Ehrlich ascites cells) 37, 56, 303
Ascosterol, 43
Asparaginase, 58
Aspartate aminotransferase, 150
Aspergillus nidulans, 237
Auxotrophs, 113
Azalomycin, 204

Bacillus cereus, 156
Bacillus licheniformis, 52, 156

Bacillus medusa, 156
Bacillus megaterium, 37, 55, 107, 134, 136, 156, 235, 296
Bacillus polymyxa, 134
Bacillus subtilis, 107, 155, 156, 206, 233, 235
Bacillus thuringiensis, 156
Bacitracin, 138, 204
Bacteriorhodopsin, 101
Baker's yeast (see also *Saccharomyces cerevisiae*), 37, 42, 47, 56, 133, 217
Barium, 146, 289
Beauvericin, 297—298
Beryllium, 289
Bilayer, lipid, 86ff, 93, 95, 96, 98, 101
Bile fronts, 56
Binding proteins, 58, 100
— nonelectrolyte, 107, 238
— ion, 107, 294
Bladder (urinary, frog or toad), 303
Blood group substances, 42
Boltzmann's statistical law, 251—252
Brassicasterol, 43
Bromostearic acid, 114—115
Brownian movement, 188
Brush border, intestinal, 235, 237, 318
Butyribacterium rettgeri, 137

Cadmium, 146
Calcium, 86, 107, 132, 145—146, 289, 293ff
Calorimetry, 82
Campestrol, 43
Candida beverwijkii, 224
Candida utilis, 37
C-antigen, 141
Capacitance, 179
Capacity, electrical, 244, 247
Cardiolipin, 39, 46, 48, 109, 148, 206
Carnitine palmitoyltransferase, 150
Carotene, 44, 154
Carrier, 183, 207ff, 214, 220, 227, 229, 231, 238, 295, 317
Carrier lipids, 106, 137, 140
Catalase, 125
Catecholamines, 146
Cathepsin, 133

Cell fusion, 115
Cell walls, 133ff
Cellulose, 143
Centrifugation, 121
Cephalins, 39—40
Cephalosporin, 138
Ceramide, 41
Cerebrosides, 41
Cesium, 289
Cetyltrimethylammonium bromide, 59
Chaos chaos, 145
Chaotropic agents, 58
Chara ceratophylla, 195, 197
Chemical potential, 170—171, 176, 179, 185, 263
Chemical shift, 70, 72
Chemotactic proteins, 238
Chitin, 142
Chloramphenicol, 204
Chlorella pyrenoidosa, 52
Chlorophyll, 45, 153—154
Chloroplast, 37, 98, 99, 115, 117, 152ff
Cholate, 59
Cholesterol, 32, 43, 48, 49, 51, 63, 74, 87, 93, 98, 106, 205
Choline dehydrogenase, 125
Cholinephosphate cytidylyltransferase, 109
Cholinephosphotransferase, 150
Circular dichroism, 64, 67ff, 99
Citrate (*si*)-synthase, 150
Clathrate, 34
Clostridium perfringens, 136
Coagel, 31
Cobalt, 146, 289
Coenzyme Q_{10}, 110
Colcemid, 116
Colchicin, 116
Colicins, 147
Colistins, 206
Comamonas, 236
Conalbumin, 114
Concanavalin A, 116, 126, 131
Constant-field assumption, 278ff
Cooperativity, 220—221
Corynebacterium, 137
Cotransport, 222
Cotton effect, 69

Countertransport, 207, 216—217, 222
Coupled transport, 229
Cratic entropy, 34
Crowns, 298$f\!f$
Cupric ion, 145—146, 289
Curie (Curie—Prigogine) principle, 19, 267
Cyanocobalamine, 237
$2':3'$:-Cyclic-nucleoside monophosphate phosphodiesterase, 58
Cycloartenol, 43
Cystine, 236
Cytochrome a, 150, 151
Cytochrome a_3, 151
Cytochrome b, 110, 150, 151
Cytochrome b_5, 100, 148, 158
Cytochrome b_5 reductase, 113, 150
Cytochrome c, 100, 107, 110, 114, 150, 151
Cytochrome c oxidase, 108, 110, 125, 133, 150
Cytochrome c reductase, 149
Cytochrome c_1, 110, 150, 151
Cytochrome P_{450}, 158
Cytolipin (globoside), 42

Debye thickness (length), 253, 278
Density gradient centrifugation, 158
Deoxycholate, 59
2-Deoxyglucose, 91, 92, 233
3-Deoxy-D-manno-octulosonic acid, 139
Deoxyribonuclease, 58, 125
Deoxyribose-phosphate aldolase, 58
Depsipeptides, 297$f\!f$
Desmosome, 128, 130
Desmosterol, 43
Detergents, 59
Diaminopimelic acid, 136
Diazosulfanilate, 82
Dictyosome, 159
Dielectric constant, 251
Differential scanning calorimetry (DSC), 82$f\!f$
Differential thermal analysis (DTA), 82, 83
Diffusion, 183$f\!f$
— exchange, 172
— lateral, 94, 100, 115, 116, 147, 167
— mediated (or facilitated), 203, 207$f\!f$, 239—241
Diffusion coefficient, 181, 184$f\!f$, 194, 197, 200, 217, 259
Diffusion equation, 186, 189
Dihydrosterculic acid, 92
Dilatometry, 85
1-Dimethylaminonaphthalene-5-sulfonyl chloride, 82, 124
Diphosphatidyl glycerol (cardiolipin), 39, 111, 112
Dissipation function, 174, 263
Dodecyl sulfate, 59, 113
Donnan phase, 248
Donnan ratio, 249—250
Double layer, electrical, 250$f\!f$, 278
Duysens' effect, 70

Effector, 146—147
Effort variables, 178
Ehrlich ascites cells, 37, 56, 69, 240, 303
Elaidic acid, 92
Electrical analogues, membrane, 286
Electrical potential, 244$f\!f$, 250, 257
Electrochemical potential, 244, 259, 261
Electrodiffusion, 256$f\!f$
Electron microscopy, 35, 117$f\!f$
Electron spin resonance, 75
Electrophorus, 146
Electroplax, 147
Ellipticity, 67
Endomyces, 142
Endoplasmic reticulum, 115, 148, 152, 156$f\!f$, 159—160
Enniatins, 297—298, 300
Enterobactin, 295
Enterochelin, 295
Epithelia, 115, 117, 257
Equilibrium, membrane, 243
Equivalent circuits, membrane, 286
Ergastoplasm, 158
Ergosterol, 43, 48, 205
Erucic acid, 88
Erythrocyte, 46, 48—50, 54, 56, 64, 74, 88, 98, 106, 119, 126, 132—133, 145—146, 220, 287, 317

Escherichia coli, 52, 54, 58, 82, 90, 92, 94, 99, 106*ff,* 115, 134, 156, 233—237, 294—295, 317—318
Esterase, 158, 163
Etruscomycin, 204
Euglena gracilis, 47
Exchange diffusion, 172

Fatty acids, 49, 52, 88, 89, 109, 156
Fatty acyl transferases, 139
Ferredoxin-$NADP^+$ reductase, 125
Ferrochelatase, 150
Fick's first law, 184, 186, 192
Fick's second law, 186, 192
Filipin, 204—205, 301
Flavin-containing amine oxidase, 107, 125
Flip-flop mechanism, 290*ff*
Flow variables, 178
Fluorescence, 79—80, 89, 90, 106
Flux, 172*ff,* 185, 193
— unidirectional, 172, 209—212
Flux ratio, 214
Folch—Lees protein, 106
Force, conjugate, 173
— generalized, 173—174
Force variables, 178
Formylmethionylsulfone methyl phosphate, 82, 124
Forssman glycolipid, 42
Freeze-etching, 118—119
Freeze-fracturing, 118—119
β-Fructofuranosidase, 133
Fructose, 233
β-Fructosidase, 59
Fucose, 60, 141
Fucosterol, 43
Fumarate hydratase, 150

Gaffkya, 136
Galactose, 14, 60, 143, 217, 233, 235, 236
β-Galactosides, 92, 114, 237
Galactosyldiglyceride, 154
Galactosyltransferase, 110
Galacturonic acid, 141
Gangliosides, 41, 48, 51
Gibbs—Donnan equilibrium, 247*ff*

Gibbs—Duhem equation, 27
Gibbs surface isotherm, 26, 28
Globoside (cytolipin), 42
Glucagon, 106, 146, 147
Glucan, 141*ff*
Glucomannan, 143
Gluconate, 235
Glucosamine, 233
Glucose, 60, 141, 143, 233, 236—237
Glucose-6-phosphatase, 109, 125, 158
Glucose-1-phosphate, 237
Glucose-1-phosphate adenyltransferase, 58
Glucose-6-phosphate, 235
β-Glucosides, 92, 233, 235
Glucuronic acid, 141, 235
Glutamate dehydrogenase, 125, 150
Glutamic acid, 236
Glutamine, 235, 236
γ-Glutamyltranspeptidase, 234
Glutathione, 234
Glycerides, 38, 46, 49, 109
Glycerol, 233
Glycerolphosphate acyltransferase, 110, 150
Glycocalyx, 127
Glycolipids, 41, 46, 48, 126, 130*ff,* 156
Glycoproteins, 60, 61, 104, 105, 130*ff*
Glycosyl transferase, 132
Goldman's equation (assumption), 261, 278*ff,* 287
Golgi apparatus, 102, 117, 122, 158*ff*
Gramicidin, 92, 296—297, 301
Granum lamella, 153
Granum stack, 153—154
Group translocation, 233, 241
Gyromagnetic ratio, 70

Haemophilus, 141
Halobacterium halobium, 55, 101, 134
Halosulfolipids, 45
α-Helix, 63, 66, 69, 99
Hematoside, 42
Henderson's equation (assumption), 261, 269—270
Hexokinase, 107
Hill coefficient, 106, 220
Histidine, 236, 237

Hodgkin—Horowicz equation, 286*ff*
Hormone receptors, 100
Hormones, 60, 146, 304
HPr (heat-stable protein), 100, 234
Hydrocarbons, 46
Hydrogen hydrogenase, 107
Hydrogen ion, 233, 304
Hydrogenosomes, 165
Hydrostatic pressure, 248, 307, 309*ff*
3-Hydroxybutyrate dehydrogenase, 59, 108, 150
Hydroxymethylglutaryl-CoA reductase, 113

Infrared spectroscopy, 63
Initial rate, 210—211, 230
Insulin, 56, 146, 147, 304
Interferon, 322
Intergranum lamella, 153
Internal conversion, 79—80
Ionophores, 296*ff*
Iron, 289
Isocitrate dehydrogenase, 125, 150
Isoprenoid-alcohol kinase, 108
Isopycnic density gradient, 123

Jejunum, 127, 128, 237
KDO (3-deoxy-D-mannooctulosonic acid), 139, 140
Kidney, 46, 47, 92, 146, 236, 303
Kronig—Kramers transform, 67
Kynureninase, 150

Lactate dehydrogenase, 133
Lactobacillus casei, 136, 156
Lactobacillus corinoides, 156
Lactobacillus plantarum, 233
Lactobacillus viridescens, 136
Lactose, 233
Langmuir adsorption isotherm, 28
Langmuir trough, 22, 25
Lanosterol, 43
Lanthanum, 145, 213
Laplace equation, 276
Larmor precession frequency, 77
Le Chatelier principle, 302
Lecithin, 32, 40, 84, 91, 93, 94, 112
Lectin receptors, 100

Leptodactyllus ocellatus, 303
Leucine, 236
Leucine aminopeptidase, 133
Leukoviruses, 37
Lightscattering, 98
Lignins, 143
Linoleic acid, 92
Linolenic acid, 106
Lipid A, 139—140
Lipid exchange proteins, 111—112
Lipid—protein interactions, 101
Lipoamide dehydrogenase, 133, 163
Lipoamino acids, 40
Lipopolysaccharides, 45, 60, 116, 134, 137, 139, 140
Lipoproteins, 103, 134, 158
Liposomes, 96—97
Liquid junctions, 247
Lithium, 91, 289
Lorentz correction factor, 66—67
Lubrols, 59
Luminescence, 80
Lymphocyte, 147, 319
Lysine, 236
Lysins, 147
Lysolecithin acyltransferase, 150
Lysophosphatides, 111
Lysophosphatidyl choline, 46, 109
Lysophosphatidyl ethanolamine, 109
Lysophosphatidyl glycerol, 109
Lysosome, 116, 122, 161, 163
Lysozyme, 57, 110

Macrolides, 204
Magnesium, 91, 107, 140, 291*ff*
Malate dehydrogenase, 150
Malate oxidase, 108
Maltose, 233, 237
Manganese, 146, 289
Mannan, 141*ff*
Mannitol, 233
Mannoheptoses, 140
Mannose, 60, 141, 143, 233
Melezitose, 233
Melibiose, 233
Membrane, artificial, 95*ff,* 287
— black, 95—96
— cytoplasmic, 134, 152

— inner, 148, 151, 152
— outer, 134, 140, 148, 151
Membrane potential, 231—232, 238 to 239, 243, 247, 256*ff*, 286
Membron, 158
Mercury, 146
Mersalyl, 59
Mesomorphic organization, 29
Mesosome, 155*ff*
Methionine, 235
α-Methyl-D-glucoside, 131
Micelles, 28, 103
Michaelis—Menten equation, 241
Micrococcus denitrificans, 235
Micrococcus lysodeikticus, 37, 47, 52, 107, 136, 156
Micrococcus radiodurans, 134
Micrococcus roseus, 135
Microelectrodes, 247
Microscopic reversibility, 174—175
Microsomes, 37, 47, 54, 108, 109, 122, 124
Microtubules, 116
Miniproteins, 54
Mitochondria, 37, 47, 48, 54, 98, 99, 105, 106, 108, 109, 115, 117, 122, 124, 132, 147*ff*
Mixed function oxidase, 158
Mobility, 185, 261
— lateral, 103
Monensin, 299, 301
Monoacylglycerolphosphate acyltransferase, 110
Monolayers, 22—23, 28, 98, 110
Mucopeptide, 134
Multi-site carriers, 220
Murein, 134
Mycobacteria, 46
Mycobacterium phlei, 156, 235
Mycobacterium smegmatis, 156, 295
Mycobactin, 295
Mycoplasma, 88, 99, 128, 130, 233
Myelin, 32, 37, 49, 55, 64, 83, 99
Myelin figures, 32
Myristic acid, 88
Myxoviruses, 37

N-acetylcandidin, 204
N-acetylgalactosamine, 41, 60, 131
N-acetyl-D-glucosamine, 60, 131, 134, 135, 138, 139, 143, 233
N-acetylmuramic acid, 134, 135, 138
N-acetylneuraminic (sialic) acid, 41
NAD (nicotinamide-adenine dinucleotide), 150
NAD^+ nucleosidase, 113, 133
NAD^+ pyrophosphatase, 133
NADH dehydrogenase, 59, 113, 125, 133, 150
NADP (nicotinamide-adenine dinucleotide phosphate), 150
NADPH-cytochrome reductase, 158
$NAD(P)^+$ transhydrogenase, 125
Nadsonia, 142
β-Naphthylamidase, 59
Nectin, 100
Necturus, 198
Negative staining, 22
Nematic state, 29
Nernst—Donnan potential, 245
Nernst—Planck equation, 260, 270
Nerve fibers, 130
Nerve impulse (see also action potential), 95
Network thermodynamics, 177
Neuromuscular junction, 321
Neurospora crassa, 42, 236
Nickel, 145
Nigericin, 299—300
Nitella translucens, 304
o-Nitrophenyl-β-galactoside, 92
Nitroxide, 75
N-(3-mercuri-5-methoxypropyl)poly-D,L-alanyl amide, 125
NMR (nuclear magnetic resonance), 70, 315
Nonactin, 92
Noradrenaline, 106
N-phenyl-1-naphthylamine (NPN), 82
Nucleic acids, 22, 320
Nucleosidediphosphatase, 133
Nucleosidediphosphate kinase, 150
Nucleoside phosphatase, 125
5'-Nucleotidase, 58, 125, 133
Nucleus, 122, 163
Nystatin, 204—205

O-antigen, 140—141
Oleic acid, 88, 92, 114
Oligomycin-sensitivity conferring protein, 100
Onsager's law, 174, 308, 310
Opsonins, 147
Optical density, 65
Optical rotatory dispersion, 64, 66$f\!f$, 99
Order parameter, 93
Ornithine, 235
Ornithine carbamoyltransferase, 150
Orthomyxoviruses, 61
Osmium tetroxide, 78, 99, 117, 127
Osmosis, 307$f\!f$
Osmotic pressure, 249, 310$f\!f$
Osmotic shock, 57, 238
Ostruthin, 82
Ouabain, 288, 292
Oxidoreductive system, 235, 239
Oxoglutarate dehydrogenase, 133, 150
Oxytocin, 146, 147

Packing areas, 88
Palmitelaidic acid, 114
Palmitic acid, 88
Pancreas, 112, 117
Pancreatic acinar cells, 319
Paramyxoviruses, 61
Paratose, 139
Particle shape factor, 121
Partition coefficient, 193, 283
Penicillinase, 58
Penicillins, 57, 138, 204
Peptide a, 54
Peptidoglycan, 134, 135, 137, 138, 140
Permeability coefficient (constant), 193$f\!f$, 199—200, 283, 312
Permeases, oligopeptide, 317
Permeation, 192
Peroxisomes, 122, 165
Phagocytosis, 116
Phagosomes, 123
Phallotoxins, 300
Phase transition, 103, 106
Phenylalanine, 236
pH gradient, 238—239
Phosphatase, alkaline, 58
Phosphatase, acid, 58, 125, 133

Phosphatidate phosphatase, 109, 150
Phosphatidic acid, 39, 46, 48, 49, 90, 110, 111
Phosphatidyl choline (lecithin), 40, 46, 48, 49, 87, 90, 98, 109, 111
Phosphatidyl ethanolamine, 39, 46, 48, 49, 87, 90, 98, 109, 111, 112, 140
Phosphatidyl glycerol, 39, 46, 48, 109, 111, 112
Phosphatidyl inositol, 39, 46, 48, 51, 106, 109, 111, 112, 148
Phosphatidyl inositol kinase, 133
Phosphatidyl serine, 40, 46—49, 87, 106, 109, 111
Phosphodiesterase, 133
Phosphoenolpyruvate carboxylase, 150
Phospho-enol-pyruvate-HPr-phosphotransferase, 109
Phospholipases, 53, 133, 150
Phospholipids, 22, 32, 34, 46—49, 51, 52, 74, 78, 85—86, 88, 90, 98, 104—106, 110—112, 134, 140, 154
Phospho-N-acetylmuramoyl-pentapeptidetransferase, 109
Phosphonolipid, 40, 46
Phosphonoplasmalogen, 40
Phosphopentomutase, 58
Phosphorescence, 79—80
Phosphotransferases, 110, 233, 239, 241
Phycoproteins, 100
Phytagglutinins, 126
Phytoglycolipids, 42
Phytohemagglutinins, 60
Phytosphingosine, 41
Pimaricin, 204
Pinifolic acid, 89
Pinocytosis, 145, 161, 183, 238, 318$f\!f$
Planck's equation, 261—262, 271$f\!f$
Plasma membrane, 105, 109, 116, 117, 122, 124$f\!f$, 130$f\!f$, 138, 141, 149, 155
Plasmalemma (see also Plasma membrane), 144
Plasmalogens, 40, 46
Pleated sheet, 63, 66, 99
Poisson's electrostatic equation, 251, 276—277

Polygalacturonic acids, 143
Polyglucose, 123
Poly-L-lysine, 68—69
Polymyxins, 206
Polyols, 239
Polyphosphoinositol, 51
Polysaccharides, 22
Polysomes, 158
Polysucrose, 123
Pores or channels, 198, 205, 243, 301
Potassium, 16, 86, 91, 95, 235, 282ff, 288ff, 304
Potassium permanganate, 117, 126
Potential, chemical, 170—171, 176, 179, 185, 263
— electrical, 244, 250, 257
— electrochemical, 244, 259, 261
— membrane, 231—232, 238—239, 243, 247, 256ff, 286
Potter—Elvehjem homogenizer, 120
Pressure diffusion, 261
Proteases, 163
Protein kinase, 133
Proteins, extrinsic, 57
— integral, 100, 103, 105, 106
— intrinsic, 56
— peripheral, 100, 103, 105, 106
— periplasmic, 57
Proteus mirabilis, 235
Proton, 235
Proton-motive force, 152, 235
Protozoa, 46
Pseudomonas aeruginosa, 37
Pseudomonas putida, 235
Pump-and-leak model, 227
Purine-deoxyribonucleoside phosphorylase, 58
Purines, 239
Puromycin, 204
Purple membrane, 55
PV peptide, 296—297
Pyrene-3-sulfonate, 82
Pyrimidines, 239
Pyruvate carboxylase, 150
Pyruvate oxidase, 108

Quantum yield, 81
Quinones, 45

Random coil, 67, 69, 99
R-antigens, 141
Rate zonal centrifugation, 123, 158
Receptor, 110
Reflection coefficient, 197, 201, 312
Relaxation time, 72, 181
Repeat distance, 30, 32, 35
Resonant frequency, 72
Retinal, 165
Retinal rod, 37, 55, 60, 62, 100, 101, 165—166, 117—118
Retinol, 82
Rhabdoviruses, 61
Rhamnose, 141
Rhodopsin, 55, 60, 100, 101, 165
Rhodospirillum rubrum, 233
Rhodotorula, 142
Rhodotorula glutinis, 164, 224
Riboflavin, 237
Ribonuclease, 58, 110, 125, 163
Ribose, 237
Ribosephosphate isomerase, 133
Ribosomes, 152, 156
Rifamycins, 204
Ristocetin, 138
Rubidium, 235, 289

Saccharomyces, 141—142
Saccharomyces cerevisiae (see also Baker's yeast), 47, 52, 128, 130, 162, 164, 224, 236
Saccharomyces pombé, 160
Saccharomycodes ludwigii, 149
S-adenosyl methionine, 111
Salmonella, 137, 139, 140
Salmonella typhimurium, 108, 233, 235 to 237, 294, 296, 317
Salt bridges, 247
Sarcoplasmic reticulum, 55
Schlögl's equation, 260, 262
Schwann cell, 130
Sciatic nerve, 87
Secretin, 147
Sedimentation coefficient, 121
Serratia marcescens, 52
Shigella, 141
Short-circuit current technique, 303
Sialic acid, 41, 60, 141

Siderophilin, 296
Sieve effect, 70
Silver, 146
β-Sitosterol, 43, 48
Smectic state, 29
Soaps, 31
Sodium, 16, 86, 91, 95, 233, 239, 283ff, 288ff
Sodium pump (see also Adenosinetriphosphatase, Na,K), 302
— electrogenic, 284
— electroneutral, 285
Sodium theory, 284, 302
Sonication, 59, 107, 121
Sorbitol, 233
Spectrin, 54, 100
Spectroscopy, infrared, 63
— ultraviolet, 64
Sphingolipids, 48
Sphingomyelins, 41, 45, 46, 48, 49, 87
Sphingosine, 40
Spinasterol, 43
Spirillum, 134
Spirillum serpens, 156
Squalene, 44
Staphylococcus, 137
Staphylococcus aureus, 108, 135, 156, 233—235
Staphylococcus epidermidis, 135
Steady-state thermodynamics, 172ff, 307ff
Stearic acid, 88, 89, 94
Sterols, 43, 45, 46
Streptococcus, 134
Streptococcus faecalis, 58, 107, 136, 156, 294
Streptococcus lactis, 233
Streptococcus pyogenes, 136
Streptomyces, 206, 240
Streptomyces albus, 136
Streptomycin, 204
Strontium, 146, 289
Succinate dehydrogenase, 108, 110, 125, 133, 150, 151
Sucrase-isomaltase, 99, 110
Sucrose, 233, 235
Sucrose gradient, 114, 124, 158
Sugar esters, 44

Sulfanilate, 124
Sulfatides, 31, 42
Sulfo acids, 49
Sulfolipids, 45, 46
Surface-active agents, 28
Surface pressure, 26
Surface tension, 25—26
Surfactin, 206
Svedberg unit, 121
Symport, 239
Synaptic cleft, 319—321
Synaptic vesicles, 165

T-antigens, 140
Teichoic acid, 134, 137, 139
Tektin A, 54
Tempol, 75, 92
Teorell's formula (equation), 185, 259
Thallium, 289
Thermodynamic equilibrium, 169ff
Thermodynamics, network, 177ff
— steady state, 172ff, 307ff
Thiamine, 237
Thiomethyl-β-galactoside, 92
Thorium, 145—146, 213
Thylakoids, 105, 152, 154
Thyreocalcitonin, 147
Tight junction, 128, 130
Titanium, 146
T-layer, 134
α-Tocopherol, 98
1-Toluidinonaphthalene-5-sulfonate (TNS), 82
Tonoplast, 163
Torpedo, 146
Tracer equilibration, 203, 214
Trans acceleration, 212
Transducer, 146—147
Transferrin, 296
Transition temperature, 85, 89ff, 106, 113, 114
Translational entropy, 34
Transport, active, 170, 222, 239—241
— passive, 170—171
— secondary active, 170, 177, 241, 303
Trehalose, 233
Trigonopsis variabilis, 162
2,4,6-Trinitrobenzene, 124

Triton, 59, 113
Tryptophan, 236
Tubulin, 116
Turnover, membrane, 112, 113
Tweens, 59
Tyrocidines, 206, 296—297
Tyvelose, 139

Ubiquinone, 150
UDPgalactose-lipopolysaccharide galactosyltransferase, 108
UDPgalactose-N-acetylglucosamine-galactosyltransferase, 125
UDPglucose-lipopolysaccharide glucosyltransferase I, 108
UDPglucosyltransferases, 140
UDPglucuronosyltransferase, 108
Ultraviolet spectroscopy, 64
Umbelliferone, 82
Unidirectional fluxes, 172, 209—212
Unstirred layers, 198ff, 217—220, 314
Uranyl ion, 145, 213
Urografin, 123

cis-Vaccenic acid, 92
Vacuoles, 161, 163
— condensing, 160
Vagus nerve, 112

Valinomycin, 92, 235, 297
Vancomycin, 138
Van der Waals forces, 34
van't Hoff's formula, 249
Vasopressin, 147
Vaucheria, 153
Vesicles, 95, 97, 98, 107, 165
Vibrio succinogenes, 107
Vinblastine, 116
Vitamin A, 165
Vitamin K, 44
Vitamins, 239
Volume flow, 310

Water, state of in cells, 314ff
— transport of, 307ff
Wax esters, 44, 46
Wickerhamai fluorescens, 157

X-ray diffraction, 29, 31, 34—35, 61 to 63
Xylose, 217, 224
Xylulose reductase, 150

Zeta potential, 251
Zinc, 289
Zymogen granule, 160, 320
Zymosterol, 43, 48

COMMUNICATING IN GROUPS AND TEAMS
STRATEGIC INTERACTIONS

FOURTH EDITION

Joann Keyton and Stephenson Beck

North Carolina State University | North Dakota State University

Bassim Hamadeh, CEO and Publisher
Todd R. Armstrong, Senior Specialist Acquisitions Editor
Abbey Hastings, Associate Production Editor
Miguel Macias, Senior Graphic Designer
Alexa Lucido, Licensing Coordinator
Kassie Graves, Director of Acquisitions and Sales
Jamie Giganti, Senior Managing Editor
Natalie Picotti, Senior Marketing Manager

Copyright © 2018 by Cognella, Inc. All rights reserved. No part of this publication may be reprinted, reproduced, transmitted, or utilized in any form or by any electronic, mechanical, or other means, now known or hereafter invented, including photocopying, microfilming, and recording, or in any information retrieval system without the written permission of Cognella, Inc. For inquiries regarding permissions, translations, foreign rights, audio rights, and any other forms of reproduction, please contact the Cognella Licensing Department at rights@cognella.com.

Trademark Notice: Product or corporate names may be trademarks or registered trademarks, and are used only for identification and explanation without intent to infringe.

Cover image copyright © 2013 by iStockphoto LP/kali9.
    copyright © 2013 by iStockphoto LP/kali9.
    copyright © 2014 by iStockphoto LP/PeopleImages.

Printed in the United States of America.

ISBN: 978-1-5165-1928-6 (pbk) / 978-1-5165-1929-3 (br) / 978-1-5165-4636-7 (al)

BRIEF TABLE OF CONTENTS

CHAPTER 1
SITUATING YOUR EXPERIENCES IN GROUPS 2

CHAPTER 2
GROUP COMMUNICATION FUNDAMENTALS 14

CHAPTER 3
CONTEXTUAL INFLUENCES ON GROUPS AND TEAMS 32

CHAPTER 4
HOW WE COMMUNICATE WHAT WE KNOW 54

CHAPTER 5
DECISION MAKING 70

CHAPTER 6
BUILDING RELATIONSHIPS IN GROUPS 92

CHAPTER 7
MANAGING CONFLICT IN GROUPS 114

CHAPTER 8
LEADING GROUPS 138

CHAPTER 9
FACILITATING GROUP MEETINGS 162

CONTENTS

A PREFACE FOR INSTRUCTORS x

CHAPTER 1
SITUATING YOUR EXPERIENCES IN GROUPS 2

Groups Are Fundamental to Society 4
  Types of Groups 4
Group Communication 6
  Interaction and Messages 6
Group and Individual Goals 9
Group Skills for Life 10
Format of the Book 11
Discussion Questions and Exercises 12

CHAPTER 2
GROUP COMMUNICATION FUNDAMENTALS 14

What Is a Group? 15
  Characteristics for Defining a Group 15
  Integrating the Pieces of the Puzzle 21
Interdependence of Task and Relational Dimensions 24
  Satisfying Task and Relational Dimensions of Groups 26
Groups or Teams? 28
Summary 29
Discussion Questions and Exercises 30

CHAPTER 3
CONTEXTUAL INFLUENCES ON GROUPS 32

Bona Fide Group Perspective 33
Permeable and Fluid Boundaries 33
A Group's Interdependence With Its Context 34
Influence of Time and Space 37
Time 37
Space 39
Influence of Diversity 40
Influence of Technology 44
Trends in Technology 49
Summary 51
Discussion Questions and Exercises 52

CHAPTER 4
HOW WE COMMUNICATE WHAT WE KNOW 54

Information Sharing 55
Cooperative Information Sharing 56
Motivated Information Sharing 57
Team Cognition 58
Group Communication Networks 60
Decentralized Networks 61
Centralized Networks 63
Evaluating Your Group's Network 64
Summary 66
Discussion Questions and Exercises 67

CHAPTER 5
DECISION MAKING 70

Group Decision-Making Theories 71
Functional Theory of Decision-Making 71

Majority/Minority Influence 76

Decision-Making Procedures 77
Brainstorming 77

Nominal Group Technique 79

Consensus 82

Voting 84

Ranking 86

Comparing Procedures 87

Decision-Making Principles 88
Summary 89
Discussion Questions and Exercises 90

CHAPTER 6
BUILDING RELATIONSHIPS IN GROUPS 92

Group Communication Climate 93
Description 94

Problem Orientation 94

Spontaneity 95

Empathy 95

Equality 95

Provisionalism 96

Changing Group Climate 96

Group Cohesiveness 97
Cohesiveness and Group Performance 99

Group Member Satisfaction 101
Improving Cohesiveness and Satisfaction 103

Trust 104
Socializing New Members Into Groups 107

Summary 110

Discussion Questions and Exercises 111

CHAPTER 7
MANAGING CONFLICT IN GROUPS 114

Defining Conflict 115
Perspectives on Conflict 116
Is Conflict Always Disruptive? 118
Is Conflict Inherent? 119
Types of Conflict 120
Power 122
Bases of Power 124
Conflict Management Strategies 126
Collaborating 127
Competing 129
Accommodating 129
Avoiding 130
Compromising 130
Which Strategy Should You Choose? 131
Summary 133
Discussion Questions and Exercises 135

CHAPTER 8
LEADING GROUPS 138

Defining Leadership 139
A Communication Competency Approach to Leadership 139
Procedural Competencies 140
Relational Competencies 142
Technical Competencies 143
Integrating Leadership Competencies 143
Becoming a Leader 145
Appointed Versus Elected Leaders 145

Emerging as a Leader 146
Shared Leadership 148
A Discursive Approach to Leadership 150
Transformational Leadership 152
Gender Assumptions About Leadership 154
Leadership in Virtual Groups 156
Enhancing Your Leadership Ability 157
Summary 158
Discussion Questions and Exercises 159

CHAPTER 9
FACILITATING GROUP MEETINGS 162

Developing a Group Charter 165
Developing a Code of Conduct 167
Importance of Meeting Management Procedures 167
Meeting Planning 169
- Pre-Meeting Planning and Preparation 169
- Leader Pre-Meeting Responsibilities 170
- Physical Environment and Material Resources 173
- Group Member Pre-Meeting Responsibilities 173

Conducting the Meeting 174
- Taking Minutes 175
- Managing Relational Issues 177
- Using Space 177
- Using Visuals 178
- Making Assignments 180
- Ending the Meeting 180
- Meeting Virtually 180

Post-Meeting Follow-Up 181
Overcoming Meeting Obstacles 182
- Long Meetings 183

Unequal Member Involvement and Commitment 183

Formation of Cliques 183

Different Levels of Communication Skill 184

Different Communicator Styles 184

Personal Conflicts 184

Summary 185

Discussion Questions and Exercises 186

APPENDIX: Creating and Delivering Team Presentations 188
GLOSSARY 198
REFERENCES 204
AUTHOR INDEX 218
SUBJECT INDEX 224

A PREFACE FOR INSTRUCTORS

Teams and groups are a basic component of society and the economy. The overwhelming majority of college graduates will work in teams during school and in their future employment. Employees are often asked to lead and facilitate groups, which may involve managing a variety of technological advances and diverse populations. Not surprisingly, employers identify teamwork skills and collaboration as two of the most important skills they *expect* technical school, two-year and four-year college graduates to bring to the workforce. Of course, teamwork is not reserved solely for employment. Outside of their work life, college graduates are the most likely population to volunteer in their nonprofit and civic communities. Whether working in fundraising, directing or being involved in community service, or completing tasks as a member of a nonprofit board or city council, teamwork is a required skill for effective and successful outcomes.

Empirical research has demonstrated that college students are not apprehensive about working in groups and teams, but they do *dislike* working in teams. To remedy this problem, instructors can't simply talk about group concepts and characteristics. Instead, course instruction must be focused on two areas: the actual behavior, and *why* that behavior is important or effective. This is why a communication perspective is the best way to teach effective group member behavior. A communication perspective emphasizes the creation and management of messages, as well as the reception and perception of meaning. It directly examines communication as the foundation of all group activity. Additionally, it challenges students to glean explanation and interpretation from interaction. A host of individual-level outcomes, such as identity, commitment, and satisfaction, are all subjective creations based on group interaction. Thus, a direct focus on interaction is not only the foundation of group communication scholarship, but also the best way to engage students in the learning experience.

To emphasize the strong and unique perspective owned by group communication scholars, the first chapters of this book create a communicative framework for the investigation of groups and teams. Part of that framework is elaborating on the strategic and contextual nature of group interaction. This starting point positions our book differently from many others in the discipline. We do not simply define what communication is, but show how through our goals and purposes we adapt messages in desire of certain outcomes, while realizing that as soon as we utter our messages to other group members we lose all control of meaning and interpretation. In the first chapters, we also spend considerable time elaborating on how the fundamental characteristics of groups directly influence, and are influenced by, communication. In other words, we take seriously our focus on groups from a communication perspective, and spend considerable effort establishing this view.

Communicating in Groups and Teams: Strategic Interactions is based on three components to assist instructors in teaching group communication. First, the material presented is based on scholarly research findings from the field of communication and related disciplines. The book is skills based—and our skill recommendations are based on rigorous and current research. Second, to emphasize the group over the individual, the book describes and explains group communication concepts with examples based on typical group interactions. That is, dialogue in the book allows students to watch group dynamics unfold. Third, the book is structured around five key elements of groups that can be used to evaluate group effectiveness.

FEATURES TO ENHANCE LEARNING

This book contains a number of features to enhance student learning:

- *Putting the Pieces Together* **boxes.** The five core elements in defining a group are used as a structure for evaluating group effectiveness. The five elements are group size, interdependence of members, group identity, group goals, and group structure. These elements are introduced in Chapter 2 and integrated in every chapter as a special feature so that students become more aware of how communication inhibits or facilitates group success.

- **Skills grounded in a solid research base.** The best advice for communicating in groups is drawn from group research and theory, which has identified the most effective processes and results for group interaction. Thus, the skills presented and suggested in the text have been tested—many of them in the field.

- **Extensive use of realistic examples.** In addition to describing what is happening in groups through the use of extensive examples, this text provides transcripts of group dialogues so students can see the communication process unfold. Group dialogues also provide an opportunity to suggest and test different communication approaches. Using the dialogue examples in this way can help students analyze how the group's conversation might have proceeded differently if alternative communication strategies were employed.

- **A wide range of group types.** The text speaks to students' experiences by providing information about a wide variety of groups, including family and social groups, work teams and high-performance task groups, civic and community groups, and discussion and decision-making groups. Whether students' experiences are with groups that are formal or informal, personal or professional, task oriented or relationally oriented, they need communication skills to build and maintain relationships that support effective problem solving and decision making.

- **Four types of pedagogical boxes emphasizing more advanced learning**
  - *Message and Meaning* boxes display transcripts from groups and teams that have been part of our research, or transcripts that are publicly available.
  - *Theory Standout* provides an in-depth look at group communication theories that are introduced in this book. Featuring theories outside of the narrative provides a closer examination and invites questions about the theory-research link.
  - *Skill Builder* provides students an opportunity to test, develop, and practice their group communication skills through exercises and activities.
  - *Nailing It! Using Group Communication Skills for Group Presentations*: In talking with professors who do not teach the group communication course, we found that a primary concern was that students who had taken a group communication class had difficulty transferring group skills to developing group

presentations in other courses. To address this issue, we have developed this feature to help students transfer what they are learning to the development of group presentations.

- **Other in-text learning aids**
  - Chapter previews: At the beginning of each chapter, there is an overview for students about what information and skills they will be learning and practicing in the chapter
  - End-of-chapter summaries and discussion questions and exercises
  - Glossary
  - Extensive list of references for further study

ORGANIZATION OF THE BOOK

To provide a foundation, Chapters 1 through 3 describe the importance of groups and teams, and how communication is basic to our understanding of these social and task forms. At the same time, we do not dismiss that individuals in groups have individual goals that may or may not be congruent with group goals. Our model, the five-piece group puzzle, is introduced in Chapter 2 and provides a structure for the remaining chapters, as well as an analytical tool for identifying key strengths and challenges of group interaction. Chapter 3 describes and explains the contextual influences that affect group interaction across group types: boundaries, time and space, diversity, and technology. Each of these influences is present in some way in every group interaction.

Chapters 4 through 8 describe and explain the core group processes for which each group member is responsible: sharing information, participating in decision making, building relationships, managing conflict, and providing leadership. These processes are interaction opportunities and problems that are regular and dynamic aspects of group interaction. Increasing students' skills in these areas will help them maximize their group interaction efforts.

The concluding Chapter 9 helps students participate in, navigate, and facilitate group meetings. Whether in the role of leader or member, students should be able to facilitate their group's interaction to help the group stay or get back on track. Armed with specific principles, procedures, and feedback techniques, students can make more informed choices about how to help their group.

RESOURCES FOR INSTRUCTORS AND STUDENTS

As group communication scholars and teachers, we develop our own Instructor's Manual materials. While we appreciate a publisher's preference for a portal linked to their brand, we also argue that Instructor's Manual materials should be regularly updated. Thus, we offer a two-pronged approach to Instructor's Manual materials. The main portal, with the password protected, with the following components are found on the Cognella site:

- *Instructor's Manual:* Outlines of learning objectives and a teaching manual suggesting active in-class and field learning exercises, as well as methods for evaluating group communication in the classroom and group assignments. This manual also includes the teaching philosophy that was a foundation for this book, syllabus examples for the group communication course, methods of obtaining feedback from students about the course and their learning experiences and expectations, chapter-by-chapter teaching resources and exercises, and suggestions for term-long group projects.
- *Test Bank:* An extensive print and computerized test bank, including multiple-choice, matching, identification, and essay questions, and identifying the pedagogical objectives addressed by each question.
- *PowerPoint:* Effective and clearly structured PowerPoint slides that avoid the trap of summarizing the chapters in PowerPoint format, as is often typical of such slides accompanying existing textbooks.

A second Instructor's Manual website—with regularly updated resources for instructors and students—will be linked directly from the authors' personal webpages (joannkeyton.com; stephensonbeck.com). Doing so allows us to update web resources quickly, as well as take advantage of current events that are group or team focused.

ACKNOWLEDGMENTS

During the development of this text, we received excellent feedback and encouragement from three reviewers: Dr. Mary Beth Asbury, Middle Tennessee State University; Dr. Linda G. Ward, University of Texas of the Permian Basin; and Dr. Cindy Peterson, MidAmerica Nazarene University. We appreciate the time they took to thoughtfully consider our approach to teaching the group communication course. Another group of communication professors provided feedback on the completed book manuscript. These reviewers were: Dr. Michelle R. Bahr, Bellevue University; Dr. Michael P. Pagano, Fairfield University; Dr. Tennley A. Vik, Emporia State University; and Marissa L. Wiley, University of Kansas.

A PREFACE FOR INSTRUCTORS

Communication scholars who provided feedback on the earlier editions of this book include: Carolyn M. Anderson, Laurie Arliss, Dale E. Brashers, John O. Burtis, Marybeth Callison, Elizabeth M. Goering, Randy Hirokawa, Michael E. Holmes, Michele H. Jackson, Virginia Kidd, Bohn D. Lattin, M. Sean Limon, Michael E. Mayer, Mary B. McPherson, Renee A. Meyers, Marshall Scott Poole, Barbara Eakins Reed, Vanessa Sandoval, Kristi Schaller, Matthew W. Seeger, Nick Trujillo, Lyn M. Van Swol, and Clay Warren.

Studying and teaching group communication is our life's work. We thank the many group and team scholars who are part of our international network. They inspire us and hold us to a high standard in conducting research and in presenting our research findings. They are also a lively bunch and are fun to hang out with.

From Joann: I want to thank my personal network, who sustain me and allow me to say "I have a deadline" and not hold it against me. They are also the first to check to see if I need anything while trying to meet those deadlines.

From Steve: I want to thank Sarah, Joshua, and Whitney for their support and love. I'd also like to thank my colleagues who are willing to put up with me.

ABOUT THE AUTHORS

Joann Keyton (BA, Western Michigan University; MA, PhD, Ohio State University) is Professor of Communication at North Carolina State University. She specializes in group communication and organizational communication. Her current research examines the collaborative processes and relational aspects of interdisciplinary teams, participants' use of language in team meetings, the multiplicity of cultures in organizations, and how messages are manipulated in sexual harassment. Her research is field focused and she was honored with the 2011 Gerald Phillips Award for Distinguished Applied Communication Scholarship by the National Communication Association.

Her research has been published in *Business Communication Quarterly, Communication Monographs, Communication Research, Communication Studies, Communication Theory, Communication Yearbook, Journal of Applied Communication Research, Journal of Business Communication, Management Communication Quarterly, Small Group Research, Southern Communication Journal,* and numerous edited collections including the *Handbook of Group Communication Theory and Research,* the Sage *Handbook of Organizational Communication,* and the Oxford *Handbook of Organizational Climate and Culture.*

In addition to publications in scholarly journals and edited collections, she has published three textbooks for courses in group communication, research methods, and organizational culture in addition to coediting an organizational communication case book. Keyton was editor of the *Journal of Applied Communication Research,* Volumes 31–33, and founding editor of *Communication Currents,* Volumes 1–5. Currently, she is editor of *Small Group Research.* She is a founder of the Interdisciplinary Network for Group Research.

For more information, contact Joann at jkeyton@ncsu.edu or www.joannkeyton.com

Stephenson J. Beck (BA, Brigham Young University; MA, University of Illinois at Urbana-Champaign; PhD, University of Kansas) is Associate Professor of Communication at North Dakota State University. His research focuses on group communication and communication strategy. His current research investigates meeting facilitation, conflict management, and decision-making communication. His research endeavors involve studying a variety of group contexts, including breast cancer support groups, nonprofit boards, first responder teams, juries, city councils, data analyst teams, organizational teams, special education teams, and military teams.

His research has been published in *Small Group Research, Group Dynamics, Journal of Applied Communication Research, Business Communication Quarterly, Communication Yearbook, Personal Relationships, Communication Studies, Journal of Family Communication,* and *Cancer Nursing,* and he has authored several contributions to edited books. He is a member of the editorial board of *Small Group Research.*

For more information, contact Stephenson at stephenson.beck@ndsu.edu or www.stephensonbeck.com

CHAPTER 1

SITUATING YOUR EXPERIENCES IN GROUPS

After reading this chapter, you should be able to:

- Describe the role of groups in your life
- Explain the importance of groups to the functioning of society
- Describe the types of groups to which people belong
- Describe group communication
- Explain the role of interaction, messages, and meanings in group communication
- Identify individual and group goals accomplished in groups and teams

A basic part of who we are as individuals is based on our interactions with others. When young we are surrounded by parents, siblings, or other family members, and our experiences with them represent a large portion of our lives. These family relationships and interactions become the first opportunity for us to learn how to communicate with others (Socha, 1999). During this early period of life, you learn how to interact with individuals holding authority, share with those who want what you have, and persuade others to agree with you. These developmental milestones are encouraged by parents. For example, parents may arrange for you to play at the park with friends. The main function of these group experiences is to allow you to learn how to interact and get along with others. Although typically unstructured, such play dates are likely your first experiences in creating and managing relationships and sharing toys with peers.

Gradually, these social interactions expand to other types of groups. When we go to elementary school, we are surrounded by classmates. This setting was likely your introduction to task-focused work groups. For example, teachers often put children into small teams or pods to learn and practice reading and math. Desks are placed into clusters and the classroom is designed for students to gather around and learn together. These types of learning groups also facilitate social development, which help young students learn how to cooperate and help one another.

As students enter middle school, they participate in extracurricular activities, such as clubs and sports, which often involves working with other students to perform in band or choir, put on a play, or compete in robotics tournaments. In high school, working in groups centers around problem-solving tasks in the classroom and more complex performing groups, such as cheerleading, football, and ice hockey. Friends form cliques, which increases the intensity of friendship groups, as well as heightens the distinction of insiders from outsiders. At home, kids play online multiplayer games, which are designed for teams of players from around the world to work together to accomplish a quest.

Upon graduation from high school we continue to encounter groups and are likely to work in teams. Fraternities and sororities evolve around groups of students, and fast food or retail jobs are often team oriented. Grocery stores, assembly lines, and construction crews are often based on work shifts where individuals regularly interact and perform in groups. Many organizational structures are situated around working in teams, which are designed to pool resources and expertise in ways to improve product quality and quantity. When important projects or decisions are required by an organization, temporary or ad hoc committees are formed to handle the situation. As adults, we also belong to groups and teams in our personal and community lives. Book clubs, religious study groups, community service projects, and bowling leagues are all based on the concept of a team. We also depend on support groups when struggling with illness, addiction, or other personal problems. In these settings, group discussion helps us to manage these difficulties.

In addition, there are groups of which we are not members that have significant influence on our lives. Groups at local, state, regional, national, and international levels address societal issues, from local political advocacy groups to the United Nations Security Council. Government hearings and committees address a host of important issues through group debate and problem solving. Although there are certain key individuals in all levels of society who are considered leaders, each of these leaders surround themselves with teams to accomplish goals and objectives.

In groups, we decide, inform, relate, entertain, and socialize. We conduct many fundamental human behaviors and fulfill interpersonal needs with our behavior in groups. Groups are not just something we are forced in to for classroom projects. Rather, groups are so central to the human experience that we often fail to recognize our continual presence and participation in groups, even though groups influence the very norms and culture of society.

GROUPS ARE FUNDAMENTAL TO SOCIETY

In other words, groups are a basic and fundamental part of society, and greatly influence how we communicate with each other (Gastil, 2010; Poole, 1990). Group work is more than simply optional activities in which we endeavor; group interaction is the human experience. Social acts, or communication from one person to another, are a result of our need to seek out opportunities to be with and converse with others. Of course, some of us feel the need to be included much more than others. Due to pressures and anxiety associated with group communication, some of us would prefer to avoid groups. Even if this is the case, groups are so central and fundamental to how we view and participate in the world that they cannot be avoided. Whether it is through families, work, school, government, religion, nonprofit organizations, or play, we spend a good portion of our time in or preparing to be with others in groups.

Types of Groups

Despite the number of groups to which one belongs, one group is never quite like another. When groups and teams were first considered by researchers, they often focused solely on **task-oriented teams**. Common examples of these teams are work groups and management teams. The purpose of these project teams and work groups is to accomplish work tasks, and it is generally assumed that group interaction should be task focused and geared toward planning, debating, or implementing group decisions. However, more recently scholars have also focused on **relational groups** (Keyton, 1999), such as support groups, fraternities/sororities, and social groups. This focus on relational groups also highlights that all teams, whether primarily task or relationally oriented, have a relational dimension. Even members of a work team must learn how to get along, especially given conflict and debates about important task decisions. Thus, even though some groups may have an overarching task or relational focus, every group is goal directed and has both task and relational dimensions that are played out through interaction.

Some groups are created for the sole purpose of a task. For example, a student organization might determine to conduct a fundraiser, and assign a five-member group to

SITUATING YOUR EXPERIENCES IN GROUPS

create the event. These individuals may have known each other from prior interactions, but there is a chance that none of them have worked together. Once this group plans, organizes, and implements the fundraiser, the group tenure will be over and they will return to their original student organization functions. Due to the short duration of the group, relationship development may be truncated and leadership structure may be formally and quickly determined and implemented.

However, some groups are ongoing. Fire and emergency medical personnel may work together on a regular basis to handle emergency situations. Experiences among team members in the present will directly influence team behavior in the future. As a result, individuals may be more conscious of relational development, knowing that preserving good relations may lead to positive results over time. Negotiation of roles may change during the life span of the group. The hiring of additional members may require a great deal of effort to acclimate new members to the norms of the established group. Behavior in this type of group may vary drastically from groups that are quickly created for a short-term purpose. Of course, there are many group types that fall between one-time groups and long-term groups.

Due to the different group types and task, group behavior can vary greatly. A social support group functions differently than a work group, which functions differently than a family. However, there is also similarity across these various groups. Groups have fundamental characteristics that make a group a group (see Chapter 2). Groups have formal or informal leaders who influence the flow of conversation. Some degree of common understanding among group members is required for successful group decisions. Groups that work together over time develop routines and norms that guide group interaction. Although these group characteristics can be manipulated and abused to the dismay of group members (e.g., social loafing, groupthink), these same characteristics can benefit a group in terms of leadership, diversity, and relational development.

GROUP COMMUNICATION

Of course, there are many ways to investigate and learn about groups (e.g., social psychological, sociological, management). However, this book examines groups from a *communication* perspective. In order to analyze groups from a communicative perspective, it is important to know what it is we mean by communication. **Communication** is the process of symbol production, reception, and usage; conveyance and reception of messages; and the meanings that develop from those messages (Keyton, Beck, & Asbury, 2010, p. 472). A communicative approach means we are examining symbols, messages, and meanings. All communication is strategic in that we combine symbols into messages in order to accomplish a specific purpose. We use messages to convey meaning to others, and other group members use our messages as evidence of what that meaning is. Unfortunately, the exchange of messages can lead to different interpretations by group members, causing miscommunication.

Interaction and Messages

In this book, we want to investigate groups by examining the processes members undertake to accomplish their task or relational goals. Process, another name for interaction, is the focus of how we study groups. Without interaction among its members, a group would cease to exist. So, while it might be interesting to study how different compositions of group members result in ineffective group outcomes, or whether face-to-face groups outperform online groups, we take these questions one step further: How do these different aspects of groups affect how groups communicate?

Thus, interaction is the foundation of all group endeavors. But due to the multiple opportunities for miscommunication, group interaction is much more complex than one-to-one or dyadic communication. For example, let's say that Sam was charged with researching a problem his sales team was having with one of its customers. After conducting the necessary research, Sam presents his information to his team members, explaining how and why the problem is significant and that it will require group members to work together to resolve the problem by next week's deadline. Since other group members have worked with Sam before, they are aware that when Sam says they will work "many hours," he means everyone will have to work an hour later each evening to get the project done. However, Tonya is new to the group, and when she hears "many hours," she is worried that they will need to work late into the night to get the project done. She may prepare her schedule differently as a result. Thus, some group members had a high level of shared understanding about the task due to their previous experience. Unfortunately for Tonya, her lack of shared experience prevents her from understanding Sam's message as he intended. Since group interaction is heard and interpreted by several individuals simultaneously, there is an increased opportunity for miscommunication and diverging perceptions of interaction. Although you may know some members very well and other members less well, all members may hear the same messages but interpret them in different ways.

Additionally, all messages are **strategic**. When we say strategic, we are referring to how speakers adapt their messages so that they will accomplish their goals (Kellermann, 1992). If you want to go out to eat with a good friend, you could probably say, "Hey, do you want to grab something to eat?" However, if you want to ask the same question of your supervisor, you may alter how you ask the question: "I was wondering if you would like to go to Café Rio for lunch on Wednesday?" Both of these messages are designed to arrange a lunch, but based on the people involved and the context of the relationship, the messages are very different (one informal, the other formal). This is what we mean by strategic—we adapt our messages so that they will best accomplish our goals. As a result, our messages may differ in terms of formality, brevity, complexity, and sincerity. And importantly, even the most elementary of messages, for example, "Will you please pass the salt?", are adapted to accomplish a specific purpose. But since these communicative processes are largely subconscious, we tend not to overly think about these adaptions in regular conversation.

Such message adaptions are true in groups as well, although the process is a bit more complicated. With multiple individuals, our adjustments have to be more complex or conscious. If your group contains close friends and superiors, you may err on adapting your messages more formally as a way of acknowledging the greater value placed on your superiors' opinions. If a deadline is approaching, your messages may be less nuanced and diplomatic, and instead become quite brief and direct. Communicating in a group context requires members to assess the situation and adjust messages accordingly. However, the situation may be difficult to assess. There may be quite a few unknowns. For example, how are people interpreting my messages? How are most group members siding on the decision? Is group member silence a good or bad thing? If the group is meeting quite regularly, it may be easier across conversations to evaluate how members' opinions are evolving.

Fortunately, from a communication perspective, we have a powerful piece of evidence for evaluating group member messages—the message itself. We can analyze how verbal and nonverbal messages are adapted in order to understand individual intent. For example, if a group member addresses the group leaders as "sir" or "ma'am" in a polite but serious way, we can assume that the group member has interpreted the group climate as very formal. The evidence is in the message itself. Thus, one of the important implications of using a group communication perspective is that we can investigate messages to understand meaning and influence on group goals and tasks.

Meaning is also extracted from the way messages follow one another. That is, messages are interdependent (Beck & Keyton, 2012). Certainly, it is impossible to predict what

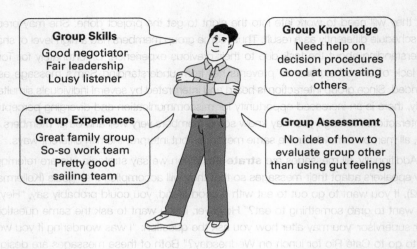

Figure 1.1 You may be good at some aspects of group communication but poor at others.

message a group member will use after another group member speaks. That's because communication is improvised in the moment of conversation. Still, the interdependence of messages that comprise group conversation allows for new ideas to be created, information to be shared, and problems to be solved.

Finally, messages are also multidimensional in that they can influence the conversation in multiple ways. At the simplest level, messages can be distinguished as task messages or relational messages. Since many groups have an overarching group goal, task messages are those messages designed to accomplish the task. This can involve sharing information, planning, coordinating, debating, deciding, and any activity behavior that is in line with accomplishing the group goal. Relational messages are those in which we show antagonism toward or friendly acceptance of others in the group. These types of messages influence the social fabric of the group. Upon completion of a project, June tells Grant that he did a good job on his part of the presentation. Such a comment shows clear support toward Grant and will benefit the relationship between the two, as well as the group as a whole. Both task and relational messages are required for a group to function successfully.

Although task and relational messages are often presented as distinct from one another, many messages contain both task and relational aspects. For example, Amber has been working on a project that she needs to report on to her boss, Katherine. When she stops by Katherine's office, Katherine says, "I don't have time to deal with that now." This message can be interpreted in multiple ways. It can be interpreted in terms of task (i.e., Katherine is busy) or relational (i.e., Katherine doesn't care about me). Even though the intent of the Katherine's message may be very task oriented (she may have a very busy day or an important deadline that she needs to meet), Amber may not use a task-oriented frame to interpret the message and may be very upset by Katherine's comment as a result. The multidimensionality of messages provides listeners several ways to interpret interaction and for group member interpretations to differ.

SITUATING YOUR EXPERIENCES IN GROUPS

GROUP AND INDIVIDUAL GOALS

Since group members are adapting their messages to accomplish an array of individual and group goals, all in consideration of context, groups tend to be a very complex and messy forum for communication. Each group member has multiple individual goals (Hollingshead, Jacobsohn, & Beck, 2007), and goals vary across participants. Additionally, group members' perception of the group goal may also be different, especially if the goal has not been discussed in depth.

Importantly, individuals are often trying to accomplish more than one goal—some of these goals may be group oriented, others may be individual oriented (Wittenbaum, Hollingshead, & Botero, 2004). Because there are many goals, not all of them will align. For example, Eliza has a great idea for a fundraising project her team has been assigned. This idea will help the team accomplish their purpose in a high-quality manner with low costs (group goal). In addition, Eliza would really like to be promoted to group leader when the current leader retires next month (individual goal). She knows that if her idea were selected, it would help her appear very leader-like and increase her chances of getting the promotion. To accomplish both her individual and group goal, Eliza successfully persuades the rest of the team to support her idea. In this example, the individual and group goals were aligned with each other, and thus Eliza was not required to prioritize her goals. However, when goals diverge, such prioritization may be complex and difficult.

GROUP SKILLS FOR LIFE

Group interaction is the mechanism that creates many of the tensions and dynamics of groups. As will be outlined in this book, group interaction is the foundation for group identity, interdependence, and structure. It creates and fixes conflicts, leads to group decision making, and is the foundation on which members develop group relationships. Since groups are brought together to take advantage of group members' skills and talents, the ability to work in groups is essential to accomplishing the group's purpose.

As you proceed through this book, you will learn skills associated with being an effective group member. According to the National Association of Colleges and Employers (2015), teamwork is the one of the top skills desired by employers. Being able to work effectively with others in a group will also be useful in your social and community life. Our hope is that by reading this book and applying the concepts described, you will be a more productive employee and member of society. Importantly, since groups are commonplace throughout society, becoming a functional and dynamic group member will increase satisfaction and enjoyment in many aspects of your life.

In addition to learning skills, another objective of this book is to help you analyze and understand the groups or teams of which you are a member or for which you are responsible. By being able to analyze group activities, the interactions among members, and the environment in which a group operates, you will discover what's unique about a particular group and what are the most effective ways to participate in the group.

There are no magic formulas for group interaction or set procedures that work in every group setting. Each group is different, so your approach to each group must differ as well. This book is a guide for your exploration of groups, based on theory and research studies. From the theories contained in this book, we can identify the skills needed for effective and successful group interaction. By developing your group communication skills, you will be able to develop practical and viable solutions to the group interaction problems you encounter.

SITUATING YOUR EXPERIENCES IN GROUPS

FORMAT OF THE BOOK

Groups are like individuals—no two are alike. The better equipped you are to analyze what is happening in and around the group, the more successful and satisfying your group experiences will be. To guide you through your investigation of group communication, this book is organized into three parts.

Chapters 1 through 3 provide the foundation for the remainder of the book. These chapters describe the importance of groups and teams and how communication is basic to our understanding of these social and task forms. At the same time, we do not dismiss that individuals in groups have individual goals that may or may not be congruent with group goals. Our model, the five-piece group puzzle, is introduced in Chapter 2 and provides a structure for the remaining chapters, as well as an analytical tool for identifying key strengths and challenges of group interaction. Chapter 3 describes and explains the contextual influences that affect group interaction across group types: boundaries, time and space, diversity, and technology. Each of these influences is present in some way in every group interaction.

Core group processes that occur in every group are described and explained in Chapters 4 through 8. These core processes are sharing information, participating in decision making, building relationships, managing conflict, and providing leadership. Each of these processes is an interaction opportunity, and each is a regular and dynamic aspect of group interaction. At the same time, these processes can create interaction problems that group members must address. Increasing your skills in these core processes will help you be more effective in group interactions.

Many groups do their work in meetings. So, in conclusion, Chapter 9 will help you participate in, navigate, and facilitate group meetings. Whether in the role of leader or member, you should be able to facilitate group interactions to help the group stay or get back on track. Armed with specific principles, procedures, and feedback techniques, you will be able to make more informed choices about how to help your group.

At the beginning of each chapter, a bulleted list presents the essential ideas for that chapter. Previewing these ideas before you read the chapter can help you organize your learning. In preparation for evaluations (tests or applications), reviewing these points can help remind you what you have learned. At the end of each chapter, you will find a numbered list of discussion questions and exercises. Answering these questions and reflecting on the exercises are an effective way of remembering and applying what you have learned. It is also a helpful way to determine if your learning was meaningful while reading the chapter. Although the material in this textbook is useful we all sometimes read a little too quickly or casually when studying for classes. Use the questions at the end of the book as a way of assessing your reading and preparation for the next class.

Throughout the book, you will notice words or phrases that are set in boldface. These words or phrases are part of the vocabulary unique to the study of communication or group communication. Each word or phrase is defined the first time it is used in a chapter. These definitions are also compiled in a glossary at the end of the book. Throughout the book, you will also find in-text citations that point you to the reference or scholarly source from which we developed our ideas. These citations comprise the References, which are at the end of the book.

Additionally, we use four features to develop your expertise about group communication. First, **Message and Meaning** boxes display transcripts from groups and teams that have been part of our research, or transcripts that are publicly available. Group interaction is not scripted in advance like a television show. Group conversation takes twists and turns based who says what, and these transcripts will draw your attention to the complexity of messages and meanings in group interaction. The second feature, **Theory Standout**, provides an in-depth look at group communication theories that are introduced in this book. Perhaps you will use one of these featured theories as a basis for a classroom assignment. The third feature, **Skill Builder**, describes how to develop a group communication skill. Whether you believe you are a novice or an expert, this boxed feature will help you identify ways to evaluate and further develop your mastery of this skill. The fourth feature, **Nailing It!**, helps you apply what you're learning to group situations you'll experience while in college. Some of these will be social groups; others will be task or decision-making groups. And, important to your classwork, some of these groups will be required to develop and give a group presentation.

DISCUSSION QUESTIONS AND EXERCISES

1 Think of a group you belong to now. Using the concepts presented in this chapter, how would you describe that group to someone unfamiliar with that group?

2 Reflect on one of your childhood groups. Compare that experience with one of your adult group experiences. What has changed? What is similar? Why?

3 Review the table of contents for this book. Thinking about your current level of group skills, and develop three lists: (a) group skills and knowledge that you have now, (b) group skills and knowledge that you'd like to learn, and (c) group skills and knowledge that you've mastered and could share with others.

4 Thinking to the future as you pursue your education or a career, what types of groups do you expect to be a member of? Be a leader?

Image Credits

Photo 1.1: U.S. Army Corps of Engineers Sacramento District. Copyright in the Public Domain.

Photo 1.2: Copyright © uwdigitalcollections (CC by 2.0) at https://commons.wikimedia.org/wiki/File:Cheerleader_tossed_aloft.jpg.

CHAPTER 2

GROUP COMMUNICATION FUNDAMENTALS

After reading this chapter, you should be able to:

- Name the characteristics essential for defining a group
- Recognize the ways in which members socially construct the group through their interactions
- Develop an understanding for thinking about a group as a process
- Explain ways in which group communication is complex and messy
- Explain why both task and relational communication are required in group interactions

WHAT IS A GROUP?

We all have different past group experiences, which lead us to have different expectations and desires for group work. In fact, individuals may disagree on what a group is and is not. Are two people having dinner together a group? What about five people waiting for a bus on a street corner? Would you consider the 50,000 fans in a football stadium a group? And what about your friends who text or send photos to stay in touch during the day? How exactly is a group differentiated from other forms or contexts of interaction? First, let's consider what a group is.

Characteristics for Defining a Group

Five characteristics are central to the definition of groups: group size, interdependence of members, group identity, group goal, and group structure. In addition to defining what a group is, these characteristics are a good place to start developing an understanding about how members of a group interact effectively. These characteristics can help you isolate group interaction problems and understand why they develop. As you read in more detail about each of the characteristics, you will come to understand why a **group** is defined as three or more people who work together interdependently on an agreed-upon activity or goal. Each person identifies themselves as members of the group, and together they develop structure and roles, based on norms and rules, as they interact and work toward their goal.

Group Size

One of the primary characteristics of a group is **group size**. The minimum number of members in a group is three; the maximum number depends on the other characteristics, discussed shortly. Communication among two people, or a dyad, is labeled as interpersonal communication, and different from group communication. The interaction of three people differs significantly from the interaction of two, because the introduction of the third person sets up the opportunity to form coalitions. As an example, **coalition formation** occurs when one member takes sides with another against a third member of the group. This type of 2-to-1 subgrouping creates an imbalance of power, one that can only occur when at least three group members are present. A coalition creates interaction dynamics that cannot occur with two people. Of course, once a group expands beyond three members, different types of coalitions can occur. Three members can contest the ideas of one member who refuses to be persuaded, or a group can break into multiple subgroups or coalitions, each promoting its own view.

Introducing a third group member also allows **hidden communication** to take place. These hidden interactions are often attempts to build alliances, which underscore the role of relationship building as groups work on tasks. For example, let's say that Nancy and Michelle meet on their way to a meeting; Jeff is waiting for them in the conference room. Nancy takes this opportunity to brief Michelle on the background of the project and provide feedback and evaluations on her previous interactions with Jeff. Jeff does not have access to this hidden interaction, but Nancy's musings to Michelle will certainly affect the interaction among these group members. In this case, there was no strategic attempt to manipulate Jeff, but Nancy and Michelle's interaction still affected the group. Naturally, the larger the group, the more these hidden interactions are likely to occur. The size of a group has an impact not only on how members interact with one another but also on how roles are assumed (or assigned) within the group and how interactions are regulated.

What happens when a group is too large? It may be more difficult for members of larger groups to decide who takes what role because many members may have the skills necessary for various roles. Also, larger groups typically have more difficulty in scheduling time to meet with one another as a group.

Research has demonstrated that increased size can produce diminishing returns. In other words, bigger is not always better (Bettenhausen, 1991; Hare, 1982; Wheelan & McKeage, 1993). Although the addition of group members can expand the pool of skills and talents from which to choose, it can also increase problems with coordination and motivation. There is a point at which groups become too large and members become dissatisfied, feel less cohesive with one another, and perceive less identification with the group. Why? The larger the group, the fewer opportunities each member has to talk, and as group size increases, logistical problems in coordinating so many people may negatively influence what a group can accomplish. Thus, increased group size affects group productivity because members have less opportunity to participate. Because of their size, large groups require more attention to group norms and group roles. Even more problematic is the fact that group members are more likely to accept the illusion that someone else is responsible for accomplishing the group's task and so fail to do their own part. As a result, there are greater demands on group leadership in large groups. But large groups can be effective—if the goal is clearly identified for all group members, if members share a consensus about the goal, and if they recognize and fulfill their roles.

Relationships also are affected when a group is too large. For example, twenty members are probably too many to deliberate on a problem and make recommendations in one written report. Having twenty people write together is difficult! Additionally, when group members feel as if they are not needed to produce the group's outcome, or if their individual efforts are not recognized, they become apathetic and feel distant from the group. This form of detachment is known as **social loafing** (Comer, 1995). Social loafers are group members who do not perform to their maximum level of potential contribution. Rather, they use group members as a shield they can hide behind while still reaping the same benefits as other group members who work to make the group a success. The opportunity for social loafing increases as the size of the group increases (Lam, 2015).

What happens when a group is too small? Members of smaller groups may find that no one in the group possesses a critical skill or certain knowledge essential to the group's activity. For complex tasks, three members may be too few to effectively reach group goals. When there are too few members and there is too much work to do, group members are likely to become frustrated, and even angry, about the task and toward the group. In smaller groups, members may also feel pressured to talk more than they feel comfortable.

As you can see, there are problems when there are too many or too few group members. Thus, group size is most appropriately determined by the group's task or activity. In essence, group size should be restricted to the number of members necessary to effectively accomplish the group's goal based on the nature of the task, especially because smaller groups develop into more productive groups more quickly (Wheelan, 2009).

The maximum number of group members depends on the other four characteristics of groups. Rather than limiting or expanding group membership to some arbitrary number, we need to consider issues such as group goals, task complexity, and interaction opportunities to identify the appropriate number of group members. Of course, some groups and teams (juries, sports teams) have specific size limits or standards, which group members cannot change.

Interdependence of Members

A second critical characteristic of a group is the interdependence of group members. **Interdependence** means that both group and individual outcomes are influenced by what other group members do (Bonito, 2002). Members must rely on and cooperate with one another to complete the group activity, because they are attempting to accomplish something that would be difficult or impossible for members working as individuals to achieve. Through their interdependence, group members mutually influence one another.

For example, members of a softball team are a group. It's impossible to play effectively without members in the roles of catcher, pitcher, and shortstop. Each member of the softball team fills a specific role that functions interdependently with those of other players. Moreover, how well one role is fulfilled affects how another role is fulfilled. Even if the team has one outstanding hitter, the team will not win very often if other members don't also hit well or if there aren't any members who specialize in defense. Not only do these team members have to fulfill their specialized roles and depend on one another, but they have to communicate with one another. It is not enough to identify the necessary roles and to assign members to them; individuals in these roles have to be actively engaged and interacting with one another.

As another example, consider a project team at a digital gaming company that has been given the task of developing a multiplayer online game. This task can be seen as a **superordinate goal**—that is, a task or goal that is so complex, difficult, or time consuming that it is beyond the capacity of one person. To be successful, group members with different skills must work interdependently to achieve the group's goal. Team members are interdependent as they share ideas in the early stages of the project; later they can test various ideas with one another before engaging expensive resources. Such interdependence is likely to save their organization time, energy, effort, and money; it is also likely to create a better game.

The communication within groups also illustrates the interdependence of group members. Let's look at a student group concerned about course and faculty evaluations. Jennetta asks the group to think of ways to improve the evaluation process. Her question

MESSAGE AND MEANING

As is typical of many support groups, members of the Bosom Buddy breast cancer support group went around the circle to check in with one another. Here is Claire's check in and how other support group member responded.

CLAIRE: My sister, who lives in Michigan, called, and her middle daughter is the one that I told you had breast cancer—and now her husband has come up with a lump in his breast. And, they're going to do an ultrasound and they're trying to figure out what's going on. But, I thought, God, that's kind of a double-whammy. You know, he's gone through this with his wife and now he's facing the possibility that this is happening to him, too. And, you know, you don't generally think in terms of men having that, so you're not looking for it. My sister said it doesn't appear to be attached to anything. I said, "Well, that's good, because, you know, frequently it goes to the bones, the rib cage." And women have a little more breast tissue, so it can, you know, be growing there more before it gets to the rib cage than for men. She was asking me if I had heard of that, and I said, "Oh, yeah." While I was doing Reach to Recovery, I worked with three different men who had had breast cancer. And all of them, at that point, were successfully treated, but—so it does occur.

MAXI: Terrible, terrible.

TINA: Do you know the percentage?

FAYE: Is the rate for cure for men as high as the rate for women?

CLAIRE: You know, that's a really good question, and I'm not sure about that.

TINA: But I suspect it might not necessarily be high, because … you're not looking for it.

MAXI: And they don't get mammograms …

TINA: That's right, and also because you're right there on the rib cage … you know, you're right there on the rib cage, so it could go to the bones and then into the lung. So, I don't know, but that's a really good question. We should find that out.

Return to Maxi's comment, "Terrible, terrible." Who is she empathizing with? Claire? Claire sister's daughter? Claire sister's daughter's husband? What is she empathizing about? Next, notice how Tina asks a question to begin the informational phase of the discussion. Does her informational query create a different type of empathy?

prompts group members to respond with ideas that she writes on the board. When they finish, Jackson comments about one trend he sees in the list. Sara asks him to elaborate. As Jackson and Sara continue their conversation, Jennetta circles the ideas they are talking about and links them together while she gives affirming nods to indicate that they should continue talking. Pamela, who said very little during the idea generation process, now says, "But the ideas you are circling are ones we as students can do little about. What about working through student government to develop an independent evaluation process that could be published in the student newspaper?" Jennetta, Jackson, and Sara turn to Pamela expectantly. Their silence encourages Pamela to continue talking: "What I'm saying is that the ideas on the board are attempts to fix a system that is not under our control. So, why not develop an independent system that students control?" Jackson replies enthusiastically, "Great idea, Pam."

Notice how the verbal and nonverbal messages in this group depend on one another to make sense. Jennetta first invites members' participation, and they all generate ideas. The list they generate motivates Jackson to make an analytical comment, which is further encouraged by Sara's question. Although Pamela initially says little, her action has an impact on other group members' communication by giving Jackson and Sara more opportunities to talk. Pamela's interjection into the conversation startles the others, and their conversation stops. Her acute observation reminds them that she has not been ignoring what's going on; rather, her assessment helps them see that they may be wasting their time.

In this example, the communication itself was interdependent. One statement can only make sense when it is placed before and after other strings of the conversation. Each individual in the group is influenced by what others say (and don't say). The group's success depends on the extent to which the verbal and nonverbal messages make sense together.

Group Identity

A third defining characteristic for a group is group identity. Group members must know and act as if they are members of this particular group. In essence, **group identity** means that individuals identify themselves with other group members and the group goal. Group identity is fully achieved when members behave as a group, believe they belong to a group, and come to like the group—both its members and its tasks (Henry, Arrow, & Carini, 1999). Group identity fosters cooperation (Jackson, 2008). Without this type of identification, group focus and interdependence will weaken.

Unfortunately, many times people are identified as a group when they have little or no expectation that group interaction will occur. Such gatherings or collections of people are more appropriately called **groupings**. Throughout our lives, we are constantly identified by the groupings people assign to us. For example, others often label us by where we live (e.g., Midwesterner, in a big city), what type of work we do (e.g., lawyer, franchise owner), and the hobbies we pursue. But these characteristics do not necessarily put us into groups where we interact and interdependently work on tasks and activities with others. At the same time, individuals may join particular groups because they want to be identified and interact as members of the group (e.g., a fraternity or community chorus). Simply doing so will not result in group interaction opportunities unless the individuals are motivated to talk to others.

Remember that, just because individuals have some reason to be together or some surface connection seems to exist among them, group interaction may not occur. Simply being identified with others who share similar characteristics doesn't create a group. Group members can only develop identity with one another, and distinguish their membership in this group from memberships in other groups, when they communicate with one another in the beginning of the group and throughout the group's history (Zanin, Hoelscher, & Kramer, 2016). When group members identify with one another and the group's goal, they adopt the norms and values of the group, increasing group members' motivations and abilities to work together effectively.

Group Goal

Identity, then, is a necessary but not sufficient characteristic for a group. We also need a fourth characteristic: group goal. A **group goal** is an agreed-upon task or activity that the group is to complete or accomplish. This goal may be long term and process oriented (such as a family functioning as a social and economic unit), or it may be short term with specific boundaries and parameters (such as a youth group holding a car wash to raise money). Regardless of the duration or type of goal, group members must agree on the group's goal to be effective (Larson & LaFasto, 1989). That does not mean that all group members have to like the goal, but it does mean that there is clarity on what the goal is and that it is perceived by members as being worthwhile.

Having a group goal gives the group direction and provides members with motivation for completing their tasks. A group's goal should be cooperative. This means that, as one member moves toward goal attainment, so do other group members. A group goal is cooperative when it integrates the self-interests of all group members. Groups that are having trouble have often lost sight of their goals—sometimes because of distractions and other times because of external forces (a change in deadline or objectives). Groups that cannot identify why they exist and what they are trying to achieve are doomed to failure.

For example, as a student in the class for which you are reading this textbook, your goal is probably to get a good grade. But getting a good grade is your individual goal, not a group goal. Each student in the class may have this same goal, but it is not a shared, consensual goal that motivates interaction and activity. If it were, everything you did in preparation for class would be designed to help you, as well as other students, achieve the *group good grade* goal. Thus, agreement on a common goal among individuals, not similarity in individual goals, defines individuals as members of the same group. Group goals create cooperation, whereas individual goals often create competition (Van Mierlo & Kleingeld, 2010).

Group Structure

The final defining characteristic of a group is its structure, or how the team members are organized along with the tasks and resources needed to accomplish the group goal (Lafond, Jobidon, Aube, & Tremblay, 2011). Whether informal (a group of friends) or formal (a parent-teacher organization), some type of structure must develop. **Group structure** tends to develop along with, or to emerge from, group rules and **norms**—patterns of behavior that others come to expect and rely on. For example, a group of friends—Pat, Emily, Donna, and Greg—meet for social activities every Friday night. If the group does not set plans for the next week, Pat takes it upon himself to text everyone to get suggestions. No one has appointed him to this role; he does it naturally in reaction to the other group members' lack of initiative. Pat has assumed the role of the group's social organizer, and the group has come to depend on him to play that role. His role playing has created a certain structure in the group, and that structure has become a norm. Group members expect him to take the lead concerning their weekly activities.

In more formal settings, a group may elect someone to record what happens in the meetings as a way of tracking the group's progress and keeping an account of details. Again, the person taking on the recorder or secretary role is providing structure for the group, as well as behaving in a normative pattern. Thus, both the recorder's actions and the record of the meeting provide structure for the group. Anytime a group member takes on a formal or informal role, group structure is created. Likewise, any discussion or outcome that provides direction for the group is considered group structure. Suppose your family decides to visit Disneyland on vacation. That decision creates structure for your family discussions in the future because now your interactions will center around the logistics of traveling to California and planning your vacation.

To be viable, groups must have some form of structure, but the structure does not have to remain constant throughout the life of the group. Much of a group's structure is provided by **group roles**, or the functions group members assume through their interactions. Roles are not necessarily fixed. Formal roles—those filled through appointment, assignment, or election—are likely to be more permanent. Informal roles—functions that emerge spontaneously from the group's interaction (such as the group member who eases tension in the group)—may change as the talents of group members become apparent or are needed by others.

Integrating the Pieces of the Puzzle

The five characteristics that define groups—group size, interdependence, group identity, group goal, and group structure—can provide a foundation for analyzing the effectiveness of a group (see Figure 2.1). The following example will lead you through this analysis.

Like most students assigned a group project, Gayle, Rebecca, Sean, Jim, and Sonya wait too long to begin work on their assignment. Now, pressed for time, each member has other obligations and, quite frankly, more pressing interests and motivations. Still, the group has to produce what the professor expects in order to receive 20 percent of

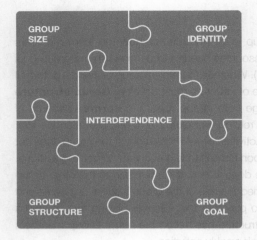

Figure 2.1

their course grade. Meeting once to get organized, Rebecca, Sean, Jim, and Sonya each assume responsibility for one area of the project, and Gayle agrees to take responsibility for integrating these parts. The group gives itself two weeks before reconvening to turn in finished materials to Gayle, who will pull it all together before the oral presentation to the class. Due to the members' late start, there will be only a few days between the group's second meeting and the oral presentation, putting extreme pressure on Gayle to integrate the project's parts and get it back to the other members so they can perform effectively during the presentation. These members are juniors and seniors, and they have done this type of group project many times in the past. They know they can pull it off.

This group is probably similar to other groups you've been in. Given their deadline, the decisions they have made about dividing up the workload may make sense. However, there may be some problems with these decisions as well. Consider the following questions:

Group Size

1 If Gayle asks her roommate for help, is her roommate part of the group? Why or why not?

2 If Sonya becomes sick and cannot perform her part of the project, is she still part of the group?

3 Would it help if the group had more members?

Interdependence

1 Does it make sense for one individual to integrate everyone's work?

2 What are the disadvantages of breaking up work in this manner? What if Sean's part of the project conflicts with the other members' parts?

3 Is this a superordinate goal? Why or why not?

Group Identity

1 Does this group have an identity? How would they know?

2 As the leader of the group, what can Jim do to enhance group identity?

3 Do the members even need a group identity, given their time constraints?

Group Goal

1 The goal was given to the group by someone external to the group. How will this affect group members' perceptions of the goal?

2 What are the boundaries of this goal? Does it align with individual goals?

Group Structure

1 Are these the only work roles the group needs to consider?

2 Is the group structure developed ideal for the project?

3 Does this project require a more formal structure to be successful?

Using the five defining characteristics of a group can help you understand what factors may be inhibiting your group. If you were in charge of this group, which of the five characteristics would be important for you to work on? Knowing, for example, that identity is weak in your group, you might want to suggest that group members spend some time getting to know one another before beginning work on the task. Or if the group goal is not clear and agreed upon by everyone, it will be helpful to spend a few minutes talking specifically about what the group is trying to accomplish. When one or several of the defining characteristics are weak or missing, the sense of groupness may be too fragile for the individuals to function effectively as a group.

To summarize: We have defined a group as three or more people who work together interdependently on an agreed-upon activity or goal. They identify themselves as members of the group, and they develop structure and roles, based on norms and rules, as they interact and work toward their goal.

Due to the defining characteristics of groups, each group takes on a life of its own. Each is unique. What we as individuals bring to group interactions is a unique compilation of all our past group experiences, good and bad. Your set of expectations resembles no one else's set of expectations, and members of the group bring different expectations to the same group experience. As a result, we live in a world of constant ebb and flow of group interactions in our personal, social, and professional lives that overlap and affect one another. It is to our benefit to understand these interactions and influence them. Not only are groups charged with completing tasks and activities, they also provide us with opportunities to develop and maintain relationships, to learn about ourselves, and to enhance our personal and professional skills. And all of this is accomplished through communication.

INTERDEPENDENCE OF TASK AND RELATIONAL DIMENSIONS

At this point, it should be obvious that a group is really a process. A group is not simply its number of members, its effectiveness, or its type of task or activity. A group is created through interaction among group members as they establish roles and relationships while they work toward their mutual goal or activity—and that process is both complex and messy. For example, group members need to meet together in order to construct role relationships and generate group identity, but most groups seldom have all members meeting together for all interactions. Likewise, in any group or team, there are likely to be multiple goals at multiple levels—group, subgroup, and individual. Finally, all group interactions are not likely to be friendly or produce positive outcomes.

Despite our best efforts, groups are not always effective, fun, or productive. When a difficulty occurs, it is often a signal that a core group characteristic—group size, interdependence of members, group identity, group goal, or group structure—is deficient, at risk, missing, or out of balance for that group's task or activity. That is, the group's size is too large or too small, group identity is too weak or too strong, interdependence of members is too loose or too strong, agreement and enactment of the group goal is under- or overemphasized, or the group structure is too rigid or too loose. Moreover, groups are not objects in containers. A group's relationship to its context is fluid and complex, which can create challenges for group members to address and resolve.

Part of the challenge is to balance the **task dimension**, or what a group does, with its **relational dimension**, or the social and emotional interactions of group members from which roles and relationships emerge. All groups have both task and relational dimensions. Groups with a strong task focus must also pay attention to the relationships that develop among group members, and even groups that are primarily social or relational have some task to perform. In these types of groups, the task may be as simple as members being there for one another, or it may be more specific, such as providing a place for members to explore their feelings. Regardless of a group's primary focus, both task and social dimensions are present, and they are inseparably interdependent (Fisher, 1971).

 THEORY STANDOUT

Task and Relational Messages

Early in the study of groups and teams, Bales (1950) identified both task and relational dimensions as the two central dimensions of group process. Watzlawick, Beavin, and Jackson (1967) went one step further and posited that all messages have both relational and task content. How can one message have both task and relational content? Think about a recent group interaction you have had. Did you use a short phrase or one word, such as "okay," to indicate your agreement? Did your "okay" mean that you agreed with your team members? Did the nonverbals that accompanied your "okay" indicate how you felt about your agreement?

Messages are complex. They can have more than one level of meaning. Most of the time, our verbal and nonverbal messages are in agreement. But sometimes, we use a verbal message to indicate agreement, and at the same time use a nonverbal message to indicate that we're not happy about having to agree. Moreover, how a message is evaluated by other group members depends on what messages occurred before and after the focal message.

Theoretical contributions by Bales and Watzlawick, Beavin, and Jackson encourage us to carefully consider what messages are and what messages do. Both task and relational messages are central to the study of group communication.

This two-dimensional aspect of groups is important because a group that concentrates solely on work without attending to its members' social or relational needs becomes boring and ineffective. Likewise, a group that focuses solely on having a good time can become tiresome if that social interaction does not lead to new information or provide opportunities to perform meaningful activities. The most effective groups are those that keep each of these dimensions in balance relative to their purposes and the needs of their members (Tse & Dasborough, 2008).

The task and relational dimensions are interdependent—that is, they work together. But this does not mean that group members use task and relational messages equally. Rather, all groups use more task messages—even groups with primarily relational or social goals. Why is that? Simply, talk does something; it is a **performative act** (Grice, 1999). For any group to achieve its goal, group members must exchange task messages which are interspersed with relational messages. For relational messages, we can analyze the frequency of positive relational messages to negative relational messages to understand the emotional tone or social climate of the group. For example, in community theatre groups, members of the casts and crews rated their relational communication with their peers as more important than task-oriented communication from the theater-group leaders (Kramer, 2005). However, a theater production would not have occurred if task messages were not also exchanged among group members.

Satisfying Task and Relational Dimensions of Groups

The balancing of task and relational group needs may depend on the nature of the group. A group's level of formality may determine the necessary frequency of task and relational messages. Some groups form deliberately; others emerge from spontaneous interaction. When groups form deliberately (e.g., work groups, neighborhood council), someone decides that a collection of individuals should accomplish a purpose or goal. Most problem-solving or decision-making groups (such as city councils) and social action groups (for instance, Stop Hunger Now) are examples of deliberately formed groups—it would be impossible for fewer people to accomplish their goals. These groups tend to have many task-oriented messages.

Other groups form spontaneously. Generally, individuals come together in these groups because of the satisfaction they expect to gain from associating with one another. A group of friends at work is a good example of a spontaneous group. In these cases, group membership is by mutual consent—each member wants to be in the group, and each is accepted as a group member. Typically, these groups form when individuals communicate frequently and voluntarily with one another. Thus, group membership is based on attraction.

However, all groups must balance task and relational dimensions. For example, a group of activists (such as a local affiliate of Habitat for Humanity) needs members with the technical skills of recruiting new volunteers and seeking and obtaining funding, as well as members who are willing to teach others the skills needed for building houses. Members also need the relational skills of motivating members to continue to work on behalf of the organization and the ability to create a supportive environment for members. The challenge is to find the appropriate balance between the two sets of skills. The balance will vary depending on the type of group and its activities and goals.

A group deliberately formed for a short period of intense work on a complex project may prefer members with a balance favoring technical skills over relational skills. For example, the technical skills of a team of doctors, nurses, and medical technicians delivering quintuplets are more important to the success of the group's task than team members' interpersonal relations. The team works together for a very short time and then disbands. Roles and responsibilities within the team are highly defined, which helps the team work effectively in the absence of well-developed personal relationships.

In contrast, a team that expects to stay together for a long period may initially favor a balance toward personal and relational skills. This is because, over time, group members can help one another increase their technical proficiency if the relationships among group members are well developed. Let's say that a project team with members representing different operations of a food manufacturer is assigned to develop prototypes for new market initiatives. With representatives from manufacturing, marketing, quality control, and food sourcing, the new product development team has six months to develop at least four products for consumer testing. Since the project is long-term and complex, it will be important for workers to work well together. If members possess the ability to work well with one another, they can also rely on one another to help fill in the technical expertise they may lack as individuals.

For instance, Jerry, the representative from manufacturing, knows very little about marketing. Initially, he relies heavily on Shanna's marketing expertise. Jerry asks Shanna lots of questions, requests marketing reports to read, and talks with her over lunch about marketing initiatives that have worked for other products. As the team works on product development, Jerry learns enough about marketing from Shanna to give informed opinions and ask appropriate questions. This process is enhanced because Jerry finds it easy to approach Shanna, and Shanna appreciates Jerry's willingness to learn about marketing.

Most spontaneously formed groups favor relational over task skills. Because group membership is based primarily on individuals' personal attraction to one another, relational skills are more important. If group members cannot get along and form a cohesive group, attraction will decrease, and members will leave the group voluntarily. This does not mean that task skills are not important—merely that relational skills are more essential in these groups.

For example, Rea's running group started over lunch when the four women discovered their common hesitance to try running. The decision first to hire a coach for the group and then to run three times a week was a natural outgrowth of the women liking one another; forming the group was not based on anyone's technical skill in running. As the group continues to work with its coach, members become confident enough to give one another friendly advice about selecting running shoes, stride length, and breathing techniques. However, if one member consistently gives poor advice or advice that detracts from another's running performance or enjoyment of running, this member's technical skill or motivation will come into question and may even disturb the relational balance of the group.

Simply put, group members will find it necessary to talk to one another to accomplish their task or activity. However, groups cannot accomplish their objectives only through task-related communication. Even if members try to constrain their messages to task issues, they are delivering implicit messages about relational issues in the group. All messages have both task and relational ramifications for group members, even if a specific message is predominantly task or relational in nature. A group is a social context, and social influence will occur whenever members are communicating. In whichever way the task and relational dimensions are balanced, the messages sent create the climate within which group members accomplish their tasks and activities (Keyton, 1999).

SKILL BUILDER

Interdependence is a defining characteristic of a group. But where does interdependence come from? Often interdependence is built into the task of the group. Sometimes being interdependent means performing similar activities, and other times quite different activities. For example, a softball team requires many different sets of skill expertise. Each member takes a turn at bat, a skill all players need. But when the team is playing defensively in the field, members play a unique position that works together with other positions.

Relational interdependence is created directly through your interaction with other group members. Providing nonverbal support (e.g., smiling; head nods; direct eye contact) and giving verbal support (e.g., "good play" or "nice hit") can help you develop or deepen relationships with other group members. Think of two different groups you belong to. What nonverbal and verbal communication behaviors could you use to build or increase relational interdependence with members of each group?

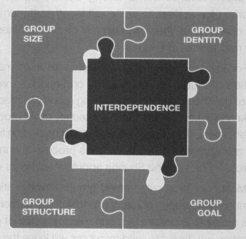

GROUPS OR TEAMS?

You have probably noticed that we do not distinguish groups from teams. That is purposeful. There is no rule for when *team* is more appropriate than *group*. These labels are often interchangeable, and there are other nouns that can be used as labels. For example, you are part of a family and a group of friends. You might belong to a book club, a soccer team, a rowing crew, or be a member of massively multimember online game (MMOG). You interact with others as a group if you are a member of a theatre troupe, a bowling team, or a cheer squad. Professionally, you may be a member of a sales team or work

crew. In professional, civic, and community settings, you may be a member of leadership council, or a ticket committee or subcommittee.

It does not matter what label is used. It does not matter if you communicate face-to-face or online. What does matter is the degree of interdependence among group members who create a group identity and group structure to support their group goal or activity. The size of a group matters to the degree that it enhances or impedes interaction among group members in accomplishing their goal.

SUMMARY

A group is defined as three or more individuals who identify themselves as a group and who can identify with the activity of the group. Five characteristics define groups: group size, interdependence of members, group identity, group goal, and group structure. Using these defining characteristics as avenues of analysis can help us understand the uniqueness of each group and the complexity of group interaction.

Clearly, it's more accurate to talk about a group as a process rather than defining a group by its number of members, its effectiveness, or its type of task or activity. The group interaction process is both complex and messy. Part of the challenge is to balance a group's task dimension, or what a group does, with its relational dimension, or the social and emotional interactions of group members from which roles and relationships emerge. All groups have both task and relational dimensions that should be balanced, based on the group's purpose and the needs of its members. Groups cannot accomplish their objectives only through task-related communication. A group is a social context, and social influence and messages about relational issues will occur whenever members are communicating.

DISCUSSION QUESTIONS AND EXERCISES

1. Think of a group you belong to in which members only communicate through technology. Now identify the group you belong to that is most diverse in its membership. Analyze these groups according to the five characteristics for defining groups.

2. Reflect on one of your childhood groups. Compare that experience with a current group of friends. What has changed? What is similar? What do you believe accounts for the differences and similarities?

3. Think of some past classroom group projects. What group characteristics made them effective? What group characteristics made them unbearable?

4. When groups are large, some members may think that their individual contributions will not be noticed and, as a result, decrease their level of activity in the group. When circumstances dictate that a group has many members, what strategies can group members use to control social loafing?

5. Think back to a group to which you belonged that was primarily task oriented. Compare that group with a group that is primarily relationally oriented. What are the similarities and differences in your assessments? Which group characteristics explain this?

Using Group Communication Skills for Group Presentations

This chapter focuses on the five defining features of a group. Let's examine how these features can be used as your group develops its presentation. Your manager wants your team to give her a presentation about new ways to motivate customers to spend more in the store. The presentation is the group goal, or group task. Including you, there are five group members and you have worked together on the weekends for more than a year. You are accustomed to working together, and those working relationships are more positive than negative, so you have developed a moderate level of group identity. You know which team members you can rely on to help you close a sale, and which members are better at explaining the technical aspects of what you sell. This structure, and the interdependencies it creates, may work for selling. But will this same structure be the most beneficial for creating and delivering the presentation?

Let's start with the group goal and work backwards. You and your work group members agree that a good presentation has several components: developing the ideas to present, designing the presentation, and delivering the presentation. The group meets to brainstorm strategies for increasing sales, and everyone participates. Now those ideas need to be developed into a presentation. Two members have expertise in computer graphics, and one member is a skillful writer. These three group members work together to make a draft presentation for the group to approve. Along the way, the remaining two group members look at drafts and provide feedback. Now the presentation is completed and it's time to deliver it. One member wants the two members who did not directly participate in the design of the presentation to deliver it. Will it be effective for these two to share the presentation task? Or is it better to have the group member with the most effective presentation skills deliver the presentation to your manager? What if the most effective presenter is also one of the members who worked on the design?

As you can see, while group size remained constant throughout the process, group identity, group structure, and interdependence changed as group members worked on different aspects of this task. Groups that are flexible, yet involve all members in the group task, are likely to be the most effective.

Image Credits

Figure 2.1: Adapted from: Copyright © Depositphotos/megastocker.
Photo 2.1: Johnny Bivera/U.S. Navy. Copyright in the Public Domain.
Figure 2.2: Adapted from: Copyright © Depositphotos/megastocker.
Photo 2.2: Copyright © Depositphotos/monkeybusiness.

CHAPTER 3

CONTEXTUAL INFLUENCES ON GROUPS

After reading this chapter, you should be able to:

- Use examples to describe how the bona fide group perspective illuminates the contextual aspects of groups and teams

- Distinguish between the contextual influences of time and space on group communication

- Describe different types of diversity found in groups and teams

- Explain ways in which individuals can overcome the challenges of diversity in group communication

- Explain the positive and potentially negative influences of technology on group and team interactions

BONA FIDE GROUP PERSPECTIVE

The five characteristics introduced in Chapter 2 (group size, interdependence of members, group identity, group goal, and group structure) are essential to defining a group. But these characteristics all occur within the group's context. The **bona fide group perspective** (Putnam & Stohl, 1990, 1996) illuminates the relationship of the group to its context or environment. The term *bona fide* was used to identify this perspective because many groups and teams studied by early scholars were ad hoc groups, or one-time groups. Group members had no history and no relationships with other members of the group.

Unfortunately, this is not a realistic view of groups or group dynamics. Rather, group members are influenced by their existing and previous group experiences, other members in these groups, the role and the importance of the group in members' lives, the way in which groups are set up by organizations or institutions, and what members expect from their group experiences. Thus, the bona fide group perspective focuses our attention on the way in which (a) groups have permeable and fluid boundaries and (b) the interdependencies between the group and its context (Putnam & Stohl, 1996; Stohl & Putnam, 2003), as well as the messages group members use in constructing and negotiating those boundaries and contexts in which a group works (SunWolf, 2008).

Permeable and Fluid Boundaries

The bona fide group perspective recognizes that group boundaries are generally stable but also permeable. In reality, a group's membership is seldom fixed. Additions are made to family groups through marriage, divorce, adoption, and death. Changes in organizational teams occur when employees leave the organization and new ones are hired. Even though we expect jury membership to remain stable, alternates are frequently required to step in when other jurors must be excused. Thus, while we often think of group membership as being static, it can be dynamic when group members are replaced, exchanged, added, or removed.

Thinking of group membership and its resulting boundaries in this way, it is easy to understand that groups are socially constructed through communication (Frey & SunWolf, 2005). Let's look at two examples. First, juries are the size they are because legal authorities debated the issues and made recommendations that became state and federal law. However, being named to a jury is not the defining feature. Rather, it is the interactions among jury members that move them from being *a* jury to being *this* jury. In essence, the jury as a group emerges through the interactions of members, not simply because they were assigned to the jury. When these jury members reflect upon and describe their experiences to others, they will point to specific interactions and specific relationships among jury members that caused them to agree that that defendant was guilty (or not).

In a second example, family members may agree that Sandra is a member of the family even though she is not related to any member. Rather, she was your mother's best friend who now lives on her own. Your family includes her as a member, inviting her to all family functions because they appreciate Sandra's thoughtfulness and helpfulness during your mother's illness. To signify her relationship to your family, you've taken to calling her Aunt Sandra. Communication between your family members and Sandra established a connection that encouraged your family to identify Sandra as a part of it and encouraged Sandra to think of herself as a member of your family.

Thus, group membership is perceived to be stable when you can identify who is in, and who is not a member of, the group. These identifications are made based upon who is communicating with whom and to what degree that interaction results in individuals identifying with a particular group. That is, individuals negotiate their identity with a group as their interactions construct the group. Of course, membership can change or be altered, permanently or temporarily.

PUTTING THE PIECES TOGETHER

Select two groups you belong to now or belonged to in the past. One group should have a relationally focused goal; the other group should have a task-oriented goal. For each group, answer these questions: How would you describe the task messages group members exchanged? How would you describe the relational messages group members exchanged? Which type of messages were more important? Why? Finally, how did the task and relational messages of group members help or hinder the group in developing its identity?

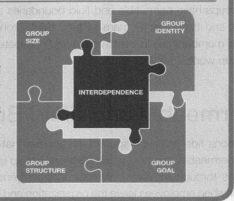

A Group's Interdependence With Its Context

Despite identifying with a particular group, members also participate in other groups. As a result, groups can interact with and influence one another. For example, if you are a member of several groups at work, you're likely to pass information from one group to another. Information gained in one group is taken—sometimes intentionally, sometimes unintentionally—to another group.

From the bona fide group perspective, a group is not a distinct entity with an environment that separates it from all other groups. Rather, groups are located within a fluid social context. The group is continually influenced by the environment in which it completes its tasks and by the social ties members have with other groups. The concepts of connectivity and embeddedness further explain how a group interacts with its larger social environment.

CONTEXTUAL INFLUENCES ON GROUPS

Connectivity is the degree to which several groups share overlapping tasks or goals. The more tightly coupled the groups, the more likely that change in one group will alter activities in others. For example, organizations are composed of many groups. A policy change developed and recommended by the human resources team is likely to affect the leadership teams of each division. A financial team may determine how much money the marketing team can use in their upcoming campaign. Although the teams have unique and specific goals, all teams function to meet the ultimate goal of producing the organization's products. When change occurs in one team, it is likely to affect other teams to which it is most tightly connected.

Connectivity increases in complexity when individuals participate in multiple groups. For instance, as a student taking several classes, it is likely that during any one term you are a member of several classroom groups. Although each group has unique membership and a goal specific to its particular course, you can use the information learned in one group in another. This information transfer is possible due to group members' multiple group memberships. Although information transfer is often viewed as a positive characteristic, it can be a negative when information learned in one group is used to the detriment of another. For example, Bryce is not thrilled to find himself in another group with Katerina.

In their statistics group, Katerina comes unprepared and seldom contributes anything meaningful. When Bryce learns that both he and Katerina have been assigned to the same group in their persuasive campaigns class, he immediately tells other group members about Katerina's substandard performance. Clearly this would influence how these members approach Katerina in their first team meeting

Another factor that contributes to complexity is **embeddedness**, which reflects the centrality of the group to its larger organizational structure. A group's position within the informal power structure or formal hierarchical structure affects its ability to obtain information and retain resources. Its position with respect to its environment also determines its degree of impact on the larger organization. For example, the student government group of your university is more deeply embedded within your university than any other student club or organization. Members of the student government have direct access to university officials; in fact, university officials may look to your student government as a primary source of student feedback and input. In contrast, a club such as Lambda Pi Eta (the communication students' honors organization) is affiliated with both the university and its national organization. To both the university and the national organization, the club is merely one student organization that competes with other organizations for attention and resources. Thus, its level of embeddedness in either the university or the national organization is shallower. In the university system, the student government group deals with issues more relevant to the university than the local Lambda Pi Eta chapter. In the Lambda Pi Eta system, one chapter is not likely to have more influence than any other local chapter.

When a group is characterized by high connectivity and high embeddedness, its boundaries are fluid. Information flows easily into and out of the group, making connections with other groups possible. Actually, it is the placement of a group within its environmental context that contributes to challenges, conflicts, and stresses group members are likely to face (Lammers & Krikorian, 1997). The more connected and the more embedded the group, the more pressures and influences it faces. When a group becomes highly embedded and connected, it may be difficult to clearly identify the group's membership.

For example, let's look at a biology study group. Five classmates meet every Thursday night to prepare for a biology test. But their interactions with one another are not limited to the Thursday night meetings. After biology class on Tuesday, the group meets quickly so Brandon can introduce a problem that involves the use of a specific lab instrument. No one in the group is sure how to use the instrument. But another student, Tom, overhears the conversation and offers to help. Because Tom is the instructor's lab assistant, members of the group consider Tom a reliable source of information. As the conversation about the lab instrument continues, Emily invites Tom to join this Thursday night's study group. Getting nonverbal agreement from the group, Tom says he will be there. Depending on how well the group interacts with Tom, he may become a regular member, and not simply a visitor. Suppose group members come to like Tom and value his contributions to the group, but Alex struggles in deciding whether to meet with the group or to play in a fraternity basketball league. Even though Alex was an original member of the group, his attendance depends on when basketball games are scheduled.

To clarify the concepts of flux and ambiguous boundaries, let's continue with the example. By the end of the semester, only two of the initial group members are left in the biology study group. Tom comes regularly now. Alex seldom studies with the group, but admits he could use an extra night of study before the final. He joins the group for this last session. To his amazement, almost half of the class is there, sitting in small subgroups going over different parts of the test material. Who is in this group? Who is not? What boundary separates this group from its environment?

INFLUENCE OF TIME AND SPACE

A part of a group's relationship to its context is the time and space of its interactions. The space a group works in, whether it is the physical space of a meeting room or the cognitive space provided by technology, influences how team members interact. Likewise, the time of day team members meet and how they use their time together also influences how members interact. Let's turn our attention first to time.

Time

Groups can have considerable histories or be of limited duration. While it is easy to think of groups that have short histories (e.g., jury, emergency task force) and long histories, (e.g., a book club that has been meeting for many years; a standing committee in a religious organization), not all group histories are this simplistic. For example, elections for city commissioners in some towns are held every two years, with three of the five commission seats up for reelection. If we measured the history of the city commission as a group, its duration would be two years. However, sometimes commissioners who receive the most votes are elected for four-year terms, while the commissioner receiving the least votes is elected for only two years. Thus, some commissioners can have considerably more experience in the group than others. This effect becomes particularly pronounced when a commissioner wins reelection many times. As another example, think of the board of directors of your local United Way. This group has considerable influence in your community, making decisions about how money is raised and how nonprofit organizations are funded. Yet any individual board member serves only a three-year term.

Another influence of time on groups is the frequency and duration of its tasks and activities (Lammers & Krikorian, 1997). One breast cancer support group, whose members are over 65 years old, has met every Monday night for one hour since 1989, while another group for younger women with breast cancer was established in 1999 and meets only once a month, but for two hours each meeting. The older women in the first group rely upon the companionship and social support of the group's members, as many are widowed. Their weekly meetings are highly social, with conversation turning to talk about vacations and grandchildren, not just methods of coping with their illness. The younger women, many of whom have children, meet less regularly but for a longer time so they can invite guest speakers to keep informed of the latest advances in breast cancer treatment and minimize the need for babysitters. Thus, frequency, or how often a group meets, and duration, how long a group meets, depends on the context of the group, the needs of members, and the task or activity the group undertakes. In turn, the length of time group members interact and the time between those interactions will influence group member relationships. Other dimensions of time influence how communication is structured in groups and teams (Ballard & Seibold, 2000, 2004). These include flexibility, linearity, pace, punctuality, delay, separation, urgency, and scarcity. **Flexibility**, or how rigidly time is structured, is apparent in how group members set deadlines. Flexibility is also seen when group members avoid setting a firm meeting schedule when they first meet. In some groups, the task they are working on will not allow a great deal of flexibility. A group's degree of flexibility can be discovered by asking these types of questions: Do group members set a rigid structure of deadlines? Or do group members allow each other to get the work done on their own time schedules? **Separation**, another dimension of time important to groups and teams, is the degree to which group members isolate their meetings from other interactions. To assess this dimension, we can ask these types of questions: How do groups and their members compartmentalize, or section off, their tasks? Do group members remove themselves from distractions by being in their own space? Or do group members welcome changes to their tasks? **Concurrency**, a third dimension of time, is based on how many tasks group members engage in. Do group members try to tackle several tasks at once? Or do group members take one task at a time and finish each task before starting another?

Time is not often talked about with respect to how groups work and group members interact. But communicating in groups does take time. Moreover, groups can develop a variety of ways of handling and expressing time as they work toward their goals. The three temporal dimensions described are just a few of the dimensions that provide context for group interaction. Groups and their members also vary by the following temporal dimensions:

- **Linearity**: Do groups create unique or special times for some events over others? Time is not only used to sequence events, it is also used to separate events. Teams expressing linearity will identify that their first meeting is to set ground rules, and the second team meeting is to focus on understanding their task.
- **Pace**: How fast are group members working? Is the tempo or rate of group activity fast or slow?
- **Punctuality and Delay**: Do group tasks have deadlines? Do group members respond promptly to each other's request for help?
- **Urgency**: Are group members preoccupied with task completion and task deadlines? Does the group treat every task as an emergency?
- **Scarcity**: How limited are the resources available to the group? Or does the group have adequate or unlimited access to resources?

CONTEXTUAL INFLUENCES ON GROUPS

- **Time Perspective**: Do group members talk about what needs to be done today, which is a present time perspective? Or do group members talk about upcoming activities and their long-term plans, which is a future time perspective?

Space

Like time, space is a contextual feature of and a contextual influence on group interaction. Some case studies drawn from research describe the different ways in which space influences group communication. In this first case, we can see how an interdisciplinary medical team at a geriatric oncology center engaged in two different types of conversation in the same work setting (Ellingson, 2003). In their formal, patient-centered conversations, team members requested information and shared impressions of patients. In this case, the presence of patients in examination rooms and the requirements of their tasks necessitated task-oriented messages. But when there was time between patients, conversations were used to build relationships among team members by talking about outside interests, such as families and vacations. Team members also bonded by complaining about their work schedules, the overbooking of patients, and the behavior of other clinic staff. Both the task-oriented and relationship-building conversations occurred in the same location—in the work space separate from the examination rooms. Although the space was the same, the presence of a patient in an examination room created a different perception of the context and, as a result, required a different type of communication among team members.

A second case describes the way in which interaction can become structured by the space in which it works. A twelve-person jury in Ohio was tasked with deciding the guilt or innocence of a defendant on two murder charges and thirty other offenses, mostly drug charges (SunWolf, 2010). After many days in the courtroom hearing testimony, the jurors were moved to a private, and very small, meeting room for their deliberations. With the other jurors seated around a long and narrow rectangular table, the jury foreperson sat at the head of the table. This seating arrangement in this small, enclosed physical space emphasized that the jurors look toward the foreperson during their deliberations. Imagine yourself in this situation. How would this physical space and close proximity to other jurors affect your interaction? Your display of physical and facial nonverbal cues? Your willingness to continue to deliberate when the others on the jury disagreed with you?

A third case of a group's use of space demonstrates how space influenced the interactions of multidisciplinary teams (Li & Robertson, 2011). If meeting rooms are outfitted with technology that is permanently placed, then presenters feel obligated to present from that fixed location even if it is not optimal to do so. When meetings are attended by many people, members who sit at the end, or at the periphery, of the meeting space may not receive eye contact from those seated more centrally in the space. As a result, the members sitting at the periphery of the space may not be called on for questions or for their opinions. When possible, space should be used so that all members have good visual access to all participants. Doing so promotes interaction among them.

Looking at groups from the bona fide group perspective, we are reminded that groups have permeable and fluid boundaries, are interdependent with their context, and are influenced by the time and space of their interactions. Most importantly, these characteristics exist because individuals construct any group and their identity in it through their interactions. If we were members of only one group at a time, then life would be simple and we could focus our attention on that group. But we are members of multiple groups, requiring that we negotiate multiple identities and manage many interaction relationships simultaneously.

As a result, multiple group memberships may create conflicting group identities for an individual. Or individuals may experience different interaction patterns or interaction roles when new members join an existing group. Moreover, the entrance and interaction of a new or temporary member can even cause group membership to change. Because individuals are members of multiple groups, an individual can serve as an implicit or explicit boundary spanner by taking information from one group to another, with the potential of creating communication exchanges between or among groups. When relationships between or among groups become established through interaction, they must coordinate or negotiate their actions and, at the same time, negotiate how they are different from one another. Thus, a member's sense of identity with a group, or sense of belonging, can shift, depending on the fluidity and permeability of the group's boundaries, the way in which the group is interdependent with its context, and the way in which group members use time and space to create a context for its interactions (Waldeck, Shepard, Teitelbaum, Farrar, & Seibold, 2002).

INFLUENCE OF DIVERSITY

Gender diversity and cultural diversity (racial, ethnic, nationality, language)—and even diversity based on profession, age, or length of membership in the group—are the primary ways in which group members distinguish themselves from one another. Individuals of any culture share common symbols, values, and norms, which result in a particular communication style with its own rules and meanings. When interaction styles are shared among group members, they perceive themselves similarly and as belonging to the same cultural group. However, when group members have different interaction styles, they are likely to attribute their differences to culture differences. Not only does this influence a group's member self-identity, it can also cause the group's identity to weaken or for subgroups to emerge (Larkey, 1996).

It is important to recognize that diversity is evident on many levels (Artiz & Walker, 2014). Obviously, team members may speak different languages or use different terminology and nonverbal symbols. Even when team members speak the same language, the cultural backgrounds of members can influence their language proficiency and language choices in both sending and receiving messages (Du-Babcock & Tanaka, 2013). Diversity can also be evident in how language is used interactively. For example, how group members initiate and respond to others, take turns, or shift to topics is influenced by the diversity of members in the group. Finally, diversity matters in how group members make decisions and assert leadership. For example, members who are from the United States display leadership by being decisive and task oriented. Alternately, group members from other cultures display leadership by being procedural or involving others in decision making. Thus, diversity matters in what is said, how something is said, and what processes develop from interaction.

The influence of diversity on groups is fairly complex for two reasons. First, while individuals from the same group (sex, gender, race, nationality) can share many interaction characteristics, there is also variation within any cultural group. Second, diversity influences individuals and the group as a whole. An individual group member's attitudes and cultural values can directly affect other members' communication behavior. In this way, each member of the group influences the perceived and real diversity in the group. There can also be a group-level influence when group members hold different cultural values. Thus, **heterogeneity**, or differences, in cultural values can influence a group's interaction processes and their performance as a group (Oetzel, McDermott, Torres, & Sanchez, 2012). We shouldn't be surprised that team members from different cultures bring different understandings of and practices for interacting in a group. That is, norms for communicating in a group vary across cultural groups. This is especially true for how group members share information and make decisions (Janssens & Brett, 2006).

 THEORY STANDOUT

More Than Cultural Differences?

Oetzel's (2005) theory of effective intercultural workgroup communication explores how self-construal, or how one defines oneself relative to others, is influential in intercultural communication. While Oetzel theorized about cultural differences as differences in geography, could the theory be applied to other types of cultural differences? Kirschbaum, Rask, Fortner, Kulesher, Nelson, Yen, and Brennan (2014) argue that cultural differences exist in groups of physicians in operating rooms. For example, when delivering a child, three physicians would be present: an anesthesiologist, a surgeon, and an obstetrician. Each physician has different tasks during the operation, and each relies on different professional norms for communicating with one another. After training that included the physicians practicing message strategies that were inclusive and responses that were inviting, physicians increased their scores on interdependent self-construal and reduced their scores on independent self-construal. As Oetzel et al. (2012) found, higher interdependence is related to a stronger positive interaction climate. And that was the case in the study of these interdisciplinary physician teams. In what other team settings can you imagine testing Oetzel's theory of effective intercultural workgroup communication?

These influences can be both positive and negative. For example, greater diversity among group members can result in changes in group membership, lower cohesiveness among members, and reduced problem-solving effectiveness by the group. Culturally diverse groups can benefit from the different perspectives group members bring, or they can allow their differences to fuel conflicts and prevent cohesion from forming (Watson, Johnson, & Merritt, 1998; Watson, Kumar, & Michaelsen, 1993). Thus, how a heterogeneous group handles its diversity is a key factor in task success.

The **theory of effective intercultural workgroup communication** (Oetzel, 2005) explains how the cultural differences among group members affect a group's task and relational communication, such as decision making and satisfaction, respectively. Let's look at the input factors of the model. First are the situational, or the contextual, features of the group. These include any unresolved conflict among members in the group or any unresolved conflict that exists historically among culturally different groups, the in-group/out-group balance among members, and members' status relative to one another. Together, these input factors can help or hinder the group in creating a common frame of reference for working together, as they represent deep-level diversity concerns. Individual differences among group members are the second input factor. These factors are self-construed, or how one defines oneself relative to others, and face concerns, which are an individual's beliefs about their image, reputation, and integrity. The third input factor is the composition of the group, or the group's diversity. These are issues of surface-level diversity, which are in contrast to deep-level diversity, which is based on broad differences among cultures and nationalities.

This theory reminds us of two valuable points. First, while there may be cultural differences, there are also differences within cultures. Second, culturally diverse groups can be more effective when encouraging equal participation, practicing consensus decision making, addressing conflict as cooperative rather than competitive, and engaging in respectful communication. These communication behaviors are the interaction climate, or the general tone, of the group that ultimately affects how well the group works together.

Although it is clear that cultural differences can influence group interaction, it is not apparent why this occurs (Oetzel, 2002). One explanation is that diverse groups can result in status differences related to ethnicity, nationality, sex, tenure, knowledge, and organizational position. Status anchored on these characteristics can affect member participation and result in negative group interactions because observable differences are used to assign group members to hierarchical positions within the group. As a result, group members are more likely to use biases, prejudices, or stereotypes in communicating with one another (Milliken & Martins, 1996). In this case, status is assigned to individuals simply because they possess or represent certain attributes.

Another explanation is that diverse groups must manage cultural differences, or the patterns of values, attitudes, and communication behaviors associated with specific groups of individuals. This explanation focuses on how differences are created through a group's interactions. These can influence relational communication, as well as member participation and turn taking, which ultimately can influence how well group members work together on their task.

Remember, though, that diversity issues go far beyond gender, race, ethnicity, and culture to include social and professional attributes and other demographic categories. In groups, we must be careful not to assume that diversity is based on a single dimension that seems to be the most obvious (Poncini, 2002). Not all of these types of diversity are equal, yet it is difficult to completely isolate one element of diversity from another. Thus, cultural diversity is really a combination of differences rather than a difference on any one dimension.

The type of group or group activity in which you are engaged creates a unique context in which diversity issues become salient. For example, your work group may be more sensitive to diversity in educational-level and political differences than to diversity in race and gender, especially if all group members are from the same department and have similar lengths of service with the organization. In this case, the differences that might exist due to race and gender are not as influential in the group because the group members know one another well and work on tasks regularly and effectively. In contrast, the cultural distance that can be created by race and gender differences may be maximized when members also represent different departments and are new to the group and its task. Group members in this situation have not had the chance to explore their differences and similarities or to develop as a group.

However, as group members gain more experience interacting with diverse others, group member participation is more equal, cooperation among group members is higher, and group members are more satisfied, which leads to fewer intercultural conflicts and prejudice (Larkey, 1996; Oetzel, Burtis, Sanchez, & Perez, 2001). As a result, task performance is enhanced when group members move from obvious surface-level or easily detectable differences to discussing different ideas or concerns relative to the task (Harrison, Price, & Bell, 1998). Moreover, there is evidence that groups with high levels of diversity have stronger beliefs about their abilities to complete the task. That is, when there are group members from many different racial and ethnic backgrounds, the group tends to avoid falling into majority/minority subgroups. As a result, communication is more effective, which facilitates group members' work on the task (Sargent & Sue-Chan, 2001).

INFLUENCE OF TECHNOLOGY

The increasing availability and variety of technology has resulted in more types of groups using technology as their primary method of communicating to accomplish tasks and other group activities. Some technologies can be used synchronously. For example, videoconferencing requires that group members coordinate their interaction as they meet in real time, although they do not need to be in one location. Alternately, other technologies are asynchronous. For example, email is asynchronous, as there are time intervals between messages among group members. Still other technologies can be used in both ways. For example, text messaging can be synchronous if all group members are attending to the interaction, or it can be asynchronous if group members only message one another infrequently on an as-needed basis. Technologies are appealing because they allow group members to work across distances—in both time and space. Today, some group members, like those in project teams at work who regularly complete complex projects, have never met in person. Interestingly, whether group members are situated around the corner or across an ocean, geographical distance among group members is not the prevailing reason for using technology. Rather, people use technologies, especially SMS (or text messaging), to enhance feelings of connectedness (Reid & Reid, 2010). As a result of these connections, people can create better working relationships, which lead to additional opportunities to become a group.

The use of technology does influence communication among group members. Technology mediates the group's interaction (Poole & Zhang, 2005). Simply, communicating through technology is not the same as communicating face-to-face. Why? Because technology leaves out or modifies social cues that we depend upon in face-to-face group interaction. Let's explore several issues group members should be aware of when communicating through technology. First, technology can eliminate or skew contextual information important to understanding messages. Second, technology can also impede the salience of information exchanged among group members. Third, groups that use technology for their interaction often do not share the cultural, social, team, or organizational norms that are more easily developed in face-to-face settings.

Technologies support a wide variety of group activities, including discussion, planning, generating ideas, and making choices, as well as collaborative document creation and

CONTEXTUAL INFLUENCES ON GROUPS

 SKILL BUILDER

How Are You Using Technology in Groups?

Research (Lira, Ripoll, Peiro, & Zornoza, 2008) has demonstrated that when a group or team meets using technology rather than meeting face-to-face, group members become more effective at using the technology and develop new strategies for accomplishing tasks. Reflecting on your online group experiences, first identify which technologies you believe you use effectively; then, describe what new technique or strategies you've discovered for helping the group accomplish its task online. How would you teach others about using one of those techniques or strategies the next time you are a member of an online work group?

editing. Moreover, the use of asynchronous technology has spread from family and social groups to work and task groups.

Let's look at the use of SMS, or text messaging, in college classroom groups (Lam, 2012). Students were randomly assigned to a SMS-only or no-SMS group. The no-SMS group could use any technology except text messaging on their cell phones; overwhelmingly they chose email for 98% of their communication with one another. Here's what happened over the eight weeks in which students had to complete their group assignment: Members of the SMS-only group, who were to only use text messaging on their cell, communicated 40% more than students in the no-SMS group, especially in the first few weeks of working together. In the first four weeks of their interaction, the SMS-only group reported higher feelings of connectedness than members of the no-SMS group. However, as the groups continued to develop across time, that difference went away. With respect to how attracted students felt to their group, there was no difference after four weeks or after eight weeks. So, members of the SMS-only group communicated more, but this greater level of communication did not result in higher levels of feeling connected to other group members or more attraction to the group than it did with members of the no-SMS group. Another interesting difference is that members of the SMS-only group asked more questions and provided more answers; however, they also sent more off-topic messages to one another than the no-SMS group. The length limitations many people follow when they are texting likely created the increase in communication quantity, as well as the increase in the conversational question and answer sequences.

Could we consider groups like the student group above a virtual team? Yes. Members of the student groups were dispersed in time and space (Timmerman & Scott, 2006). Of course, group members are not limited to SMS or email technologies. Organizations routinely provide employees with teleconferencing, videoconferencing, and web conferencing systems, internal social networking tools, and project management systems in which team members can share documents and manage schedules to help team members communicate and collaborate. Each technology provides greater flexibility for team members in how they communicate. Each technology also creates a permanent, and often searchable, record of what group members communicated to one another. Some

MESSAGE AND MEANING

A few years ago, Nancy, Genevieve, Frankie, Kyle, and John worked as a group to develop a recycling center for five small rural communities. Each community wasn't large enough to warrant its own center. This five-person team developed the concept, presented it to each community, and received approval from each county to initiate the recycling center. In its 10th year, the center operates at a profit and keeps recyclable trash out of refuse stream. However, none of the team members are currently involved in the day-to-day operations.

A state environmental agency contacted Nancy asking her to check with the other team members to see if the team would present a panel discussion at a state recycling conference. The conference organizer wanted the team to explain how their idea came about, how the center was started, and what lessons could they pass along to other communities. Nancy sent this initial email:

> From: Nancy
> To: Genevieve, Frankie, and John
>
> Hi. The state environmental agency wants to know if we would put together a panel discussion for the state conference—some sort of retrospective. What do you think? I don't know if I have the bandwidth to take the lead on this, but I do think it could be of interest to the participants.
>
> Also, I don't have contact info for Kyle. Do any of you?

A few hours later, in a separate email, John lets Nancy know that he has Kyle's email and will see if it works, and then get back to her.

Another few hours later:

> From: Nancy
> To: Genevieve, Frankie, Kyle, and John
>
> Now we have Kyle in the loop (Hi Kyle!)! It's great to have us all together again! I have very fond memories of our time together. It seems like we all are interested in putting together a panel session for WeRecycle! on the history of the recycling center. I personally would love to participate, but with my new job, I'd prefer not to take the lead on this. Would any of you like to drive it?

An hour later:

> From: John
> To: Genevieve, Frankie, Kyle, and Nancy

Why not ask one the current volunteers to organize the panel and be the facilitator of questions to us?

Three days later:

> From: Genevieve
> To: John, Frankie, Kyle, and Nancy
>
> Unfortunately, I'm in a similar situation as Nancy. I also have a new job. So, I can't take the lead on this. But, I do have some thoughts. To merit a presentation, I think there needs to be some new information that we could present. What data are there that demonstrate the impact the recycling center has had on the communities?

Another six days later:

> From: Nancy
> To: Genevieve, Frankie, Kyle, and John
>
> Hi everyone. I conferred with organizer. He is interested in a fun retrospective, not an information session. He sees it as being during a lunch, or something like that.
>
> Here are some of my ideas . . . we could do a slide show with pics, etc. Get creative. Perhaps we each could take the lead on a different aspect of the history of the center. Like the initial idea and meetings, the first years in operation, the formation of the board and nonprofit business, the meaning it has given our lives, and potential effects on the communities.
>
> Your thoughts . . .

A day later:

> From: John
> To: Genevieve, Frankie, Kyle, and Nancy
>
> I'm okay with that.

Another five days later:

> From: Nancy
> To: Genevieve, Frankie, Kyle, and John
>
> Sounds like we're all in. Just need to hear back from Frankie!
>
> Here's what we have so far . . .

> Kyle: "I would be happy to do a short bit on the origins and what came before, and the barriers we had to overcome."
>
> Genevieve: "I can talk about planting the seeds of what would become the recycling center. The "meaning in our lives" question is a significant one for me. My involvement changed me personally and professionally in so many ways . . . still seeing the impact today.
>
> John: "When Citizens Try to Run a Nonprofit Organization" 😀
>
> Frankie: (with a quizzical look on his face)
>
> Nancy: I think it would be really cool to do a network map of who uses the recycling, including our community partners, but that would be work, which I have absolutely no time for. So instead, I could talk about the challenges of running the organization in terms of all we tried to do to develop a culture of collaboration. I could also talk about our winter board meeting where we all got snowed in!
>
> After seven years of not meeting regularly, this team tries to reestablish itself to make a presentation about how they got the recycling center started. In the beginning, they regularly used email to stay in touch. Is email still the best technology to reunite the team? Some types of group communication are better suited for some technologies rather than others. What would recommend for this team?

technologies, like project email and management systems, also let group members send, receive, and store documents. As a result, reviewing the performance of virtual teams is easier.

The obvious advantage is that groups using technology are able to communicate, collaborate, and work on their goals regardless of temporal or geographic constraints. Simply, the group—and its members—have greater flexibility in determining when they work and meet. Group members may work asynchronously, as group members will not read or send messages at the same time. Group members can log on and communicate according to their own preferences and needs, and within their own timeframe. This type of asynchronous group communication has several advantages. Members do not have to compete for talking time when using technology that is textually based. They also have the opportunity to reflect on what is posted before responding. In general, group members have a greater opportunity to participate, as they are not closed out by powerful or talkative members. The asynchronous mode also allows any group member to introduce ideas into the discussion, as there are fewer opportunities for members to control the group's task or activity. These benefits of asynchronous communication are especially useful when a group is working on a less complex task, such as idea generation (Bell & Kozlowski, 2002).

Other groups use technology for synchronous communication, interacting at the same time even though they cannot be face-to-face. This type of group technology is effective for more complex tasks, as they require greater coordination. Increased coordination relies on feedback among group members. Thus, synchronous communication, especially technology with richer media, for example videoconferencing, is more effective for tasks that rely on greater interdependencies among group members (Bell & Kozlowski, 2002).

Trends in Technology

As new technologies develop, two trends are appearing (Darics, 2014). First, the line between synchronous and asynchronous communication is becoming blurred. When group members create norms for using technology such that time lags or gaps between messages are minimal or when the gap is only the few milliseconds required for the technology to send and receive messages, technology-based communication is more like talking over the telephone. The conversation is in real time; only the group members are distributed.

The second trend is that most group members are comfortable using many different technologies. Work or project groups, for example, will use instant messaging, videoconference, and email in addition to face-to-face meetings. Thus, visual, vocal, and physical cues available in the face-to-face setting are assumed to be presented when communicating through technology. As a result, a sense of copresence is present whether group members are meeting face-to-face or not. **Copresence** is the idea that a group member's behavior and messages are shaped by others in the group. That is, the sense of *being there* with the other group members when interacting face-to-face bleeds over to communication in which group members are not colocated (Flordi, 2005; Merolla, 2010).

The one disadvantage shared by all communication technologies is that each—some to a lesser degree than others—lack nonverbal cues. Some technologies lack visual cues; other technologies lack verbal cues. Video-based technology does not resolve this, as visual and verbal cues are degraded (e.g., the video is not clear, not all members can be heard when they talk, view of interactants is narrow or limited). Without these cues, it is more difficult for group members to appropriately contextualize their interaction (Baron, 2010). Many of the nonverbal cues available in face-to-face interaction are missing when group members communicate through technology. When nonverbal cues are missing, it is more difficult to clarify ambiguities and to accurately determine other members' interpersonal needs. As a result, group members who communicate through technology can find it more difficult to develop shared meanings with other group members (Berry, 2011).

Anonymity is a feature of some group-oriented technologies. Some technology-based groups can also work on their tasks without members knowing their identity. Interacting anonymously is common in online support groups and web forums in which participants work on solving problems. Regardless, groups that use technology are distanced by space, even if they are in the same building. Moreover, group members' communication is mediated by the technology (Bell & Kozlowski, 2002). This is not a moot point. Members must use and rely on technology-mediated communication to link them together as a group. In addition, groups must take the time to adjust to or learn about their mediated environment. Not doing so will negatively influence task completion and harm member relationships. Thus, groups who expect to work together on a task over a long period of time are better suited to using technology than groups for which the time horizon is very short (Walther, 2002).

Generally, a more complex task requires greater coordination among group members and more immediate feedback to ensure interdependence among team members. While there is little difference in task effectiveness on less complex tasks between groups that meet face-to-face and groups that meet virtually, groups perform better on complex tasks when they meet face-to-face.

Thus, two recommendations can be made when groups must work virtually. First, groups will have greater task success if the technology is rich enough to include both asynchronous and synchronous media, as well as easy methods for sharing documents, drawings, and other graphical information (Bell & Kozlowski, 2002). Second, technology can limit the development of relational ties among members of more formal groups. If possible, virtual groups should consider meeting initially face-to-face to help group members create a group identity with one another and develop agreement about the group's goal or task (Meier, 2003).

Another influence of the use of technology is that virtual teams often cross functional, organizational, and cultural boundaries. Because technology makes it easier for groups of people to meet, virtual groups often include members with different backgrounds, knowledge sets, motivations, and communication styles. Technology often makes communication in these groups more difficult because members have more trouble conveying meaning and in knowing when they are not understood. It is also difficult for groups using technology to form relationships and organize themselves (Kiesler & Cummings, 2002). To overcome these potential difficulties, groups can develop procedures or rules for interacting and creating linkages among members by assigning tasks to subgroups of members with different backgrounds (Bell & Kozlowski, 2002). Group members also need to address relationship development because groups using technology whose members have developed collective positive attitudes and beliefs about their group's ability to perform are more effective at their tasks (Pescosolido, 2003).

While groups that use technology to communicate have some special challenges, they are more like groups that meet face-to-face than they are different. Both types of groups have defined and limited membership, create a structure and group identity to work interdependently towards a shared goal, and need to manage their task as well as their relationships. The primary difference is that virtual teams are geographically dispersed for some or all of their interaction, making them rely on technology rather than face-to-face interaction (Berry, 2011). It is likely that face-to-face groups also use technology to some degree. Thus, virtualness is a matter of degree.

SUMMARY

The bona fide group perspective illuminates the relationship of the group to its context or environment. The perspective recognizes that group boundaries are generally stable, but also permeable and fluid when group members are replaced, exchanged, added, or removed. As a result, as group members communicate with one another, they socially construct or negotiate the group's boundaries. This perspective also acknowledges that a group is not a distinct entity within an environment, but is connected to or embedded in other groups in a fluid social context.

Time and space are two other parts of a group's context. Issues of time include: the length of a group's history, the frequency and duration of its tasks, how flexibly the group treats time, the degree to which group members are separated from distractions while they work, and the number the tasks worked on at one time. Issues of space include: group members meeting in formal or informal spaces, how members are seated relative to one another, and the group's use of technology to bring members together from around the block or from another country.

Diversity is a major influence on group member interactions. Some degree of heterogeneity exists in every group. Diversity can be based on gender, sex, race, ethnicity, nationality, language, age, profession, and length of membership in the group. These types of diversity can create both positive and negative influences. The theory of effective intercultural workgroup communication explains how diversity influences group decision making, as well as members' satisfaction with the group. It is important to remember that there may be differences within cultures, as well as diversity between cultures.

Some groups rely completely on technology to complete their goals. Even groups whose members can meet face-to-face will use some form of technology to work together. Some technologies are used synchronously; others are used asynchronously. Research has demonstrated that communicating through technology is not the same as communicating face-to-face. Technology can eliminate some contextual information that is important to creating shared understanding, skew information making it more difficult for all members to identify the important information, and diminish members developing and sharing relational and task norms. Technology continues to change, which creates new opportunities for group members to interact. Experts recommend that groups use technology that includes both synchronous and asynchronous channels of communication. Groups that will work virtually are encouraged to meet face-to-face the first time. Doing so will give group members an opportunity to create group identity and develop agreement about the group's goals.

DISCUSSION QUESTIONS AND EXERCISES

1. Watch a situation comedy. Identify the external factors that influenced the group of characters. For example, why are these characters interacting? What's their goal? Is the goal easy or difficult? Is the group working under time pressure? How well do these characters know one another? What other groups are the characters connected with? Use the bona fide group perspective to identify what made that interaction effective or ineffective.

2. How would you describe the influence of time when you are interacting in a group of friends? In a required classroom group? In your work team? In a sports team? Is the influence of time consistent or inconsistent? Explain the differences or similarities.

3. In a public space (e.g., library, shopping center, restaurant, park), observe a group conversation. How did that group mark their space boundaries? Use their space? Expand their space?

4. Recall a group situation in which there was cultural diversity. What evidence do you have that the distinctions you noted about group members were based on cultural differences? Evaluate the steps you took (or could take) to help the group overcome any cultural obstacles.

5. List the communication technologies that you are comfortable using when talking with a group of friends. Doing a group project at school? How does technology help or hinder those group interactions?

NAILING IT!

Using Group Communication Skills for Group Presentations

Reflect on an experience of developing and delivering a group presentation. How was time a liability as your group worked on the presentation? In what ways could technology have helped you overcome these time constraints? Is communicating through technology always a time saver over face-to-face group interaction? Why or why not? When developing a group presentation, encourage your group to develop a timeline, or a chronological list, of tasks that need to be accomplished in developing your presentation. Whether drawn by hand or with created with software, a timeline should include:

- a detailed description of each task,
- the group members responsible for completing the task, and
- the date each task must be completed.

These items are organized from beginning to end; as group members take ownership of different tasks, some tasks can be done in parallel. Each time the group meets, the timeline should be reviewed and revised as necessary.

Image Credits

Figure 3.1: Adapted from: Copyright © Depositphotos/megastocker.

Photo 3.1: Copyright © Depositphotos/AndreyPopov.

Photo 3.2: Copyright © hepingting (CC BY-SA 2.0) at https://commons.wikimedia.org/wiki/File:CB106492_(5299266966).jpg.

CHAPTER 4

HOW WE COMMUNICATE WHAT WE KNOW

After reading this chapter, you should be able to:

- Explain why sharing information is central to group communication
- Describe how the alignment of individual and group goals may impede sharing information in groups and teams
- Describe team cognition
- Identify different types of ties in a communication network
- Differentiate between centralized and decentralized communication networks
- Identify when a faultline influences the creation of subgroups

Sharing information is a primary way groups complete their work. Members may share information to inform, to persuade, or to create common ground. In this chapter, we explore how and why information sharing takes place in groups. First, we investigate information sharing itself, examining two theoretical frameworks that differ in their views of how groups communicate. Second, we investigate how sharing information influences team cognition, or the level of understanding shared across team members. Oftentimes team members believe they think about an issue the same way, when in fact they do not. Unfortunately, this may not be discovered by the group until it is too late. Last, we consider how communication networks form to facilitate the flow of information within and across groups.

INFORMATION SHARING

Oftentimes the information needed to make a decision is spread across group members. During discussion, information may be shared with fellow members in the hope that it will aid the group in making an effective decision. Not only are there differences in the distribution of information across the group, but the various resources and expertise of group members may lead to different interpretations of the information. For example, a marketing team receives information from a group member who is charged with identifying vendors. He learns that it will cost $100,000 to advertise a product. The marketing team member in charge of the budget looks at this information and worries about the high cost. The member in charge of advertising looks at this amount and believes it is not enough based on budgets from similar projects. Other members of the team look at this number and doubt its accuracy. Even though everyone is looking at the same information, different interpretations abound. This is not uncommon in teams. However, this does make it more difficult to create a common understanding of an issue.

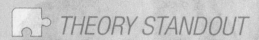

THEORY STANDOUT

Motivated Information Sharing Paradigm

Research aligned with the motivated information sharing paradigm suggests that group members have both individual and group goals that drive their interaction (Hollingshead et al., 2007; Wittenbaum, Hollingshead, & Botero, 2004). Although the individual and group goals may contradict (what is best for me vs. what is best for the team), that isn't necessarily the case. Sometimes individual goals naturally align with group goals, or at least oftentimes they don't directly conflict. Additionally, group members may be able to work toward an individual goal while disguising the attempt as a move toward the group goal. The recognition of both individual and group goals creates an environment where group members are strategic in how they communicate to their groups.

Thus, groups often try to align individual and group goals so that members are incentivized to work for the betterment of the group. If group members see other members as threats to their individual goals, then they may not communicate in a way best for the group. Joint group reward is one such way that this is accomplished. What are other ways that leaders can align individual and group goals?

Hence, the sharing of information is not always straightforward. As mentioned in Chapter 1, group members communicate in a strategic way, in hopes of accomplishing their purpose. Group members who possess important information potentially hold power and influence over the group (see Chapter 7). The way information is presented may make one option more appealing than another. At the same time, the sharing of information is essential to group functioning. How groups share information has been debated over the years, and two perspectives on information sharing have emerged.

Cooperative Information Sharing

The **cooperative information sharing paradigm** investigates how group members share information when trying to make a decision (Stasser & Titus, 1985, 1987). Central to this model is the **hidden profile**, which is a research technique used to study information sharing. In the hidden profile experiment, information is distributed among members of the group. Some of the information is shared by all group members, meaning that every member has the same information. Additional information is unshared, in that only one group member has a specific set of information. If all members decide to share the information that they uniquely hold, then the correct choice should be discovered and selected. However, if all the unshared information isn't shared, then the group won't make the right decision. To uncover the best decision (or hidden profile), all the information must be shared among group members.

Unfortunately, research has shown time and time again that team members rarely discover the hidden profile. In fact, group members readily share *shared* information, but rarely share *unshared* information (Reimer, Reimer, & Czienskowski, 2010). For example, a group is discussing whether to select option A or B. Cindy has some unique information about option B that would discourage the group from selecting it, and some positive information about option A that everyone already knows. Research suggests that Cindy is more likely to share the positive information than the negative information because it is already shared by others. Perhaps Cindy is doing this because she is afraid that no one else will support her unique information. Perhaps Cindy shares common information so that she appears supportive and the information can be confirmed by others, making her look credible. Either way, relying on information that everyone already knows can prove problematic when trying to determine the correct decision.

Researchers have explored a variety of ways to encourage group members to share unique information, including considering various group member characteristics, such as apprehension, (Henningsen & Henningsen, 2004), altering the group structure (van Swol, 2009), and altering when group members receive information (Reimer, Reimer, & Hinsz, 2010). However, in general, groups are often ineffective when it comes to discovering the hidden profile that exists in the information that is distributed among them.

Motivated Information Sharing

After more than two decades of focusing on the cooperative information sharing paradigm, Wittenbaum, Hollingshead, and Botero (2004) critiqued the model and offered an alternative: the **motivated information sharing paradigm**. This paradigm uses a communicative lens to explain why groups are ineffective at sharing necessary information. First, these scholars argued that some of the premises behind the cooperative information sharing paradigm were faulty. In other words, groups and group members really do not act in a way consistent with the assumptions of the model. For example, they argued that Stasser and Titus' approach did not consider contextual factors. The hidden profile assumed that group members had all the information necessary to make a decision, all the information was objectively and concretely understood, and group members were all united in their focus on the group goal. In fact, this rarely happens. Most of the time, groups do not know if they have all the information necessary to make a decision, and sometimes must engage in information searches to discover more information. Even though members may do their best to share information objectively, this is very hard to do. Why is this the case?

Group members consider individual goals simultaneously with the group goal, and sometimes these two types of goals do not align. For example, let's say that Stephanie has a good idea for an upcoming project. During a meeting, she introduces the idea and shares information pertaining to it so the group can successfully accomplish the group goal. However, Stephanie may also have a desire to be promoted within the group, and wants to be featured in the group's presentation. This may change her delivery of the message; she may be more prone to deliver it in a way so that she receives more credit for it. She may even be more critical of other group members' ideas so that her idea looks better. Of course, individual goals may be in harmony with or contradictory to the group goal. However, since individuals adapt their communication in accordance with both individual and group goals, Stephanie's creation of her message may be quite complex, and could be at odds with her group goal. Individuals may have more than one goal, which can cause individual team members to be distracted from the team goal (Hollingshead, Jacobsohn, & Beck, 2007).

Our tendency to not share unique information and our complication of balancing multiple individual and group goals when communicating creates a potentially complex social situation for sharing information. Effective group facilitators will look for opportunities to not only encourage members to share unique information, but to minimize any perceived risks of doing so. In addition, group members can use the messages of other members as evidence for their personal goals. For example, if someone becomes defensive when a decision option is critiqued, it may be that an individual goal (e.g., to be promoted, to be right) has been hindered. Analysis of the interaction may lead a group member to recognize other group members' interests and adapt their messages accordingly.

TEAM COGNITION

One reason group members share information is to increase the level of understanding across group members. **Team cognition** refers to the level of knowledge or information commonly shared and expressed by group members (Beck & Keyton, 2012). An easy way of conceptualizing this is to think of yourself in a new group. Although you may not know anyone in the group, it is still probable that you share a basic level of understanding of what behavior is appropriate. For example, you would probably not start yelling at another group member or start sharing very intimate information about your life. These discussions would be awkward and surprising, and would likely alienate you from other group members. Even if groups or group members are new, there is a common understanding about decorum and meeting behavior based on culture and context.

In addition, creating a common understanding about issues is important for groups. One of the reasons we share information is to create common talking points for group members. When sharing information, discrepancies may arise. This is because when groups share information, their communication becomes evidence of how they are thinking (Keyton, Beck, & Asbury, 2010). It is not only the message content that suggests differences in perceptions, but also nonverbal behavior. For example, Brett may come to a meeting believing her sales team should aggressively seek customers for their new product. She sees it as opportunity to move the company forward and help the team make a name for itself. However, when she arrives at her team meeting she remains silent for the first part of group discussion, where several of the more experienced group members give reasons why an aggressive sales strategy would be risky. With this new information, Brett reconsiders her view of the product. Since she hasn't spoken up, many might interpret her silence as support for what the experienced group members are saying. She may reassess her position based on explicit messages from these team members—namely, that it's not worth the risk, or by more subtle cues, such as a questionable facial expression, sarcasm in a team member's voice, or even silence. Brett will use these communicative cues from others as a way to understand how the group is viewing an issue.

SKILL BUILDER

Having all team members on the same page is vital for group functioning. But oftentimes team cognition is difficult to assess. How does a group know if all members understand an issue the same way?

Unfortunately, group members usually don't know until it is too late. For example, in an undergraduate class a group was assigned to present on a topic in the next class period. All group members thought it would be beneficial to arrive early to class to make sure they were prepared. Most of the group showed up 30 minutes before class time, expecting to have a rehearsal of the entire presentation. However, one class member showed up five minutes before class and said, "I'm glad everyone showed up a few minutes early to make sure we are all ready and prepared to present. I've been practicing all day and am ready to go. How is everyone else doing?" Clearly, the last class member had a different understanding of what "showing up early" meant than the others. How could the group have been more aware of their expectation differences ahead of time? What is the best way for groups to assess team cognition?

What can you do to help your team members share information ? Importantly, the only ways that group members can know what other group members are thinking is through communication (de Vries, Van den Hooff, & de Ridder, 2006; Park, 2008). First, examine the attitude you display in the team. Do you willingly share information? If you do, you can create a norm of reciprocity. That is, your willingness to share information can encourage other team members to share information. Second, do you display enthusiasm for the group and what the group is working on? When you are enthusiastic, this will enhance other members' eagerness to share information. Third, are you willingly sharing information? If you remain silent in a conversation, members of the group tend to assume that you don't have a belief that severely contradicts the course of conversation. This is how **false consensus**, or the belief that group members are all in agreement when they are not, can emerge. When team members do not encourage and promote opportunities for everyone to share their beliefs and information, groups can be dominated by a few very vocal members.

GROUP COMMUNICATION NETWORKS

As group members regularly communicate about issues, certain channels of communication become routinized within the group. This can also occur for communication that moves across groups. This **communication network**, or a structure of who talks to whom, is the interaction pattern or flow of messages between and among group members. A network creates structure for the group because the network facilitates or constrains who can (or will) talk to whom.

A network is a social structure that consists of group members and the relationships or ties among them. While we generally think of who talks to whom, or communication ties, there are other types of ties among members (Katz, Lazer, Arrow, & Contractor, 2005). **Formal ties** describe who reports to whom or any other power-laden relationship. **Affective ties** describe who likes or trusts whom. **Material ties** describe who gives resources to whom. **Proximity ties** describe who is spatially close or electronically linked to whom. Finally, **cognitive ties** describe who knows whom. In a group, it is likely that describing the network among group members would look different based on which set of ties or relationships you were examining. Thus, the terms *communication, formal, affective, material, proximity*, and *cognitive* describe the nature of ties or relationships you have with other group members.

Ties or relationships among members in a network can further be described by their communication attributes. For example, ties can vary in direction. That is, communication may flow one way, from one group member to another group member. Or communication can flow between group members, with each person sending and receiving of messages. Ties can also be described based on the content of the communication, the frequency of the interaction, and the medium or channel used for communication among group members.

Network ties are described as strong or weak, based on the intensity and reciprocity in the relationship. You probably have stronger ties with your family and friends than you do with members of your work group because ties in groups where your relational needs are met require a greater amount of trust than ties in work groups, which may be temporary or situational. Of course, it may be helpful to have some weak ties, which require less relationship maintenance and effort but still allow you access to a variety of resources. Similarly, ties can be described as being positive or negative. Identifying a tie as strong or weak and positive or negative is really your evaluation of the importance of the tie to you.

Networks in groups and teams are multiplex, meaning that relationships among group members can be described or evaluated across a number of these dimensions. Moreover, describing group relationships according to one type (i.e., formal ties) will reveal a different network than describing relationships according to a second type (i.e., proximity ties). Creating a **sociogram**, a type of visual representation of the group members' relationships, for any particular type of tie reveals the number and pattern of network ties among group members. Because a sociogram reveals the network among group members, it is easy to determine if subgroups or cliques exist. Figures 4.1 displays two sociograms for a baseball team based on formal and affective ties. Notice both the similarities and the differences in the interdependence of member relationships in the two different networks.

Network theory can help us understand the importance of thinking of groups as networks (Wellman, 1988). One proposition of network theory proposes that group members' behavior can be predicted by the nature of their ties to one another. That is, group members tend to use their existing networks to get information, seek resources, and request support. Thus, a network presents a set of opportunities if a group member is well connected, but can also act as a constraint if the member is connected only weakly to just a few other members. While networks are comprised of sets of relationships between two members, another proposition suggests that every two-member relationship must be considered relative to how that particular relationship is situated within the larger network of the group. In other words, the pattern of all of the relationships are important, not just a specific relationship between two group members. How members are connected influences the flow of information and resources. That is, a member who is connected to only one other group member could still receive most of the information distributed in the group if the person he or she is connected to has relationships with all of the other group members.

Networks emerge among group members because they are working interdependently on a task or activity. Ties between members are based on both relational (i.e., affective) and task (i.e., formal) dimensions. Simply put, what group members talk about and how group members talk to one another creates structure for the group, and different structures will emerge based on the nature of the ties among members. In the case of the baseball team, the group's activity influences who talks to whom in a group. If you are the catcher for your softball team, you will talk to the pitcher and infielders more frequently than you will talk to the outfielders. By virtue of your position on the team, you must talk to the pitcher to plan your approach to opposing batters and to the infielders to coordinate your team's defense. A primary way to evaluate networks is to examine their centralized or decentralized structure.

Decentralized Networks

Most groups use a decentralized network that allows each group member to talk to every other member. There are no restrictions as to who can talk to whom. This pattern is decentralized because group members communicate without restrictions, and it is typical of most group interactions. Although a **decentralized communication**

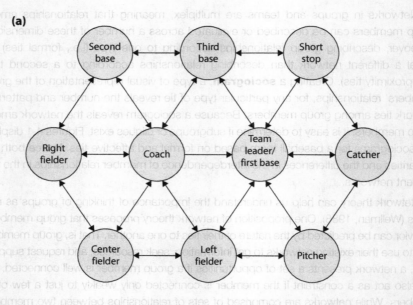

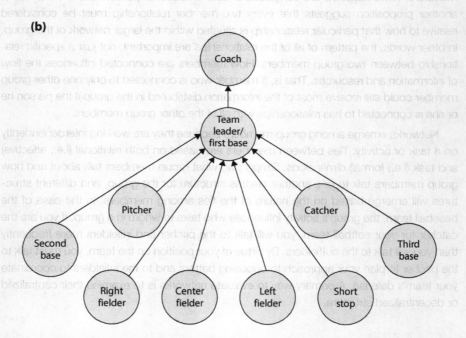

Figure 4.1 (a) Decentralized network of baseball team based on affective ties; (b) Centralized network of baseball team based on formal ties

network is helpful for certain group tasks, such as discussions, problem solving, and decision making, it may slow down other types of group activities. A decentralized network is good for building group and team cohesiveness. For example, the affective network of the baseball team in Figure 4.1 is decentralized. The team has been together for a several years. Although there are formal positions of coach and team leader, nearly everyone is the same age, as the team comprises individuals who played baseball in college and now play as a team in an amateur adult league. Team members take frequent road trips together, and for many players the team has become their primary social outlet. Thus, team members get along well with one another and communicate frequently about topics other than baseball.

An open, or decentralized, network provides the most input, but it can also produce **communication overload**—too much or too complex communication from too many sources. When overload occurs, messages may compete or conflict, causing stress and confusion. Sometimes in discussions, groups may need a facilitator or coordinator to monitor turn taking so that everyone has a chance to be heard. Still, when a group works on a complex task or activity, a decentralized pattern is more effective (Brown & Miller, 2000).

Centralized Networks

Any type of network that imposes restrictions on who can talk to whom is **centralized**. The constraints might be real or perceived. For example, a real constraint is that members of the baseball team are asked to bring issues about uniform repair to the team leader. If they can't be resolved there, then the team leader takes the issue to the coach. A perceived constraint is that a relatively new member of the team feels uneasy about asking the team leader for new equipment. There is no rule or policy to suggest that he can't make this request, but he doesn't because he believes other players are not making similar requests.

When networks are centralized, some members may experience **communication underload**—too infrequent or too simple messages. Group members in an underload situation often feel disconnected from the group. A centralized network can develop if the leader of the group controls the distribution and sharing of messages in the group. From this central and controlling position, the member in the role of leader talks to other group members individually. Group members do not talk with one another; they communicate only with the leader. This type of pattern often develops when there is a strong, domineering leader. If this is the only communication pattern within the group, members are likely to be dissatisfied with the group experience. This pattern also restricts the development of a group identity and weakens the interdependence of group members.

In the case of the baseball team, a formal centralized network may be appropriate for the team's task. Usually, one person (i.e., the manager) is in charge of making decisions for the team. Of course, players can provide feedback to their manager, but during a game a manager cannot take into consideration everyone's viewpoints. Instead a manager may limit conversations to his coaching staff when making decisions. This allows the manager to respond more quickly to situations that require immediate attention. Centralized networks prevent communication overload, especially when time is of the essence.

Evaluating Your Group's Network

Most groups think they use a decentralized or open network in which group members are free to talk to whomever they want. But as roles and norms develop in groups, a structure may be created that affects who talks to whom and who talks most frequently. Status and

 MESSAGE AND MEANING

Instead of holding a meeting, Jeff has decided to text his team requesting feedback on how large the marketing budget should be for the project. Within 10 minutes, he received the following responses:

JIM: $200,000

FRED: It really depends on whether you want to prioritize this project over Project B.

SARA: Shouldn't we first see how much the ads will cost?

JIM: Or we could probably run it at $150,000.

SARA: Jim also brings up a good question. Which project are we prioritizing? I meant to say Fred. Fred brought up the question.

FRED: $150,000 is way too low.

CAL: What do you think, Jeff?

JIM: I don't know anything about Project B. I think $150,000 may be what the clients are thinking.

SARA: You should check out the last marketing report from the downtown office.

ALLIE: I'm a little behind on this conversation. Is Project A our next project?

JEFF: Let's hold off on this topic and meet sometime next week.

Within a few minutes Jeff had received 11 texts about several different aspects of his question. He was overwhelmed with the information and couldn't control the flow of conversation. What type of communication network did they have for this team? What characteristics of their communication helped you determine the network type? Will holding a team meeting resolve the network problem? When is texting an appropriate communication medium for groups?

power differences among members also affect a group's communication network. As a result, some group members will end up talking more, some will be talked to less, and some will talk only to specific other members.

One way to examine the effectiveness of a group's network is to examine which members hold which knowledge or expertise. This is a first and necessary step, but simply knowing who has what knowledge is not sufficient. A group's knowledge network is only beneficial if group members have developed strong communication ties to share that information. Finally, member expertise and their communication ties must match the task interdependency of group members (Yuan, Fulk, Monge, & Contractor, 2010).

A second way to evaluate group network is look for **faultlines**, or characteristics or attributes of diversity that are salient for a particular group and its task (Lau & Murnighan, 1998). Faultlines can divide members into subgroups because group members commonly communicate more frequently with members who they perceive to be similar. For exam-

PUTTING THE PIECES TOGETHER

Group Size, Interdependence, and Group Structure

Reflect on a group or team you belong to now. How much of the overall group communication do you contribute? Twenty percent? Fifty percent? How does group size influence your amount of communication in this group? The way that group members interact influences the roles that group members take. For example, is the most talkative member of your group the leader? Why or why not? If you wanted to become the leader of the group, how would you have to interact?

Some group members may have expertise in different areas of the group task. How do you know which members are experts? How do members communicate their expertise?

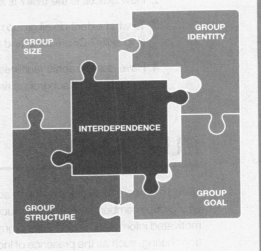

ple, a work group is discussing their organization's new early retirement and family leave policies and the impact of these new policies on the workplace. Older members or female members may create subgroups in the larger discussion, as age and gender are salient to the topic because older group members will easily identify with the potential of retirement and female employees may identify with the potential of taking family leave. When faultlines like these occur in a group discussion, subgroups can emerge. One way to avoid or decrease these types of demographic faultlines is to create relational ties among members on other dimensions. In this case, providing employees with a common goal for their unit that requires the skills and talents of all group members will encourage them to create a network that minimizes these differences.

So, which network should your group use? This depends on several factors. Although the task or activity of the group is often the primary determinant (Hirokawa, Erbert, & Hurst, 1996), do not forget about the effects of a communication network on a group's social or relational development. On the one hand, centralized networks place a heavy burden on the person at the center of the network. At the same time, a centralized pattern limits the opportunity for group members to get to know one another, to develop relationships within the group setting, and to create a group identity.

On the other hand, decentralized, open communication networks may slow the group's work on the task. Yet members communicating in this fashion are generally more satisfied with the group and its activity and are more committed to the group. You can ask yourself these questions to determine which communication network will work best for your group situation. More than likely, multiple networks will be required to satisfy both the relational and task dimensions of your group.

1. What is more important to the group right now—working on this task or developing relationships and commitment to the group?
2. How difficult is the task? Is it simple or complex?
3. To what extent do all group members need to develop leadership and followership skills? Or are roles and functions specifically set in this group?
4. Have demographic faultlines created subgroups? What other networks could be facilitated if subgroups are hindering group success?

SUMMARY

Group members share information to accomplish group tasks. Early research investigated why group members were often unsuccessful in their attempts to share information. The motivated information sharing paradigm explains that other factors also influence information sharing, such as the presence of individual goals or not having access to all necessary information. These other factors may lead group members to adjust their messages in ways that inhibit pursuit of the group goal.

One of the reasons members share information is so that members form a common understanding of issues. There may be differences in how members view an issue, but these differences are unlikely to be made known unless the group discusses them. Thus, member interaction is a great way of determining if other members have the same understanding. Without interaction, there is no way of knowing if agreement across the group is really false consensus.

HOW WE COMMUNICATE WHAT WE KNOW

As group members communicate within and outside the group, patterns of interaction are formed. These patterns may be based on a variety of characteristics, such as through hierarchy, proximity, or past relationships. The type of network a group needs is based upon the nature of the task they are trying to accomplish. For example, groups that have tasks requiring flexibility and creativity (i.e., design team in an advertising agency) are best managed by the use of decentralized networks. However, groups that have tasks that must be efficient or are urgent (i.e., emergency medical providers) may prefer centralized networks. Group member knowledge expertise or faultlines may also influence group communication networks.

DISCUSSION QUESTIONS AND EXERCISES

1. Have a conversation with your friends or coworkers. Ask them about instances in which their groups or teams had problems sharing information. Can they identify why such a problem existed? How did they become aware that information known to some group members was not being shared with everyone in the group? Did the unshared information have a positive or negative effect on the group?

2. Team cognition is a difficult concept to explain. Can you draw a picture to help explain this concept to others?

3. Observe a group in action, for example, a civic or governmental group, or a student, sorority, or fraternity council, or watch a situation comedy that features

 NAILING IT!

Using Group Communication Skills for Group Presentations

Developing and delivering a group presentation is a complex group goal. How well you and other group members share information during the developmental process will make a difference between a well-developed and cohesive presentation versus one that is disjointed.

First, to share information effectively, it is necessary to meet as a team—and meeting face-to-face may be helpful to communicate both verbal messages and nonverbal cues. Second, make a list of the type of information the group will need for the presentation. It is highly unlikely that one group member will have all the expertise or information the team needs. Assigning information-gathering tasks to all members helps to ensure that the group will have the information required for the presentation. Perhaps, the group will make information-gathering assignments based on types or sources of information (e.g., information from the web, scholarly journals, popular press books, or interviews). Or, perhaps your presentation topic is easily divided into different topics (i.e., environmental concerns about water quality, air quality, and soil quality).

Third, as you discovered in this chapter, there is a tendency for group members to not share information each member uniquely holds. To overcome that tendency, the group should make a plan for sharing information with one another. One way is to hold group meeting where the only item on the agenda is to make short informational presentations to one another. Holding a group meeting for this purpose only is a good strategy, as it allows group members to identify gaps, overlaps, or inconsistencies in the information gathered. Further, it allows group members to ask questions, which can help to uncover potentially unshared information.

group interaction, and create their communication network. Identify the type of communication network that emerges in the group interaction. After evaluating the strengths and weakness of the network, what suggestions would you make to the group and to individual members for changing the network structure?

Image Credits

Photo 4.1: Copyright © Depositphotos/nd3000.
Figure 4.2: Adapted from: Copyright © Depositphotos/megastocker.

CHAPTER 5

DECISION MAKING

After reading this chapter, you should be able to:

- Explain why decision making is a core group activity
- Describe the steps of functional decision making and identify them when they occur in group interaction
- Distinguish between majority and minority group behaviors
- Describe and explain these decision-making techniques to others: brainstorming, nominal group technique, consensus, voting, and ranking
- Compare these decision-making techniques and select the most appropriate for a particular group decision

In order to accomplish a group goal, members must make decisions about a variety of issues. Often, the first decision is to decide on a meeting time and location. Making decisions about group composition and leadership structure are also important. Of course, decisions pertinent to the purpose of the group can be quite difficult. In fact, groups and teams are often created because individuals do not want to make difficult decisions.

Thus, decision making is central to group and team interaction. In order to explore this topic, we first investigate two theories of decision making. Then we consider several techniques that have proven beneficial to improving communication during decision making. Last, we investigate several decision-making principles that are the foundation of successful group interaction.

GROUP DECISION-MAKING THEORIES

Functional Theory of Decision Making

When group communication scholars use a functional approach, they are speaking of three primary functional assumptions (Wittenbaum, Hollingshead, Paulus, Hirokawa, Ancona, Peterson, Jehn, & Yoon, 2004). The first assumption is that groups are goal oriented, something emphasized in Chapter 1. The second assumption is that group performance can vary and be evaluated; in other words, groups do not always succeed given the difficulty of tasks. This assumption also points out that a group's success can be measured. Third, group performance (or its effectiveness at decision making) is a product of interaction among group members, which is based on (a) internal inputs, such as group size and member composition, and (b) external circumstances, such as time pressure. A group's interaction reflects both of these. Obviously, the goal of task-oriented groups is to find the solution or decision. According to the **functional theory of group decision making**, these are the five critical functions in decision-making and problem-solving activities (Gouran & Hirokawa, 1983; Hirokawa, 1982):

1 thoroughly discuss the problem,

2 examine the criteria of an acceptable solution before discussing specific solutions,

3 propose a set of realistic alternative solutions,

4 assess the positive aspects of each proposed solution, and

5 assess the negative aspects of each proposed solution.

A function is not just a step or a procedure. Each function represents a type of interaction required among group members to make a decision. The functions do not have to be completed in order, but when the five functions are not accomplished, a group diminishes its chances of making good decision.

For the first step, group members need to achieve an understanding of the problem they are trying to solve. The group should deliberate until all members understand the nature and

MESSAGE AND MEANING

The jury has deliberated all day. After making decisions about a defendant's guilt or innocence of double murder, the jury still has 25 counts to deliberate. Each of these counts is about drug possession or drug trafficking.

JUROR 1: Can I say something . . . I am willing to work here until 10 o'clock tonight if we have to, to finish this. Maybe in another hour or so, we can have some food brought in. Pizza or Chinese—I don't care. Whatever anybody wants. Let's just grind away at this.

JUROR 2: I mean I think we need to listen to the tape. It's gonna take time.

JUROR 3: Hopefully, they're [the audio tapes] readily accessible.

JUROR 2: You know I mean . . .

JUROR 3: The simple fact is that if we won't get into the hotel room till 11. Then we're gonna have to get up again at 7 o'clock and I don't want to do this. Perhaps we should wait until tomorrow.

JUROR 4: I don't think it's gonna take till 10. Obviously, the sooner we get to these tapes, the better off we are. So instead of debating this, let's just get these tapes going. You have to take a break, go take a break for 15 minutes. It will, it will take them 12 minutes to get the tape ready for us. And then we're back in here.

JUROR 1: As far as I'm concerned, I'll stay here all night.

[SEVERAL JURORS SPEAK AT ONCE]

JUROR 1: Because then by the time they order the food,

[SEVERAL JURORS SPEAK AT ONCE]

JUROR 5: I want to leave.

[SEVERAL JURORS SPEAK AT ONCE]

JUROR 2: . . . it will be 8:30. I do not want to stay here any longer.

JUROR 3: I don't want to stay here till 9 o'clock at night, for 12 hours. I don't want to do it, period.

JUROR 6: Let's eat dinner here and leave at 7 and resume tomorrow.

JUROR 1: I don't care if we stay later than that late. But I don't want to stay until 12 o'clock at night, 10 o'clock at night.

> [SEVERAL JURORS SPEAK AT ONCE]
>
> **JUROR 4:** By 8 o'clock, if there're two verdicts left, I'll stay until they're done just so we can get it over with. But if they're still several verdicts to go, I'm not staying here, no way.
>
> **JUROR 1:** Well, you know we're all reasonable people here. Let's do the best we can and assess it when we get closer to a decision. It's only 5 o'clock. We're reasonable . . . listen, nobody wants to drag this out, obviously. But let's be reasonable about it.
>
> Using the functional theory of group decision making, how effectively are jury members making this procedural decision about whether to stop deliberating for the day, or to stop for dinner and then continue to deliberate into the evening. How would you evaluate their abilities to 1) thoroughly discuss the problem, 2) examine the criteria of an acceptable solution before discussing specific solutions, 3) propose a set of realistic alternative solutions, 4) assess the positive aspects of each proposed solution, and 5) assess the negative aspects of each proposed solution? Is the functional theory of decision making appropriate for procedural decisions? If the jury were to effectively deliberate their current problem (whether to stop, or to eat and then deliberate further), what influence might that have on the remaining decisions they have to make?

significance of the problem, its possible causes, and the consequences that could develop if the problem is not dealt with effectively. For example, parking is generally a problem on most campuses. But a group of students, faculty, staff, and administrators addressing the parking problem without having an adequate understanding of the entire issue is likely to suggest solutions that will not actually solve the problem. The parking problem on your campus may be that there are not enough parking spaces. Or it may be that there are not enough parking spaces where people want to park. Or perhaps the parking problem exists at only certain times of the day. Another type of parking problem exists when students do not want to pay to park and park their cars illegally on campus and in the surrounding community. Each parking problem is different and so requires different solutions. When group members address this function—understanding the nature of the problem before trying to solve it—their decision-making efforts result in higher-quality decisions (Hirokawa, 1983).

Second, the group needs to develop an understanding of what constitutes an acceptable resolution of the problem. In this critical function, group members come to understand the specific standards that must be satisfied for the solution to be acceptable. Groups must develop criteria by which to evaluate each proposed alternative. Let's go back to the parking problem. In this step, group members need to consider how much students and employees will be willing to pay for parking. Group members also need to identify and discuss the type of solutions campus administrators and campus police will find acceptable. The group should consider if the local police need to agree with the recommendation. In other words, the group has to decide on the objectives and standards that must be satisfied in selecting an appropriate solution. Any evaluation of alternatives must be based on known and agreed-upon criteria (Graham, Papa, & McPherson, 1997).

Third, the group needs to seek and develop a set of realistic and acceptable alternatives. With respect to the parking problem, groups frequently stop generating alternatives when the first plausible solution is suggested. Look at the following dialogue:

MARTY: Okay, I think we should think about building a parking garage.

LINDSEY: Where would it go?

MARTY: I don't know. But there are all kinds of empty lots around campus.

HELEN: What about parking in the church parking lots?

LINDSEY: That's an idea, but I like the idea of our own parking garage better.

TODD: I like that, too. It would be good to know that whatever time I go to campus a parking spot would be waiting for me.

MARTY: Any other ideas, besides the parking garage?

LINDSEY: No, I can't think of any. I think we need to work on the parking garage idea.

TODD: Me, too.

HELEN: Shouldn't we consider something else in case the parking garage idea falls through?

MARTY: Why? We all like the idea, don't we?

If a group gets stuck in generating alternatives, as our parking group does, a brainstorming session or nominal group technique (discussed later in the chapter) may help. A group cannot choose the best alternative if all the alternatives are not known.

Fourth, group members need to assess the positive qualities of each of the alternatives they find attractive. This step helps the group recognize the relative merits of each alternative. Once again, let's turn to the parking problem. Students and employees probably will cheer for a solution to the parking problem that does not cost them more money. Certainly no-cost or low-cost parking will be attractive to everyone. But if this is the only positive quality of an alternative, it is probably not the best choice. For example, to provide no-cost or low-cost parking, your recommendation is that during the daytime, students park in the parking lots of churches and at night they park in the parking lots of office buildings. Although the group has satisfied concerns about cost, it is doubtful that those who manage church and office building properties will find this alternative attractive. Fifth, group members need to assess the negative qualities of alternative choices. By assessing positive and negative qualities separately, group members can avoid the tendency to provide an overall positive or negative evaluation when assessing an option.

When group members communicate to fulfill these five functions, they increase the chance that their decision making will be effective. This is because group members have worked together to pool their information resources, avoid errors in individual judgment, and create opportunities to persuade other group members (Gouran, Hirokawa, Julian, & Leatham, 1993). For example, the members of the parking group bring different information to the discussion because they come to school at different times of the day. Those who come early or late in the day have a harder time finding a place to park than those who come early in the afternoon. By pooling what each participant knows about the parking situation, the group avoids becoming biased or choosing a solution that will resolve only one type of parking problem.

In addition, as the group discusses the problem, members can identify and remedy errors in individual judgment. It is easy to think that parking is not a problem when you come in for one class in the early afternoon and leave immediately after. In your experience, the parking lot has some empty spaces because you come at a time when others have left for lunch. And when you leave 2 hours later, the lot is even emptier, making you wonder what the fuss is about in the first place!

Discussion also provides an opportunity to persuade others or to be persuaded. Discussion allows alternatives to be presented that might not occur to others and allows for reevaluation of alternatives that initially seem unattractive. Let's go back to the group discussing the parking problem:

MARTY: Okay, where are we?

HELEN: Well, I think we've pretty much discussed parking alternatives. I'm not sure.

LINDSEY: What about using the bus?

TODD: You've got to be kidding.

LINDSEY: Why not? The bus line goes right by campus and the fare is only 50 cents.

MARTY: Well, it's an idea.

HELEN: Well, what if the bus doesn't have a route where I live?

LINDSEY: Well, that may be the case for you, Helen, but I bet many students and employees live on or near a bus line.

MARTY: I wonder how many?

LINDSEY: Let's go online and get a copy of the entire routing system.

MARTY: Good idea, Lindsey. We were looking for parking alternatives and hadn't thought about other modes of transportation.

Groups that successfully achieve each of the five critical functions of decision making make higher-quality decisions than groups that do not (Hirokawa, 1988). However, the functional perspective is not a procedure for making decisions, because there is no prescribed order to the five functions. Rather, it is the failure of the group to perform one of the five functions that has a profound effect on the quality of the group's decision making. But do the five functions contribute equally to group decision-making effectiveness? An analysis across hundreds of groups indicates that the most important function is group members' assessment of the negative consequences of proposed alternatives.

Next in importance were thorough discussion and analysis of the problem, and the establishment of criteria for evaluating proposals (Orlitzky & Hirokawa, 2001). The functional theory of decision making continues to be the central theory of decision-making communication.

Majority/Minority Influence

Another avenue of group communication research is to investigate how different subgroups argue for their respective positions in a group. One way to do this is to look at majority and minority subgroups and how they communicate to accomplish their respective goals. Early efforts on such influence focused on the differing purposes of each group. In general, majorities try to effect compliance through their control and use of resources, and by using their numerical advantage to enforce public compliance (Moscovici, 1976). Such influence may be used in a variety of different ways. For example, majority subgroup members could try to communicate more than other groups since there are more members to speak. They may try to threaten to be unhappy with or noncompliant to others' ideas, knowing that without the majority members supporting a position, it will likely fail, no matter if the leader feels differently.

On the other hand, minorities try to provoke or stimulate thought to encourage majority members to reflect on their attitudes and beliefs (Nemeth, Swedlund, & Kanki, 1974, Nemeth, 1986). Simply put, minority subgroups may be stubborn about their position, and attempt to make bold statements to lead others to second guess their beliefs. They can also make it very clear that if the majority wins, the minority will be very difficult throughout the rest of the process.

In terms of group communication, majority/minority influence has found three primary results (Meyers, Brashers, & Hanner, 2000). First, the majority subgroup is often victorious in winning the final result. This is not surprising given that most decisions are majority-rule, but it is an important finding nonetheless. Minorities know from the outset that they have a difficult mountain to climb to have any effect on the proceedings, let alone to convince others of their position. Second, argument consistency is important for both majorities and minorities to be influential (Bazarova, Walther, & McLeod, 2012). If either group moves away from their initial position, they are often considered weak or wishy-washy. For minority subgroups, this stubbornness is especially important when seeking concessions from the other side.

Third, majorities and minorities tend to trigger different thought processes in group members. Majority subgroups tend to encourage integration and convergence from others. Majority subgroups members may attempt to control the conversation and perhaps even minimize communication if it is clear that they hold the winning position. Minorities, on the other hand, try to spark discussion or conflicts, and sometimes the anxiety created by such behavior may lead others to want to reduce it, even if that means conceding on some points or ceasing to seek out helpful information Meyers et al., 2000.

A direct study of argument interaction (Meyers et al., 2000) found that majority subgroups used more convergence-seeking messages, or messages that encouraged group members to unite with a point of view. For example, members of the majority were more likely to agree with another person's position, or show support and understanding of others' viewpoints. On the other hand, minority subgroups used more disagreeing messages in their argumentation. If, for example, another group member said something that a minority subgroup member disagreed with, the minority member would be more prone to vocalize that disagreement. The findings by Meyers et al. (2000) support earlier findings, that majority and minority group members strive to accomplish different purposes in decision making.

DECISION-MAKING PROCEDURES

In order to facilitate the decision-making process, there are variety of communicative techniques that group members use. These procedures differ in many ways, including their level of formality (Schweiger & Leana, 1986), forcefulness, and participation. Procedures help members balance task and relational dimensions of their group and develop agreement among members (SunWolf & Seibold, 1999). In this section, we focus on five: brainstorming, nominal group technique, consensus, voting, and ranking.

Brainstorming

Brainstorming is an idea generation technique designed to improve productivity and creativity (Osborn, 1963). Thus, the brainstorming procedure helps a group to function creatively. In a brainstorming session, group members first state as many alternatives as possible to a given problem. Creative ideas are encouraged; ideas do not have to be traditional or unoriginal. A central component to brainstorming is that all ideas be accepted without criticism—verbal or nonverbal—from other group members. Next, ideas that have been presented can be improved upon or combined with other ideas. Finally, the group evaluates ideas after the idea generation phase is complete. The group should record all ideas for future consideration, even those that are initially discarded. A group member can act as the facilitator of the brainstorming session, but research has shown that someone external to the group may be more effective in this role. The facilitator helps the group maintain momentum and helps members remain neutral by not stopping to criticize ideas (Kramer, Fleming, & Mannis, 2001).

This brainstorming procedure helps groups generate as many ideas as possible from which to select a solution. Generally, as the number of ideas increases, so does idea quality. Members may experience periods of silence during idea generation, but research has shown that good ideas can come after moments of silence while members reflect and think individually (Ruback, Dabbs, & Hopper, 1984). So, it may be premature to end idea generation the first time all members become quiet.

When should a group use brainstorming? Brainstorming is best used when the problem is specific rather than general. Why? If the problem is too generally stated, suggestions will not be focused or helpful. For example, brainstorming can be effective for identifying ways to attract minority employees to an organization. But the problem—What does a group hope to accomplish in the next 5 years?—is too broad. Use a brainstorming session to break issues down into subproblems, and then devote a further session to each one. Brainstorming works best with smaller rather than larger groups. Brainstorming

THEORY STANDOUT

Persuasive Arguments Theory

In order to demonstrate the benefits and strengths of a communication perspective, it is also important to consider a noncommunicative theory on group decision making. Early research assumed that the process of group decision making largely consisted of group members making up their minds prior to meetings, meaning that communication during meetings was largely unimportant.

For example, **persuasive arguments theory** posits that instead of interaction being important, it was rather the cognitive processes of individual group members prior to meetings that led to interaction outcomes. In other words, group members make up their minds prior to a meeting. Thus, the real purpose of meetings is not to develop the correct solution, but to simply convince others to support premeeting decisions by group members. Since group members have made up their minds prior to group discussion, it becomes increasingly difficult to persuade other group members. One approach may be to present extreme arguments in order to shock others into supporting the prevailing view. This persuasive technique for group decision making, however, leads to **group polarization**, or the tendency for groups to gravitate toward decisions more extreme than any of the members would prefer individually.

Although a provocative way to view group decision making, group communication research suggests there is little data to support it (Meyers, 1989). Persuasive arguments theory minimizes the influence created in group discussion specifically, and communication, more generally.

Questions:

1. Do you make up your mind on issues prior to your group's discussion? Is this beneficial or harmful to group decision making?
2. Is suggesting extreme views to win a point ethical?
3. Are there differences in how group members talk when trying to discover the best solution as opposed to simply persuading other group members to agree with them?

also works best when group members are diverse in the knowledge they hold about the issue (Wittenbaum, Hollingshead, & Botero, 2004). Finally, members are more likely to generate a greater number of unique ideas if they write their ideas down before presenting them to the group (Mullen, Johnson, & Salas, 1991).

Brainstorming can help increase group cohesiveness because it encourages all members to participate. It also helps group members realize that they can work together productively (Pavitt, 1993). In addition, group members report that they like having an opportunity to be creative and to build upon one another's ideas (Kramer, Kuo, & Dailey, 1997), and they usually find brainstorming fun. However, groups do better if they have a chance to warm up or to practice the process (Firestien, 1990). The practice session should be unrelated to the subject of the actual brainstorming session. Practice sessions are beneficial because they reinforce the procedure and reassure participants that the idea generation and evaluation steps will not be integrated. Posting the five brainstorming steps so they are visible during the session helps remind participants of the procedure's rules.

Brainstorming can be done effectively in groups that meet face-to-face, as well as in groups that use technology to facilitate the process (Barki & Pinsonneault, 2001). Also note that groups do better at brainstorming—both in quantity and quality of ideas—if they have been through a practice round. One way to do this is to have group members use the brainstorming technique as a way to get to know one another better. For example, group members can be asked to "generate as many ideas as you can related to your area of expertise, age, sex, major, ethnicity, and geographical location" (Baruah & Paulus, 2008, p. 530). Practicing on the topic of major, groups members could brainstorm the types of jobs graduates with communication degrees could apply for. Notice that brainstorming is a procedure for generating ideas, and not for making decisions. As a result, brainstorming by itself cannot satisfy the five critical functions of group decision making. However, it is especially effective in helping a group seek and develop a set of realistic and acceptable alternatives and in coming to an understanding of what constitutes an acceptable resolution, and moderately effective in helping group members achieve an understanding of the problem.

Nominal Group Technique

The same basic principles of brainstorming are also applied in the **nominal group technique** (NGT) except that group members work both independently as individuals and interdependently in the group. Thus, the nominal group technique is an idea generation process in which individual group members generate ideas on their own before interacting as a group to discuss the ideas. The unique aspect of this procedure is that the group temporarily suspends interaction to take advantage of independent thinking and reflection (Delbecq, Van de Ven, & Gustafson, 1975). NGT is based on two principles: (a) individuals think more creatively and generate more alternatives working alone, and (b) group discussion is best used for refining and clarifying alternatives.

NGT is a six-step linear process, with each step focusing on different aspects of the problem-solving process. In Step 1, group members silently generate as many ideas as possible, writing down each idea. It's sensible to give members a few minutes after everyone appears to be finished, as some of our best ideas occur to us after we think we are finished.

In Step 2, the ideas are recorded on a flip chart by a facilitator. Generally, it is best to invite someone outside the group to help facilitate the process so all group members can participate. Members take turns, giving one idea at a time to be written on the flip chart. Duplicate ideas do not need to be recorded, but ideas that are slightly different from those already posted should be listed. Ideas are not discussed during this step. The person recording the group's ideas on the flip chart should summarize and shorten lengthy ideas into a phrase. But first, this person should check with the member who originated the idea to make sure that editorializing did not occur. When a member runs out of ideas, the member simply says "pass," and the facilitator moves on to the next person. When all members have passed, the recording step is over.

In Step 3, group interaction resumes. Taking one idea at a time, group members discuss each idea for clarification. If an idea needs no clarification, then the group moves on to the next one. Rather than asking only the group member who contributed the idea to clarify it, the facilitator should ask if any group member has questions about the idea. By including everyone in the clarification process, group ownership of the idea increases.

In Step 4, group members vote on the ideas they believe are most important. For instance, if your group generates 40 ideas, consider asking group members to vote for their top five. By not narrowing the number of choices too severely or too quickly, group members have a chance to discuss the ideas they most prefer. If time permits, let group members come to the flip charts and select their most important ideas themselves. This helps ensure that members select the ideas that are important to them without the influence of peer pressure.

SKILL BUILDER

Which Procedures Will Help Your Group?

Think of three recent group experiences in which decision making was the focus of your group's activity. Which procedures do you believe might have been most beneficial for each group? Why? Could the groups have benefited from using more than one procedure? How might you have initiated the use of procedures in your groups? Would group members have welcomed this type of procedural assistance or resisted it? What strategy or strategies could you have used to get your groups to adopt decision-making procedures? What communication skills could you have relied on to help the groups adopt these procedures? Which communication skills will you use in your next group meetings to encourage the groups to adopt the procedures?

In Step 5, the group discusses the vote just taken. Suppose that, from the 40 ideas presented, 11 receive two or more votes. Now is the time for group members to further elaborate on each of these ideas. Direct the discussion according to the order of ideas as they appear on the flip chart, rather than starting with the idea that received the most votes. Beginning the discussion in a neutral or randomly selected place encourages discussion on each item, not just on the one that appears most popular at this point in the procedure.

With that discussion complete, Step 6 requires that group members repeat Steps 4 and 5. That is, once again, members vote on the importance of the remaining ideas. With 11 ideas left, you might ask members to select their top three choices. After members vote, the group discusses the three ideas that received the most votes. Now it is time for the final vote. This time, group members select the idea they most favor.

The greatest advantage of NGT is that the independent idea generation steps encourage equal participation of group members regardless of power or status. The views of more silent members are treated the same as the views of dominant members (Van de Ven & Delbecq, 1974). In fact, NGT groups develop more proposals and higher-quality proposals than groups using other procedures (Green, 1975; Kramer et al., 1997). Another advantage of NGT is that its specified structure helps bring a sense of closure and accomplishment to group problem solving (Van de Ven & Delbecq, 1974). When the meeting is finished, members have a firm grasp of what the group decided and a feeling of satisfaction because they helped the group reach that decision.

When is it best to use NGT? Several group situations can be enhanced by the NGT process (Pavitt, 1993). NGT is most helpful when proposal generation is crucial. For example, suppose your softball team needs to find new and creative ways to raise funds. Your team has already tried most of the traditional approaches to raising money, and members' enthusiasm for selling door-to-door is low. NGT can help the team identify alternatives because it encourages participation from all group members without surrendering to the ideas of only the coach or the most vocal players. NGT also can be very helpful for groups that are not very cohesive. When a group's culture is unhealthy and cohesiveness is low but the group's work must be done, NGT can help the group overcome its relationship problems and allow it to continue with its tasks. The minimized interaction in the idea generation phase of NGT gives everyone a chance to participate, increasing the likelihood that members will be satisfied with the group's final choice. Finally, NGT is particularly helpful when the problem facing the group is particularly volatile—for example, when organizational groups have to make difficult decisions about which items or projects to cut

PUTTING THE PIECES TOGETHER

Group Goal and Interdependence

Think about a group decision in which you participated and consensus was the decision-making procedure chosen by the group. To what extent did consensus decision making help the group achieve its goal? How well did consensus decision making reflect interdependence among group members? Was consensus the most appropriate decision-making procedure for this group and this decision? In what way did the group's practice of consensus match the description of consensus given in this chapter? Considering what you know now about decision-making procedures, what three pieces of advice would you give to this group about its use of consensus?

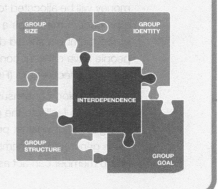

from the budget. The conflict that is likely to occur through more interactive procedures or unstructured processes can be destructive. The structured process of NGT helps group members focus on the task because turn taking is controlled.

With respect to the five critical functions of group decision making, NGT satisfies four. Because interaction is limited, especially in the idea generation phase, group members are not likely to achieve understanding of the problem. The discussion phase of NGT, however, should be effective in helping group members come to understand what constitutes an acceptable resolution to the problem, develop realistic and acceptable alternatives, and assess the positive and negative qualities of alternatives considered.

Consensus

Consensus means that each group member agrees with the decision or that group members' individual positions are close enough that they can support the group's decision (DeStephen & Hirokawa, 1988; Hoffman & Kleinman, 1994). In the latter case, even if members do not totally agree with the decision, they choose to support the group by supporting the decision. Consensus is achieved through discussion. Through members' interactions, alternatives emerge and are tested. In their interaction, group members consult with one another and weigh various alternatives. Eventually, one idea emerges as the decision that group members can support.

To the extent that group members feel they have participated in the decision-making process, they are satisfied with the group's interaction. That satisfaction is then extended to the consensus decision. Thus, when all group members can give verbal support, consensus has been achieved. To develop consensus, a group uses discussion to combine the best insights of all members to find a solution that incorporates all points of view. For example, juries that award damages in lawsuits must make consensus decisions—everyone must agree on the amount of money to be awarded.

Too frequently, consensus building is seen as a freewheeling discussion without any sort of process, plan, or procedure. But there are guidelines a group can use to achieve consensus. This procedure is especially useful for groups that must make highly subjective decisions (e.g., a panel of judges deciding which undergraduate should be selected as the commencement speaker, or the local United Way board of directors deciding how much money will be allocated to community service agencies) (Hare, 1982). Thus, consensus is a procedure that helps a group reach agreement. However, for consensus to work well, group members should discuss what they mean when they use the word *consensus*, as people have different conceptualizations of what consensus means and different feelings about its effectiveness (Renz, 2006).

To develop consensus, the leader or another group member takes on the role of coordinator to facilitate the group's discussion. This coordinator does not express opinions or argue for or against proposals suggested by the group. Rather, the coordinator uses ideas generated by members to formulate proposals acceptable to all members. Another group member can act as a recorder to document each of the proposals. Throughout the discussion, the recorder should read back statements that reflect the initial agreements of the group. This ensures that the agreement is real. When the group feels it has reached consensus, the recorder should read aloud this decision so members can give approval or modify the proposal.

In addition to following these steps, all group members need to be aware of a few basic discussion rules. First, the goal of the group's discussion is to find a solution that incorporates all points of view. Second, group members should not only give their opinions on the issue, but also seek out the opinions of other members. The coordinator should make an extra effort to include less talkative members in the discussion. Third, group members should address their opinions and remarks to the group as a whole, and not to the coordinator. Finally, group members should avoid calling for a vote, which has the effect of stopping discussion.

Consensus can only be reached through interaction. Although each group member should be encouraged to give an opinion, group members should avoid arguing for their personal ideas. It is better to state your ideas and give supporting reasons. Arguing about whose idea is better or whose idea is more correct will not help the group achieve consensus. If other group members express opinions that differ from yours, avoid confrontation and criticism. Rather, ask questions that can help you understand their points of view.

As the group interacts, it can be tempting to change your mind just so the group can reach consensus and move on to other activities. Be careful! Changing your mind only to reach agreement will make you less satisfied with the process and the decision. If the group has trouble reaching consensus, it is better to postpone the decision until another meeting. Pressing for a solution because time is short will not help group members understand and commit to the decision. If a decision is postponed, assigning group members to gather more information can help the next discussion session.

How well does consensus achieve the five critical functions of decision making? As a decision procedure, it is very effective in helping group members achieve an understanding of the problem they are trying to resolve, identify what constitutes an acceptable resolution, and develop a set of realistic and acceptable alternatives. Discussion leading to consensus allows more viewpoints to be discussed, so members are made aware of issues and facts they did not previously know. As a result, group members become more knowledgeable about the problem. Consensus discussions involve everyone, which results in a high degree of integration, as at least part of everyone's point of view is represented in the final decision. Thus, consensus can help achieve the first three critical functions. However, it is less effective in helping groups assess the positive and negative qualities of the alternatives presented.

There are a few disadvantages to using consensus. First, this procedure takes time. When not enough time is allotted, some members may opt out of the discussion process, allowing the group to come to a **false consensus**—agreeing to a decision simply to be done with the task. Thus, the extent to which consensus is effective depends on the voluntary and effective participation of group members. Second, consensus is usually not effective when controversial or complex decisions must be made. A group charged with making a decision that heightens emotional issues for members is likely to make a better decision with a more standardized approach that structures group inquiry. This is why the consensus procedure is not always effective in assessing the positive and negative qualities of the alternatives presented.

Voting

Voting, another decision-making procedure, is simply the process of casting written, verbal, or online ballots in support of or against a specific proposal. Many organizational groups rely on the outcomes of majority voting to elect officers or pass resolutions. A group that votes needs to decide on three procedural issues before a vote is taken.

The first procedural issue centers on the discussion the group should have before members vote. Members do not simply walk into a meeting and vote. Voting should be on clear proposals, and only after substantial group discussion. Here is a suggested procedure to follow in voting (Hare, 1982). Members bring items to the attention of the group by making proposals in the form of motions. Let's say that your communication students' association is making decisions about its budget. Karen says, "I move that we set aside part of our budget for community activities." But subsequent discussion among group members reveals two ambiguities. What does Karen mean by "part of our budget"? Twenty percent? Forty

percent? And what are "community activities"? Do they include teaching junior high students how to give speeches? With other members' help, Karen's proposal is made more specific: "I move that we set aside 20 percent of our budget for community intervention activities that help children appreciate the value of communicating effectively." Now, with a specific motion, Karen can argue for her proposal by stating its merits. Even with a specific proposal, she is going to receive some opposition or face more questions. That is okay because it helps all group members understand her motion more clearly. During this discussion, the group leader makes sure that all those who want to be heard get a chance to talk. However, the leader does not argue for or against any particular motion. To do so would put undue influence on the group. The group's secretary or recorder keeps track of the motions and identifies which ones receive approval from the group.

The second procedural issue is to decide how the vote will be taken. When sensitive issues are being voted on, it is better to use a written ballot. Similar ballots or pieces of paper are given to each group member. Written ballots allow group members to vote their conscience and retain their anonymity. Two group members should count the votes and verify the decision before announcing it to the group.

A verbal vote, or a show of hands, is more efficient when it is necessary only to document the approval or disapproval apparent in the group's discussion. For example, suppose your communication students' association has several items of business to take care of at the next meeting. Specifically, the association needs to elect officers, approve the budget, and select a faculty member for the outstanding professor award. The budget was read to members at the last meeting and then discussed. Although members will ask some questions before the vote, the group basically needs to approve or disapprove the budget. Because there is nothing out of the ordinary about the budget and little controversy is expected, it is okay to use a show of hands in this case.

However, electing officers and voting for one professor to receive an award can bring up conflicting emotions among group members. Both of these matters are better handled with written ballots. This ensures that group members can freely support the candidates and the professor they desire without fear of intimidation or retaliation.

The final procedural issue that needs to be agreed on before taking a vote is how many votes are needed to win or decide an issue. Most of the time, a simple majority vote (one more than half of the members) is satisfactory. However, if a group is changing its constitution or taking some type of legal action, a two-thirds or three-fourths majority may be preferable. Both the method of voting and the majority required for a decision need to be agreed upon before any voting takes place.

Voting can be efficient, but it can also arbitrarily limit a group's choices. Many times, motions considered for a vote take on an either/or quality that limits the choice to two alternatives. And a decision made by voting is seen as final—groups seldom revote. This is why having an adequate discussion period before voting is necessary. As you can see, voting is not the best choice when complex decisions must be made.

To summarize, the procedures for voting include the following:

1. Hold discussions to generate a clear proposal.
2. Decide how the vote will be taken—written ballot, verbal vote, or show of hands.
3. Decide how many votes are needed to win or decide an issue.
4. Restate the proposal before voting.

How well a group develops the discussion before voting determines how well the group satisfies the five critical functions of group decision making. Although voting is often perceived as a way of providing a quick decision, inadequate time for group discussion can severely limit the appropriateness or effectiveness of the proposals to be voted on.

Ranking

Ranking is the process of assigning a numerical value to each decision alternative so that group members' preferences are revealed. Groups often use a ranking process when there are many viable alternatives from which to choose, but the group must select the preferred alternative or a set of preferred alternatives. There are two steps to the ranking process.

First, each member individually assigns a numerical value to each decision alternative. In effect, rankings position each alternative from highest to lowest, as well as relative to one another. Usually, 1 is assigned to the most valued choice, 2 to the next most valued choice, and so on. These rankings may be based on a set of criteria developed by the group, for instance, how well the alternative fixes the problem or if the alternative is possible within the time frame allotted the project.

Second, after group members complete their individual rankings, the values for each alternative are summed and totaled. Now the group has a score for each alternative. The alternative with the lowest total is the group's first-ranked alternative. The alternative with the second-lowest score is the group's second-ranked alternative, and so on. This procedure, which helps group members come to agreement, can be done publicly so group members can see or hear the ranking of one another's alternatives, or the process can be done on paper so individual rankings are anonymous.

Just as with voting, the ranking procedure is most effective when the group has adequate time to develop and discuss the alternatives to be ranked. Compared to groups instructed to "choose the best alternative," groups that rank-order their alternatives do a better job, as all alternatives must be discussed for members to perform the ranking task (Hollingshead, 1996). Thus, the extent to which this procedure satisfies the five critical functions of group decision making depends on the quality of the group's discussion.

Although ranking decreases group members' feelings of personal involvement or participation, groups using this procedure report little negativity in decision making. Fewer arguments or conflicts are reported when ranking is used because it is more difficult for one or two individual members to alter a group's decision-making process. All members get to indicate their preference, and all preferences are treated equally. Thus, group members report feeling satisfied with the outcome (Green & Taber, 1980). Group members usually prefer ranking to voting for making a decision when more than two alternatives exist.

Comparing Procedures

Procedures help groups by managing their discussions and decision-making processes. In turn, this enhances the quality of decision making in the group by coordinating members' thinking and communication, providing a set of ground rules all members can and must follow, balancing member participation, managing conflicts, and improving group climate (Jarboe, 1996; Poole, 1991; SunWolf & Seibold, 1999). Most importantly, procedures help groups avoid becoming solution minded too quickly.

But which procedure is best? Sometimes the group leader or facilitator selects a procedure. Other times the group relies on familiarity—selecting the procedure it used last time regardless of its effectiveness. Rather than select a procedure arbitrarily, groups should select a procedure or a combination of procedures that best suits their needs and satisfies the five critical functions of group decision making. Table 5.1 summarizes the ways in which each procedure satisfies the five functions.

Table 5.1 The Ways in Which Various Procedures Satisfy Problem-Solving and Decision-Making Functions

	Understand the Problem	Understand What Constitutes Acceptable Resolution	Develop Realistic and Acceptable Alternatives	Assess the Positive Qualities of Alternatives	Assess the Negative Qualities of Alternatives
Brainstorming	Somewhat	Yes	Yes	No	No
NGT	No	Yes	Yes	Yes	Yes
Consensus	Yes	Yes	Yes	No	No
Voting	Depends on quality of group discussion before voting	Depends on quality of group discussion before voting	Depends on quality of group discussion before voting	Depends on quality of group discussion before voting	Depends on quality of group discussion before voting
Ranking	Depends on quality of group discussion before ranking	Depends on quality of group discussion before ranking	Depends on quality of group discussion before ranking	Depends on quality of group discussion before ranking	Depends on quality of group discussion before ranking

Before you select a procedure, you should analyze the type of task before your group. If the task is easy—for example, the group has all of the necessary information to make effective choices—the type of procedure you select will have less influence on the group's ability to resolve the problem or reach a decision. However, if the group task or decision is difficult—for example, members' decision-making skills vary, the group needs to consult with people outside the group, or the decision has multiple parts—the decision procedure selected will have a greater impact on the group's decision-making abilities. Generally, in these situations, the procedure that encourages vigilant and systematic face-to-face interaction will result in higher-quality outcomes (Hirokawa et al., 1996).

Regardless of which procedure your group selects, all members must agree to use the procedure if any benefits are to be achieved. Also remember that the procedure itself does not ensure that all members will be motivated and willing to participate. Decision procedures cannot replace group cohesiveness.

DECISION-MAKING PRINCIPLES

Regardless of the procedure or process your group uses, four principles seem to fit most group problem-solving and decision-making situations (Hirokawa & Johnston, 1989). First, group decision making is an evolutionary process. The final decision of the group emerges over time as a result of the clarification, modification, and integration of ideas that group members express in their interaction. A student government group may know that it needs to make a decision about how to provide child care for university students, but the final decision results from the group bringing new information to meetings and other group members asking for clarification and developing ideas. Thus, a group will have a general idea about a decision that needs to be made, but not necessarily its specifics.

Relatedly, the second principle is that group decision making is a circular rather than a linear process. Even when they try, it is difficult for group members to follow a step-by-step approach to group decision making. Group decision making is circular because group members seldom bring all the needed information into the group's discussion at the same time. For example, let's say that your group decides to hold a fundraiser on May 3, close to the end of the spring semester. In discussion, your group needs to make this decision first to secure a date on your university's student activities calendar. Now that the date is settled, your group can concentrate on what type of fundraiser might be best. But you have to take into consideration that it is late in the semester. Not only will students have limited time because of term papers and final exams, but their funds likely will be depleted. That information will affect the type of fundraiser you will plan. But wait! At that point in the semester, students really enjoy having coffee and doughnuts available in the early morning, after all-night study sessions. And your group can sell lots of coffee and doughnuts to many students for very little money. As you can see, this example shows how ideas and questions do not simply emerge in a linear fashion. Instead interaction evolves in its own way, and group decisions will continually adjust for these changes.

The third principle is that many different types of influence affect a group's decision making. Group members' moods, motivations, competencies, and communication skills are individual-level variables that affect the group's final decision. These are individual-level variables because each member brings a unique set of influences to the group. The dynamics of the interpersonal relationships that result in group member cohesiveness and satisfaction also affect a group's decision making. Finally, the communication structure or network developed in the group impacts information flows among group members. The quality of information exchanged by group members affects a group's decision outcomes.

Additionally, forces outside a group can influence decision making. An example of this type of external influence is the generally accepted societal rule about making decisions quickly and cost-effectively.

The fourth principle of group decision making is that decisions are made within a system of external and internal constraints. Few groups have as much freedom of choice as they would like. Groups are constrained by external forces, such as deadlines or budgets imposed by outsiders and the preferences of the people who will evaluate or use the group's decision. Internal constraints are the values, morals, and ethics that individual members bring to the group setting. These values guide what the group does and how it does it.

These four principles reveal that decision making may be part of a larger problem-solving process. Problem solving is the communication group members engage in when there is a need to address an unsatisfactory situation or overcome some obstacle. Decision making and problem solving are often used interchangeably, but they are different. Decision making involves a choice between alternatives; problem solving represents the group's attempts to analyze a problem in detail so that effective decisions can be made (SunWolf & Seibold, 1999). Hence, this problem: Groups often make decisions without engaging in the analysis associated with problem solving. For anything but the simplest matters, groups are more likely to make faulty decisions when they do not take advantage of the problem-solving process to address contextual details.

Research has demonstrated that groups using formal discussion procedures generally develop higher member satisfaction and greater commitment to the decision. Yet many groups try to avoid using procedures. This is because discussion and decision-making procedures take time, and groups must plan their meetings accordingly. Group members often are reluctant to use procedures because they are unaccustomed to using them or initially find them too restrictive. It is often difficult for groups to stick with a procedure once it has been initiated.

Procedures help group members resist sloppy thinking and ineffective group habits (Poole, 1991). When procedures seem unnatural, it is often because group members have had little practice with them. When all group members know the procedure, it keeps the leader from assuming too much power and swaying the decision process. In addition, procedures help coordinate members' thinking and interaction, making it less likely that a group will go off topic. As member participation becomes more balanced, more voices are heard, and more ideas are deliberated.

SUMMARY

Decision making is a primary activity of many groups. To a large extent, our society depends on groups to make decisions—decisions that affect governmental and organizational policies, long-term policies, and day-to-day activities. Family and other social groups also make decisions. Across the variety of group decision-making situations, group members need task, relational, and procedural skills. Generally, groups are better decision makers than individuals when the problem is complex, when the problem requires input from diverse perspectives, and when people need to identify with and commit to the decision.

Functional theory of group decision making advocates five functions as necessary for effective decision making: understanding the problem, understanding what constitutes an acceptable choice, generating realistic and acceptable alternatives, assessing the positive qualities of each alternative, and assessing the negative qualities of each alternative. Groups whose communication fulfills all five functions are more effective in their decision making because information has been pooled and evaluated.

Decision-making procedures can help groups stay on track, equalize participation among members, and balance emotional and social aspects with task issues. Groups can choose from a variety of decision-making or discussion procedures. Brainstorming is an idea generation procedure that can help groups be creative in thinking of alternatives. The nominal group technique also assists the idea generation process, although it controls the timing of communication among group members. Consensus is a technique with wide application in group decision making. In this procedure, one group member helps facilitate the discussion and makes sure that all members have the opportunity to express their points of view. Voting is a popular procedure when groups must make their final selection from a set of alternatives. Like ranking, voting allows each member to equally affect the outcome.

Each procedure can help groups be more effective, but some procedures are better suited to different aspects of the decision-making process. Groups can compare procedures for their ability to satisfy the five critical functions according to functional theory.

Regardless of the procedures or process your group uses, four principles seem to fit most group problem-solving and decision-making situations. Acknowledging that group decision making is evolutionary and circular, and that there are multiple influences on decisions made within a larger context of constraints, allows us to embrace rather than fight the process. No one person or procedure will make group decision making effective. It is the appropriate selection and combination of people, talents, procedures, and structures that strengthens group decision making. By monitoring your own performance and the performance of the group as a whole, you will be able to select the most appropriate decision-making or discussion procedures.

DISCUSSION QUESTIONS AND EXERCISES

1 For one week, keep a diary or journal of all the group decisions in which you participate. Identify who in the group is making the decisions, what the decisions are about, how long the group spends on decision making, and what strengths or weaknesses exist in the decision-making process. Come to class ready to discuss your experiences and to identify procedures that could have helped your group be more effective.

DECISION MAKING

2. Using the data from your diary or journal (from the first item), write a paper that analyzes your role in the problem-solving and decision-making procedures your group uses and the ways in which your communication skills in that role influence your group's decision-making effectiveness.

3. Select someone you know who works full time in a profession you aspire to, and ask this person to participate in an interview about group or team decision making at work. Before the interview, develop a list of questions to guide the interview. You might include questions like these: How many decision-making groups or teams are you a part of? What is your role and what are your responsibilities in those groups and teams? How would you assess the effectiveness of your decision making? Is there something unusual (good or bad) that helps or hinders your groups' decision-making abilities? If you could change one thing about how your groups make decisions, what would it be?

4. Reflect on situations in which your family or group of friends made decisions. Using the five key components of a group—group size, interdependence of members, group identity, group goal, and group structure—describe how this social group was similar to or different from a work group or sports group to which you belonged. Selecting one decision event from both types of groups, compare the decision process in which the groups engaged. What task, relational, and procedural skills did you use to help the groups with their decision making? Which of these skills do you wish you would have used? Why?

 NAILING IT!

Using Group Communication Skills for Group Presentations

When deciding on how to present a group project, it will be important to incorporate everyone's input. There are many aspects of a group project that should be decided upon (e.g., who should present, should PowerPoint® be used, how do we grab the audience's attention). With all of these options to consider, what is the best approach for choosing a presentation format?

With help from Table 5.1, analyze which decision-making procedure would be most helpful when determining the format for your group's presentation. What is your rationale for making the selection? Is it possible that all procedures could be helpful?

Image Credits

Photo 5.1: Copyright © Sebastian ter Burg (CC by 2.0) at https://commons.wikimedia.org/wiki/File:GLAM_people_vote_on_the_next_GLAMwiki_conference_location_The_Netherlands.jpg.

CHAPTER 6

BUILDING RELATIONSHIPS IN GROUPS

After reading this chapter, you should be able to:

- Contribute to developing a supportive group communication climate
- Assist your group in developing cohesiveness
- Maximize your satisfaction in a group and help create satisfaction for other group members
- Maintain the trust of group members
- Socialize new members into a group

A group is a social context in which people come together to perform some task or activity. Just as group members work interdependently on a task or goal, they also create interdependent relationships. This chapter will explore the communication relationships among group members that help to form a group's communication climate and the degree to which communication contributes to cohesiveness and satisfaction among group members. The strength and resiliency of group member relationships is a critical factor in the equity and trust group members perceive in the group. Thus, how you build and maintain relationships with other group members will influence whether and how well the group completes its task.

GROUP COMMUNICATION CLIMATE

A group's **communication climate**, or the social atmosphere group members create, results from group members' use of verbal and nonverbal communication and their listening skills. Another way to think about group communication climate is to ask yourself: How do you feel about being a group member? Are you glad you are a member of this group or team? Do other group members communicate with you in a friendly manner? Assessing your group's communication climate will help you present your ideas and opinions in ways that make them more likely to be accepted. By paying attention to your presentation strategies, you are also likely to strengthen your relationships with others in the group and avoid unnecessary confrontation. Thus, a group's communication climate is the tone, mood, or character of the group that develops from the way group members interact and listen to one another.

 *PUTTING THE PIECES TOGETHER*

Group Identity and Group Structure

A group's communication climate is a critical factor in how members identify with and structure their group. Of the six supportive communication climate dimensions—description, problem orientation, spontaneity, empathy, equality, and provisionalism—which of these is most important to developing group identity? Which of these are more important to how norms, roles, and communication networks develop in the group? How have the dimensions you identified been instrumental to the development of your relationships with other members in your groups?

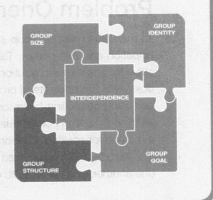

Early research on groups (Gibb, 1961) demonstrated that climates in groups range on a continuum from defensive to supportive. Not many of us would want to be in a group that has a **defensive climate**—a climate based on negative or threatening group interaction that discourages other members from communicating. Rather, a **supportive climate**, or a positive or encouraging environment, is more effective. But, not all groups and teams need the same level of supportiveness. For instance, counseling or support groups need a more supportive climate since they often include members who need encouragement to participate. An arbitration or legal group may not need as supportive of an environment, as lawyers are trained to stand up for their clients even if others disagree. All groups require a supportive climate, but groups do vary in the degree of supportiveness needed.

A group's supportive or positive communication climate is built on group members' use of six different types, or categories, of communication behavior: description, problem orientation, spontaneity, empathy, equality, and provisionalism.

Description

To arrive at a good idea, it is typical for group members to suggest many options for consideration. Not every idea is great, and when a group member suggests an idea that is considered less than ideal by other members, there is a tendency to evaluate the person rather than the idea. A more positive approach is to use **description**. Describing what is wrong with an idea gives the group member who introduced it the opportunity to clarify its presentation or to amend the idea for consideration by the group. Describing the idea may also encourage other group members to join in and help transform a poor idea into a better one. Thus, description is almost always preferable to evaluation because group members benefit when they know why an idea is rejected. **Evaluation**, or using language that criticizes others and their ideas, simply humiliates the person who offered the idea, and is likely to result in a defensive relationship and a defensive group climate.

Problem Orientation

To make the most of group situations, members need to exhibit the spirit of group participation and democracy. Taking a **problem orientation** approach, group members strive for answers and solutions that will benefit all group members and satisfy the group's objective. The opposite of problem orientation is **controlling behavior**. A group member using controlling behavior assumes there is a predetermined solution to be found. Alternately, asking honest questions like "Is the budget we're working with $250,000?" is a method for seeking collaboration on solving the problem. When group members adopt this attitude and it is reflected in their interactions, it is easier for them to cooperate with one another, and, as a result, develop group cohesion while completing their task.

Spontaneity

Group members who act with **spontaneity** are open and honest with other group members. These group members are known for their immediacy in the group and willingness to deal with issues as they come up. If a member is sincere and straightforward with others, the group members are likely to reciprocate that honesty and openness, which creates a more supportive communication climate for completing the group's task or activity. When a group member fakes sincerity or tries to hide motivations that could be hurtful to the group, the member is likely to be accused of being **strategic** or manipulative.

Empathy

We have all had bad days and not performed at our best in group situations. In a supportive group climate, other group members express empathy for our situation because we have expressed empathy for them. **Empathy** does not mean that other group members are excused from doing their assigned tasks; rather, it means that group members express genuine concern and are helpful if their help is requested. Empathic communication conveys members' respect for and reassurance of the receiver. Nonverbal behaviors are especially good at conveying empathy. For example, a smile, a kind gesture, and respect for someone's privacy are ways group members can express empathy for one another. However, when a group member reacts in a detached or unemotional way, the member is demonstrating **neutrality**. When group members react with a lack of warmth, other members often feel as if they are not important.

Equality

Groups are more likely to create supportive communication climates when **equality** is stressed. This does not mean that everyone does the same thing. Instead, it is important for group members to perform equitable work and to treat each other accordingly. Remember that trust and respect are earned and given incrementally. It is every group member's responsibility to work for trust and respect, and to give trust and respect when these are due. And because each group member must establish a relationship of trust with each other group member, creating a sense of trust within the group is a long-term, complex process. Using a respectful, polite tone and using the same kind of language to talk to different members will help in establishing equality. If members try to reinforce their **superiority**, or the belief that they are better than other group members, equality can be quickly diminished.

Provisionalism

To act with **provisionalism** is to be flexible. Rather than taking sides, provisional group members want to hear all ideas so they can make better, more informed choices. Provisionalism encourages group members to experiment with and explore ideas in the group. This creates an opportunity for group members to ask more questions, which can diffuse the dominance of one or two group members. Alternately, group members who believe they have all the answers or who *know* what another group member is going to say or do are communicating **certainty**. For instance, certainty is revealed by a group member who cuts off the attempts of other members to provide more information on a certain alternative.

Why does creating a positive or supportive group communication climate matter? First, when a group has a more positive or constructive communication climate, group members are more willing to seek and share information, which in turn creates members' commitment to the group (van den Hooff & de Ridder, 2004). A positive group communication climate is, simply, more open, which encourages group members to be more creative in working on their tasks or goals (Shin, 2014). It's not surprising then that a supportive communication climate is positively related to a group's performance (Liu, Hartel, & Sun, 2014).

Changing Group Climate

How can you help your group develop a more supportive climate? The important first step is to monitor your behavior and adopt more positive behaviors based on the six previously mentioned categories. The behaviors that create a supportive communication climate require you to assess the group situation, think about what you want to say, and evaluate your statements for their potential impact on other group members. Of course, you must follow through. Good intentions alone do not lead to a supportive communication climate. Use description rather than evaluation to assess the input of others. Create opportunities for all group members to participate. Be open and honest, but tactful. Express empathy for others in the group. Create a sense of equality through equitable assignments and responsibilities. Finally, be flexible and open. If you adopt some of these supportive interaction strategies, you will strengthen your relationships in groups and help your groups achieve more effective outcomes.

A second way to help your group develop a more supportive climate is to monitor your reaction to the interaction of other group members and respond to them in a more productive fashion. For example, an effective way to neutralize superiority is to respond with equality. It is every group member's responsibility to break destructive cycles in group interaction, and the development and maintenance of a defensive group climate is one such destructive cycle. Communication climate is changeable, but first group members must recognize what communication habits have been established and which ones should be changed. By using supportive communication, group members can create a more positive climate. When defensive strategies are present in a group, tolerating those behaviors can indirectly reinforce their use.

GROUP COHESIVENESS

Cohesiveness is the degree to which members desire to remain in the group. When the desire to be a group member is strong across all members, they are more likely to demonstrate cohesion and be committed to the group's task (Whitton & Fletcher, 2014). But cohesiveness is an elusive concept. Most people can sense when it exists within a group. But cohesion can manifest differently depending on the group or team context.

 MESSAGE AND MEANING

A nonprofit organization, FreeInternetForAll.net, was having their weekly team meeting. At the end of the meeting, Fred, the team leader, went around the room as he often did, asking everyone for a quick update on their work.

FRED: Alright. Joel, where are you at?

JOEL: Let's see. What did I do this week? Well, the big thing I did this week and that no one noticed, and, well, that's fine, I'll just deal with it, but ...

CHRISTY: Hey, if you do your job right, we shouldn't notice, right?

JOEL: Exactly, exactly. [laughter]

FRED: Yeah, if he's doing his job right, we shouldn't notice [laughter].

JOEL: That's right, not a big deal. But this week I created ...

Following this exchange, Joel proceeded to give the details on his accomplishment during the week. However, no one had previously seemed to notice that Joel had done something of significance during the week, and he decided to use a joke to inform the group. Do you think that Joel wanted others to recognize his work from last week? Was his use of joking and laughter a way of receiving recognition, or was it simply a funny way of giving his work report? Group members may have to think of creative ways to receive recognition for one's efforts, because by simply announcing a good deed the group member may come off as arrogant. However, the other group members communicatively interpreted it as a joke. Do you think they recognized that Joel wanted people to notice his work? The use of joking and laughter can be a tricky type of relational communication (Keyton & Beck, 2010).

There are three general ways that cohesiveness develops (Carron & Brawley, 2000). Some types of group are based on belonging. Family and friendship groups are good examples. Likewise, sorority and fraternity groups also display high cohesion because membership is controlled or invited. The sense of belonging that is driven by the feeling or desire to be a member creates cohesion. Second, cohesion can develop when groups are completing highly interdependent work that requires coordination. Volunteer groups building houses for Habitat for Humanity are a good example of this, as are work groups that meet regularly to finish complex projects on a tight deadline and within a tight budget. It takes everyone's skills and expertise to finish these tasks successfully. Third, cohesiveness can develop when individuals are attracted to a group. For example, individuals are drawn to community action groups or groups that espouse certain values or beliefs. Members are attracted because being a member of this group provides an opportunity to interact with others who share their beliefs and values.

Thus, group cohesion is a complex, multidimensional construct. Cohesion in one group may not look or feel like cohesion in other groups. As Pescosolido and Saavedra (2012) describe, "members of a hockey team may develop greater understanding of shared strategies and tactics over a season, members of a group therapy session may develop feelings of acceptance and a sense of belonging, and members of a business unit may develop a sense of shared responsibility and success" (p. 753). Thus, we distinguish between social cohesion, which is based on the quality of relationships among group members, and task cohesion, which is based on commitment to the group's task (Hackman, 1992).

Importantly, balancing social and task cohesion may be tricky in certain types of groups. Although online groups are commonplace, it may be more difficult to create social cohesion by communicating online. This is not to say that it can't happen, but it may require extra effort or longer periods of time to develop. Intentional efforts to communicate relationally (which may be more natural in face-to-face groups) may be required, such as an online session specifically devoted to getting to know one another or communicating via social media (e.g., tweeting with one another; becoming a Facebook friend). Online platforms may be better geared toward social (Facebook) or task (Google Drive) purposes, and combined, these platforms may improve both types of cohesion.

How group members communicate with one another is a good indicator as to whether cohesion exists. For example, when group members use the same or similar types of words at the same rate, especially words that provide structure for conversation, cohesiveness among members is higher. Prepositions (e.g., at, for, into), conjunctions (e.g., and, but, or), and pronouns (e.g., we, us) help speakers organize and present their thoughts in ways that generalize across content, context, or channel of communicating. Why is this the case? Communication requires coordination, and one measure of coordination is the degree to which group members use the same or similar words. In essence, when group members like one another they tend to use the same rates of prepositions, conjunctions, and articles. When a group member mimics the style of speaking of other group members, cohesiveness tends to be higher (Gonzales, Hancock, & Pennebaker, 2010).

One group member cannot build cohesiveness alone, but one member's actions can destroy the cohesiveness of the group. Although we say that the group is cohesive, cohesiveness actually results from the psychological closeness individual group members feel toward the group. Cohesiveness can be built around interpersonal attraction to other members, attraction to the task, coordination of effort, and member motivation to work on behalf of the group (Golembiewski, 1962). That is, when group members believe that their task and relational needs are being fulfilled, they perceive the group as cohesive (Carron et al., 2004).

Sometimes people refer to cohesiveness as the glue that keeps the group together; others describe cohesiveness as the morale of the group. In either case, cohesiveness serves to keep group members together because of their attraction to the group. There are three specific advantages to building and maintaining a cohesive group. First, members feel that they are a part of the group. Second, cohesiveness acts as a bonding agent for group members. Members of cohesive groups are more likely to stick with the group throughout the duration of its task (Spink & Carron, 1994). This, in turn, creates more opportunities for norms to be developed and followed (Shaw, 1981). Third, cohesive groups develop a "we" climate, not an "I" climate. In fact, one way to know if group members are close is whether they start using plural pronouns like "we" and "us."

Cohesiveness and Group Performance

Research has reportedly demonstrated that communication is the central binding force of team and group activities (Weick, 1969). When group and team members use cooperative communication practices, such as information exchange, opinion sharing, and agreement seeking, group cohesion is strongest (Bakar & Sheer, 2013). Simply stated: The more cohesive the group, the more likely the group is to perform effectively (Cohen & Bailey, 1997). This relationship between cohesiveness and performance is often reciprocal (Greene, 1989). When an ongoing group performs well, it is also likely that its members will generate additional cohesive feelings for one another or at least maintain the current level of cohesiveness in the group. Cohesiveness in a group can also affect individual group member performance. Members of groups with high task cohesiveness put more energy into working with and for the group (Prapavessis & Carron, 1997). In other words, individual members have greater adherence to the team task when the group's task is attractive to them. Thus, group members are successful because they have helped the group become successful.

However, the relationship between cohesiveness and performance is not a straightforward one. The degree of interdependence needed to perform a group's task affects the cohesiveness-performance relationship (Gully, Devine, & Whitney, 1995). When a group task requires coordination, high levels of interaction, and joint performances from group members, the cohesiveness-performance relationship is stronger. But when task interdependence is low, the cohesiveness-performance relationship is much weaker. An emergency response team of firefighters and medical personnel is an example of a group that requires a high degree of interdependence to successfully complete its task. The coordination and communication efforts during a rescue are very high. Thus, the more cohesive the group, the more effective the team will be in its tasks, because high cohesiveness also motivates individual group members to perform well.

Alternately, tasks with low interdependence provide less opportunity for members to communicate and to coordinate their actions. Even if cohesiveness develops, there is

less opportunity for it to be demonstrated and for it to affect group performance. A sales team is a good example of how low interdependence and low cohesiveness-performance interact. The salesperson approaches a client while the service manager makes the follow-up call. Back at the office, two administrative assistants talk the new client through the initial steps of filing forms and preparing documents. Once the goods have been delivered, the salesperson steps back in to say hello. This task has low interdependence because the steps of the task are not done simultaneously and because team members are responsible for unique tasks. Thus, cohesiveness is likely to be low in these situations.

What benefits can cohesive groups expect? Members of cohesive groups are less likely to leave to join other groups. For groups with long-term goals, this can be an especially important benefit because the group does not have to spend time finding, attracting, and integrating new members. More cohesive groups also exert greater influence over their members. Thus, norms are less likely to be violated because cohesiveness exists. In addition, this level of influence encourages group members to more readily accept group goals and tasks. Generally, there is greater equality in participation in cohesive groups because members want to express their identity and solidarity with the group (Cartwright, 1968).

Can a group ever be too cohesive? Yes. Some groups can develop such a high level of social cohesiveness that new problems appear. For example, giving constructive criticism to one another becomes difficult when social cohesion is high. Other problems associated with high social cohesion are: (a) the group's focus on socializing overshadows their work on their task or goal, (b) group members feeling isolated because they feel left out of a social clique that is developing in the group, and (c) a general reduction in commitment to the group's task.

When a group develops very high levels of task cohesion, other problems develop. These include: (a) working on the task overshadows the development of group member relationships; (b) communication among group members can be taken the wrong way, as social relationships are not well developed; (c) focusing on the goal or tasks diminishes enjoyment normally associated with it; (d) the development of perceptions that some members are overly serious about the task, and (e) group members feeling too much pressure about their performance and the group's goal achievement (Hardy, Eys, & Carron, 2005).

What group members consider to be the optimum level of cohesiveness will vary depending on the group's task or activity and the type of attraction group members hold for the group. When cohesiveness is based on interpersonal attraction, groups are more susceptible to **groupthink**, a condition that occurs when group members' desires for harmony and conformity results in dysfunctional and ineffective decision making. When cohesiveness is based on task attraction, groups are less susceptible to these deficiencies (Mullen, Anthony, Salas, & Driskell, 1994). When your group has strong interpersonal relationships and cohesiveness is high, the group leader or facilitator may want to take extra precautions to prevent groupthink from developing.

GROUP MEMBER SATISFACTION

Closely related to cohesiveness is satisfaction with the group. **Satisfaction** is the degree to which you feel fulfilled or gratified as a group member; it is an attitude you express based on your experiences in the group. When your satisfaction is high, you are likely to feel content with the group situation. As an individual group member, you perceive some things about the group as satisfying (such as being assigned to the role you requested), but you may also perceive some group elements as dissatisfying (for instance, having to meet too frequently).

The types of things that satisfy individuals in group settings are quite different from those that cause dissatisfaction (Keyton, 1991). As long as the group is moving along its expected path, group members are likely to be satisfied. This occurs, for example, when group members feel free to participate in the group, when they feel that their time is well spent, and when their group interaction is comfortable and effective. Alternatively, dissatisfaction develops when group members spend too much time on activities that do not contribute to task completion, when the group lacks organization, and when members display little patience. Thus, dissatisfaction is more likely to result from negative

THEORY STANDOUT

Satisfaction and Dissatisfaction

The word *satisfaction* is frequently used as group members' evaluation of what is happening in their group or team. But satisfaction is more complicated than that. According to Keyton (1991), satisfaction is a global factor, as well as situationally bound. When a group is moving along as expected, satisfaction is a positive and global affect evaluation. When a group is not moving along very well, then group members start to anchor on specific negative things that dissatisfy them. Thus, what dissatisfies us in a group or team is not simply the opposite of what satisfies us. Take a look at these statements and see if you agree. For example, Keyton suggests that group member participation is expected. Therefore, participation does not help in developing satisfaction. But, when members do not participate, dissatisfaction occurs.

Global Satisfiers
1. Everyone seems genuinely interested in getting something accomplished.
2. Group members can provide constructive criticism to others.
3. Group members interact well with one another.

Global Dissatisfiers
4. Not everyone in my group is participating.
5. It is difficult for my group to come to a decision.
6. My group gets sidetracked by distractions.

Situational Satisfiers
7. There is diversity of ideas among my group members.
8. The individual personalities of the group members do not clash.

Situational Dissatisfiers
9. My group has too many people making the decisions.
10. Interaction roles have not been established.
11. My group can come to an agreement, but getting there is frustrating.

Think of two very different groups and evaluate each group on the 11 items above. Do you see evidence that some things satisfy or dissatisfy you regardless of the group? Do you see evidence that what satisfies and dissatisfies you in one group does not in the other?

assessments you make about the group as a whole (such as the perception that the group is in chaos) than from an evaluation of your individual interaction opportunities.

As with cohesiveness, your satisfaction with the group is partially based on communication behaviors, task elements, or a combination of both (Witteman, 1991). The use of verbal immediacy behaviors by group members results in higher group member satisfaction (Turman, 2008). Examples of verbal immediacy behaviors include: asking questions that invite other group members to give their opinion, encouraging others to respond, using other members' first names, and engaging in interpersonal interaction before and after the group meets. With respect to group tasks, when you are satisfied with the activity of the group, you try harder to communicate more effectively. As a result, you are satisfied with communication within the group.

How a group handles its conflict also affects member satisfaction. Members of groups that identify viable solutions to conflict have greater satisfaction than members of groups that avoid conflict. Thus, even groups that experience conflict can have a satisfying group experience, especially when group members use these behaviors to: (a) make direct statements about the conflict rather than try to avoid it, (b) work to find a solution by integrating the ideas of all group members, and (c) demonstrate flexibility, which demonstrates goodwill toward other group members. Note that it is how group members handle conflict that influences group member satisfaction, not its absence. Finally, group members who are satisfied are more likely to attribute credit to other team members rather than to themselves (Behfar, Friedman, & Oh, 2016).

Improving Cohesiveness and Satisfaction

Building cohesiveness and satisfaction in a group cannot be accomplished alone. But there are some tactics you can undertake as an individual to help the process along. For example, you can adapt and monitor your communication so that your interaction encourages positive climate building in the group. Your interaction should be more supportive than defensive. You can encourage your group to celebrate its successes; this creates a history and tradition for the group. But do not wait until the project is over; each time the group accomplishes some part of the task, recognize the achievement. If you facilitate or lead the group, adopt a reward system that encourages all members, and not just a few, to participate. Basing rewards on group output rather than individual output builds cohesiveness. To make this work, however, group goals must truly be group goals. Additionally, group members should have input in developing goals. Each time the group gets together, group members should be aware of how their communication and activities contribute to the pursuit of these goals.

There are three cautions in developing closeness in groups. First, groups that are not cohesive and in which members are not satisfied are unlikely to produce positive outcomes. But high levels of cohesiveness and satisfaction among group members do not always lead to acceptable output (McGrath, 1984). Cohesiveness in a group can be so high that members overlook tasks in favor of having fun. Cohesiveness can also insulate a group, making it less able to fully explore its task or options. Instead of making a group more vigilant, overly high cohesiveness among group members can make the group susceptible to faulty thinking. With respect to satisfaction, group members may be satisfied because they like one another and as a result become focused on the relational aspects of the group while minimizing their attention to tasks.

Many people want to believe that cohesiveness and satisfaction are so tightly related that, as one increases, so must the other. This is the second caution. If group members become overly cohesive, they may start to reject or ignore their task. For instance, if you are attracted to a genealogy group because you want to learn more about how to research your family's history, you will probably not be very satisfied if the group regularly focuses its conversations on other topics. Although the cohesiveness of the group might enhance the discussion of any topic, your satisfaction with the group may actually decrease because you are not accomplishing your goal.

The third caution involves group size. When a group becomes too large, creating greater complexity than can be handled by the communication structure of the group, there are fewer interaction opportunities among group members. This diminishes cohesiveness and satisfaction. Both cohesiveness and satisfaction develop from the opportunity for members to interact on a regular basis. Moreover, developing cohesion and satisfaction is worth the effort. When group members are comfortable enough with one another, they perceive each other as friendly peers and colleagues, and, as a result, disclose more personal information with one another. In turn, these group members experience greater cohesion and satisfaction. As importantly, they are more favorable toward the group and learn more as they work on their task (Myers et al., 2010).

TRUST

You have probably heard that trust is earned, that trust is not automatic or given freely to others. **Trust** is a group member's positive expectation of another group member, or a group member's willingness to rely on another's actions in a risky situation (Lewicki, McAllister, & Bies, 1998). In other words, when we trust someone, we expect that they will be helpful, or at least not harmful. Thus, trust resides in one group member's relationship with another group member, is based on previous experience with one another, and develops over time as relationships unfold and confidence builds. When you trust another group member, it helps you predict how this group member will behave or react.

Trust can develop early in a group's interaction, especially when group members have some history or familiarity with one another. At that point, however, trust is not very complex. As group members interact more, two other types of trust emerge: affective trust and cognitive trust (Webber, 2008). **Affective trust** is based on the interpersonal concern group members show for one another outside of the group task. This type of trust is enduring and significantly influences how interpersonal relationships among group members develop. Not surprisingly, affective trust is demonstrated through group members' commitment and communication to and with one another. Alternatively, **cognitive trust** is based on group members' assessments about the reliability, dependability, and competence of other group members while they are working on the task. Interestingly, cognitive trust only develops if group members develop a significant level of early trust and group members demonstrate reliability in their task performances for the group.

BUILDING RELATIONSHIPS IN GROUPS

As you can see, trust is not easy to establish. Moreover, as a group member, trust is established with each group member, which is why establishing trust in group settings takes such a long time. Trust is also extended slowly and incrementally. For example, if Russ trusts Sonya a little, this will often lead Sonya to trust Russ a little in return. If Russ begins to feel comfortable with how his relationship with Sonya is developing (and Sonya is not explicitly communicating anything to the contrary), then Russ may extend a deeper level of trust to Sonya. Extending trust is risky, which is why we are unwilling to give full trust to new people in new settings. An individual's level of trust for all group members will collectively determine how the individual feels about the group as a whole.

 SKILL BUILDER

How Fragile Is Trust?

Reflect on one of your current group experiences. What is the level of trust in that group? Specifically recall behaviors you have used to help develop or maintain trust within the group. Have you done enough to reinforce trust among group members? What else could you do to strengthen the level of trust among group members? Consider the following: Do you always follow through on commitments you make to the group? Have you ever broken a confidence inside or outside the group? Do you ever withhold information from other members? Have you ever indicated that you would do one thing for the group and then did something else? Identify what you would do and communicate to sustain or enhance trust among group members. What communication skills are most central to sustaining and maintaining trust?

Trust is also multifaceted, based on honesty, openness, consistency, and respect (Larson & LaFasto, 1989). As you might suspect, it is difficult to trust a group member who is not honest. It is also difficult to trust a group member who is not open. Sharing part of yourself with the group by revealing personal or professional information helps others to get to know you. Not only must you share with others, you must also be receptive to receiving personal and professional information about other group members. Openness cannot be one-sided. Being consistent in your group's interactions helps others understand you as well. When you are inconsistent in your interactions with other group members, they could become hesitant around you because they are never quite sure how you will react.

Finally, trust is based upon respect. It is hard to trust someone who does not respect others or who acts or communicates in ways that do not deserve respect. In a group, our behaviors and interactions are always being evaluated by other group members. If you tell offensive jokes about someone when that person is not present, other group members might assume that you tell jokes about them when they are not around too.

How can you build trust in groups? First, be aware of your communication style in the group and work to minimize your apprehensiveness. This will increase your ability to develop positive interactions with others. Second, use the supportive climate interaction

characteristics discussed earlier. If you interact in a defensive manner, it is unlikely that others will extend trust to you. Third, use appropriate self-disclosure. We have all met people who tell us more than we want to know about them in our initial meeting with them. Extend only the personal and professional information about yourself that you believe will be perceived by others as positive contributions to the group. As group members warm up to one another, self-disclosure often becomes more personal. Remember, however, that in decision-making and other task groups, revealing too much personal information may be considered unprofessional. Moreover, personal information can be used against you—once you reveal it, you lose power over the information. Fourth, focus on developing a positive and collaborative climate with *all* group members. For instance, Devon may resist extending his trust to you because you treat Maggie more favorably than other group members. Finally, monitor your interaction behavior to ensure that you are not overusing defensive behaviors or behaving in such a way that other members label you as dysfunctional. Trust is seldom extended to group members who are perceived negatively. "Skill Builder Feature: How Fragile Is Trust?" explores the effect of lack of trust on group interaction. These recommendations will help you build affective trust with your team members. But it is not enough. You must also build cognitive trust by performing well and completing the tasks you agreed to in order to help the group be successful.

Socializing New Members Into Groups

Socialization is the process that new and existing group members go through when a new member joins a group. New members want to both fit in and to adapt to the group and its goal. New members also need feedback from other group members so they can reduce the uncertainties that being new in a group brings (Riddle, Anderson, & Martin, 2000). Even if the new members are joining a group for which they have task familiarity, they must be introduced to the other group members and become comfortable with their role in this particular group (Kramer, 2011).

Sometimes membership changes are temporary, as when a factory worker takes a 2-week vacation and is replaced by a utility person who floats among team assignments. At other times, membership changes are permanent. Members may leave the group for a variety of reasons (e.g., quitting, retiring, transferring), or they may simply grow tired of the group and drop out. Each of these membership changes is member initiated.

Membership change in groups can be critical because a change in the composition of the group also changes the cluster of knowledge, skills, and abilities within the group (McGrath, Berdahl, & Arrow, 1995) as well as the relationships among group members. Let's explore some of the issues surrounding membership change to assess its impact on groups (Arrow & McGrath, 1993). First, when membership change increases or decreases the number of members in the group, other aspects of the group—group roles, norms, and communication networks—must change as well.

For example, Kara's honors study group meets with its honors advisor to work on a research project. For over a year, the group—the professor and four students—has met in the professor's office. After much discussion about whether to add other students, the group decides that adding one more student will help lighten the workload of running the experiments. Unfortunately, the group does not think about how an additional group member will affect the group's meetings. The professor's office is crowded with stacks of books and computer printouts. There is a couch that can seat only three people, as well as a visitor's chair and the professor's chair. When the group with its new member meet for the first time, someone has to sit on the arm of the couch. This seating arrangement creates awkward dynamics among group members and impedes the group's ability to work together coding data. Certainly, the group could move to a new meeting location, but that would cause another disruption in the group. Although adding a member seems to be a gain for the group because it will lighten the workload, it also creates an unexpected negative consequence.

Second, it is important to know why there is a change in group membership. Group members react differently to situations in which membership change is member initiated, and not controlled by someone outside the group. You are more likely to accept a new member who persuades you to let her join than a member who joins because your supervisor says she must. Sometimes groups actually recruit new members because they need skills that group members lack. Does a group ever purposely change members? Yes. Sometimes a group member creates a logjam, making it difficult for the group to accomplish anything (Cohen, 1990). Groups can get stuck when strong, self-oriented individuals are in the leadership role, and groups can fail to meet their potential if there is a weak link. Changes to replace ineffective members are made deliberately to help the group out of its entrenched patterns. We expect professional sports teams to use this strategy, and we should want our groups to do the same.

The timing of a change in group membership is also important. Most of you expect that your membership in classroom groups will remain stable throughout the term or the course. What will happen to your group and how will you respond if 3 weeks after your group forms, your instructor adds a student to your group? Making changes after your group has formed and while members are developing a group identity and structure will likely be disruptive.

The frequency with which groups change membership influences the group's ability to chart a course for itself, and can reflect poorly on the group's leadership. For example, an executive team that cannot hold any person in the position of executive assistant for longer than 6 months makes you wonder more about the executive team and less about the individuals rotating through the administrative role. Groups that have regular turnover like this are often questioned about members' ability to work together. The assumption is that their inability to do so effectively drives off the executive assistants.

Fourth, it is important to know which members are leaving, because group members are not interchangeable. A group that loses an effective leader will experience disruption and frustration. Sometimes another group member can assume that role and take on new responsibilities; other times, however, a group role must be filled. For example, your relay swim team relies on the swimmer who can assess how the other teams are doing in relationship to your team and then really kick in for a quick finish. Thus, if you lose the member who normally swims the anchor or final leg of the relay, your group has a specific role to fill.

What is affected by changes in membership (Arrow & McGrath, 1993)? Obviously, membership dynamics and relationships will change. Any change in group membership will alter to some degree the interactions among group members. Not only does the structure and the process of the group change, but members' performance is also likely to be affected. The more interdependence among members of the team, the more the team will feel the effects of membership changes. And the more central the member is to the team, the more the team will feel the effects of the change.

Group members build a history together, and each group develops a memory of how and why it does certain things in certain ways. At the very minimum, a new group member will be unfamiliar with a group's habits and routines (Gersick & Hackman, 1990). These need to be explained, or the new group member will feel left out. And the new group member cannot share in the memory of the group—there is simply no way for the new group member to know what it feels like to be a part of this group (McGrath et al., 1995). Assumptions or existing knowledge that other group members use in making decisions simply are not available or do not make sense to the new group member.

Also, there is no guarantee that current members will be able to make the behavioral adjustments needed when a new group member enters. However, a group can overcome these effects by realizing that membership change actually creates a new group. The group must allow members time to resocialize and to reidentify the role structure. The challenges of membership change are (a) to initiate the new members into the team, (b) to learn from the new member's fresh perspective, and (c) not to sacrifice the pace and focus of the team (Katzenbach & Smith, 1993).

When group membership changes, socialization, or the reciprocal process of social influence and change by both newcomers and established members, occurs (Anderson, Riddle, & Martin, 1999). Group socialization is an ongoing process, but it becomes especially salient to group members anytime an established group finds itself creating or re-creating itself or its activities.

Obviously, group members do not join a group as a blank slate. Individuals bring knowledge, beliefs, and attitudes about group work with them, just as they bring their own motivations for joining the group and their own sets of communication skills. The socialization process really commences as an individual begins to anticipate what it will be like to be a member of a particular group. For example, you are thrilled that you were selected to be a member of your university's ambassador team but also a little concerned about how you will blend in with students who have already served on this team for several years. And even as you are anticipating joining the team, the existing team members also have anticipatory expectations about you. They might be wondering if you'll be able to devote the time these group activities demand. They might also be hoping that you'll live up to the performance level you claimed you could achieve in the interview.

The first day you meet with the ambassador team, you and the other members begin to adjust to one another, and to negotiate the formal and informal roles, norms, and communication networks of the group. There are many ways to introduce yourself to other group members. If you get there early, you can greet the other members as they arrive. Always be sure to introduce yourself to anyone you do not know well. When you do this, you help establish a friendly and supportive climate and create a sense of openness to which others will respond.

At your first meeting and in subsequent encounters with other group members, you need to continue to assimilate yourself into the group. One way to get to know others is to sit next to someone you do not know at all or do not know well. You can start a conversation by bringing up an easy topic (weekend activities, hobbies, your role in this group). The objective here is to create an opportunity for interaction that will help you get to know the other person. But the topic of conversation should not be threatening or invasive.

Now, as a member of the team, you are influencing the group, just as the group and its other members are influencing you. You are fully integrated, or assimilated, into the group culture when you and the other team members establish a shared group identity by working effectively and interdependently within the group's structures toward a common group goal.

Of course, current members should also help to welcome and socialize new members into a group. During a meeting, you can assimilate new group members by asking them to comment on what the group is talking about ("Michael, what do you think of this plan?"). This is especially important because as new members they may believe their opinions are not welcomed, given that they have little or no history with the group. Creating opportunities for new members to contribute to the group's task or activity also provides a mechanism for them to create interdependence with other group members, from which relationships flow.

SUMMARY

Working in groups means building relationships with others. You are likely to develop dependent relationships with some group members, but be careful. Dependent relationships create superior-subordinate interaction. Alternatively, interdependent relationships are those that are more equally balanced. When group members perceive interdependence with one another, they can use communication to resolve differences. Interdependence also creates a seamless communication flow among group members.

Group communication climate is the atmosphere group members create from the content, tone, mood, and character of their verbal and nonverbal messages. A group will develop a defensive or negative climate if group members send messages that are based on evaluation, control, strategy, neutrality, superiority, and certainty. Alternatively, a group will develop a supportive or positive climate if group members send messages that are based on description, problem orientation, spontaneity, empathy, equality, and provisionalism. While groups will vary in their need for a supportive climate, all groups require some level of supportiveness to maintain interdependence among members to accomplish the group task or activity.

When a group is cohesive, members want to remain in the group, whether they are attracted to the group's task or to group members. Group cohesiveness contributes to the development of interdependence. Cohesive groups often perform more effectively, but cohesiveness does not ensure good group performance. In fact, too much cohesiveness can actually interfere with the group's ability to critically examine alternatives. Like cohesiveness, satisfaction develops from interdependent relationships. Members are satisfied when they are fulfilled or gratified by the group, particularly when the group is moving along its expected path. When group members are satisfied, they are willing to work harder and be more committed to the group.

Trust is a group member's positive expectation of another group member, or a group member's willingness to rely on another's actions in a risky situation. Thus, trust resides in group member relationships, and is developed as relationships unfold over time and individuals build confidence in each other. Trust is based on honesty, openness, consistency, and respect. Group members must earn trust in one another through their interactions. Trust is fragile; once broken, it is hard to reinstate.

Building relationships with group members is particularly important when a new member joins an established group, thereby changing the group's cluster of knowledge, skills, and abilities, as well as the relationships among group members. Socialization, or the reciprocal process of social influence and change by both newcomers and established members, occurs any time membership changes in a group. Both new and established members are responsible for making the transition a smooth one.

BUILDING RELATIONSHIPS IN GROUPS 111

DISCUSSION QUESTIONS AND EXERCISES

1. Think of a recent group experience, and write a short essay analyzing the communication climate within the group. Give specific examples to support your conclusions. Provide three recommendations for maintaining or increasing a supportive climate.

2. You've been asked to be a consultant to a team at work that is experiencing relational disharmony. What advice will you give this group for developing and maintaining positive and productive group member relationships?

3. Describe three things that you see as evidence of group member cohesiveness and satisfaction in one of your groups. Do these descriptions apply to all groups, only to groups like yours, or only to your group?

4. Think back to one group in which trust was high and to another group in which trust was low. Write a short paper responding to these questions: What accounted for the difference in level of trust? How did the level of trust affect the group's communication? How did one group build trust? What happened in the other group to erode trust?

5. Based upon your experiences in joining groups and teams, what advice would you give to a friend who is joining an established sports team? Changing teams at work? Welcoming a new member into the family? Would your advice be similar or different across these three contexts? Why or why not?

 NAILING IT!

Using Group Communication Skills for Group Presentations

Your group presentation is not fully developed because a group member has dropped out of the project and your supervisor recommends that Emma join your group. Because you are the only member of the group who knows Emma, you make a mental list of behaviors to welcome your new group member and to help her become socialized into your group. As other members arrive, you facilitate introductions between them and Emma. After introductions, you briefly describe the objectives, timeline, and audience for the group's presentation. Then you invite Emma to ask questions. You are careful not to answer her questions, which allows Emma to begin relationship building with the other group members. Finally, you ask Emma to talk about her strengths in developing and giving presentations. To your delight, Emma describes her experience with different types of data visualization tools—a skill no one else in the group possesses. As the meeting winds down, you feel confident that your group handled this membership transition well, and that rather than hurt the group's presentation, adding Emma is a bonus!

Image Credits

Figure 6.1: Adapted from: Copyright © Depositphotos/megastocker.
Photo 6.1: Copyright © Depositphotos/vilevi.
Photo 6.2: Copyright © Depositphotos/Blulz60.

CHAPTER 7

MANAGING CONFLICT IN GROUPS

After reading this chapter, you should be able to:

- Explain the advantages and disadvantages of group conflict
- Identify types and sources of conflict
- Recognize the relationship between power and conflict
- Identify five conflict management strategies
- Help your group engage in an effective conflict management strategy
- Avoid or respond to nonproductive conflict management strategies

Seldom do individuals—each with unique experiences, perspectives, knowledge, skills, values, and expectations—develop into a group or team without experiencing conflict. In fact, the more complex the group task or activity, the more likely conflict is to occur. Although most people don't like conflict, nearly every group experiences it. Conflict is inherent in group situations because incompatibilities exist among group members. At its worst, conflict interferes with task coordination, creates opportunities for dysfunctional power struggles, and disrupts the social network and communication climate among group members. At its best, conflict can help groups find creative solutions to problems and increase rationality in decision making.

DEFINING CONFLICT

In current society, we use the word *conflict* regularly, often not thinking about what it really means. Conflict results from incompatible activities (Deutsch, 1969). To be in a **conflict** means that at least two interdependent parties (individuals or groups) capable of invoking sanctions oppose each other. In other words, one party believes that the other party has some real or perceived power over it or can threaten it in some way, or that the other party will use that power, real or perceived, to keep it from reaching its goal. Generally, conflicting parties have different value systems or perceive the same issue differently, thus creating incompatibilities (Jehn, 1995).

Communication is central to conflict as it is through group members' verbal messages and nonverbal expressions that other group members understand that conflict exists. As a process, "communication shapes the very nature of conflict through the evolution of social interaction" (Putnam, 2006, p. 13). Thus, (a) communication within a group can cause conflict, (b) conflict develops from patterns of interactions among group members, and (c) communication among group members can resolve conflict.

PERSPECTIVES ON CONFLICT

There are three perspectives for understanding why conflict occurs in groups (Poole & Garner, 2006). An **instrumental perspective on conflict** is about a group's performance and outcomes. From this perspective, conflict can be productive or destructive. Productive conflict is that which helps group members figure out what the goals are and how to accomplish its tasks. Destructive conflict impedes the group and keeps the members from reaching their goals, as this type of conflict stems from relational problems, or the individual goals of group members. A **developmental perspective on conflict** treats conflict among group members as a natural part of the group's development and results from the typical challenges and dilemmas any group must work through to achieve their goal. In this perspective, conflict is productive if its emergence and the group's resolution helps the group move forward. Alternatively, conflict is destructive if a group does not allow conflict to emerge. Conflict is also destructive from this perspective if conflict does emerge, but group members get stuck in the conflict and cannot move forward. A **political perspective on conflict** situates a struggle for power as the source of conflict. When these struggles allow alternative points of view to surface, then conflict can be productive. However, if a conflict only reaffirms or strengthens the dominant member or members in the group, then the conflict is destructive. These three perspectives provide different vantage points for describing and analyzing conflict in a group.

Regardless of the perspective, conflict is a process that occurs over time or in a sequence of events (Thomas, 1992). A conflict starts when one or all group members realize that an incompatibility exists. Group members may then frame conflict by identifying what the conflict is over, who the conflict is with, and what the odds are for succeeding. The conflict continues with group members interacting, each trying to obtain his or her goal. Here is where a conflict can become really interesting, because no matter how well you think through a conflict strategy or how well you rehearse what you are going to say, you can never fully predict what the other person will say or do. Thus, you have to adjust your intentions and behaviors during the conflict management interaction.

Fortunately, most conflicts have an end. This occurs when each party is satisfied with what it won or lost or when those involved believe that the costs of continuing the conflict outweigh the benefits of continuing the conflict. Regardless of the outcome, it's important to realize that one conflict episode is connected to the next one. How you evaluate the outcome of the first conflict episode will affect your awareness of the next (potential) conflict. Suppose you are in a conflict with another person and believe that the two of you have reached a mutually agreeable decision. The next time you engage in conflict with that person, you will expect to again reach an agreeable outcome. Alternatively, if you believe the decision was unfair, you may harbor negative emotions that will influence your interaction the next time you have a conflict. In other words, your awareness of conflict in subsequent episodes is heightened by your sense of success or loss in the first one.

The feelings that result from conflict are the **conflict aftermath** (Pondy, 1967). In other words, each conflict leaves a legacy. If the conflict results in a positive outcome, members are likely to feel motivated and enthusiastic about the group, as they recognize that conflict does not necessarily destroy group relationships. However, if the conflict is resolved with some group members feeling as if they have lost, they may have negative feelings for other group members and even become hostile toward the group.

How important is conflict aftermath? Conflict aftermath affects group members' perceptions of their ability to work together. When the aftermath is negative, members are less likely to embrace conflict in the future. After all, they believe that they lost this time and do not want to lose again. For instance, Theo is distraught over his group's recent conflict episode because he feels as if he really didn't get a chance to express his viewpoint during the argument. "Man, that wasn't a good situation," he concludes. "Now nobody in the group gets along, and we were just given another project to complete." Theo's feelings—his conflict aftermath—will affect his interaction in the group. His motivation to work with group members, his trust of other members, and his interpersonal relationships with other members are all negatively affected.

But conflict can also create positive outcomes (Wall, Galanes, & Love, 1987). A moderate amount of conflict can actually increase the quality of group outcomes. Conflict can motivate members to participate and pay attention and can strengthen a group's ability to solve problems. When group members manage the conflict by satisfying their own concerns, in addition to the concerns of others, group outcomes are of a higher quality (Wall & Galanes, 1986; Wall & Nolan, 1986). Furthermore, when conflict is managed effectively by the group, high levels of trust and respect are generated (Jehn & Mannix, 2001).

Thus, conflict is an emotionally driven process, and group members experience emotion in positive and negative ways. When individuals are vested in the group or its activities, group members identify with the group. Emotions are more salient because group members' identities are affected when conflict occurs and threatens members' positions, ideas, or points of view. This, of course, can influence their relationships. Thus, conflict is emotionally defined, valenced as positive or negative, identity-based, and relationally oriented (Jones, 2000).

Is Conflict Always Disruptive?

Although we often think of conflict as being disruptive, conflict can be productive for groups (Deutsch, 1969). When conflict exists, groups are engaged and talking with one another. This means that stagnation is prevented and that interest in and curiosity about the group and its activities are stimulated. Conflict provides an opportunity for members to test and improve their abilities and to create solutions. This is not to say that groups seek conflict. Rather, conflict naturally occurs, and it is an important part of group functioning. Conflict about the group's task can keep members from prematurely accepting or agreeing on solutions. In fact, a moderate amount of task conflict improves a team's problem solving (De Dreu, 2006).

 ## SKILL BUILDER

Did I Do That?

Think about the last group conflict in which you were involved. At what point did you know that a conflict had developed? What cues led you to this conclusion? What was your role in helping to develop or establish the conflict? Did you say something that someone found offensive, inaccurate, or personal? Or did you neglect to say something when you should have spoken up? Did you behave in a way that demonstrated lack of interest in the group or in what the group was doing? Could you have changed your communication or behavior in any way to help the group avoid or minimize the conflict?

Although conflict can help groups generate creative and innovative solutions, many groups try to avoid conflict at all costs. Other groups deny that conflict exists and continue with their interactions as if nothing is wrong. Still other groups believe that conflict is disruptive and detrimental. Why do groups hesitate to engage in conflict, given that positive outcomes can be achieved? One reason has to do with the anxiety associated with conflict. When you compare the language and interaction of groups in conflict with that of groups not in conflict, distinct differences emerge. Group members in conflict actually change their verbal patterns. For instance, they become more repetitive and use simpler forms of language, speaking in habitual ways and repeating phrases without adding anything new to the conversation. In addition, their anxiety rises, which affects their ability to take the perspectives of others in the group. Look at the following example:

LUCY: Can we just get on with it?

JIM: Sure. I want us to vote for the incorporation.

ELLA: Right, as if the incorporation will do us any good.

JIM: Well, it will . . . the incorporation, I mean.

LUCY: Can you explain more about the incorporation plan, Jim?

JIM: Well, as you know, the incorporation plan will incorporate all of the surrounding towns into Plainview.

ELLA: If we incorporate, we'll be just like Plainview. No different, but just like Plainview.

LUCY: I've figured out you're against incorporation, Ella. But could someone please tell me what incorporation means?

In this case, the low levels of language diversity or redundancy—such as the repeated use of the word *incorporation* with no explanation of what it means—reinforce the disruptive nature of conflict (Bell, 1983). Anxiety is increased in conflict situations like this, affecting group members' abilities to take the perspective of others in the group. Although group members are using the same word—incorporation—there is no evidence that they have similar meanings for it. Because members take positions, they are not likely to have common ground, and misperceive what others in the group are saying (Krauss & Morsella, 2000).

Is Conflict Inherent?

Conflict occurs because one of three things happens (Smith & Berg, 1987). First, groups often need people with different skills, interests, and values to accomplish their goals. These differences alone can create conflict. Although group member differences may be necessary, they can also threaten a group's capacity to function effectively. To benefit from diversity, group members must become interdependent in a way that provides unity while preserving differences.

Second, groups have a natural tendency to polarize members in terms of how they communicate (Bales & Cohen, 1979). Friendly group members often feel that they are in conflict with members who are negative and unfriendly, and submissive group members sometimes feel opposed to members who are dominant, outgoing, and assertive. When differences in levels of group member dominance are great, conflict is likely to occur (Wall & Galanes, 1986). For example, both Yvonne and Joe are talkative, bold, and expressive. They usually initiate group conversations and occupy most of the group's talking time. In comparison, Marc and Wendy are more submissive. They really do not like to talk much, preferring to follow Yvonne and Joe's lead. But this does not necessarily mean that Marc and Wendy will go along with everything the other two suggest. Wendy, in particular, may become angry—but will not show it—when Yvonne and Joe decide what the group will do. As this example suggests, differences in ideas are often exacerbated when there are differences in levels of group member dominance. More dominant members take responsibility for the group, often without asking other members for their input or for agreement. More submissive members are less likely to take a vocal or overt stand against ideas, making it appear that they agree with more dominant members.

In addition, conflict is also likely to occur when there are differences in group members' orientation (Wall & Galanes, 1986). Some members may have a higher task orientation, whereas other group members have a higher relationship orientation. This difference in orientation affects how individuals perceive their group membership and the primary function and goal of the group. The task-oriented members think the other members are slowing them down, and the relationship-oriented members think the task-oriented members need to relax to allow group members to develop stronger relationships. Frequently, these group members will find themselves locked in a distributive conflict in which one side will win and the other will lose. Try the "Skill Builder Feature: Did I Do That?" to reveal your role in a group conflict.

Thus, group life is filled with many opportunities for oppositional forces to exist and many instances in which members perceive that opposition exists. This means that individuals in groups and groups as a whole will always be managing differences even as they are seeking a certain level of homogeneity (Smith & Berg, 1987).

Third, group members can experience feelings of ambivalence about their group membership, which can cause conflict. You may want to identify with others and be part of a group, but you also want to retain your individuality and be different. Thus, you feel both drawn toward the group and pushed away from it: "I'm like them; I'm not like them." The desire to be both separate from and connected to the group can result in individual-to-group conflict. For example, Jones is a member of a fraternity, and he values his affiliation with his fraternity brothers. He wears their logo proudly on a cap and a sweatshirt, and plays on their soccer team and captains their softball team. But Jones's fraternity brothers are notorious for waiting until the last minute to fulfill their service work as campus safety escorts. Because Jones lives in the frat house, he is called on frequently to take shifts when others do not show up. Lately, he has become resentful of others relying on him to take their shifts. "After all," Jones complains, "don't they realize I have a life of my own?"

Conflict can even be inherent in the most common of group tasks. For example, in a group brainstorming session, members produced 64 unique ideas for consideration. When group members voted on their top 5 ideas, 33 different ideas received at least one top-5 vote (Warfield, 1993). As you can see from this example, group members' different skills, interests, and values generated a substantial number of solutions for consideration. This is certainly one advantage of working in groups. However, with so many ideas capturing the interests of group members, conflict will surely arise as the different ideas are debated and discussed. According to the **law of inherent conflict**, no matter the issue or group, there will be significant conflict stemming from different perceptions of relevant factors (Warfield, 1993).

TYPES OF CONFLICT

As mentioned previously, conflict is not necessarily destructive. It may seem that way when you are involved in a conflict situation, but conflict actually can be productive for the group. Conflict is destructive if it completely consumes the group's energy and time, prevents members from working together, or escalates into violence. Alternatively, conflict can be productive if it exposes new ideas, helps clarify an issue, or alerts the group to a concern that needs to be addressed.

Is the conflict personal? **Relational conflict** is rooted in interpersonal relationships, emotions, or personalities (Jehn, 1995). Even when group members agree about the group's goals and procedures, this type of relational conflict can keep a group from accomplishing its task. For instance, when Angie refuses to listen to what Scott has to say because she thinks he is arrogant and that he thinks he is better than the rest of the group, the conflict is relational. This type of relational conflict is based on social or relational issues like status, power, perceived competence, cooperation, and friendliness, and it generally increases emotional responses. Effective groups are those that can experience conflict over ideas without tying the conflict to particular group members. When conflict is linked to a particular group member, it is personalized. This type of conflict is likely to be more dysfunctional because it is like a deep current running through the group, and it is often subtle. Typically, relational conflict is disruptive and can negatively influence team members' satisfaction with the group. Also important, the greater the relational conflict among team members, the less likely members will seek information from one another (Meng, Fulk, & Yuan, 2015).

Second, **task conflict** is rooted in issues or ideas, or disagreement about some aspect of the group's task (Jehn, 1995). For example, members disagreeing about the appropriateness of two alternatives or about the scope of their responsibilities are having task, or substantive, conflicts. Managed effectively, task conflicts help groups to be creative, improve their problem-solving abilities, and generate member satisfaction with decision making (De Dreu & Weingart, 2003; Yong, Sauer & Mannix, 2014). However, if group members perceive or are told that the group's performance is inadequate or ineffective, relational conflict is likely to develop (Guenter et al., 2016).

Third, **process conflict** occurs when group members have disagreements about which group member is assigned duties or given access to resources (Jehn, 1997). Coordination among group members is central to process conflict. If group members are having difficulty with time management, how work and responsibility are distributed among members, or how group members decide to approach their work, they are experiencing process conflict (Behfar, Mannix, Peterson, & Trochim, 2011).

When conflict erupts in a group, it is seldom the case that group members will be able to agree about the source or type of conflict. To be effective, groups must express some agreement about the type of conflict that is occurring (Pace, 1990) before they can move toward managing it. What can group members do when task or relational conflict occurs? Sharing more information among group members appears to strengthen group process and lead to better decision making (Moye & Langfred, 2004).

Another distinction concerning conflict is whether the conflict exists in a competitive or cooperative environment (Guetzkow & Gyr, 1954). **Competitive conflict** polarizes groups, with one side winning and the other side losing. When this happens, group members are likely to escalate the conflict and become defensive or even hostile toward one another. Alternatively, **cooperative conflict** occurs when the disagreement actually helps move the group along with its task or activities. In this case, the climate surrounding the conflict is supportive or positive. As a result, the group is more likely to find a mutually beneficial resolution to the conflict.

How important is the conflict? A third distinction is the centrality of the conflict to the group (Guetzkow & Gyr, 1954). How important is the issue to the members who are in disagreement or to the group as a whole? If the conflict is about a trivial matter (e.g., what type of paper to copy the agenda on; which room to meet in), then the conflict is not as salient, or important, to the group's objective. If group members are arguing about a critical feature of the group's project (e.g., the scope of the project or its budget), then the conflict is salient and has the capacity to create more dysfunction in the group.

Is the conflict over information? A **cognitive conflict** exists when group members disagree about information or the analysis of that information (Knutson & Kowitz, 1977). For example, when discussing a group project, Jacob says, "the instructor said that the most important part of the task is the class presentation." However, Dante says, "No, the teacher said the most important part of the project is the portfolio." One team member (or potentially even both members) has heard the information incorrectly or made an incorrect assumption about the information provided. In most cases like this, the conflict can be resolved when group members get more or better information. But it must be evidence or data from a valid source. It does not work to say "because I say so." To resolve a conflict over information, data must be available to all group members.

Is the conflict about expectations? **Normative conflict** occurs when one party has expectations about another party's behavior (Thomas, 1992). In other words, conflict occurs when someone evaluates your behavior against what that person thought you should have done. Sororities and fraternities often deal with normative conflict when a member ignores or breaks one of the house rules. For instance, Heidi turns in her sorority sister because she violated sorority house rules by bringing her pet to the house over the weekend. Heidi expected her sorority sister to know the rules and abide by them. Normative conflict can evoke an emotional response like blame, anger, or disapproval, and it is usually followed by sanctions intended to produce conformity to the formal rules or implied standards.

When group members disagree about the nature of the conflict, such issues need to be addressed before the group can resolve the primary conflict. Disagreements over these issues allow group members to perceive that incompatible goals exist. The more disagreement there is about the nature of the conflict, the more strain there will be on the group's interpersonal relationships. Group members who can come to agreement through interaction on the nature of conflict are more likely to build consensus and cohesiveness (Pace, 1990).

POWER

Central to all types of conflict is the use of power among group members. Why? It is unlikely that all group members are equally skilled or knowledgeable, and even more unlikely that all group members perceive that they have equal status. **Power** is the capacity to produce effects in others or influence the behavior of others (Burgoon & Dunbar, 2006). This influence, which results from social interactions or is created by the possession of or access to resources, is an issue in all types of groups (Lovaglia, Mannix, Samuelson, Sell, & Wilson, 2005). Think for a moment of your family group. Who has the most power? Your mom? Your dad? Your little brother? Now think of your work group. Who has the most power? The leader?

MESSAGE AND MEANING

To use power, others must recognize and endorse your power (Clegg, 1989). If others do not, then whether you actually have power is debatable. The city commission in a medium-sized town was debating a rezoning issue amongst themselves. One of the commissioners was arguing that the city had made a plan for development, and that the city commission ought to adhere to it.

SALLY: Jack, do you have any comments on the proposed rezoning for this store?

JACK: I think we are missing a very important piece of information as we talk about this issue. Five years ago, we created a resource plan. This was an important plan that was supposed to guide the city commission in decision making over the next decade of development. And I've heard no one on this commission referring to it as we discuss this issue.

SAUL: Jack, that plan is outdated. We all know this. We aren't referring to it because it is worthless.

JACK: But I think it is still helpful . . .

In this circumstance, Jack presented information that no one was using. He hoped that this new information, a formal report that the city commission had funded a few years earlier, would hold considerable weight in the debate. After Saul's comment, can Jack have any influence by using this plan? Can someone simply refuse to endorse someone's comments (like Saul did) and reduce an argument to nothing? Why? What must Jack do to be persuasive along this line of argumentation?

Power is not inherently good or bad. Having access to power or knowing that others see you as powerful helps you feel confident in group settings. At the same time, we are all familiar with the misuse of power in group settings and the relational damage it can cause. Positive or negative, power resides in the relationships among group members. Some types of power facilitate conflict among group members, and other types prevent conflict from developing. It's how power is communicated and used that determines its influence on and for group members.

Bases of Power

Power exists in relationships among group members. When a group member has power, that member has interpersonal influence over other group members because they have accepted or allowed the attempt at power to be successful. If you perceive the influence of another group member and alter your behavior because of that influence, power exists. Although power traditionally has been seen as residing primarily in group leaders, any member of the group can develop power and use it in relationships with other group members. Six power bases have been identified in research: reward, coercive, legitimate, referent, expert, and informational (French & Raven, 1968; Raven, 1993).

You are probably most familiar with **reward power**. Rewards can be relationally oriented, such as attention, friendship, or favors. They can also take on tangible forms, such as gifts or money. Group members behave and communicate in a certain way because they are rewarded when they do so. In contrast to the positive influence of rewards, threats represent negative or **coercive power**. In group settings, coercive power results from the expectation that you can or will be punished by another group member. Coercion can take the form of denying a group member the opportunity to participate or threatening to take something of importance away from a group member.

Legitimate power is the inherent influence associated with a position or role in the group. Leaders or facilitators often have legitimate power—they can call meetings and make assignments. Group members allow the member with legitimate power to do these things because the power is formal, or inherent in the role the person has within the group. This type of power exists within the role, not within the person. Without another power base, a leader relying solely on legitimate power will have little real influence in a group.

Expert power is influence based on what a group member knows or can do. Group members develop expert power when they offer their unique skills to help the group, and their behavior matches the expectations they have created. Suppose Amber says she can use computer-aided design software to design the team's new office space. Her team members will reward her with this expert power only when she demonstrates this skill and the office layout is approved by the group. Saying you can do something is not enough; your performance must match the expectations you create.

Informational power is persuasion or influence based on what information a group member possesses or the logical arguments the member presents to the group. In a mountaineering team, everyone must work together to safely complete the trek. Team members who possess a diversity of expertise can strengthen the team. For example, one member is responsible for checking the weather forecasts and briefing the others before the climb, another member is certified in first aid and CPR, and a third member has the most hours of climbing experience. Each of these members has more legitimate informational power in one area than the others. If one member is hurt, the member who is certified in first aid should have the most influence on how to treat the injury. Finally, **referent power** is influence given by you to another group member based on your desire to build a relationship with the member. In other words, you admire or want to be like another group member. Thus, you allow yourself to be influenced by this person. For instance, if Wes admires Henry and wants to be like him, Wes will allow himself to be influenced by Henry and will follow his suggestions and recommendations. Anyone in the group can possess referent power, which is often based on charisma. And members can have referent power over others without intending to do so. A group member with a pleasant or stimulating communicator style often develops referent power with others, which gives this person additional opportunities to develop power bases with these same group members.

Of these types of power, all except coercive power are essential for effective group process. Group members want someone to be in control (legitimate power); like it when others compliment them on their contributions (reward power); find it beneficial when someone is the group's motivator, cheerleader, or contact person (referent power); and expect others to contribute their skills and knowledge (expert, informational, and legitimate power).

A group member can hold little power or can develop power in many areas. But to be influential, the base of power must be essential to the functioning of the group. Position power is not powerful if the position is with another group. For example, the leader of your work team will not necessarily have power on the company basketball team.

For a group member to have power, other members must support or endorse that power (Clegg, 1989). Oftentimes this happens only after we communicate in a way that demonstrates our power. For example, if Andy does not profess his expertise in creating graphics for the group's final report, other group members cannot create this power relationship with him. Andy's power arises only if the others are aware of his knowledge and skill. Thus, power emerges through interaction. Although the group leader or facilitator typically holds more power than other group members, all members should develop and demonstrate some base of power to augment their credibility in and worth to the group. When power is distributed among group members, participation is more balanced, and cohesiveness and satisfaction are enhanced.

How important is it that you develop a power base as a group member? Very. Power used positively creates attraction from other group members. We tend to like to be associated with powerful people. In fact, power has a greater effect on other group members than status (Bradley, 1978). This is because power is developed within the group's relationships, whereas status is generally brought into the group. You may have status or prestige because of where you live, what your parents do for a living, what car you drive, and so on. But these status issues may not be salient or relevant to the group and its activity. Would it really matter to another group member that you drove an expensive car if you did not follow through with your group assignments? Would it matter that last semester you worked for a prestigious law firm if you did not share your knowledge with the group? Try the Putting It All Together feature "Group Structure and Group Size" to better understand the power bases of a group experience.

A power source that is often overlooked is the control of resources—real or imagined. As groups work on their tasks, information and materials from outside the group frequently are needed. For example, when a group member volunteers to use his or her connections to obtain permission to use the dean's conference room for the group's meetings, he or she is exerting power over needed resources. Group members are often thankful that someone in the group has access to these resources and impressed that the individual can obtain what they cannot.

However, overusing these connections can create a defensive climate in the group. Here power can be perceived as strategic manipulation. For instance, if a team member volunteers to reserve a conference room, then it is possible that she will schedule it to fit her schedule.

If power is neither inherently good nor bad, how does power relate to conflict? There are two primary ways (Sillince, 2000). First, those with power use the communication resources of a group differently than those with less power. Powerful members of a group talk more, respond to questions more, issue more challenges, and introduce more new topics into the group than do less powerful members. Using these communication strategies, a powerful member is more likely to set the agenda for the group, and that can cause conflict. Second, less powerful group members often want more power and try to find ways to increase their power base. This can cause conflict because the more powerful members may feel threatened or unappreciated.

 PUTTING THE PIECES TOGETHER

Group Structure and Group Size

Think of a recent group experience in which you took on a relatively minor role in the group. How was legitimate power used by the member in charge or the leader? To what extent did the leader use other bases of power to influence group members? How did this use of power affect the structure of the group? What bases of power did other members have or develop? Was power easily shared among members?

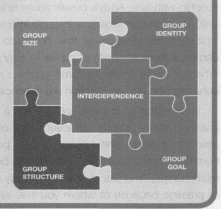

CONFLICT MANAGEMENT STRATEGIES

You may think that there is little difference between the terms *conflict resolution* and *conflict management*, but the two terms represent widely divergent views of conflict. Resolving conflict requires that you view conflict as a destructive phenomenon or as a disruption that you need to eliminate. Thus, the only kind of *good* conflict is the absence of conflict. In contrast, **conflict management**, implies that conflict is a normal and inevitable situation that groups must handle. Groups that manage their conflicts take advantage of conflict situations to solicit alternative views not previously addressed. Managing conflict results in creative and innovative solutions.

There are five general conflict management strategies: collaborating, competing, accommodating, avoiding, and compromising (see Figure 7.1). Although you probably have a primary orientation toward managing conflict, group members can learn new skills and approaches as well. Not all conflict situations are alike and the conflict management strategy you're most comfortable with may not be the best strategy for all conflicts.

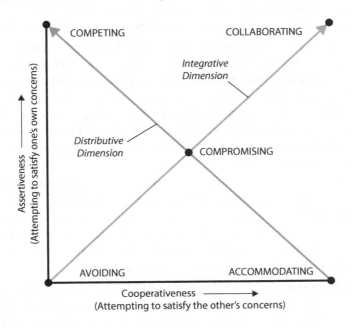

Figure 7.1 Five Approaches to Managing Conflict

Collaborating

Which is the most effective conflict management strategy? Many people would argue that using **collaboration** helps a group achieve a win-win outcome—an outcome with which everyone can agree—because information sharing and collaboration are promoted. Collaboration is an **integrative conflict management strategy** because you attempt to maximize the gains of all conflicting parties. This promotes rather than inhibits relationships among group members (Canary & Spitzberg, 1989).

In collaborating, the parties replace their incompatible goals with the superordinate goal of solving the problem even though their initial ideas for how to solve the problem differ. For example, even though conflict partners might have different ideas for resolving the problem, their common interest is in finding a solution. Thus, parties must communicate with one another to redefine the situation so that they can identify a shared or mutually acceptable interest or goal. Besides sharing the goal, the opposing parties must develop or build a common language in order to create a shared frame from which to view the problem.

Ultimately, this means that opposing parties who see each other as enemies must be able to move from that framing device to one in which the other parties are viewed as partners. To effectively work in the problem-solving mode, a new "we" must be created to include all parties involved. Framing others as adversaries must give way to a frame of allies.

Collaboration is an integrative strategy because it relies on parties communicating with one another. How can you initiate this strategy? You can ask other group members how they feel about the problem or let others know that something about the group or its task is bothering you (Jarboe & Witteman, 1996). Or you can self-disclose—a good way to get others to self-disclose and open channels of communication (Sillars & Wilmot, 1994).

As communication between parties begins, look for objective criteria by which to evaluate potential solutions. This cooperative strategy allows both parties to find and develop common ground, and diminishes the emotional and subjective aspects of the conflict. Groups that can achieve integrative conflict management produce higher-quality outcomes and generate higher group member satisfaction (Wall & Galanes, 1986; Wall et al., 1987). Because collaboration requires a good-faith effort from everyone and open communication among everyone, it is easier to initiate and sustain when group identity and task interdependence is high.

There are some difficulties with using a collaborative approach. First, it takes the most time and effort. Collaborators must question and listen to the other side in order to truly understand them. It may be difficult to use a collaborative approach in situations with a deadline or requiring efficiency. Second, collaboration works best if both sides are using the same style. It is hard to be collaborative with someone who does not reciprocate a desire to find an outcome that will leave both sides satisfied.

Although collaboration can take time and energy, it is a constructive way of managing differences, and it is the conflict management style people report using most often (Farmer & Roth, 1998). Openly discussing differences, having access to a variety of opinions, and carefully critiquing assumptions help create commitment to the group, as well as trust among and respect for group members (Thomas, 1992). Collaboration also can offset problems caused by initial differences among members or unequal participation caused by diversity within the group (Kirchmeyer & Cohen, 1992). More importantly, collaboration can produce higher-quality decisions and solutions.

Additionally, collaborating should be used when you view the issue as too important to compromise, when you have a long-term relationship with the other party and other conflicts are inevitable, when you could learn from merging your insights with the insights of others, and when you need to build a sense of community and commitment. Although the other strategies may seem effective in certain situations, they also have costs or risks associated with them.

What are the alternatives to the integrative strategy of collaboration? There are several, but none can create the same win-win outcome as collaboration. Two strategies—competing and accommodating—are **distributive conflict management strategies,** meaning that they are characterized by a win-lose orientation, or an outcome that satisfies one party at the expense of the other. Group members who use one of the distributive strategies often show anger and use sarcasm. One of these strategies may settle the conflict, but the relational aspects of the group will be damaged because of the strategy's win-lose orientation. When win-lose strategies are used, the quality of the outcome is also lower (Wall et al., 1987).

Competing

Competing as a conflict management strategy emphasizes your own triumph at the other person's expense. In a sense, you take from the conflict but give very little. In using a competing strategy and wielding power over others, you are being assertive and uncooperative. Group members who use a competing style are likely to force the issue and dominate interaction. They will also believe that they are right and the other person is wrong.

When group members compete, communication channels close down (Deutsch, 1969). In fact, you might go out of your way not to talk to other group members as part of the competition. And even when communication is taking place, you may be suspicious of the information you receive from other parties. Thus, error and misinformation abound. Competing also contributes to the view that the solution to the conflict can be of only one type—the type that is imposed by one side on the other.

Competing can be used when decisive action is vital, as in an emergency, or when you need to implement an unpopular action. Competing can be effective when the conflict is with others who would take advantage of you if you used a noncompetitive strategy or when you know you are right (just be sure you are!).

Accommodating

The other distributive conflict management strategy is **accommodating**, in which you give everything and take very little from the conflict. Here you are cooperative, yet unassertive. You focus on trying to satisfy the other's concerns rather than your own. You try to smooth over issues and relationships by being obliging and yielding. For example, each time Henry brings up a sensitive issue, Nell becomes submissive and quiet. Even when Henry tries to force Nell to talk about the problem, she lowers her head and says something like "Whatever you say, Henry. I'm sure you must be right. You've never led us down the wrong path before." In essence, Nell accommodates Henry and his viewpoint to end the conflict.

Although the accommodation style focuses on the other person, there still may be advantages for the speaker. Oftentimes when someone gets their way, there is an expectation that the next time around the other will reciprocate. In other words, accommodators may be building up credit for future interactions. Accommodators may also believe that more competing styles could potentially make the problem worst, and thus it is simpler and better in the long term for the other side to get their way.

Accommodating is especially effective if you discover that you are wrong. By accommodating, you demonstrate that you are willing to learn from others. Accommodating can also be effective when you need to satisfy others to retain their cooperation; in other words, the issue is more important to them than it is to you. And, because you value the long-term nature of the relationship, it is important to be reasonable and to demonstrate harmony in and loyalty to the relationship. Thus, accommodation can be an effective strategy for managing minor internal group disagreements while the group is resolving more significant problems.

Avoiding

Neither integrative nor distributive, the **avoiding** conflict management strategy is nonconfrontive. Group members choosing this strategy try to sidestep the conflict by changing the topic or shifting the focus to other issues. They hope that if they ignore it or do not draw attention to it, the conflict will disappear. How is this strategy used to end a conflict? Generally, people who use the avoiding conflict management strategy verbally withdraw from the conversation. They can also physically withdraw by not showing up. Members of work groups often use the avoiding strategy. One member will acknowledge a problem he has with another group member and then back off (e.g., "I get pretty ticked off sometimes, but it's not really a problem"; "That's not how I would do it"). This type of denial is to be expected, given that members are assigned to these types of task groups and often do not have a choice in group member selection. This type of forced intimacy can be characterized by nervous tension and denial; thus, one party to the conflict withdraws and the conflict ceases to exist. Although not recommended as a strategy for managing most conflicts, avoidance can be the most effective strategy when it allows group members who must work together to focus on their task (De Dreu & Van Vianen, 2001).

Avoiding is effective when the issue is trivial and you can let go of it or when other matters are more pressing. Sometimes it makes sense simply to walk away from a conflict. Avoiding can be an effective strategy when you perceive no chance of satisfying your concerns. Why fight over something you cannot have? As a teenager, you probably used avoiding with your parents to let everyone cool down and gain perspective before dealing with an important but sensitive issue again.

Compromising

A fifth strategy, compromising, also deserves attention. **Compromising** is an intermediate strategy between cooperativeness and assertiveness. Although compromising may settle the problem, it also will offer incomplete satisfaction for both parties. You have given up something, but you are still holding out for something better. Although a compromise may be easier to obtain than collaboration, it is at best a temporary fix (Putnam & Wilson, 1983; Thomas, 1977). Groups tend to manage conflict with compromise when they feel time pressures to reach an agreement. Unfortunately, giving in to a compromise may mean that not all ideas or concerns have been heard.

You have probably compromised with a roommate over who will perform household tasks, such as taking out the trash. These compromises are okay at first because they solve the immediate problem by ending the fight about who will take the trash out right now. But over the long term, compromises tend not to hold. Eventually, your roommate will forget to take out the trash when it is his turn, and you will blow up and start looking for a new living arrangement.

Compromising may be effective if you and the other party are willing to accept a temporary settlement. Also, a compromise may be the best you can achieve when both sides are adamantly fixed on opposing or mutually exclusive goals. For example, suppose members of management and union representatives are meeting as a group to discuss a potential strike. What management offers to forestall a strike is likely to be different from what the union representatives want. The only type of settlement likely to be achieved is a compromise. Generally, these two groups are fundamentally opposed to each other's view, so anything more than a compromise is unlikely. Recognize, however, that the resolution achieved through compromise is only temporary. Parties with fundamentally conflicting views are likely to reinitiate the conflict or start another one.

Which Strategy Should You Choose?

In conflict situations, you are managing three views of the situation: (a) your view of the conflict, (b) your belief of what the other person's view is, and (c) your evaluation of the relationship between you and the other person. When conflicts occur, it is typical to start with a strategy that emphasizes your view of the problem. Even if you enter the conflict with little concern about your view, that quickly changes. You would not be in the conflict

---

THEORY STANDOUT

Conflict in Virtual Teams

Poole and Garner (2006) summarize the research on conflict in virtual teams. Studies that examine how face-to-face communication is different than digital communication are conducted using the instrumental perspective of conflict. This perspective is useful as researchers are attempting to identify how technology influences team members in their management of conflict. One consistent finding is that it is not easy for team members to resolve conflict when they are not face-to-face. Why is that the case? The lean channels of digital communication can hide many nonverbal cues associated with conflict. This makes it more difficult to confront conflict (Poole & Garner, 2006). If your team meets using technology, what advice or practices could you introduce to prevent conflict from being hidden?

if your own view of the problem were not important to you. And, as you might expect, your attention to the other person's view often diminishes throughout the interaction. Thus, a pattern dominates most conflict situations: People enhance their own view of the situation while minimizing the view of the other person (Nicotera, 1994).

Initially, it seems reasonable to put your view of the conflict first. Attempts to minimize the view of the other person also seem reasonable given that you don't want to lose in conflict situations. And you probably have some interest in continuing a relationship with the other person, as people generally have conflicts with others who matter to them or with members of groups that matter to them. But remember that your conflict partner has the same orientation to the conflict. Your partner views the issue as important, is attempting to minimize your view, doesn't want to lose in the conflict, and sees you or the group as matter of personal importance. Essentially, both parties have the same, but incompatible, goals.

How you manage conflicts provides others with a means for evaluating your communication competence. Your conflict messages are assessed according to their appropriateness and effectiveness. In turn, these assessments act as a filter through which evaluations of your competence are made. One conflict episode can influence how your relationship with someone develops. Certainly a pattern of conflict over time will influence your ability to maintain relationships with others. Generally, you will be perceived more negatively and as being less competent if you manage conflicts with competing, accommodating, or avoiding strategies. In contrast, group members who use collaboration to focus on the issue or content of the conflict are perceived more positively (McKinney, Kelly, & Duran, 1997).

Most conflict situations, however, are complex enough that groups change their conflict management strategies over the course of a discussion (Poole & Dobosh, 2010). What combinations of styles are effective? Generally, group members who use some combination of collaborating, competing, and compromising are judged to be effective at conflict management because they are best at creating mutually acceptable outcomes and maintaining relationships. Once a group develops a pattern for managing its conflicts, it tends to use the same strategies in subsequent conflicts. Such a pattern can have far-reaching effects. For example, groups that manage conflict collaboratively increase group efficacy, which in turn encourages members to believe that they can also handle subsequent conflicts (Alper, Tjosvold, & Law, 2000). Moreover, groups that master collaboration as a conflict management strategy incorporate it into their decision making more than groups that rely on confrontation or avoidance (Kuhn & Poole, 2000).

How you manage conflict depends both on what you say and how you say it. Which conflict management strategy is best? It depends upon you and your involvement in the conflict situation (Thomas, 1977). Because conflict creates emotional reactions, the nonverbal messages you communicate (consciously and unconsciously) and the words you choose influence other group members' responses in conflict interactions. For example, suppose you say, "That's a good point, but I disagree." Said in a polite and respectful tone, other group members might interpret this as indicating a willingness on your part to develop a collaborative solution. But if you say, "That's the dumbest idea I ever heard" as you roll your eyes, you are likely to get a very different response from other group members. Even though the content of the two messages is very similar—that you disagree—the interpretative frames through which the messages are sent and received are quite different.

To select the strategy that you believe will be most effective, it can be helpful to analyze the conflict you're experiencing according to four dimensions: (a) the level of emotionality in the conflict, (b) the importance of the conflict, (c) the degree to which there are group norms for handling conflict, and (d) the conflict's resolution potential (Jehn, 1997).

First, identify how many negative emotions are being expressed during the conflict. These emotions might include anger, rage, annoyance, frustration, resentment, or simple discomfort. In your group, yelling, crying, banging fists, or talking in an angry tone are clear signs that negative emotions exist. Second, identify the importance or centrality of the conflict to the group. In other words, is this conflict a big deal or of little importance? The importance of a conflict is often tied to the perceived consequences of being in the conflict. If the outcome of a conflict will greatly influence the identity of a group (e.g., splitting a larger team into two separate teams; some group members stop attending), the consequences can be considerable.

The next step is to identify the norms this group has about conflict, particularly the degree to which group members find conflict acceptable. Do they regularly engage in conflict? Do they perceive it as normal for conflict to occur? For groups in which conflict is not acceptable, members often try to downplay or avoid conflict. For groups in which conflict is a regular occurrence, members may view conflict as a healthy and constructive part of the group process.

Finally, assess the conflict for its resolution potential. The key question is, do group members *believe* this conflict can be resolved? What's important here is group members' perceptions, not how an outsider might assess the resolution potential of a conflict. Some group conflicts are easy to resolve by gathering additional information or by giving group members additional time to work through difficult issues. However, other conflicts, especially personal or relational conflicts, can be difficult to resolve. Issues that influence members' perceptions of resolution potential include the degree of group member interdependence, the group's history with conflict, uncertainty about the group and its activities, and status and power differences. Regardless of the outcome, one conflict episode is connected to the next one. The feelings that result from the first conflict, or conflict aftermath, influence your awareness of the subsequent conflict and your interaction in it.

SUMMARY

The word *conflict* generally conjures up negative emotions. However, in our society and in working groups, conflict is a normal part of day-to-day activities. When parties are in conflict, they have mutually exclusive goals. Because the parties are interdependent, they cannot all have things their way simultaneously. Conflict can be productive for groups by stimulating interest and providing opportunities to evaluate alternatives.

Conflict is inherent in group interactions because different skills, values, and talents of members are needed to complete complex activities and goals. These differences, although necessary, also allow conflict to occur. Effective communication can help groups manage their differences and find solutions to conflict problems.

Conflicts can be over relational issues or substantive issues, and they can be cooperative or competitive. Conflicts can occur over judgment or cognitive tasks, over the use of (or lack of) procedures, or over incompatible personalities. Conflicts can arise over differing goals or interests, and they can develop when one party evaluates another in terms of what should have been done or accomplished. Because gender and culture are primary ways in which we identify ourselves and others, differences attributed to these characteristics can be salient in conflict interactions. Despite gender stereotypes, there is little evidence that one gender communicates differently than another in groups. Still, when gender distinctiveness is relevant to the group's task, conflict can occur. Likewise differing cultural orientations can also cause conflict. In both instances, however, we should remember that gender and intercultural differences cannot explain all instances of conflict. Many conflicts are based on other affective or substantive issues.

When relationships and interdependence develop among group members, power issues are inevitable. Power is created through communication and can be based upon rewards, coercion, role or position, charisma, expertise, or information. Power is best analyzed contextually. The extent to which power develops may be based on formalized power structures and the degree to which a struggle over power occurs. Power is fluid, not static; power in relationships changes frequently. Power can also be created when a group member has control over real or imagined resources.

Conflict between groups is common in our society. We create enemies by talking about our adversaries. We belong to multiple groups and sometimes find ourselves caught in the middle. When our group is in conflict with another, intragroup cohesiveness and commitment build as we distinguish ourselves from the out-group members.

There are five types of conflict management strategies. Collaborating, or problem solving, is an integrative strategy that can produce high-quality solutions and decisions for the group. Through discussion, all parties contribute their ideas to find one solution that satisfies everyone's concerns. The distributive conflict management strategies of competing and accommodating are characterized by a win-lose orientation. In competing, you win and the other party loses. In accommodating, you allow yourself to lose to let the other party win. Avoiding is characterized by verbal or physical withdrawal from the conflict situation. Compromising is an intermediate strategy in that it may settle the problem but also offer incomplete satisfaction for both parties.

Usually, you will select a conflict management strategy that emphasizes your view over those of others. But as the discussion continues, you are likely to change your strategy. In any case, the strategy you use affects how others judge your communication competence. Although each strategy has advantages and disadvantages, most people prefer the integrative strategy of collaborating.

DISCUSSION QUESTIONS AND EXERCISES

1. Most television shows revolve around conflict—even situation comedies. As you watch television this week, make a list of the group conflicts you see. Label each conflict according to whether it is relational, task, or process, as well as competitive or cooperative and cognitive or normative. Also note how the conflicts are managed. What conflict management strategies did characters use? What conflict management strategies would have been more effective?

2. Keep a journal for one week of the groups you are involved with and the conflicts they are experiencing. In addition to classroom groups, include your family or living group and any work groups. Describe and analyze each conflict in terms of the following characteristics: Who was involved in the conflict? When did you become aware that you were involved in the conflict? How did you communicate with the other person(s)? Did your plan for managing the conflict change as you communicated with the other person(s)? How long did the conflict last? What was its outcome?

3. You probably can remember at least one group conflict that did not turn out as you expected or wished. Think back to that conflict and reflect on which members displayed what types of power. What type of power did you display? What other power options did you have that you did not use?

NAILING IT!

Using Group Communication Skills for Group Presentations

When group members are preparing to create a group presentation, there may be many opportunities for conflict. Certain types of conflict may be more common than others. Normative conflict is one potential type that may arise as group members practice their presentation. Even though the presentation may be well planned out, the actual practicing of the presentation may reveal differences in expectations. For example, the first presenter may be asked to provide a brief overview of the presentation. But if Elissa thinks a brief overview is 10 seconds and Shawn thinks a brief overview is 2 minutes, a normative conflict may arise. This is because we often don't know what our norms or expectations are until seeing the actual behavior. Normative conflict may be a common conflict type during group presentation preparations, but it also may be very valuable. Becoming aware of violated expectations before the final presentation allows the group to adapt accordingly.

Image Credits

Photo 7.1: Copyright © Depositphotos/Farina6000.

Figure 7.0: Adapted from: Copyright © Depositphotos/megastocker.

Figure 7.1: Adapted from: Ralph H. Kilmann and Kenneth W. Thomas, "Interpersonal Conflict-Handling Behavior as Reflections of Jungian Personality Dimensions," *Psychological Reports*, vol. 37, no. 3, p. 210. Copyright © 1975 by SAGE Publications.

CHAPTER 8

LEADING GROUPS

After reading this chapter, you should be able to:

- Describe the relationship of communication to leadership
- Describe the communication behaviors associated with group members who emerge as leaders
- Explain why and how leadership can be shared
- Develop the qualities needed to be a transformational leader
- Identify when gender assumptions inhibit or facilitate who becomes a leader
- Take steps to enhance your leadership capacity

In person or online, leadership is a communicative process of influence (Fairhust, 2007). Whether appointed or elected to the formal leadership role, the person who influences other group members is the leader of the group. Leaders influence what groups do or talk about. They also influence how groups perform their activities and achieve their goals. Because leadership roles are based on influence, it's not unusual for a group to have members in both formal and informal leadership roles.

Society's conceptualization of what a leader is has changed over time. Today, leaders must be skilled in team building and in helping team members collaborate (Martin, 2007). So, it's not surprising that leadership theory has focused on the leader as a motivator—someone who can provide the group with energy. But there is one thing you should keep in mind as you read about leadership: Research on leadership has focused almost exclusively on groups that make decisions in formal, hierarchical, or task-oriented settings. Thus, some of the findings presented here may need to be adapted for groups in which initiating, developing, and maintaining relationships are primary goals.

DEFINING LEADERSHIP

We can define leadership in a number of ways. In its broadest sense, **leadership** is the process of using interpersonal influence to help a group attain a goal (Northouse, 2017). As a process, leadership is the way a person uses noncoercive influence to direct and coordinate group activities in pursuit of group goals. Leadership in groups and teams is complicated because the leader must address the individual needs of all group members as well as facilitate the processes (i.e., information sharing, decision making, and conflict management) that arise from group interaction (Shuffler, Burke, Kramer, & Salas, 2013).

A COMMUNICATION COMPETENCY APPROACH TO LEADERSHIP

This functional view of leadership focuses on how group members communicate with one another to identify who is displaying leader-relevant actions and how others are responding to those actions. Generally, the content of leadership-relevant actions can be categorized as (a) procedural, or how the task is performed; (b) relational, or how the relationships among members of the group are being managed; and (c) technical, or the substantive content of the task (Pavitt, High, Tressler, & Winslow, 2007).

> ### SKILL BUILDER
>
> If leadership relies on task, relational, and technical competencies, how would you develop strengths in one or more of these? Would it be easier to develop task competencies in a setting where the task is known and structured? Or when the task is ambiguous? Would it be easier to develop relational competencies in a group for which the focus was more social or relational? Or in a group for which there was a clear task orientation? Finally, would it be easier to develop technical competencies while you were the leader? Or while you were not? Think through these questions and develop a strategy for strengthening the competency that you believe needs the most attention.

Regardless of how leadership develops in the group, the member or members who take on leadership responsibilities must communicate procedural, relational, or technical competency. The **communication competency approach to leadership** (Barge & Hirokawa, 1989; Johansson, Miller, & Hamrin, 2014) is based on the principle that a leader's communication must be competent communication. Simply communicating with group members is not enough. Competence is not synonymous with quantity of communication. To be competent, a leader must be able to adapt to the differing needs of group members and the group task. Moreover, group leaders must be flexible, possessing the ability to change to or adopt other competencies when the group needs it.

Three assumptions are the foundation of this approach. First, leadership is action that helps group members overcome the barriers or obstacles they face in achieving their goals or completing their tasks. This means that a leader must take active steps to reduce ambiguity and manage the complexity faced by the group. In other words, the leader helps the group create a system for working together and accomplishing its goals. Second, leadership occurs through communication. Thus, the relationships established and maintained between leader and group members through verbal and nonverbal communication are central to defining the nature of leadership. Third, individuals use a set of skills or competencies to exercise leadership in groups (Barge, 1994; Carter, Seely, Dagosta, DeChurch, & Zacarro, 2015; Johansson, Miller, & Hamrin, 2014).

Procedural Competencies

Procedural leadership competency is displayed when one or more group members coordinate group activities and help members function as a group. Procedures often help a group achieve its goal, and group members look to others in the group for procedural aid. The person who facilitates procedures best is likely to be selected as the group's leader (Ketrow, 1991). Thus, leaders provide team coordination; that is, they (a) successfully focus the group on its goals; (b) coordinate the skills, abilities, and resources available in the group; and (c) facilitate decision making. To do that, leaders must be organized and responsible, and knowledgeable about the project (Lambertz-Berndt & Blight, 2016). Indeed, groups perform better when their interaction is explicitly directed (Pavitt, High, Tressler, & Winslow, 2007).

To help the group accomplish its tasks or activities, a leader should be able to initiate structure or establish operating procedures. That is, when a group begins its work, the leader must help the group define its mission, and set goals and expectations. In a second set of task competencies that facilitate the group's work, leaders can help group members make sense of the group's task. This requires the leader to anticipate events and the impact it may have on the team. By articulating what may happen and describing how it will affect the team, the leader helps the team adapt to a new or modified situation. Group leaders can further help the group facilitate their work by providing performance feedback, and, if needed, provide coaching or training. Of course, leaders also need the essential task-related communication skills to encourage information flow and facilitate the group's deliberations and discussions, as well as give feedback or descriptive information about how team members are working together and progressing toward their goal.

Specific to decision-making tasks, the leader should demonstrate competency in analyzing problems, generating criteria to evaluate potential solutions, identifying those criteria for solutions or actions under consideration, and selecting the best solution or activity. Some group leaders must also demonstrate competency by coordinating the activities of their groups, especially in competition or performance tasks.

Procedural leadership behaviors are those that help the group assess and evaluate its discussions or expected outcomes. Group members recognize the need for someone to display task-relevant behaviors, such as initiating topics of discussion, giving information, and summarizing, and believe that the member who displays this type of behavior is the most influential person in the group. Groups can perform better when a leader focuses other members' attention on what information is critical for understanding the task and for obtaining information that is accurate (Pavitt, High, Tressler, & Winslow, 2007).

Relational Competencies

Relational leadership competency is displayed one or more group members help other group members cooperate with and express support for one another. In this function, leaders develop and maintain relationships with group members to foster interpersonal ties, increase motivation and goal activity, build commitment and cohesiveness among members, and create perceptions of fairness within the group. Thus, a leader must be able to demonstrate a wide repertoire of communicative behaviors, including being respectful and honest, being confident or assertive, being a good listener, sharing information and being open to information provided by others, and being competent and outgoing (Lambertz-Berndt & Blight, 2016).

There are two distinct sets of **relational competencies**: those that help group members with their internal relational dynamics, and those that connect the group to those outside the group. As individuals work together to accomplish the group task or activity, it is natural that miscommunication and conflicts will surface, challenging interpersonal relationships among group members. So, in relation to internal dynamics, effective leaders provide four types of relational assistance—interaction management, expressiveness, other-orientation, and relaxation—to help group members maintain, manage, and modify relationships within the group. Effective leaders assist the group in managing its conversations by clarifying and summarizing the comments of group members. Interaction management is also visible when the leader balances participation among group members. Managing conflicts and building consensus are further examples of the types of interaction management assistance leaders can provide.

Relational assistance with expressiveness helps groups avoid ambiguity. An effective leader encourages group members to express themselves clearly by identifying undocumented opinions and irrelevant remarks. In providing an other-orientation, the leader displays concern for and interest in other members, which helps the group develop a climate of trust and respect. Anxiety is a natural state in a group and occurs because individuals are often hesitant to express their ideas for evaluation. An effective leader reduces the amount of social anxiety in the group by creating a relaxed atmosphere of involvement and participation.

Thus, a leader has primary responsibility for maintaining a positive social climate among team members. To be effective in facilitating positive internal dynamics, the leader must be open or approachable and trustworthy. The leader must also be a good listener.

A leader must also have relational competencies in representing or connecting the group to others. The leader is often the connecting link between groups. In that role, the leader must monitor the external environment for opportunities or threats. To do this competently, the leader must manage the relational boundaries of the group and utilize the group network effectively.

Technical Competencies

Technical leadership competency should not be overlooked. Leaders must demonstrate technical competence relative to the technical demands of the group's activity (Bass, 1981). Leaders who cannot express or share their expertise, or who are unwilling to learn new skills on behalf of other group members, will be disregarded by group members. This does not mean that the leader of, say, a softball team must be the best fielder and hitter, or that the chairperson of the budget and finance subcommittee must be a gifted accountant and a tax law expert. It does mean, however, that the leader must possess enough technical competence to help other group members and to know when outside expertise is needed. Generally, we expect leaders to be qualified or technically competent in at least one area relevant to the group's problem or activity.

Integrating Leadership Competencies

Let's see how a leader can manage procedural, relational, and technical competencies. Kia is supervisor of an engineering team, and her team members have between 1 and 3 years of experience at this organization. Her team is charged with developing all of the materials and procedures a new customer needs when they purchase the organization's computer system. As the installation team leader, Kia's team handles three to five projects simultaneously.

> KIA: Before we start on this installation, I need to explain what's unique about this project. This is our first system installation in this company. So, besides what we normally do, we must also be brand ambassadors. Management is counting on us with the hopes that this installation goes so well that the company purchases another system.
>
> AARON: Okay, I'll get started by getting the specifications from the manufacturing team and I'll break them up in way that seems best for us.
>
> KIA: Great, thanks Aaron. Let me know if you need help from me or another member of the team. [*Pause.*] Aaron, would it be possible for Zack to follow you on this task? [*Pause.*] Zack, would you like to do that?
>
> ZACK: Sure . . . I've learned a lot from Aaron already. It would be great to work with him again.
>
> AARON: Won't Zack have his own tasks?
>
> KIA: Yes, Zack will have his regular responsibilities of keeping the timeline and budget. [*To Zack.*] I believe you can do both. Right?
>
> ZACK: Yes.

KIA: Frankly, I'm expecting that Zack following Aaron's work will allow Zack to improve our timeline and budget tracking.

LING: I guess I do what I always do?

KIA: Right, Ling, I want you to be the installation liaison again. You're doing a good job keeping the customers happy. [*To everyone.*] Remember, if you have a problem with the customer in anyway, contact Ling first. I'll take a look at how this project will fit with our other deadlines and get back to you by the end of the work day if I see that anything needs to shift.

First, notice how Kia demonstrates task competencies. She initiates structure for the group and helps members identify their roles for the task. Since satisfying customers is a primary goal, Kia praises Ling for her ability to do this, and reminds others that Ling will help them. With respect to relational competencies, Kia identifies the tension and ambiguity she has caused when she assigns Zack to work with Aaron. At first mention of this, Aaron seems defensive, perhaps because he believes Zack is not taking on as much responsibility as other team members. Kia provides an explanation to reduce this relational tension. Third, notice Kia's technical competencies. She expresses familiarity with the installation task and the integration of this task with others.

From a communication competency view of leadership, group members do not have to view leadership as residing in one person. Rather, many group members can provide leadership. By defining leadership as a process, we make communication central to the discussion (Clifton, 2006). Leadership is a social phenomenon, as group members in the roles of leader and follower need interaction with one another for leadership to occur. Leadership vividly demonstrates the type of interdependence found in group situations. Recall that one of the defining elements for a group, given in Chapter 2, is that members must have agreement on a goal. The interdependence created by group members sharing a collective goal forces issues of leadership to surface.

What competencies will your group require? It depends on two factors: (a) the type of goal your group is working toward and (b) the situational complexity of the group's environment. When the group's goal is primarily task-oriented (such as a sales team developing a marketing plan), the leader needs more procedural competencies. When the group's goal is primarily relation-oriented (e.g., maintaining solidarity among a fraternity group or a neighborhood association), the leader will need more relational competencies. In either case, a group leader must have some degree of technical competency. It would be difficult for other group members to look up to or follow a leader who was not at least moderately skilled in or knowledgeable about the content of the group's task. The degree of situational complexity—goal complexity, group climate, and role ambiguity—also affects the degree to which the leader needs to demonstrate these two types of competencies.

Four caveats are worth mentioning here. First, you are not exhibiting leadership if others are not following. If group members do not respond to your leadership attempts, you are not the leader. Second, being appointed as head, chair, or leader does not guarantee that you will influence others. Simply, having a title does not make anyone a leader. Group members will follow the member or members who exhibit influence in a positive manner to help them achieve their group and individual goals. Thus, leadership influence is not inherent in a position. Third, leadership and power are not synonymous. Leadership may be infused with power (Hollander, 1985), but other group members also control power in the group. Finally, the leader cannot do everything (Hollander, 1985). There are limits to everyone's capacities, knowledge, skills, and motivation in performing this role. As a result, many followers perform leadership roles in groups. Thus, the distinction between leader and follower may not be as clear as you might initially believe.

Based on expectations created by societal standards and by experiences in other group situations, members have expectations about how leaders should behave (Pavitt & Sackaroff, 1990). First, group members expect the leader to encourage participation by others. Second, they expect that the leader will keep the group organized by talking about the procedures the group will use, summarizing the group's discussion, and facilitating group discussion. Third, group members expect that the leader will work to develop and maintain harmony in the group by managing group conflicts. Finally, they expect the leader to play the role of devil's advocate or critical advisor.

BECOMING A LEADER

When we enter a new group situation, often one of the first things we want to know after identifying the group's task is who is going to be the leader. Leaders come to their positions in one of three ways: (a) they are appointed, (b) they are elected, or (c) they emerge from the group's interaction.

Appointed Versus Elected Leaders

An authority outside of the group can appoint leaders, or group members can elect their own leader. How a leader is selected affects the group environment (Hollander, 1985). Each method of leader selection validates one person as leader, and each creates a different reality for testing a leader's legitimacy.

When leaders are elected by group members—usually by a simple majority vote—members have a stronger investment in and more motivation to follow the leader than when the leader is appointed by outsiders. When things are going poorly for the group, elected leaders are more likely to be rejected by group members. Thus, elected leaders may have a greater sense of responsibility and face higher expectations for leader success than appointed leaders.

For example, suppose your group elects Jason as chairperson. You expect him to take responsibility for the group, yet you will blame him if he fails. One way to interpret this is in terms of the group giving a reward to one group member in advance, with the other group members then expecting the elected leader to "pay back" the group by producing favorable outcomes (Jacobs, 1970). Now let's examine what happens if Jason is appointed leader of your group. Your evaluation of Jason as a leader depends on his performance as leader and your confidence in whoever appointed him. If he does not perform well, you may attribute the group's failure to Jason. You can also attribute the group's failure to whoever appointed him, and you will be more likely to do so if Jason is well liked in the group. Although it may be more efficient to elect or appoint a leader, these procedures do not guarantee that the leader will be an effective communicator or that group members will perceive this person as leader of the group. Generally, a leader who is elected by the group after a process of allowing leaders to emerge and be tested is in the strongest position to get things done (Hollander, 1978).

Emerging as a Leader

Some groups rely on **emergent leadership**, whereby a leader who is not appointed or elected emerges as a result of the group's interaction. That is, a group member becomes the leader because other members accept and recognize an individual as a leader in this situation and in these specific interactions (Emery, Daniloski, & Hamby, 2011). Often, other group members assess a group member's ability to be their leader by the size, as well as the quality, of their contributions (Jones & Kelly, 2007).

Emergent leadership often occurs in groups in which no leader was appointed or in groups with ineffective leadership. At the start, group members assess the trustworthiness and authoritativeness of members to see who might be leader-worthy (Baker, 1990). The group member most likely to gain influence over other group members is the one with these characteristics:

- Is not hesitant to speak and speaks frequently
- Uses nonverbal movement to communicate a sense of dynamism, alertness, involvement, and participation
- Is supportive of and concerned with the welfare of others
- Says and does the things that others in the group want to hear
- Is charismatic
- Does not control resources to demonstrate power
- Contributes procedural and task-relevant messages

Thus, those members who take an active role and have a wide repertoire of communication skills in the group are most likely to end up in the leadership role (Hill, 2013; Pavitt, 1999). As equally important, group members who perceive themselves as leaders are more likely to become leaders (Emery, Daniloski, & Hamby, 2011).

Emergent leaders are generally those group members who demonstrate interpersonal understanding and trust of others. Using open and supportive communication, emergent leaders increase members' engagement with the task (Druskat & Pescosolidio, 2006). When one group member possesses this type of social and task awareness, other group members are likely to look to this person as the natural leader of the group. Leadership emergence can occur in two different ways. Let's use Ava as an example. Group members may willingly support the emergence of Ava as leader and encourage her to take on the leadership role. Or they may allow Ava to emerge as leader because they are passive and do not want to assume any of the group's leadership functions. In either case, a leader can only emerge through the sanctioning behavior of other group members. Thus, Ava emerges as a leader when other group members perceive her as leader and act as if she is leading them, and when her attempts to initiate action or structure the group's interaction are successful.

How you communicate within a group is important because other group members are evaluating your potential for leadership by assessing your communication skills (Schultz, 1986). In particular, your ability to communicate clear goals, give directions, and summarize will either identify you as a potential leader or eliminate you from consideration. If several members are competing for the leadership role, the degree to which you communicate in a self-assured manner contributes to your selection as leader. The member most likely to emerge as the group's leader is the one who can identify sources of differences or conflicts within the group and then develop and present a compelling rhetorical vision that can transcend those differences (Sharf, 1978).

Let's see how these principles are revealed in the following group:

NANCY: I'm glad I'm in your group. This should be fun.

QUINN: Me, too. It'll give me a chance to get to know Andrea better.

JOEL: Yeah.

ANDREA: Uh . . . what's your name, again?

NANCY: I'm Nancy, and that's Quinn and Joel.

QUINN: Can we get started? I've got another meeting in an hour.

NANCY: Sure. Where should we start?

ANDREA: I'm not sure I know enough at this point to really help out.

JOEL: Me either.

QUINN: Let's try getting started by identifying what each of us knows about the registration problem.

NANCY: Good idea, Quinn. For me, my enrollment time slot is when I'm in class. It just doesn't make sense to me. The university's enrollment system knows I'm registered for classes this semester. Why am I assigned an enrollment time that conflicts with my schedule?

QUINN: Joel, what do you think the problem is?

JOEL: I, uh, . . . don't really know. I just know it doesn't work.

QUINN: Andrea?

ANDREA: Well, it seems that . . . maybe I shouldn't say since this is my first semester.

QUINN: Okay. This is my fourth time to register this way. I agree with you, Joel, that it doesn't work. One thing I've noticed is that the registration form my advisor signs doesn't follow the registration prompts on the computer.

NANCY: Right. That sure makes it confusing.

QUINN: And I've had trouble trying to give another option when my first course selection is closed out. Well, it sounds like we've had different problems, but it also seems that we believe a different system for registering could be developed. Do you agree?

Who do you believe will emerge as leader of this group? Nancy and Quinn are certainly more assertive, and both are contributing ideas for the group to consider. Joel is both vague and tentative. Andrea bases her hesitancy to help the group on her limited experience at the university. But does that mean she could not be a good leader? If the conversation continues in a similar way, we can expect that Nancy or Quinn will emerge as leader.

SHARED LEADERSHIP

Most frequently, when we think of leadership, we think about one person in the leader role. However, given the complexity of group tasks and the distribution of group members across geography and time, it may be more effective for leadership to be shared. **Shared leadership**—sometimes called collective or distributive leadership—is the notion that there may be more than one leader or that leadership rotates among members across their work on their task (Contractor, DeChurch, Carson, Carter, & Keegan, 2012).

One principle of shared leadership is that leadership should be enacted by individuals at many or all levels. This is very different than traditional leadership models in which the leader has more formal power or status. A second principle is that leadership is located in a network of members-as-leaders who are interdependent. Third, shared leadership emerges from social interactions that are fluid and multidirectional. Thus, shared leadership focuses on collective achievement, mutual learning, and shared responsibility making it a valid alternative for team leadership (Fletcher & Kaufer, 2003).

Let's look at an example that demonstrates why shared leadership of both types is necessary. The town of Cary, North Carolina, in the United States, embarked on a multiyear process of creating a community plan as a guide for the town's growth. Across time, the town engaged in collecting data from citizens about existing conditions, generating and testing ideas for potential implementation, drafting policies, and getting feedback from the community. Across these processes, the town's council and town employees worked with a consulting firm. Citizens were invited to sessions at which they provided feedback in focus groups. Later, citizens were invited to evaluate the draft plan using online surveys.

Although a consulting group provided leadership for the entirety of the project, citizens were also invited to join self-managed groups that had the responsibilities of publicizing the process, speaking to community groups about the process, conducting focus group research, and talking about the process and the draft plan with their neighbors. Each type of group needed leadership, and there needed to be coordination across groups. Thus, leadership requirements changed across time as the project unfolded. Consulting group members were the official leaders of the project, but would have less influence than community members. Thus, citizens were invited to apply for the leadership positions that required direct contact in the community. The groups they led required specific, but different, sets of task expertise (i.e., public relations skills, public speaking skills, focus group moderator skills, knowledge of the town's history) in addition to the relational skills that foster citizen motivation and engagement.

PUTTING THE PIECES TOGETHER

Group Goal, Group Structure, and Interdependence

After reading this chapter, you should have developed some idea of your leadership effectiveness. Think about one of your group leadership experiences. How would you describe or characterize your leadership? Specifically, what communication strategies did you use? To what extent did these strategies help the group achieve its goal? In what ways did your leadership enhance or inhibit interdependence among group members? Did other group members find it easier or more difficult to work together? How did your leadership affect or alter the group's structure or its use of decision procedures? Were you the only leader? Were additional leaders required? Did additional leaders emerge? If so, why were other leaders needed? To what extent did each group member exhibit leadership to help the group?

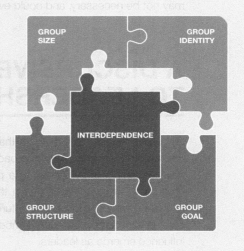

Unfortunately, for these groups the leader was always changing. Sometimes all members of the consulting group were present; at other times, only one of the consultants was in town. Likewise, citizens had work and family responsibilities that prevented them from being at all events. Despite the changing memberships, each group required leadership. This meant that a community member had to step in to provide coordination or directions for completing a task if others were not available. In this multiyear process, it would be unreasonable to expect one person, or even one set of people, to lead every aspect of the project. The leadership skills also needed to change as the project developed across time from exploration to data gathering to presentation of ideas.

This example illustrates several principles (Kramer, 2006). First, shared leadership is an ongoing process of balancing leadership roles over time. Different parts of the project required different skills from leaders and members. Second, the inconsistent availability of group members demonstrates that shared leadership must be a fluid process. Third, sharing leadership may be temporary. In this case, sharing leadership was effective and appropriate, as there was (a) a compelling vision that provided direction, (b) enough shared or easily available information among group members to complete required tasks, and (c) coordination among various groups loosely coordinated by the consulting group. Particularly in this example of community volunteers, participating in shared leadership boosted their confidence, satisfaction, and ownership—which ultimately benefited the overall project (Solansky, 2008).

Shared leadership can be a strength for a group or team, especially at the beginning of a new task. Shared leadership helps a group as members orient themselves to one another and the task. However, when a group reaches its midpoint, shared leadership may become an ineffective model of leadership for a group or team to use (Wang, Han, Fisher, & Pan, 2017). Why? Research has demonstrated that when a group nears the midpoint, group members have greater task familiarity, as the group has established behaviors and procedures for accomplishing its task. At this point, continuing with shared leadership may not be necessary, and could even be harmful.

A DISCURSIVE APPROACH TO LEADERSHIP

Discursive leadership considers that leadership develops through talk and is managed through conversation. This approach to leadership asks how a conversation functions, how leadership is evidenced in a particular conversation, and what kind of leadership emerges (Fairhurst, 2007). From the discursive leadership approach, decision-making talk both frames and defines future action of a group that needs be resolved. In the conversation among group members, influence is negotiated and those with the most influence emerge as leaders.

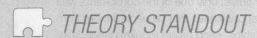

THEORY STANDOUT

Discursive Leadership Perspective

Examining interaction among group members from a discursive leadership perspective illustrates how leadership occurs (Aritz & Walker, 2014). Comparing transcripts of leaders using three different styles of communication also illuminates how messages delivered by leaders facilitate or hinder cross-cultural communication.

Aritz and Walker collected interaction data from multicultural groups whose members were from East Asian and American cultures, and another set of data from homogenous groups of American participants. All groups participated in the Subarctic Survival in which group members are placed in the role of airplane crash survivors who must discuss and agree on the ranking of items to be salvaged from the aircraft.

In Case 1, the group comprised one male English speaker (Speaker 1), two male Asian speakers (Speaker 2 and Speaker 5), and two female English speakers (Speakers 3 and Speaker 4). Speaker 1 communicates with a directive style of leadership by speaking first and identifying his item to be salvaged. Then Speaker 1 uses questioning to direct others to select his preferred option.

S1: I figure you can use fire, otherwise you're screwed.

S4: Okay, so let's—

S3: But if you, but if you just have matches, what are you going to do with them?

S4: Yeah.

S1: Well, at least you can start a fire though, don't you think? I mean it could be one or two, it doesn't matter.

This directive style of leaders was the one most commonly observed when groups comprised American and Asian participants. Member contributions were not balanced. This style of leadership was likely responsible for Asian participants reporting lower satisfaction with respect to feeling included, valued, or supported within their groups.

Case 2 illustrates how a leader emerges in group interaction by demonstrating a cooperative and inclusive leadership style. This group comprised one female English speaker (Speaker 1), two female Asian speakers (Speaker 2 and Speaker 3), 1 male English speaker (Speaker 4), and two male Asian speakers (Speakers 5 and Speaker 6). Although Speaker 1 does not begin as the leader, she emerges as the leader by using yes/no questions and open-ended questions to solicit information.

Speaker 1 does this by asking questions of the two Asian females who had not spoken, thus giving them a chance to join the group (i.e., "What did you guys put as the number one?" Later, Speaker 1 recaps the group discussion by summarizing and listing the items in order, which elicits an affirmative confirmation by other speakers. This case illustrates a more cooperative and inclusive leadership style which results in greater balance among member contributions. Unfortunately, this type of leadership occurred only a few times in these intercultural groups consisting of East Asian and U.S. participants.

In Case 3, a collaborative style of leadership develops among the five American participants who begin the discussion by using questions to establish the collaborative nature of interaction in the group. The first questions used by several members in the group frame the type of discussion that will follow. The leadership style is collaborative because all the group members are actively engaged in co-constructing the rules and the process for discussion. For example:

S2: Do we wanna go around and just give, like, our top five?

S1: What's the best, what's the least.

S5: Sure.

This type of distributed leadership among group members is more likely to occur when all group members communicate with a more aggressive and direct style that may not be appropriate in all cultures.

From this perspective, we wouldn't ask *who* is the leader. Rather, we'd ask *how* is leadership being negotiated and talked out in this particular conversation. "Consequently, leadership is not necessarily the property of any one person; it can be distributed and it is open to challenge" (Clifton, 2012, p. 150). And, as you would expect, group members who emerge as leaders are those who have powerful discursive resources. Thus, in leaderless groups, and even groups with a formal leader, leadership is negotiated as group members work on their task. Moreover, leadership may be distributed among many members, it is often performed collaboratively, and it is always contested in the group's conversation.

Because a discursive approach to leadership focuses on the interaction among group members, we must remember that how leadership is discursively created in one team can be different in another team (Baxter, 2015). We should also be sensitive to how difficult it can be for minority members of a group to use discursive practices this way (Walker & Aritz, 2015).

TRANSFORMATIONAL LEADERSHIP

A **transformational leader** is an exceptionally expressive person who communicates in such a way as to persuade, influence, and mobilize others. According to this theory, acting as a role model, the transformational leader sets an example for group members to follow. This type of leader uses rhetorical skills to build a vision with which members can identify. That vision creates a sense of connection with group members and motivates them toward goal completion (Bass, 1985, 1990). Although transformational leaders are perceived by group members to be powerful, they do not rely on their position of power or the use of organizational rewards. Rather, they communicate a sense of urgency and utility—a group vision—that members find appealing.

This type of leader creates power through the use of dramatic and inspirational messages. Thus, you can find transformational leaders in all levels of organizations and in group settings in which motivating people and providing services are more important than monetary rewards. For instance, your soccer coach might be a transformational leader, and your church group may be empowered by a transformational leader. Many civic and community groups, particularly grass roots organizations, are led by transformational leaders. As you might guess, transformational leaders are successful at recruiting group members and helping them achieve high-quality performance.

Transformational leadership occurs when leaders broaden and elevate the interests of group members, when they generate awareness and acceptance of the group's purpose and mission, and when they encourage group members to look beyond their own self-interests and work for the good of the group. Thus, group members are encouraged to take on more challenges and greater responsibility. In doing so, a transformational leader can help a group that has considerable diversity develop an integrative social identity, which in turn can help a group minimize group differences, especially those based on demographics (Kunze & Bruch, 2010; Wu, Tsui, & Kinicki, 2010).

Transformational leaders have charisma. This means that they have confidence in their communication competence and conviction in their beliefs and ideals. Such a spirit generates feelings of faith, trust, and respect from other group members. But, more importantly, transformational leaders inspire others by communicating high expectations. These leaders are animated, which arouses others and heightens their motivation. Transformational leaders are intellectually stimulating, helping group members to be more aware of problems and to pay more attention to problem solving. Most importantly, and the key reason for their success, transformational leaders give special attention to each group member, treating members as individuals. Thus, each group member is treated differently according to the members' needs and capabilities.

As you can see, the way in which a transformational leader interacts with group members can be very powerful. Other studies have demonstrated that when a leader uses a transformational style, group members are better at problem solving and use fewer counterproductive messages, such as going off topic, criticizing, or complaining.

Transformational leaders are particularly good at getting group members to perform the extra work that is often necessary to achieve performance goals (Avolio, Waldman, & Einstein, 1988; Gardner & Avolio, 1998). How is this accomplished? Transformational leaders inspire their followers. Together, leader and followers create a larger collective with which members identify. As identification with the leader increases, so does commitment. Despite time and energy pressures that threaten to keep group members from contributing to group activities, transformational leaders are able to persuade group members to do whatever it takes to achieve group goals. The confidence and inspirational qualities of transformational leaders are the motivating factors for group members. Let's see how Deanna uses transformational leadership:

> I know you don't need one more thing to relearn, but there needs to be a change in pharmacy procedure. When I heard about this, I suggested that our team create the new protocol. Why? Because we're trend setters, and I know we can perform this task effectively and efficiently! Here are the details about handling patient medication. I think this new procedure has lots of merit. [*Deanna hands out the information sheet.*] As you work with patients today, please consider what recommendations you would make about handling patient medications. After the shift tonight, let's take 10 minutes to discuss your ideas. I'm sure that collectively we can produce a draft to be considered.

How can you become a transformational leader? First, you must assess the working climate and task of the group. You might rearrange or restructure work on group tasks to provide more stimulating activities. By knowing the current state of affairs in your group, you can then address what you would like the group climate to be like. Ultimately, these assessments will lead you to strategies that can help group members recognize their individuality and creativity, and their responsibility to the group. What you will find as a transformational leader is that you are valuing group members differently. Together your group will have been transformed from "what is" to "what is desirable" and "what ought to be" (Rosenthal & Buchholz, 1995).

Second, transformational leadership is based in communication ability (Levine, Muenchen, & Brooks, 2010). If you can answer yes to the following questions, you may have the communication skills necessary to be a transformational leader:

1. Does your communication act as a role model for group members?
2. Can you define and articulate a vision for the group?
3. Do you earn the trust and respect of others in the group?
4. Can you influence and inspire other group members to excel?
5. Can you stimulate group members to think in new ways?
6. Do you avoid criticizing group members in their attempts to try new things?
7. Do you consider and recognize each group member as an individual?
8. Are you willing to listen to and empathize with others?
9. Can you coach or mentor group members?

If you consistently answered yes, you likely possess the four traits necessary to be a transformational leader. First, a transformational leader demonstrates idealized influence by acting as a role model and articulating a vision and goal for the group. Second, a transformational leader creates inspirational motivation by communicating high expectations. Third, a transformational leader creates intellectual stimulation by challenging members to think creatively. Finally, a transformational leader practices individualized consideration by providing a supportive climate that helps each group member develop and reach the member's full potential.

GENDER ASSUMPTIONS ABOUT LEADERSHIP

Our gender assumptions about leadership are firmly embedded in society (Ridgeway, 2001), and are frequently reproduced in groups and teams. One reason for this may be the way in which the male leadership assumption is entrenched in our language, which makes it difficult for us to examine our gender stereotypes about leadership (Walker & Aritz, 2015). For example, recent research demonstrates that group members continue to view male and female leadership differently in terms of the gender stereotypes leaders display, as well as the gender stereotypes group members hold (Wolfram & Gratton, 2014). One way we may reinforce this is through the language we use. When talking about leadership in general, it is common for someone to use masculine pronouns.

It can be difficult to avoid assuming that males are more likely linked to leadership roles when group members use gender-specific language, such as in "the person we elect as leader, well, he should be forceful, strong, and willing to work as hard as we do" or "the chairman will decide when the report will be due." You can avoid this assumption and encourage both men and women to consider the leadership role in your group by using gender-neutral language (for example, chairperson, not chairman) when talking about group roles.

As our societies and workplaces become more internationalized, we need to be mindful that not all cultures had such gendered views of leadership. For example, groups in Turkey, a country with more feminine and collectivist characteristics, neither sex or gender role predicted who would emerge as group leaders (Türetgen, Unsal, & Erdem, 2008). Further, research has also demonstrated that when group leaders hold strong and positive beliefs about diversity, groups have higher levels of cohesion and are less likely to create subgroups based on gender differences.

MESSAGE AND MEANING

Sung (2011) published a journal article that examined the gendered nature of leadership as displayed on the television reality show *The Apprentice*. In this show, competitors form teams; each week someone from each team is "fired." In the following excerpt from the show, Sung identifies Omarosa's performance of leadership as a mixed masculine and feminine, yet predominantly masculine discourse style. This conversation (Sung, 2011., p. 99) was held in a cab. Heidi is in the cab with Omarosa when she receives a phone call from two other team members, Jessie and Kwame, requesting that Omarosa get the phone number of the foundation they are going to work with. As Sung describes, her rejection is made in a masculine style.

Omarosa: [*Answering the call from Jessie.*] Hello?

Jessie: Hey, Omarosa, can I get the number for Katie Card, plus your contact for the foundation?

Omarosa: Okay. Why, why are we calling her?

[*Pause.*]

Omarosa: Hey, let me speak to Kwame.

Kwame: Yeah, give me the number for Katie.

Omarosa: I wanna talk with her as well cos I haven't had an opportunity to talk with her just yet.

Kwame: Right now we need the number quickly.

> **Omarosa:** [*Talking to Heidi.*] Are we here?
>
> (we are here)
>
> **Kwame:** Okay, what's the number?
>
> **Omarosa:** Let's talk when we get together.
>
> **Kwame:** Would you please give it to me?
>
> [*Omarosa hangs up her cell phone.*]
>
> **Kwame:** Hello?
>
> **Omarosa:** [*Talking to Heidi.*] I'm sorry I had to bang it on them. They're not listening to me.
>
> What meanings to you extract from this conversation? What arguments would you make to support Omarosa's leadership style as feminine? Masculine? How does her gendered leadership style influence her effectiveness as leader?

There is evidence that women and men prefer different styles of leadership (Berdahl & Anderson, 2005). When all group members are women, decentralized leadership is preferred. When all group members are men, or there are more men than women, centralized leadership is preferred. Interestingly, groups in which the number of men and women are the same, or in groups in which there are more women than men, centralized leadership is used at the beginning of a group. Over time, however, leadership becomes decentralized.

LEADERSHIP IN VIRTUAL GROUPS

Especially in organizational settings, leaders will use both face-to-face and online communication with members of their teams. If the team is communicating on video and audio platforms, we would expect that most of what is known about group and team leadership would be salient there as well. However, some teams use textual online communication, such as chat, which significantly reduces auditory and visual cues. In face-to-face settings, leaders rely on these cues to understand how team members are working together, and how they feel about the work and each other. In an online textual environment, it's easy to quickly compose a message and send it without thinking how the message will be received. As a result, leaders of teams in online environments are likely to have fewer cues by which to measure uncertainty (Gilstrap & Hendershot, 2015). This may cause leaders to over-communicate with their team members.

Research has shown, however, that communicating online frequently with team members who have not met face-to-face may not be beneficial to group performance. In other words, more online communication may not increase trust among team members or with the team leader (Chen, Wu, Ma, & Knight, 2011). A better strategy, when possible, is for team leaders to use both face-to-face and online communication, and to move to face-to-face interaction when uncertainty is high.

ENHANCING YOUR LEADERSHIP ABILITY

Effective team leadership is critical to a group's success (Hirokawa & Keyton, 1995; Larson & LaFasto, 1989). To be an effective leader, you must be in control of three factors: knowledge, performance, and impression. Together, these three factors form the basis of how group members evaluate your communication competence and leadership ability.

First, are you knowledgeable about leadership issues? Do you understand a variety of leadership styles, and can you explain why different types of leadership may be needed? Do you know if your group has a greater need for relational support or for task guidance? Can you identify the decision procedure that is most needed by your group? However, being knowledgeable about leadership isn't sufficient to make you a competent group leader. Other group members can't benefit from your knowledge unless you demonstrate your knowledge through your leadership performance.

Second, can you perform a variety of leadership behaviors and functions? Or are you stuck, having to rely on one type of leadership behavior? Is your leadership situationally appropriate? Being flexible and able to adapt to the needs and expectations of other group members is a hallmark of effective leadership (Nye, 2002). Still, there are some leadership behaviors that are effective in nearly all group situations: establishing and communicating the goal or intention of the group, keeping the group focused on its primary activities, taking steps to establish a positive group climate, monitoring or facilitating interactions among team members, and modeling competent group communication skills (Galanes, 2003). These are common leadership expectations across a variety of groups. The leadership performance you communicate to and with other group members is what other group members evaluate.

Third, what kind of impression do you make as a leader? A group leader who is generous with his or her time and energy, is willing to do favors or make sacrifices for others, shows personal interest in others, and praises others' ideas and actions will create a favorable impression with group members (Rozell & Gundersen, 2003).

In many group situations, leadership may be better expressed as facilitating group member interactions leading to goal realization. Too frequently, leadership is conceptualized as an overly directive style, with the leader arguing for a particular position, refuting information that challenges it, and advocating for a decision that supports this position.

While this style may satisfy the leader's needs, it is unlikely to satisfy other group members. Moreover, this style of leadership can have detrimental effects on a group. By imposing leader preferences, the flow of information from and among other group members is stifled (Cruz, Henningsen, & Smith, 1999).

Leadership is like walking a tightrope. You must balance task and relational concerns throughout the group process (Barge, 1996). The effectiveness of a leadership style will change as the group matures and moves from a beginning to an ending point. You must be able to anticipate and deal with unexpected problems and to regain control if the situation warrants it. Your flexibility as a leader will dictate your balance, sense of control, and confidence—and hence your success as a group leader. Try the "Putting the Pieces Together: Group Goal, Group Structure, and Interdependence" feature to assess your leadership effectiveness.

SUMMARY

Leadership is a process of influence that occurs when group members interact. According to the functionalist view of leadership, it is important for leaders to have the ability to manage several types of group situations. Procedural, task, and maintenance behaviors are all important as group leaders navigate the complex social dynamics of group interaction.

Leaders may be appointed, elected, or emerge from the group interaction. Elected leaders generally have a greater sense of responsibility and a higher level of accountability than appointed leaders. Emergent leaders are usually group members who are active and dominant in the group's conversation, are trustworthy and authoritative, and can monitor the group situation to meet the task and relational needs of members. Additionally, as a group juggles a variety of challenges and goals, shared leadership may be necessary. In this type of leadership, members use the interdependent nature of task and group interaction to allow multiple members the ability to influence the process. Shared leadership allows a group to be more flexible and adapt according to members who have skills appropriate for a given task.

The communication competency approach to leadership is based upon a leader's competence in both task and relational skills. A third competency, technical skills, enhances these other areas. A leader helps organize and manage a group's environment, facilitates members' understanding of obstacles they face, and helps members plan and select the most effective actions. The more complex the group activity, the more complex the leader's communication needs to be.

Transformational leadership theory explains why some leaders are more effective than others. A transformational leader communicates a sense of urgency and utility, which motivates group members. Group members report that transformational leaders are charismatic, inspiring, and intellectually stimulating, and that they treat each group member as an individual. Thus, this type of leader can empower group members to accomplish more than they originally thought possible.

Society holds stereotypes about leadership and gender. Sometimes these stereotypes are reinforced when masculine language is used to describe leadership approaches. Research has shown there to be differences in how females and males lead; however, both can be effective leaders.

Effective team leadership is critical to a group's success. You must be knowledgeable on leadership issues and be able to perform a variety of leadership behaviors and functions, and you should leave a favorable impression as a leader. Together, these three factors form the basis for how others evaluate your communication competence as a leader.

DISCUSSION QUESTIONS AND EXERCISES

1 Select at least two people you know who lead or direct groups, and ask them to participate in informal interviews on their views of leadership. You might select someone who (a) chairs a task force or project team in a for-profit organizational setting, (b) leads a not-for-profit group of volunteers, (c) chairs a committee for an educational or government organization, or (d) leads a religious study or self-help group. If possible, find someone who leads an online team. Develop at least five questions to guide your interaction with your two leaders. For example, how do they view their role as leader? What leadership functions do they perform for the group? How did they come to be in that particular leadership role? How do they believe other members of the group perceive them and evaluate their leadership? If there is one thing they might do to improve their leadership, what is it?

2 Think of a community, regional, or national leader who is a transformational leader. What evidence do you have to support that claim? Do others agree with your assessment? Is your evidence based on the leader's communication behavior, the communication behavior of the leader's followers, or the outcomes achieved by the group? Which of these do you believe is the best direct evidence that transformational leadership is an effective method for creating and sustaining positive leader-member relationships?

3 Identify the leadership task, relational, and technical competencies that you feel comfortable using in groups. Are there leadership competencies that you should develop further? What leadership competencies do you lack? What could you do to develop those behaviors you identified?

4 Set a timer for 3 minutes. In that time, think of as many labels as you can for "leader." In addition to your own experiences, also think about what leaders

may have been called at different points in history, in organizations, in families, in friendship groups, in community and civic groups, and so on. Compare your list with other students' lists. How did your lists differ? What labels did you overlook? How do different labels imply different approaches to group or team leadership?

Using Group Communication Skills for Group Presentations

When delivering a group presentation, it may be important to have a specific individual provide leadership or guidance. This person may be a coordinator, synthesizer, facilitator, or present a particular important part of the presentation, based on the leadership needs of the group. For example, a coordinator may introduce the members of the group and the topic of the presentation, and may also help with transitions during the presentation. A synthesizer may present the last part of the presentation, bringing all previous presenters' messages together for a final meaningful summary. A facilitator may help engage the audience with the group presentation in a meaningful way. Of course, a leader may play several of these roles at once. Importantly, a leader should ensure a clear structure to the presentation. Additionally, the leader is likely the person group members will turn to if any difficulties arise.

Image Credits

Photo 8.1: Copyright © Depositphotos/michaeljung.
Figure 8.1: Adapted from: Copyright © Depositphotos/megastocker.

CHAPTER 9

FACILITATING GROUP MEETINGS

After reading this chapter, you should be able to:

- Create a draft of a group charter to present to your group or team
- Carry out the pre-meeting responsibilities as a leader or member of a group
- Design and lead an effective group meeting
- Select and prepare appropriate visuals to help the group record what is happening
- Take effective minutes for your group meeting
- Carry out the post-meeting responsibilities as a leader or member of a group
- Assist your group in overcoming typical meeting obstacles

Most group interaction depends to some degree on meetings. Indeed, interaction among group members within a meeting is necessary for groups to be effective (Somech & Drach-Zahavy, 2007). For some groups, meetings will be the only opportunities for all group members to meet and interact. Even for groups that perform physical tasks (e.g., a fire fighting team, a cheerleading squad), a group meeting may be held before or after the task. Meetings may be held to create ideas, exchange opinions, solve problems, make decisions, negotiate agreement, or develop policy or procedures. Regardless, meetings are events at which group members make sense of themselves, their task, the group, and the environment (Schwartzman, 1989).

Essentially, group meetings are the core of group interaction. Some meetings are private, other times nonmembers are invited to attend. Other group meetings are open to the public. For some public meetings, the public serves as an audience at the meeting; in some of these meetings, members of the public can participate in the meeting. Finally, some group's meetings are recorded in detail (e.g., written notes or minutes, or digitally recorded), whereas others groups meet without documenting what happened. Regardless, group meetings are opportunities for relationships to develop among people, tasks, resources, roles, and responsibilities (Baran, Shanock, Rogelberg, & Scott, 2012; Mirivel & Tracy, 2005).

Meeting interaction is never neutral or benign. Research has demonstrated that what is said in meetings is interpreted strategically by other group members—even in meetings that are routine and follow a standardized agenda. While tasks are the focus of most meetings, group members infer relational intent or relational strategies from others' meeting interaction. Thus, meetings are also events where relationships are defined between and among group members (Beck & Keyton, 2009).

Since meetings can vary dramatically in their private or public nature, level of formality, number of attendees, and scope of activities, most people who attend meetings develop their own criteria for determining a meeting's effectiveness (Beck, Littlefield, & Weber, 2012; McComas, 2001). Thus, what you may believe is effective interaction at a meeting may be evaluated as ineffective by others.

There are, however, generally accepted principles of meeting management. This chapter explores methods of managing group interactions during meetings and describes the responsibilities of both leaders and group members. Too frequently, meetings become sites of information transmission and fail to make use of group members' knowledge and skills (Myrsiades, 2000). Thus, it is the responsibility of all group members to manage meetings to facilitate group productivity.

Traditionally, group or team leaders take on the responsibility for much of the planning associated with meetings. But members also have obligations. **Meeting citizenship behaviors** are those actions that members can take to maintain a positive social and communication climate. To support the goals of the meeting, members can be prepared to speak up when input is requested, volunteer helpful information, make suggestions to the agenda, and encourage other members to participate by asking questions. By taking on these responsibilities, members help to more effectively facilitate the meeting (Baran et al., 2012).

MESSAGE AND MEANING

Meeting agendas provide structure to group discussion. Not only do agendas list meeting topics for all members to see, but leaders can also designate amounts of time for each item. Members may gain insight into how complex or important meeting items are based on the length of time given to the respective issue. It is also common for leaders to describe the agenda as tentative, in case adjustments need to be made as new topics develop just before or during the meeting, or if one topic requires the entire meeting time. These adjustments may allow leaders to keep the meeting discussion focused on the topic at hand, as in this case during a team meeting of special education instructors at an elementary school:

JILL (leader): So, the first item on the agenda is whether we should increase the number of students we work with.

FACILITATING GROUP MEETINGS   165

> **PAM:** I think we should. If we don't do it, the district [of education] may force us to anyway. If we increase it, at least we will have a say in the matter.
>
> **CALVIN:** I agree. The writing is on the wall—let's just go ahead and do it.
>
> **TERESA:** I'm not so sure. We are already pretty busy.
>
> **AIDEN:** We should also increase the number of teachers we have. I'm afraid the next budget hasn't allotted enough for new teacher hires.
>
> **JILL:** I appreciate your point Aiden, which you have brought up a few times. To make sure we discuss that issue, I'm going to add it as agenda item number 4. Teresa, could you elaborate a bit more on your concerns about increasing the number of students?
>
> **TERESA:** Sure. My first concern is that we all have different workloads currently based on …
>
> Note that during the meeting Jill added an item to the agenda in order to discuss Aiden's point. Do you believe Jill was justified in handling Aiden's comment in this way? How does this adjustment benefit or harm the discussion? How does adding the agenda item address Aiden's concern?

One final caveat: Most of the research on meetings has been conducted on work teams or public meetings. As you know, group interaction spans a wide variety of contexts. For example, athletes on sports team have meetings before games. Some meetings in health care settings are attended by social workers, therapists, parents, and children (for examples, see Davis, 2008; Wittenberg-Lyles et al., 2013). The team meets to discuss and assess progress and to develop new goals. In this case, the health of the child is the focus, and the role of the team is to ensure that goals are met. While this type of team differs from those meetings for which members have a formal hierarchical structure and tasks assigned by outsiders, some of the principles and procedures described in this chapter can be modified and used effectively.

DEVELOPING A GROUP CHARTER

If yours is a group that will meet over an extended period, developing a charter can help your group develop cohesiveness and find unity of purpose. A **group charter** or mission statement describes the goals or mission of the group, and describes behaviors that are appropriate in this particular group. Developing a charter early in the group's history is recommended, as doing so gives team members the opportunity to discuss and agree on expectations related to member contributions, how meetings will be managed, and how work will be distributed and, perhaps, evaluated. Talking through these issues and affirming them in writing will help the group move effectively toward its goal.

To develop a group charter, the group must discuss and agree upon what members view as important and what they hope to accomplish in this group experience. Group goals are generally the primary component of a charter or mission statement (Mathieu & Rapp, 2009). Each goal should be listed individually, and both task and relational goals should be included. What individual members can expect to learn or obtain from the group can also be listed. Figure 9.1 provides an example of a group charter. As you can see, the charter is specific and clear, and provides direction for the group. However, it does not dictate how the group will meet its objectives.

Check your charter by answering the following questions:
- Is the statement understandable to all group members?
- Is it brief enough that team members can remember it and keep it in mind?
- Does it clearly specify the activities of the team?
- Does it reflect realistic goals?
- Is it in line with members' values and beliefs?
- Is it inspiring or motivating to members?

Not only does a charter or mission statement help a group define and solidify its purpose, this document can be used to stimulate group discussion at a later meeting (Heath & Sias,

---

Group Charter

Group's Mission Statement: To work interdependently as team members to identify relevant issues, resolve problems, learn new skills, and have fun.

Our group will develop a strategic plan for our organization for the next five years. The plan must: (a) be accepted by the executive committee to which this group reports, (b) be implemented by the rest of the organization, and (c) meet a set of conditions given to us by the executive committee.

In working on the strategic plan, each group member should develop skills in group facilitation, organizational forecasting, and team member effectiveness.

Our goal is to complete the first objective in 6 months from our start date.

Figure 9.1 A Sample Group Charter

1999). Posting the charter or mission and asking "Are we on track?" or "What are we doing to implement our mission?" or "Has our mission changed?" helps group members assess their progress toward their purpose and goals. Does having a team charter really make a difference? Yes, teams that do develop a group charter report greater member satisfaction and group cohesiveness, more effective communication, and greater effort by team members (Aaron, McDowell, & Herdman, 2014).

DEVELOPING A CODE OF CONDUCT

Additionally, some groups and teams also create a **code of conduct**. The purpose of this document is to help group members identify behaviors that are expected from team members and avoid behaviors that are not (Hill & Rapp, 2014). A code of conduct lists those behaviors that members feel are appropriate and will help them be effective in the group. Figure 9.2 provides an example.

Too frequently, group members do not discuss what they expect from one another in terms of behavior. Left undiscussed, members are unsure of what is appropriate or inappropriate behavior. Thus, they use behavioral norms from past group experiences to guide their behavior. Of course, norms from previous groups are not always transferrable to other group situations. Attendance and preparedness are examples of individual behavior to be included in the code of conduct. Group-level behaviors—those related to role sharing within the group, decision-making rules, meeting attendance, election procedures, and group structure—can also be described and included. Specifying both individual and group behaviors ensures that all members are aware of what is expected of them in this group.

A code of conduct provides a set of guidelines—much as rules establish the guidelines by which you participate in any sport. Guidelines provided by the code of conduct create equity in the group process because all group members share in their creation. By developing a group charter and a code of conduct, members are more likely to share perceptions about what constitutes effective group membership. Developing these documents also helps a group crystallize its identity and culture.

IMPORTANCE OF MEETING MANAGEMENT PROCEDURES

Meetings can be valuable in helping group members reach their goal, as the two most common functions of meetings are to generate and share information. But they can also

> **Code of Conduct**
> **created and agreed to by all group members,**
> **August 14, 2017**
>
> As a team, we expect all members of the team to support the team's charter. Our communication, actions, and decisions should emphasize teamwork, not individual accomplishment. Each group member is responsible for being involved in the day-to-day business of the team. Each group member expects to receive information and influence from all other group members.
>
> As a group, we will elect a leader for each month we are together as a team. By rotating the leadership role, we will help to develop each member's leader and follower skills. The leader is to provide overall direction and support for the team. Followers are to carry through their assigned responsibilities and inform the leader if they encounter any obstacles.
>
> We expect to make decisions as a team using the majority vote rule. In cases where a minority vote member is so uncomfortable with the decision outcome that he or she cannot support the group's decision, he or she may ask the group to reconsider the issue.
>
> Group members are expected to attend all meetings. When a member cannot attend, other members will expect his or her assignments to be completed and handed in to the leader before the next meeting. In the event any member finds that he or she cannot fulfill the responsibilities of being a member of this team, he or she may ask the group for a reduction in, or termination of, group responsibilities.

Figure 9.2 A Sample Code of Conduct

waste valuable time. Whether meetings are positive or negative really depends on how prepared you are for the meetings' activities and what meeting facilitation skills you can contribute to the group. Even though some organizations are replacing meetings with more frequent use of technology that allows employees to meet across divisions in time and geography, there is no technology that can fully replace face-to-face meetings.

Whether face-to-face or online, you need to be skilled in basic meeting management procedures. Even simple agendas can provide structure for groups and keep meetings running smoothly. Procedures help group members coordinate their thinking and provide a set of objective rules all members can follow. You have probably been a member of a group that did not accomplish what it intended because another topic was introduced into the discussion. As a result, the group spent most of its time on this new topic, forcing you and other group members to make important decisions in the last 10 minutes without adequate discussion.

Meeting management procedures and facilitation strategies benefit groups by balancing members' participation. When all group members share their input, higher-quality decisions result and members are more supportive of the group's output. There are many other advantages to using meeting management and facilitation strategies. These techniques help to uncover and then manage conflicts that can steal valuable resources and time from the group. They provide structure that can be revisited if a group takes a temporary detour, and they encourage group members to reflect on their meeting process and progress.

Meeting management procedures and facilitation strategies help teams develop more effectively and overcome obstacles. Being able to help your group manage its meetings and providing it with facilitation expertise are responsibilities that go along with group membership. Fulfilling each of these responsibilities allows you to participate in the group to the best of your abilities. At the same time, you are helping your group's interaction become more effective and efficient. You might think that the leader should bear these responsibilities. But when group members participate in helping the group's interaction develop effectively, the group's process is smoother and members are more likely to feel satisfied on both task and relational dimensions (Kauffeld & Lehmann-Willenbrock, 2012).

Let us first explore meeting planning. You have probably used some of these procedures in the past, but just because your group has developed an agenda does not necessarily mean that it effectively manages its time together. There is more to formal meeting planning than simply listing items of business.

MEETING PLANNING

Meetings should not just happen, but many do. When a meeting is called, most of us jot down the time and date, and then show up. But that is really not enough.

Pre-Meeting Planning and Preparation

The group's leader or facilitator should do pre-meeting planning, and every group member should do some pre-meeting preparation as well. If you need some motivation to do this extra work before your next group meeting, think about this: A typical group meeting generates somewhere between 100 and 600 speaking turns or opportunities for individuals to talk (Scheerhon, Geist, & Teboul, 1994). Can you imagine trying to make sense out of so much information without some prior knowledge beforehand?

Leader Pre-Meeting Responsibilities

Before calling any meeting, the leader should first decide if there is enough business to hold a meeting, and if so, what the meeting's purpose should be. If there is not enough business, or if a clear purpose does not emerge, do not hold a meeting. One way to make these decisions is to list the specific business items you want the group to consider or accomplish during its next meeting. Now look at the items. Can they be organized in some fashion that will make sense and move the group forward? If not, are these issues really ones that the entire group needs to discuss? Could talking individually with some group members take care of these issues?

It may seem obvious to consider the overall purpose of the meeting, but answering these questions can force you to consider why you need a meeting in the first place. Perhaps you are going to call a meeting because your boss requested the vacation schedule for your department. Is a meeting the best way to collect and coordinate this information? If you cannot identify a purpose for the meeting, do not have one!

Now consider the participants. You need to sort out who should be invited to the meeting and who should be informed about it. The two lists are not always the same. If a key person cannot attend, should the meeting be cancelled or rescheduled? Or should the meeting go on regardless of who shows up? Answering these questions can help you determine the importance of those attending the meeting, as well as the importance of the meeting itself.

Once you have decided that there is a valid reason for the group to meet, you need to consider how long the meeting should be. Everyone identifies a start time, but few groups know when they can expect to be finished. Identifying a stop time is important because it can help a group focus on its work. Knowing that time is limited is a motivator that can keep group members from delaying action or making a decision. Group members appreciate knowing when a meeting should be finished. This actually increases attendance because it allows group members to schedule around the meeting and avoid time conflicts. This is important because group members are likely to hold membership in other groups, and time devoted to meetings must be integrated with their other responsibilities.

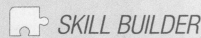

SKILL BUILDER

Developing a Meeting Agenda

For the next meeting you will participate in (for this class, at work, in your community), develop an agenda. If the group typically uses an agenda, try not to rely on a past agenda for form or substance. As you develop the agenda, use the principles described in the chapter to identify the meeting's activities, and provide enough information for the agenda to be useful to members before and during the meeting. What would you say to group members to encourage them to use an agenda to help structure group meetings? What specific advantages can your group expect from implementing this meeting procedure?

FACILITATING GROUP MEETINGS

Time limitations make an agenda that much more important. An agenda lists what the group needs to consider in detail and what the order of consideration will be. Group members should receive a copy of the agenda before the group meets and not simply at the beginning of the meeting. This way, they can plan what they want to say and collect data or information to support their point of view. When group members have an agenda before the group meets, they are better prepared to contribute effectively and efficiently. Figure 9.3 shows a sample agenda.

The **agenda** should list the meeting's start and stop times, the location of the meeting, the expected attendees, and the overall goal of the meeting, as well as the specific goal of each agenda item (e.g., to share information, to discuss a proposal, or to make a decision). Additionally, the agenda should identify or describe any preparations that group members should make, such as "Come to the meeting with ideas on how to help our

---

Agenda
Project Development Work Group
Thursday, January 29, 1:00 to 2:00 p.m. Conference Room A

Participants: Cynthia, Dan, Lu, Marquita, Tyron
Purpose: Project Update Tracking

Welcome

Introduce any guests

Preview agenda; ask for additional agenda items

Information sharing

    Review developments since January 15th meeting

    Dan, report on final numbers for December's project activity

    Cynthia, tell group about presentation to Federal Railways

Discussion items

    Progress on planning of telephone service cut-over

    Evaluation of new project tracking board

Decision item

    Need decision on feasibility of upgrading digital networks

    (bring cost estimates)

Suggestions for next meeting's agenda

Set next meeting date/time

Adjourn

Figure 9.3 A Sample Agenda

department pass its safety assessment." Notice how the agenda starts with items that are easy for the group to manage and that give group members the opportunity to contribute. Welcoming members, asking for additional agenda items, and sharing information are activities that do not take much time but that can help the group establish a positive or supportive climate. Now that the group has warmed up, it's time to take on more difficult tasks. Unless decisions are interdependent and need to be made in a particular sequence, it's best to make easy decisions first and work your way up to the more difficult ones. Wrapping up the meeting by discussing the agenda and date for the next meeting gives members an opportunity to regain their composure if the decision making was contentious. Ending on a positive note with respectful and positive messages completes the meeting cycle (Tropman, 2003). Try "Skill Builder: Developing a Meeting Agenda" to test your skills in this area.

Is creating an agenda worth the effort? Yes! Research has demonstrated that an agenda helps "guide attendees through the meeting, regulate activities, facilitate discussion, and minimize the need to backtrack" (Odermatt, Konig, & Kleinmann, 2015, p. 55). As important, when a written agenda is distributed before the meeting and used during the meeting, meeting effectiveness is improved (Leach, Rogelberg, Warr, & Burnfield, 2009). Once an agenda is complete and distributed to all group members, the group leader still has preparation work to do:

- Given the items on the agenda, what leadership style should you use?
- What decision procedures will be most appropriate?
- Will the group meeting require any equipment?
- Have space and equipment been reserved?
- Are there enough seats?
- Can the participants fit comfortably around the table?
- Does the table and the configuration of the room enable all participants to see and speak to one another easily?
- Is the room available when you need it?
- Do you need to make a reservation?
- Will you need refreshments?
- What level of documentation is needed?
- What agendas, minutes, or reports will the meeting require?
- How many copies will be needed and who will make them?
- Will it be necessary to have overheads, flip charts, chalkboards, or technology during the meeting?
- Do you have to make an equipment reservation?

After this needs assessment, you are ready to plan the meeting and invite those you identified as necessary participants and inform others who simply need to know about the meeting. Make sure to give adequate lead time and send along the agenda and any other documentation they will need prior to the meeting. If you want participants to prepare in some special way for the meeting (e.g., to bring budget requests; to prepare a work schedule), make sure to tell them that. The more completely you prepare for the meeting, the more quickly the group will be able to work on its business and complete its activities.

Physical Environment and Material Resources

Usually, it is the leader who arranges for or secures the physical environment in which a group will meet. It is important to find a quiet meeting place where the group can have privacy. This type of setting promotes relational development because group members feel more comfortable negotiating differences of opinion in private. To the extent possible, seating arrangements should emphasize equality. Circular tables are more likely to provide this perception because conversational distance between all members is about the same. Circular settings also promote an open network of communication in which each group member can easily talk to every other member or to the entire group at once. When chairs are arranged in a lecture format (all chairs facing forward toward the leader or facilitator), it encourages one-way communication and reliance on the leader. A group member who is part of the audience has to gain formal acknowledgment that it is okay and appropriate to speak. And other group members may not be able to see the speaker. These physical conditions inhibit free-flowing interaction and limit the opportunities to develop relationships with other group members.

Stand-up meetings are becoming more popular, especially for group tasks that can be addressed within 10 to 20 minutes (Odermatt et al. 2015). These types of meetings are not less effective than meetings in which members sit at a table. But, stand up meetings encourage team members to keep the meeting short!

The greater the level of connectivity and embeddedness among group members, the more difficult it is for a group to find a time to meet. As our organizations become more team-oriented, meeting time and meeting preparation time become even more serious considerations. Groups that meet formally might also require time and space to meet informally in between regularly scheduled meetings. Informal interaction further anchors group member relationships and gives members an opportunity to test ideas with others before presenting them to the entire group. You may think time is only a problem for organizational groups, but this is not so. Given the variety of demands on your schedule, your family, personal, and recreational groups may be even more pressed for adequate time.

Group Member Pre-Meeting Responsibilities

You have just put your next work group meeting on your calendar. Now what? To be an effective contributor, you should review the agenda (or ask for one if it is not provided) to determine if you need to prepare anything before the meeting. For instance, in looking at the agenda, Dan sees that his group is going to begin considering alternative work schedules at its next meeting. He has not been asked to prepare anything, but he knows that these discussions will be emotional. Even though his group members complain frequently about the schedule they work, changing the work schedule will also cause problems. First, Dan reviews the overtime records to see how much overtime each member has worked. Then he reviews the project record to see if there is any pattern to how projects flow into the department. He notices that only a few projects come in the first week of the month but that the pace picks up steadily each week until many projects must be worked on simultaneously and completed by month's end. This gives him an idea: Why not propose that everyone work flextime the first 2 weeks of the month and take some additional time off? This would balance out the overtime needed during the final 2 weeks of the month. Now, Dan has an alternative based on data to present to the group for consideration.

Preparing for a meeting may require talking with other group members. For example, after reviewing her agenda, Marta believes that she should talk with Dan about the scheduling issues. As a single mother who depends on child care, she has a special interest in changes in schedules. She must get to the child care center by 6 p.m. or face a late penalty and an anxious child. Marta talks first with other parents in the group to see how they manage their child care arrangements. Cynthia tells her about one child care center that is open until 8 p.m. and Karen tells her that the company is scheduled to open an on-site child care center within a few months. At the meeting, Marta suggests that the work team lobby the company's executive board about the importance of an on-site child care center. With that benefit, employees like Marta will be willing to work unusual schedules. Although the child care center issue is not on the agenda, child care considerations affect the group's discussion of work schedules. Without talking beforehand to Cynthia or Karen, Marta would not know of any alternative child care arrangements or of the company's plans for an on-site child care center. Without this knowledge, Marta could easily steer the group off its primary topic to more emotional issues.

CONDUCTING THE MEETING

The group's leader or facilitator should arrive at the meeting site early to make sure everything is ready. When it is time for the meeting to start, the leader can call the meeting to order, preview the agenda with the group, and ask if other topics need to be added to it. By presenting the agenda as tentative rather than firm, the leader gains group members' support when together they agree that the agenda includes all important items (Schwarz, 1994). Posting the agenda so everyone can see it will help keep the meeting moving along. Before starting the meeting, the group should agree on the ground rules (e.g., when the meeting will end, what will happen if there is a tie vote). Finally, the group should review developments since the previous meeting. These should be brief reports to bring group members up to date.

Now the group is ready to move ahead with new business. As leader, taking each agenda item in order, you can announce the item and then ask what process might be most appropriate for this item of business. You can make suggestions but should be open to the ideas of other group members. With the item of business described and the process decided upon, you proceed with the discussion or action item. Generally, the leader's role is to initiate and structure discussion, not to control the discussion content. It is normal, if not always desirable, for group members to look to the leader for approval. One way to break this pattern and encourage input from everyone is to ask one member to respond to what another member says. This keeps the group discussion from developing into a pattern in which the leader says something each time a group member speaks.

Sometimes members complain, taking up valuable group time unnecessarily. For complaining members, your job as leader is to listen carefully to the complaints for their relevance to the agenda item. If a complaint is really about another topic, ask other group members to respond so you can gauge the extent to which this is a group rather than an individual concern. If it is a group concern, suggest that this issue be made another agenda item for later in the meeting (if there is time) or for a future meeting. If it is an individual issue, let the complaining group member know that you will speak with the member about it after the meeting. Besides controlling complaining speakers, you may also have to encourage less talkative members to contribute. You can do this by asking open-ended questions, such as "David, you've worked at other companies with rotating schedules. What can you tell us about your experiences?"

When different ideas are presented, summarize these in a compare-and-contrast format. Ask group members if your summaries are complete and accurate. If group members are quiet, do not assume their apparent consensus. Ask questions until you believe that group members really do agree on the substance of the issue. When an argument or conflict begins, do not take sides. Rather, ask group members to clarify their comments and probe for alternative viewpoints. You should reveal your own viewpoint only if it differs from those already expressed. To help clarify what the conflict is really about, ask group members to write down their response to the statement "I believe our conflict is about. . . ." This technique allows all group members to identify their perceptions of the conflict. Then ask group members to read their statements to the group. You may find that there is disagreement over what the conflict actually is about. Once the conflict is identified and agreed upon by all group members, encourage joint problem solving through discussion. When it is time for the group to make a decision, consider the advantages of each of the decision-making procedures described in Chapter 5. Be sure to let group members know if the decision they are making is a binding one or if a vote is simply an opportunity to see how group members are currently thinking about an issue.

Taking Minutes

Groups need a record, or **minutes**, of what they did at each meeting. Minutes should report on who attended the meeting, what was discussed, what was decided, who agreed to take on what responsibilities, and what the group plans to do next. Generally, the group's secretary or recorder takes the minutes, finalizes them, and presents them to the group at the next meeting. Many groups prefer that the minutes be prepared and distributed prior to the next group meeting. This gives members ample opportunity to review the record for accuracy. At the next group meeting, the minutes should be reviewed and corrected, if needed, before being accepted by the group as its formal record of activity. Figure 9.4 shows the minutes that resulted from the meeting conducted with the agenda displayed in Figure 9.3.

Besides providing a review of the meeting, the most important part of the minutes is to list member assignments for the next meeting. A main reason group meetings become

---

Minutes
Project Development Work Group
Thursday, January 29, 1:00 to 1:40 p.m. Conference Room A
Participants: Cynthia, Dan, Lu, Marquita, Tyron
Guest: Jensen
Purpose: Project Update Tracking

Lu called the meeting to order at 1:00; all members of the work group were present. Jensen Clark attended the meeting at Lu's invitation.

Lu reported on three items that occurred since the January 15 meeting: (a) promotion of competing software, (b) manufacturing status of our software, and (c) results from software demonstrations. Competing software has entered the market but has received unfavorable reviews. Our software is still on target for a March 1 release date, and feedback from the demonstrations has been positive. *Decision*: The group decided that additional demonstrations were not warranted. *Action*: Dan will check manufacturing status every Friday and email an update to each member of the group. *Action*: Lu will ask the marketing department to watch consumer reaction of competing software.

Dan reported that December project activity was slightly off due to higher than anticipated vacation days. January numbers appear to be in line with estimates.

Cynthia reported that her presentation to Federal Railways...

page 1 of 2

Figure 9.4 Sample Meeting Minutes

dysfunctional is a lack of follow-up on assignments. Without a reminder, group members may forget their assignment from one meeting to the next. Meetings become pointless and frustrating if group members are not ready with their assignments. Meeting minutes are one way to combat this problem.

Managing Relational Issues

Besides conducting the meeting and helping the group accomplish what is on the agenda, the leader or facilitator is also responsible for developing and maintaining a supportive group climate. Greeting group members as they arrive and engaging them in small talk can help establish a friendly meeting environment. If group members do not know one another well, brief self-introductions, name badges, and table tents can help them learn names more quickly.

If the meeting is long, suggest taking a break. This not only gives people time to take care of personal needs or get a snack but also relieves the tensions that can develop in groups. Another way to help group members feel more comfortable is to ask for volunteers for assignments. When someone volunteers, ask other members who know and like this person to work with the volunteer.

As the leader, you also have primary responsibility for establishing and setting group norms. Group members will follow your lead. As the meeting progresses, analyze norms for their effectiveness. Just because your group has always done it a certain way does not mean that it is effective. When you speak, try to use "we" and "our team" rather than "I" language. These subtle cues help create a team atmosphere other members can accept and adopt.

Because conflict is a natural outgrowth of group discussions, watch for cues from group members that conflicting positions or hidden agendas are developing. Conflict cues include rising voices as the conversation goes back and forth between group members and the use of more dominant nonverbal behaviors, such as a visible tightening of arms and faces, forceful gestures, and averted bodies. Some group members become silent and withdraw from a group's conversation when conflict arises.

A hidden agenda may be developing when one group member dominates the conversation while dismissing input from others. A **hidden agenda** exists when a group member has an ulterior motive. Group members ask the question or raise the issue for personal gain or personal satisfaction, not to help the group. Loaded questions ("So, you don't think I would be a good chairperson?") are often associated with hidden agendas, causing other group members either to retreat or to respond with the answer the person is looking for merely to keep peace in the group. When you see cues that conflict or hidden agendas developing, deal with them. The longer you wait, the more entrenched they become, making them more difficult for the group to manage and making the group less effective. (See Chapter 7 for strategies for managing conflicts.)

Using Space

Four principles should guide the selection and use of meeting space (Schwarz, 1994). First, all group members should be able to see and hear one another. Many different room configurations can accommodate this goal, but round or rectangular tables generally are best for groups of less than 10. A U-shaped table configuration can work well for groups of 20 or so. But, members should only be seated on the outside of the U so they can easily see one another. Second, the seating arrangement should allow each member to easily view the flip chart, the white board, or the screen. Third, if nonparticipants are invited to the meeting to provide information or simply to observe, they should not sit with the group members who will discuss or vote on issues before the group. Nonparticipants can easily contribute when called upon if they sit outside or just beyond the participants. This seating arrangement keeps nonparticipants from invading the psychological or relational space of group members. As a final consideration, the space for the group meeting should fit the needs of the group but not be so large as to allow for empty seats among participants. Group members may need space for writing, and they should not feel crowded. But allowing for too much extra space between members may increase the psychological distance among them and impede group progress.

Using Visuals

Even the best meetings and the best groups can profit from keeping visual records of what is happening in the group. Although a secretary or recorder may be taking minutes, these generally reflect group outcomes or the final decisions made by group members. Graphics or visuals can be used to keep track of the group's process and progress. By keeping and posting a running record of the group's key ideas and central themes, several positive things occur. First, group members know immediately if others are accurately hearing them. As a group posts its ideas on a flip chart or white board, it is easy to determine whether another group member accurately summarized someone's 4-minute statement and to correct misperceptions as needed. Second, writing what people say makes members feel acknowledged and part of the group process. When this type of validation occurs, group members are more likely to continue to contribute to the group discussion, which increases levels of participation, cooperation, and involvement. Third, visualizing or graphing what is going on in a group helps to spark the creativity of group members. Providing a visualization of the interaction helps group members both analyze and synthesize ideas before the group. Finally, visualizing the group's interaction provides a graphic record for the group, helping to reinforce group decisions. In this case, seeing is believing. The visuals can also be used for making minutes more detailed. The graphic record can be referred to in future meetings when the group needs to revisit something it has already addressed.

What does it take to visualize a group meeting? Markers and flip chart pads or white boards are the best tools. Making visuals does not require artistic talent, but it does require that you be interested in what is happening in the group and be able to follow the interaction. Most group members can visualize or graph a group's interaction with just a little practice.

Here is how the process works. Any group member can visually track the group's interaction. The role of this person is to capture what people say, not to evaluate ideas. Record everything, as accurately as you can, in group members' words. Your job is to provide some structure or organization to what people say. You might want to use different colors—say, green for positive attributes, red for negative attributes, and purple for questions that still need to be explored. Periodically stop and ask other group members if you have captured everything accurately and clearly. Use asterisks (*), boxes and circles, and underlining to highlight important items or to indicate what has been decided upon. Use forms (stick people, smiling faces, dollar signs, check marks, question marks—anything you feel comfortable drawing) to help structure or organize the record.

Other visual techniques that can assist your group include mapping, clustering, and flow charts. In the clustering technique, you place comments together within circles as themes start to emerge in a group's discussion. Starting with just a few spread-out circles helps you cluster items together. And you can draw lines out to other ideas to connect circles (themes). The clustering technique helps groups separate and integrate ideas. The mapping technique is similar to clustering in that it separates and integrates ideas, but it adds elements to reflect the flow of the group's discussion. The group member developing the visual reflects the decisions made during the discussion by noting the questions members ask and the answers the group develops. In mapping, it's a good idea to start on the left side with general ideas and questions and to work toward the right as the group develops the answers. Arrows can be drawn to connect answers to questions and to indicate the sequence of the discussion. Look at the map of a group meeting in Figure 9.5. Can you tell what this group discussed?

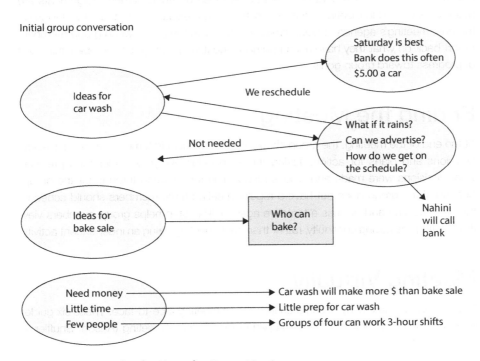

Figure 9.5 A Sample of a Map of a Group Meeting

Whatever visualizing or graphing techniques your group uses, do not throw them away. The group's secretary or recorder can use them to write more detailed minutes. Group members might want to refer back to them between meetings to see if an idea was

Making Assignments

Most group meetings reach a point at which additional information is needed. The leader may assign individual members these responsibilities, or group members may volunteer. In either case, you need to develop action statements and get agreement about what is to be completed.

For example, it becomes obvious to Terry that his group needs more information on how to use the company's videoconferencing system before this group will agree to adopt its use. Being comfortable with technology, Terry offers to find information about training for the group. Sounds good, right? But what exactly is Terry going to do? Will he find out when the training is scheduled? Will he explore what is covered in the training? Will he see if there is a training manual that can be placed near their computers? When should he report back to the group? And how? By email? By specifying what should be accomplished, Terry's expectations will parallel those of others. When assignments are made or accepted in meetings, this action should immediately be noted for inclusion on the next meeting's agenda. It helps create continuity in the group when group members report back on what they have accomplished, and it keeps all group members informed of progress toward group goals.

Ending the Meeting

At the end of the meeting, the group should do two things. First, members should review decisions and plans for action. Taking this step helps everyone understand precisely what decisions were made and who is responsible for following through for the group. Second, if the group does not have a regular meeting time, members should schedule the next meeting and discuss a tentative agenda. This step helps group members view the meetings as having continuity, rather than each meeting being an independent activity.

Meeting Virtually

Meeting virtually has many of same challenges as meeting face-to-face. These six guidelines are offered as action items in preparing for an online meeting (Allison, Shuffler, & Wallace, 2015).

1. Select a facilitator. The person in this role does not have to be the team's leader, but should have expertise in the technology being used for the meeting.

2. Select the communication technology. Choice of technology should match team members' communication preferences and meet the purpose of the meeting. (See Messersmith, 2015, for a review of group meeting technologies.)

3   Set norms for the meeting. Virtual meetings require a more formal structure and it may be necessary to determine how members will signal they want to talk.

4   Set and reinforce team roles. Clarify team goals and responsibilities at the beginning of the meeting.

5   Acknowledge both time zone and cultural differences. Rotate meetings so everyone can share in the inconvenience of a bad meeting time. If language barriers are present, add email, teleconferencing, or videoconferencing.

6   Follow up with action items. Use appropriate technology (often email) to follow up with members.

POST-MEETING FOLLOW-UP

Most leaders consider their job done when the group concludes its meeting. However, to make meetings more effective, a few follow-up steps should be performed. First, the leader should review the minutes with the person who took them and distribute them to each group member. This should be done as soon after the meeting as possible. This enables other group members to review the minutes for completeness and accuracy so that corrections can be made as soon as possible and the minutes be redistributed. Because the minutes include action statements for which group members agreed to be responsible, this reminds them of their commitment to the group. Second, if a group's actions will affect other groups or individuals, the leader should share the group's decisions with those parties. And third, when reviewing the actions to be taken from this agenda, the leader should begin preparing the initial framework for the group's next agenda.

The leader has another responsibility toward the group. After each meeting, the leader should analyze what went well and what did not work. To a great extent, the leader is responsible for making sure the group realized its goals during the meeting. Did that occur, and if not, why not? The leader should also think back over the meeting's interaction to assess whether individual group members' goals appear to be in alignment with the group's goals. If not, what could the leader do to encourage or motivate group members?

After important group business is conducted, the leader should also analyze to what extent inequality was an issue in the group. Some inequalities may stem from the leader's influence attempts. Some leaders are too assertive or dominant in their communicator style, which effectively shuts down members' contributions. In a sense, this influence pattern diminishes the need for a meeting, in that the leader is the only group member talking, giving input, and making decisions. Another type of negative influence occurs when a leader always looks to and speaks to the same group members. By consistently relying on only certain group members to answer questions and take on responsibilities for the group, the leader is implicitly saying to the others "You don't count" or "I don't trust you to do this for us." In either case, the leader's influence creates subgroups—the dominant subgroup that performs most of the group's work and a subordinate subgroup whose members are expected to follow along meekly. A leader can avoid these problems by making eye contact periodically with all group members, encouraging more silent group members to give their opinions, deferring the input of more dominant group members, and using decision-making procedures to help equalize any undue influence in the group.

Group members also have post-meeting responsibilities. If you were assigned or took on a responsibility to the group, be sure to fulfill it. If you believe the group forgot to cover something important, ask the leader to make sure it is part of the next meeting's agenda.

OVERCOMING MEETING OBSTACLES

Despite your best efforts in planning for and conducting meetings, problems can still arise. You have probably encountered one or several of six general obstacles to effective group meetings: (a) long meetings, (b) unequal member involvement and commitment, (c) the formation of cliques, (d) different levels of communication skills, (e) different communicator styles, and (f) personal conflicts (Gastil, 1993). Let's examine each obstacle and consider how you can help your group overcome each one.

PUTTING THE PIECES TOGETHER

Group Identity, Interdependence, and Group Structure

Reflect on a meeting in which your group experienced one or more obstacles, like those described in this chapter. First, identify the obstacles that arose. Now, assess each obstacle for its impact on group member identity. For example, did the length of the meeting cause members to resent being in the group? Did the formation of cliques cause negative feelings and emotions and threaten the identity of the group? To what degree did the obstacles that arose affect interdependence among group members? How could you determine that interdependence was adversely affected? Finally, what structural elements of the group—group roles, norms, and communication network—contributed to these problems? Using the principles of meeting management, what suggestions to group structure would you make to prevent these obstacles from occurring in the future?

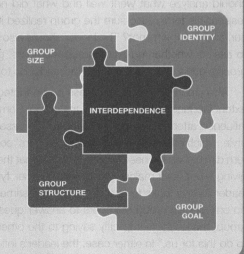

Long Meetings

No one likes long meetings, but a lack of preparation by group members actually contributes to this dilemma. Review what happened at the last meeting and what the group wants to accomplish at this meeting. At the meeting, speak in a clear but concise manner. Do not ramble, and do not let other group members do so either. If a group member gets off track, ask them to clarify their points. If your group has several long-winded talkers, you might want to consider asking group members to establish a time limit for individuals to contribute to the discussion. This can quicken the pace of the meeting. Keep side conversations to a minimum because one side conversation tends to escalate into several more. Having definite starting and stopping times for your meetings can help. And if you cannot cover all of the agenda items in your meeting, ask members for their commitment to continuing the meeting or schedule a follow-up.

Unequal Member Involvement and Commitment

You cannot be directly responsible for another member's level of involvement. However, there are ways to encourage equal participation. Asking questions to all members may help, as well as linking the interests of each member to the goal or activity of the group. Pointing out individual member benefits can help them identify with the group more strongly. Generally, when members identify with the group, they become more committed. Another strategy for increasing involvement and commitment is to allow the group to create and develop its own goals. When members help direct the activities of the group, their involvement and commitment follow. These strategies can help a group overcome social loafing, or the failure of group members to perform to their potential (Comer, 1995). When group member participation is unequal, less talkative group members may become detached from the group because they feel as if their contributions do not matter.

Formation of Cliques

Cliques, or subgroups, develop when there is a reason or need to communicate outside the group setting. When cliques develop, not all group members will have access to needed information. You probably cannot entirely avoid the formation of cliques, but you can reduce the impact of cliques on the group by having an alternative means of communicating with all group members. You can post the group's minutes, activities, or agenda online or on a bulletin board. If it is going to be a long time until the next group meeting and group members do not have access to a common area, send crucial information to all members. Ask group members to communicate important developments that occur between group meetings to all other members before the start of the next meeting. Finally, be sure to recognize personally each group member early in the group's discussion.

Using their names and asking them questions that personally involve them in the group's discussion increases each member's involvement in the full group.

Different Levels of Communication Skill

You may think there is not much you can do to enhance another group member's lack of communication skill, but there is. You might begin a group session by asking all group members to report on what they accomplished while away from the group or to reflect on what happened in the previous group meeting. The important thing here is to give each group member an opportunity to speak freely. You can help other members improve their skills by asking them questions that you know they can answer easily but that still contribute to the group. For example, you know that Marianne did a great deal of work checking out three sites for the festival. But you also know that she has some difficulty in giving detailed descriptions. Here is how you can help: Ask Marianne to tell the group about the three sites. When she pauses, ask her which site she prefers. What did she particularly like about that site? What criteria did she use in selecting sites? By asking Marianne follow up questions you know she can answer, you are helping her overcome her anxiety, as well as providing details for other group members so that they can appreciate the work Marianne did.

Different Communicator Styles

People differ in their personalities and their communicator styles. What can you do to decrease differences among communicator styles in your group? The key is to remain flexible and to accept other styles. If everyone had the same communicator style, the group's interaction could be boring and less productive. Think about maximizing the opportunities that differences offer to the group rather than negating others who communicate differently.

Personal Conflicts

Personal conflicts and personality conflicts are especially likely to happen if the group is feeling other pressures (such as time, resource, or deadline pressures). Rather than panic when these conflicts occur, use them as opportunities to learn more about other group members. Having an expectation that some conflict may occur will prevent you from being caught off guard when such situations arise. Another way to avoid personal conflicts is to create a supportive climate in which members can express their feelings in the group. Sometimes conflicts occur simply because we think someone said something other than what they did. When a conflict does arise, help members work through it by having each side express its views clearly. Finally, if an intense conflict develops, direct the group's attention to the primary conflict issue before continuing with other group activities or business. Failing to deal with the conflict when it arises will likely escalate the conflict later. The "Putting the Pieces Together: Group Identity, Interdependence, and Group Structure" feature will give you further insight into conflict and other meeting obstacles.

 THEORY STANDOUT

Strategic Meeting Interaction

Beck and Keyton's theory of strategic meeting interaction assumes that all types of interactions, whether verbal or nonverbal, are strategic (Beck & Keyton, 2009; Beck, Paskewitz, & Keyton, 2015). The theory makes three assumptions. First, team members adapt their messages to the relational and task aspects of group context. Second, strategic meeting interaction allows the examination of both messages produced and messages received by all group members in the interaction setting. Third, strategic meeting interaction provides a basis to examine the potentially different meanings that are produced from the same interactions. Their study of a regular end-of-work-week meeting revealed that, in some meeting interactions, members had similar interpretations of what had happened. At other times, members have considerably different view of what was going on. Why? When members described the strategic intent of others, they identified the *relational* intent of the messages. But when members described the strategic intent of their own messages, they identified the *task* intent of the messages. This suggests that team members are less aware of their relational strategies and, perhaps, even their relational intentions. Unfortunately, then, members may base their subsequent messages in the meeting on a foundation that other members do not share.

SUMMARY

Meeting management procedures and facilitation strategies are designed to capitalize on the strengths of group members and group processes. Meeting planning includes pre-meeting planning and preparation by group members and by the group leader. An agenda, identifying both start and stop times and all items the group will consider, should be prepared and distributed before any group meeting. Remember: If you cannot identify a specific purpose and goals for a meeting, do not have one.

The group's leader conducts the meeting according to the agenda. However, the leader's role is to initiate and structure discussion, not to control discussion content. Besides helping the group move through its business issues, the leader is also responsible for developing and maintaining a supportive group climate. Introducing members, establishing norms, and managing conflict are some of these responsibilities.

One way to help your group is to develop a group charter or mission statement and a code of conduct. A group charter describes the goals or mission of the group; a code of conduct describes behaviors that are appropriate for this particular group. Both can provide direction and clarity for group members.

A secretary or recorder should take minutes at each meeting. Minutes should include what was discussed or decided, who agreed to take on what responsibilities, and what the group plans to do next. Minutes should be distributed, and revisions made, as soon as possible.

The space in which a group meets is important. All group members should be able to see and hear one another easily. The size of the space for the meeting should fit the needs and size of the group. Members should feel neither too crowded nor too distant from others in the group. Using visuals and graphics can help a group record what is happening in the group. Listing topics of conversation, drawing a diagram of the group's conversation, clustering ideas together, creating data tables, and drawing organizational charts are just a few types of visualization that help a group capture a pictorial memory of its interaction.

In most meetings, group members volunteer for or are given assignments to be completed before the next meeting. These actions should be noted both in the minutes and on the next meeting's agenda. At the end of a meeting, the group should review decisions and plans for actions, schedule the next meeting time, and discuss future agenda items.

Most groups experience some obstacles. Long meetings, unequal member involvement and commitment, the formation of cliques, differing levels of communication skills, different communicator styles, and personal conflicts are common obstacles groups must overcome. Any group member can help a group surmount these barriers.

DISCUSSION QUESTIONS AND EXERCISES

1. Think of your most recent group experience—one in which the group will meet again. Write a three- to five-page paper analyzing your group by answering these questions:

    a. What did your group accomplish? How does that compare to what it should have accomplished?

    b. What is one aspect of the group process or procedure that was effective, and one aspect that was ineffective?

    c. Which meeting management strategies would have helped your group? How? Be specific.

    d. What did you learn about yourself as a group member that you can carry forward to the next group experience? What did you learn about the group that you can apply in the next group session?

FACILITATING GROUP MEETINGS 187

2   Gather the agendas of several different meetings. Compare and analyze them for their effectiveness. If the agendas are from meetings you attended, consider the usefulness of the agenda to the structure and purpose of the meeting. If the agendas are from meetings of other individuals, ask them to what degree the agendas helped them prepare for the meeting (if they got the agenda ahead of time) and to what degree the agenda reflected the meeting's activities.

3   Interview at least three people who have been members of organizational groups or teams, or community or civic groups. Ask each person to describe how the group or team accomplished its work at meetings. If these are not mentioned, ask each person about the group's use of agendas, graphics, and minutes.

 NAILING IT!

Using Group Communication Skills for Group Presentations

Your group is ready to make its presentation. The few hours before the presentation can be hectic. How can you use meeting management techniques to help calm everyone down, smooth out the remaining issues, and deliver a great presentation? Return to your group charter, and let your group members know that together they've met the group goal. Make a written list on a notecard of the last few things that need attention. Use this list as a way to check in with every group member. Before the presentation, take a few minutes for a group huddle. Thank members for the time, skills, and energy they put into developing the presentation. Be sure to encourage them to do their best during the presentation. After the presentation, check in with them again. Point out what they did in the presentation (or in developing the presentation) that was especially effective and the audience enjoyed. Finally, if your group borrowed equipment or supplies, make sure those are returned. Although the group presentation is not a meeting, many of the same techniques apply.

Image Credits

Photo 9.1: Copyright © Depositphotos/imtmphoto.

Figure 9.6: Adapted from: Copyright © Depositphotos/megastocker.

APPENDIX

CREATING AND DELIVERING TEAM PRESENTATIONS

You've been assigned a team presentation. Now what do you do? Whether you are a first-time or experienced presenter, presenting as a group can create some new challenges! This appendix is divided into three sections focusing on developing, preparing for, and making the presentation, respectively.

DEVELOPING YOUR PRESENTATION

Thinking About Your Presentation

- Ask yourself, what is the purpose of the presentation? To persuade? Inspire? Inform? Teach? Instruct? Entertain?
- As a group, outline or map your presentation. Decide who will be responsible for each part of the presentation and for transitions between the parts. Even though individual members might specialize in various parts of the presentation, every group member is responsible for all of the material in the presentation.
- Consider the unique talents of your team members.
- All team members should have a part in the presentation.
- Consider your audience in developing the presentation material.
- Watch the local news for examples of how to make transitions between parts of the presentation.
- If you have been given 25 minutes for the presentation, plan on filling 20 of it. Presentations seldom start on time—perhaps because latecomers are accommodated, the previous class let out late, or a few announcements are made. It's better to finish early and have time for questions than to shove 25 minutes of material into 20 minutes or to run overtime.
- Develop your material with three specific main points you want the audience to learn from your presentation. Outline or map these points before you begin, deliver the three points, and summarize these points.
- Organize your points in one of these ways: (a) from simple to complex, (b) in chronological order, (c) from general to specific, or (d) from problem to solution.

PREPARING FOR THE PRESENTATION

Setting Up the Room

- Ask your instructor if you can rearrange the room for your presentation. Changing how the room is set up creates interest and signals that something different is about to happen. If you do so, it should be done before any class members arrive.
- Try some of these shapes: a U-shape, angled tables or chairs, amphitheater style, a circle with a large opening. Or turn all of the chairs in a different direction, making another wall the temporary *front* of the classroom.
- Don't use the lectern just because one is there. If you're not going to use it, remove it. A lectern can create an unnecessary barrier between you and your audience.

Using Visual Aids

- Make a commitment to using visual aids in your presentation. They add value to your presentation and increase your effectiveness.
- Use visual aids to serve as your notes, help maintain audience interest, underscore your points, and keep you focused.
- For every visual aid you use, ask whether it is worth it and what objective it serves. You and your group members are the presentation. The visual aids are just helpers.
- Don't use too many visual aids—no more than one per minute.
- Don't turn the lights out even when projecting visuals. This will only put your audience to sleep!
- Use visuals as reminders to yourself, but don't read them word for word—the audience can do that!
- If you have a list of concepts on a slide, use the transition feature to reveal each item as you talk about it. Otherwise, your audience will read the list while you are reporting on the first item. Revealing concepts item by item also helps you control your thinking and your rate of speaking.
- Simplify. Use the simplest visual aid that gets your point across, easily and clearly.
- Think horizontally. People are used to viewing television.
- Use color and contrast; it makes a greater impact.
- Include no more than four colors per visual.
- Use dark print on a light background or light print on a dark background.
- Maintain the same background color throughout your presentation.
- Don't use red for text. Use it to highlight with bullets, arrows, and so on.

- Avoid red/green contrasts; some people are color blind.
- In planning a color scheme, use darker colors on the bottom, medium colors in the middle, and lighter colors on top.
- Use short titles.
- Use plenty of white space. Don't crowd the slide or chart.
- Use graphics, charts, pictures, audio, or video to help communicate complex ideas.
- Use pie charts to show the distribution of a whole into its component parts (e.g., budgets or times).
- Use bar charts to represent quantity by the length or height of the bar. Use color coding to help audience members interpret the bars (e.g., blue for this year and yellow for last year).
- Make sure the visual aids use a consistent form, color scheme, and type style so the result is a *group* presentation.
- Have all group members proofread all visual aids regardless of who designed or made them.
- Rehearse with the visual aids.

Preparing and Using Flip Charts

- Recognize that flip charts are one of the cheapest and most effective presentation aids (i.e., they can't break down and they allow the lights to be at their brightest). They can be prepared ahead of time or used spontaneously during the presentation.
- Don't use a flip chart if there are more than 30 audience members—not all members will be able to see it.
- Leave a blank page at the beginning of your flip chart pad.
- Staple a blank sheet behind each flip chart page you are going to use. This way writing won't bleed through, and it makes the pages easier to turn.
- Don't think you have to be an artist to use a flip chart. Well-spaced, large letters, and simple figures and diagrams work well on flip charts.
- Use very light blue lines—which you can see but the audience can't—to help you keep your lettering straight and all the same height (two-inch height is recommended).
- As a rule of thumb, include no more than six lines per flip chart. Begin each item with a bullet so audience members can tell where to stop and start.
- Use two flip charts. One can be prepared ahead of time, and the other can be used spontaneously during the presentation.
- Place flip charts to the side of the presentation space. The presenter is more important than the flip chart.

Using PowerPoint Presentations

- Ensure that your memory stick works with the computer you will be using. Also email yourself a copy of the presentation; that way you can download the file if necessary.
- Ensure that the version of PowerPoint in which you created the presentation is the version installed on the computer. Not all versions are compatible.
- Have a *welcome* slide with the title of your presentation up when the audience comes in.
- The last slide should be a prompt for questions or provide contact information for the speakers.
- Use a wireless device to advance the slides. That way, the speaker is not trapped behind equipment.
- Incorporate charts, graphics, pictures, animation, and video to help make your point. Don't, however, let these features overshadow what you have to say.
- If your presentation is more textual than graphic, try using circles, squares, or triangles to group text together for more impact.
- For text, use the 6 x 6 rule: no more than six words per line, and no more than six lines per slide.
- Use uppercase and lowercase text.
- Put a paper copy of each slide on the floor. If you can read it while standing up, the type is probably large enough. If not, the type is too small.
- As you show a new slide, allow a second or two for the visual impact to sink in. Then give the explanation.

Using Video

- Use no more than 2 to 3 minutes of video at a time.
- Know where the pause and stop buttons are on the screen.
- Test the video out on others before you show it to your audience. Is it as funny or moving or dramatic as you think?
- Check the volume before the presentation.

Using Handouts

- Check for accuracy (grammar, punctuation, spelling).
- Don't reproduce your visual aids and distribute them at the beginning of the presentation. You'll be sure to lose your audience!

- If you must use a handout because of complexity or detail, distribute it when you come to that part of the presentation.
- If you want audience members to have a complete set of handouts, tell them they'll be available after your presentation, and hand them out as they go out the door.

Using Numbers and Statistics

- Round off numbers so the audience can remember them. Which would you remember better: nearly $5 million or $4,789,187?
- Use the most recent statistics you can find. Don't present 2000 figures in 2018. What were you doing in 2000?

Preparing Your Notes

- Put your notes on 5 x 7 note cards. They are easy to hold, and they don't make noise.
- Alternatively, try putting your notes on the four sides of an empty manila file folder. You can keep your handouts or overheads in it before and after the presentation.

Working as a Team

- Act like a team. A group presentation isn't a series of individual presentations.
- Agree on who will handle questions from the audience. Generally, those who presented the material should respond to the question.
- Work out transitions between sections. Use the next speaker's name in handing off—for example, "That covers decision-making groups. Now Andrew will cover families as groups."
- Assign one member to keep track of the time during the presentation. Have preplanned cues for signaling information about time.
- Focus on the audience. Make the presentation to the audience, not your group.
- When you're not speaking, look at the person who is speaking.
- Be sure to tell the audience the agenda for your presentation. When multiple people speak, the audience needs a road map.
- Rehearse as a team.

Rehearsing Your Presentation

- Do not play it by ear. Your audience deserves a rehearsed presentation.
- Recognize that writing out the presentation is not the same as rehearsing it.
- Keep in mind that rehearsals tend to run shorter than the actual presentation. A 20-minute rehearsal usually means a 25- to 30-minute presentation.
- Rehearse using all of the visual aids and equipment you plan to include in the presentation. There is no substitute for a dress rehearsal.
- Have all members do their parts as planned. Filling in for someone at rehearsal increases the likelihood that you'll fill in for that person at the presentation. Again, there is no substitute for a dress rehearsal.
- Review all visual aids for spelling, grammar, and punctuation.
- Check the voice levels of all presenters. Ask a couple of friends (but not anyone who will be part of the audience) to help you by playing the role of audience members during your rehearsals. They can give you valuable feedback.
- If you can, rehearse in the room in which you'll give your presentation.
- Talk about what you'll wear for the presentation. You won't look like a group if four of you show up in dress clothes and one of you wears jeans. Clothing should be coordinated, but it doesn't have to match.
- Know that the best rehearsal is a video rehearsal. Get a friend to record your presentation, or set up a camera with a wide-angle shot. Video is the most powerful tool for identifying flaws. If you don't have access to video, at least audiorecord the rehearsal. You'll be amazed at what you'll hear.

DELIVERING THE PRESENTATION

Introduction

- Introduce all members of the group.
- Be sure to tell audience members why your topic is important. Give them at least three reasons they should pay attention to what you have to say.
- Capture the audience's attention by answering the question "What's in it for me?"

Communicating Verbally

- Do not start with humor or a joke.
- Do not apologize or make excuses for anything or anyone.

- Use specific language—for example, "We can increase membership by 60 percent" rather than "We can increase membership a lot."
- Use vivid language, paint word pictures, and use metaphors.
- Use action words.
- Use short, simple words.
- Eliminate clichés or overused phrases.
- Think of your presentation as a conversation with audience members.
- Anchor new information in something familiar to the audience.
- Use transitions:

    "That brings us to our next point."

    "Now that we've discussed X, let's talk about Y."

    "So far we've covered decision making and leadership. Now let's take a look at conflict."

    "In addition to consensus building, we need to address conflict."

    "To begin, let's take a look at …"

    "The next important factor is …"

    "That's the first reason to be flexible as a leader. The next reason is …"

    "Finally, let's consider …"

    "In conclusion, …"

    "To summarize, …"

    "We'd like to leave you with this thought."

Communicating Nonverbally

- Move around when you talk.
- Use hand movements and arm gestures.
- Vary the quality of your voice. It's easy to do this if you avoid the straight presentation of information. It's easier to be expressive when you use analogies, tell stories, give demonstrations, and ask questions.
- Make eye contact with your audience.
- Manage your nonverbal adaptors.
- If you're using visual aids, turn and look at the aid to get the point securely in your mind. Then turn to the audience and address the point to them. Keep looking at the audience until you are finished and are ready to address the next point on the visual aid.

Handling the Question-and-Answer Session

- While you're developing the presentation, make a list of questions you believe audience members will ask. Prepare an answer for each one.
- Be aware that you're likely to be met with silence if you ask, "Are there any questions?" Rather, ask, "Are there any questions about how leaders should be flexible?" or "Are there any questions about the difference between task and relational messages?"
- If you ask the audience a question, pause for a moment to give members time to think of one.
- If there are no questions, continue by saying "One question we had when we started our research was …" or "One of the most frequent questions asked about [the topic] is …"
- When an audience member asks you a question, start your response by restating the question so all audience members can hear it. Many times, others in the audience haven't heard the question, which means you're responding to something that doesn't make any sense to them.
- Assuming the Q & A session happens at the end of the presentation, after the last question thank the audience for the questions and then summarize and close out the presentation.

Handling Interruptions From the Audience

- If you are interrupted by a question that will be answered later as part of your planned presentation, say, "Dale, I'm coming to that in about 2 minutes" or "The answer is yes, and I'll explain it fully in just a few minutes."
- If you are interrupted by a question that will not be part of your planned presentation, say, "Dale, that's a good question. I'd like to answer it at the end of the presentation."
- If you are interrupted by a question or comment that you prefer not to answer publicly, say, "Dale, that's interesting (or "I hadn't thought of that"). Let's talk about it after the presentation."

SUGGESTED READING

Brennan, M. (2016, Nov. 29). To persuade people, trade PowerPoint for papier-mâché. *Harvard Business Review Digital Articles*, 2–5.

Duarte, N. (2008). *Slideology: The art and science of creating great presentations*. Sebastopol, CA: O'Reilly Media.

Duarte, N., & Sanchez, P. (2016). *Illuminate: Ignite change through speeches, stories, ceremonies, and symbols*. New York, NY: Portfolio/Penguin.

Glonek, K. L., & King, P. E. (2014). Listening to narratives: An experimental examination of storytelling in the classroom. *International Journal of Listening*, 28, 32–46. doi:10.1080/10904018.2014.861302

Hertz, B., van Woerkum, C., & Kerkhof, P. (2015). Why do scholars use PowerPoint the way they do? *Business and Professional Communication Quarterly*, 78, 273–291. doi:10.1177/2329490615589171

Hertz, B., Kerkhof, P., & van Woerkum, C. (2016). PowerPoint slides as speaking notes: The influence of speaking anxiety on the use of text on slides. *Business and Professional Communication Quarterly*, 79, 348–359. doi:10.1177/2329490615620416

Kernbach, S., Bresciani, S., & Eppler, M. J. (2015). Slip-sliding-away: A review of the literature on the constraining qualities of PowerPoint. *Business and Professional Communication Quarterly*, 78, 292–313. doi:10.1177/2329490615595499

Pruim, D. E. (2016). Disaster day! Integrating speech skills though impromptu group research and presentation. *Communication Teacher*, 30, 62–66. doi:10.1080/17404622.2016.1139148

Reynolds, G. (2014). *Presentation Zen: A simple visual approach to presenting in today's world* (2nd ed.). San Francisco, CA: New Riders.

Worley, R., & Dyrud, M. (Eds.). (2004). Presentation and the PowerPoint problem. *Business Communication Quarterly*, 67, 78–94.

Worley, R., & Dyrud, M. (Eds.). (2004). Presentation and the PowerPoint problem—Part II. *Business Communication Quarterly*, 67, 214–231.

GLOSSARY

accommodating A win-lose conflict management strategy exemplified by trying to satisfy the other's concerns.

affective ties In a communication network, relationships built on the liking or trusting of another group member.

affective trust A type of trust among group members based on the interpersonal concern group members show for one another outside of the group task.

agenda A list of activities or topics to be considered at a group meeting; should also include starting and stopping times, the location of the meeting, the attendees, and the overall goal of the meeting, as well as the specific goal of each agenda item and any preparations that group members should make.

avoiding conflict management strategy A nonconfrontive strategy for managing conflict, based on verbal, physical, or psychological withdrawal.

bona fide group perspective A theoretical frame that illuminates the relationship of the group to its context or environment by recognizing a group's permeable and fluid boundaries and the time and space characteristics of its interactions.

brainstorming A group procedure designed to help groups generate creative ideas.

centralized communication network Communication network that imposes restrictions on who can talk to whom and for which one or two group members control those restrictions.

certainty An attribute of a negative communication climate, group members communicate as if they possess all the answers or believe they can predict what another group member is going to say or do.

coalition formation Phenomenon that occurs when one member takes sides with another against yet another member of the group; creates an imbalance of power; can only occur with at least three group members.

code of conduct A group document that describes behaviors appropriate for the group.

coercive power A type of power resulting from the expectation that one group member can be punished by another.

cognitive conflict A type of conflict involving disagreement over interpretations or analyses of information or data; also known as *judgment conflict*.

cognitive ties In a communication network, describe who knows whom.

cognitive trust Based on group members' assessments about the reliability, dependability, and competence of other group members.

cohesiveness The degree to which members desire to remain in the group.

collaboration A conflict management strategy based on parties sharing a superordinate goal of solving the problem even though their initial ideas for how to solve it differ.

communication The process of symbol production, reception, and usage; conveyance and reception of messages; and meanings that develop from those messages; all communication is strategic in that people communicate to accomplish a specific purpose.

communication climate The atmosphere that results from group members' use of verbal and nonverbal communication and listening skills; can be defensive or supportive.

communication competency approach to leadership A model for leadership based on three principles: (1) leadership is action that helps a group overcome barriers or obstacles, (2) leadership occurs through interaction, and (3) there is a set of skills or competencies that individuals use to exercise leadership in groups.

communication network The interaction pattern or flow of messages between and among group members; creates structure for the group based on patterns of who talks to whom.

communication overload Communication that is too extensive or complex and that comes from too many sources; causes stress and confusion among group members.

communication underload Communication that is infrequent and simple; causes group members to feel disconnected from the group.

competing A distributive conflict management strategy exemplified by forcing; emphasizes one party winning at the other party's expense.

competitive conflict Polarizations; one side winning with the other side losing.

compromising A conflict management style; an intermediate strategy between cooperativeness and assertiveness; compromising may settle the problem but will also offer incomplete satisfaction for both parties.

concurrency A dimension of time; group members completing tasks simultaneously or one at a time.

conflict Situation in which at least two interdependent parties capable of invoking sanctions on each other oppose each other; based on real or perceived power; occurs because parties have mutually desired but mutually unobtainable objectives.

conflict aftermath The feelings that group members have developed as a result of a conflict episode; the legacy of the conflict interaction.

conflict management The relational skill of keeping group members focused on task-related ideas rather than personal differences.

connectivity The degree to which several groups share overlapping tasks or goals.

consensus A decision procedure in which each group member agrees with the decision or in which group members' individual positions are close enough that they can support the group's decision.

controlling behavior A dimension of defensive communication climate in which the sender assumes to know what is best for others.

cooperative conflict A type of disagreement that helps move the group along with its task or activities.

cooperative information sharing paradigm A model of how group members share information when trying to make a decision. See *hidden profile*.

copresence Group member's behavior and messages are shaped by others in the group; a sense of being with other group members even if they are not physically present.

decentralized communication network Communication network that allows each group member to talk to every other group member.

defensive climate A communication climate based on negative or threatening group interaction.

delay The manner in which group members resist setting deadlines or time to work.

description A dimension of a supportive communication climate that occurs when a group member responds to the idea instead of evaluating the group member who offered the idea.

developmental perspective on conflict Perspective that views conflict as a natural part of the group's development; can be productive if conflict emergence and resolution helps the group move forward; can be destructive if a group does not allow conflict to emerge.

distributive conflict management strategy A win-lose conflict management strategy exemplified by competitiveness or accommodation; yields an outcome that satisfies one party at the expense of the others.

embeddedness The degree to which the group is central to its larger organizational structure.

emergent leadership A type of leadership in which a group member is not appointed or elected to the leadership role; rather, leadership develops over time as a result of the group's interaction.

empathy A dimension of supportive communication climate that expresses genuine concern for other group members; conveys respect for and reassurance of the receiver.

equality A dimension of supportive communication climate in which trust and respect for all group members is expressed.

evaluation A dimension of a defensive communication climate in which a group member uses language to criticize other group members.

expert power A type of influence based on what a group member knows or can do.

false consensus A belief among group members that they all agree when they do not; agreeing to a decision only in order to be done with the task.

faultlines Demographic characteristics or other attributes salient for a particular group and its task; members are likely to communicate with similar others, which can divide a group into subgroups.

flexibility A dimension of time; how rigidly or flexibly time and deadlines are structured.

formal ties Who reports to whom or any other power-laden relationship in a communication network.

functional theory of group decision making Comprised of a set of critical functions group members should engage in for effective decision-making and problem-solving activities: (a) thoroughly discuss the problem, (b) examine the criteria of an acceptable solution before discussing specific solutions, (c) propose a set of realistic alternative solutions, (d) assess the positive aspects of each proposed solution, and (e) assess the negative aspects of each proposed solution.

group Three or more members who identify themselves as a group and who can identify the interdependent activity or goal of the group.

group charter A group document that describes the goals or mission of the group.

group goal An agreed-upon task or activity that the group is to complete or accomplish.

group identity The result when members identify themselves with other group members and the group goal.

groupings People identified as a group when they have little or no expectation that interaction will occur with one another; often based on demographic characteristics.

group polarization Tendency for groups to make decisions or choose solutions more extreme than any member would prefer individually.

group roles Interactive positions within a group; the micro components of a group's structure.

group size The number of members in the group; the minimum number of members is three; the maximum number depends primarily on the complexity of the task or activity.

group structure The patterns of behavior that group members come to rely on; develops with or emerges from group rules and norms.

groupthink A type of faulty decision making based on the tendency of highly cohesive groups to adopt faulty solutions because members failed to critically examine and analyze options while under pressure from the external environment.

heterogeneity The expressed differences in cultural values that influence group members' interactions.

hidden agenda A group member has an ulterior motive; uses the group for personal gain or personal satisfaction.

hidden communication Interaction among only some members of the group; often used to build alliances.

hidden profile In decision making, information is distributed among group members such that some information is shared by all members and other information is held by only one member.

informational power Persuasion or influence based on what information or knowledge a group member possesses or presents as arguments to group.

instrumental perspective on conflict A productive conflict, which helps group members figure out what the goals are, and how to accomplish its tasks; or destructive, which impedes the group from reaching their goals.

integrative conflict management strategy A win-win conflict management strategy based on problem solving or collaboration; produces an outcome with which all parties can agree.

interdependence Phenomenon whereby both group and individual outcomes are influenced by what other individuals in the group do; group members must rely upon and cooperate with one another to complete the group activity.

law of inherent conflict The premise that no matter the issue or group, there will be significant conflict stemming from different perceptions of relative factors.

leadership As a process, how a person uses positive influence to direct and coordinate the activities of group members toward goal.

legitimate power A type of power based on the inherent influence associated with a position or role in the group.

linearity The manner in which group members create times for some events over others; putting group tasks in some order.

material ties In a communication network describe the relationship among two members who give resources to one another.

meaning The understanding that is extracted from the way messages follow one another in interaction.

meeting citizenship behaviors Actions taken by group members to maintain a positive social and communication climate.

minutes A record of what the group did or accomplished at a meeting; should reflect who attended the meeting, what content was discussed, what was decided, who agreed to take on what responsibilities, and what the group plans to do next; usually taken by the group's secretary.

motivated information sharing paradigm An explanation for why groups are ineffective at sharing necessary information (i.e., groups do not know if they have all the information necessary to make a decision).

neutrality A dimension of defensive communication climate expressed when a group member reacts in a detached or unemotional way; a lack of warmth or caring for other members, making them feel as if they are not important.

nominal group technique (NGT) A decision-making procedure in which the group temporarily suspends interaction to take advantage of independent thinking and reflection before coming together as a group to discuss the ideas generated.

norm An expectation about behavior; an informal rule adopted by a group to regulate group members' behaviors.

normative conflict A type of conflict that occurs when one party has expectations about and evaluates another party's behavior.

pace The tempo or rate activity of group interaction.

performative act Messages exchanged that do something; for example, group interaction that introduces group members to one another.

persuasive arguments theory Posits that cognitive processes of individual group members prior to meetings lead to interaction outcomes.

political perspective on conflict Struggle for power is the source of conflict; if alternative points of view surface, conflict can be productive; if a conflict reaffirms or strengthens dominant members, conflict can be destructive.

power The influence of one person over another; the ability to get things done or to find needed resources; power bases are reward, coercion, legitimacy, expertise, information, and referent.

problem orientation A dimension of a supportive communication climate that strives for answers and solutions to benefit all group members and to satisfy the group's objective.

procedural leadership competency Behavior exhibited when group members coordinate group activities; helps members function as a group and achieve their goal.

process conflict Disagreements about coordination of group member duties, the group's time management, and/or access to resources.

provisionalism A dimension of a supportive communication climate that is committed to solving the group's problems by hearing all of the ideas; encourages the experimentation and exploration of ideas in the group.

proximity ties In a communication network describe group members who are spatially close or electronically linked to one another.

punctuality The manner in which group members set deadlines or respond to one another promptly.

ranking A decision procedure in which members assign a numerical value to each available position; rankings are then ordered.

referent power A type of influence given by a group member to another member based on a desire to build a relationship with that member.

relational competencies Skills individuals use to help manage relationships among group members and to create and maintain connections with those outside the group.

relational conflict Conflict based interpersonal relationships, emotions, or personalities.

relational dimension Group interaction that provides social and emotional support, as well as a mechanism for developing and maintaining role identities within a group.

relational groups Groups with membership based on personal or social relationships; examples include support groups, fraternities/sororities, and friendship groups.

relational leadership competency Displayed when group members cooperate with and express support for one another.

reward power A type of positive influence, relationally oriented and based on such things as attention, friendship, or favors, or materially oriented and based on tangible influence such as gifts or money.

satisfaction The degree to which a group member feels fulfilled or gratified based upon experiences in the group.

scarcity The degree to which resources are available to the group.

separation A dimension of time; how group members isolate their meetings from other interactions.

shared leadership Enacted by all or many group members; also called collective or distributive leadership; leadership rotates among members.

socialization The reciprocal process of social influence and change between new and established group members.

social loafing The idea that individual efforts decrease as the size of the group increases; a detachment from the group that occurs when group members feel as if they are not needed to produce the group's outcome or as if their individual efforts are not recognized by other members.

sociogram A visual representation of the group members' communication network or relationships; reveals the number and pattern of network ties among group members.

spontaneity A dimension of a supportive communication climate exemplified by a group member who is open and honest with other group members; creates immediacy with other group members.

strategic A dimension of a defensive communication in which senders manipulate others by placing themselves above the group or its task; or how speakers adapt their messages so that they will accomplish their goals.

superiority A dimension of a defensive communication climate exemplified when group members continually reinforce their strength or position over others.

superordinate goal A task or goal so difficult, time-consuming, and burdensome that it is beyond the capacity of one person.

supportive climate A communication climate based on positive group interaction.

task conflict Conflict based on issues or ideas, or disagreement about the group's task.

task dimension A group's interaction that focuses on its task, activity, or goal.

task-oriented team Also labeled *work groups* and *decision-making teams*; their purpose is to accomplish work tasks; it is generally assumed that group interaction should be geared toward planning, debating, or implementing group decisions.

team cognition The level of knowledge or commonly shared information (about one another or the task) among group members.

technical leadership competency Demonstrated when group members display the task or technical competencies required by the group's activity.

theory of effective intercultural workgroup communication An explanation of how group members' cultural differences their group interactions.

time perspective Group members completing their tasks with a focus on the past, present, or future.

transformational leader A type of leadership based on the premise that the leader sets an example for group members to follow; uses rhetorical skills to build a vision that members can identify with and use as a guiding force toward goal completion.

trust A group member's positive expectation of another group member; reliance on another group member in a risky situation.

urgency Group members approach their tasks as if there were a crisis or pressing deadline.

voting A decision procedure in which group members cast a written or verbal ballot in support of or against a specific proposal; generally, a majority or two-thirds vote is needed to support a proposition.

REFERENCES

Aaron, J. R., McDowell, W. C., & Herdman, A. O. (2014). The effects of a team charter on student team behaviors. *Journal of Education for Business, 89,* 90–97. doi:10.1080/08832323.2013.763753

Allison, B. B., Shuffler, M. L., & Wallace, A. M. (2015). In J. A. Allen, N. Lehmann-Willenbrock, & S. G. Rogelberg (Eds.), *The Cambridge handbook of meeting science* (pp. 680–705). New York, NY: Cambridge University Press.

Alper, S., Tjosvold, D., & Law, K. S. (2000). Conflict management, efficacy, and performance in organizational teams. *Personnel Psychology, 53,* 625–642. doi:10.1111/j.1744-6570.2000.tb00216.x

Anderson, C. M., Riddle, B. L., & Martin, M. M. (1999). Socialization processes in groups. In L. R. Frey, D. S. Gouran, & M. S. Poole (Eds.), *The handbook of group communication theory & research* (pp. 139–163). Thousand Oaks, CA: SAGE.

Aritz, J., & Walker, R. C. (2014). Leadership styles in multicultural groups: Americans and East Asians working together. *Journal of Business Communication, 51,* 72–92. doi:10.1177/2329488413516211

Arrow, H., & McGrath, J. E. (1993). Membership matters: How member change and continuity affect small group structure, process, and performance. *Small Group Research, 24,* 334–361. doi:10.1177/1046496493243004

Avolio, B. J., Waldman, D. A., & Einstein, W. O. (1988). Transformational leadership in a management game simulation. *Group and Organization Studies, 13,* 59–80. doi:10.1177/105960118801300109

Baker, D. C. (1990). A qualitative and quantitative analysis of verbal style and the elimination of potential leaders in small groups. *Communication Quarterly, 38,* 13–26. doi:10.1080/01463379009369738

Bakar, H. A., & Sheer, V. C. (2013). The mediating role of perceived cooperative communication in the relationship between interpersonal exchange relationships and perceived group cohesion. *Management Communication Quarterly, 27,* 443–465. doi:10.1177/0893318913492564

Bales, R. F. (1950). *Interaction process analysis: A method for the study of small groups.* Cambridge, MA: Addison-Wesley.

Bales, R. F., & Cohen, S. P. (1979). *SYMLOG: A system for the multiple level observation of group.* New York, NY; Free Press.

Ballard, D. I., & Seibold, D. R. (2000). Time orientation and temporal variation across work groups: Implications for group and organizational communication. *Western Journal of Communication, 64,* 218–242. doi:10.1080/10570310009374672

Ballard, D. I., & Seibold, D. R. (2004). Communication-related organizational structures and work group temporal experiences: The effects of coordination method, technology type, and feedback cycle on members' construals and enactments of time. *Communication Monographs, 71,* 1–27. doi:10.1080/03634520410001691474

Baran, B. E., Shanock, L. R., Rogelberg, S. G., & Scott, C. W. (2012). Leading group meetings: Supervisors' actions, employee behaviors, and upward perceptions. *Small Group Research, 43,* 330–355. doi:10.1177/1046496411418252

Barge, J. K. (1994). *Leadership: Communication skills for organizations and groups.* New York, NY: St. Martin's.

Barge, J. K. (1996). Leadership skills and the dialectics of leadership in group decision making. In R. Y. Hirokawa & M. S. Poole (Eds.), *Communication and group decision making* (2nd ed., pp. 301–342). Thousand Oaks, CA: SAGE.

Barge, J. K., & Hirokawa, R. Y. (1989). Toward a communication competency model of group leadership. *Small Group Behavior, 20,* 167–189. doi:10.1177/104649648902000203

Barki, H., & Pinsonneault, A. (2001). Small group brainstorming and idea quality: Is electronic brainstorming the most effective approach? *Small Group Research, 32,* 158–205. doi:10.1177/104649640103200203

Baron, N. S. (2010). Discourse structures in instant messages: The case of utterance breaks. *Language@Internet, 7.* Retrieved from http://nbn-resolving.de/urn:nbn:de:0009-7-26514

Baruah, J., & Paulus, P. B. (2008). Effects of training on idea generation in groups. *Small Group Research, 39,* 523–541. doi:10.1177/1046496408320049

Bass, B. M. (1981). *Stogdill's handbook of leadership. A survey of theory and research.* New York, NY: Free Press.

Bass, B. M. (1985). *Leadership and performance beyond expectations.* New York, NY: Free Press.

Bass, B. M. (1990). From transactional to transformational leadership: Learning to share the vision. *Organizational Dynamics, 18*(3), 19–31.

Baxter, J. (2015). Who wants to be the leader? The linguistic construction of emerging leadership in differently gendered teams. *International Journal of Business Communication, 52,* 427–451. doi:10.1177/2329488414525460

Bazarova, N. N., Walther, J. B., & McLeod, P. L. (2012). Minority Influence in virtual groups: A comparison of four theories of minority influence. *Communication Research Communication Research, 39,* 295–316. doi:10.1177/0093650211399752

Beck, S. J., & Keyton, J. (2009). Perceiving strategic meeting interaction. *Small Group Research, 40,* 223–246. doi:10.1177/1046496408330084

Beck, S. J., & Keyton, J. (2012). Team cognition, communication, and message interdependence. In E. Salas, S. F. Fiore, & M. Letsky (Eds.), *Theories of team cognition: Cross-disciplinary perspectives* (pp. 471–494). New York, NY: Routledge.

Beck, S. J., Littlefield, R. S., & Weber, A. (2012). Public meeting facilitation: A naïve theory analysis of crisis meeting interaction. *Small Group Research, 43,* 211–235. doi:10.1177/1046496411430531

Beck, S. J., Paskewitz, E. A., & Keyton, J. (2015). Toward a theory of strategic meeting interaction. In J. A. Allen, N. Lehmann-Willenbrock, & S. G. Rogelberg (Eds.), *The Cambridge handbook of meeting science* (pp. 305–324). New York, NY: Cambridge University Press.

Behfar, K. J., Friedman, R., & Oh, S. H. (2016). Impact of team (dis)satisfaction and psychological safety on performance evaluation biases. *Small Group Research, 47,* 77–107. doi:10.1177/1046496415616865

Behfar, K. J., Mannix, E. A., Peterson, R. A., & Trochim, W. M. (2011). Conflict in small groups: The meaning and consequences of process conflict. *Small Group Research, 42,* 127–176. doi:10.1177/1046496410389194

Bell, B. S., & Kozlowski, S. W. J. (2002). A typology of virtual teams: Implications for effective leadership. *Group & Organization Management, 27,* 14–49. doi:10.1177/1059601102027001003

Bell, M. A. (1983). A research note: The relationship of conflict and linguistic diversity in small groups. *Central States Speech Journal, 34,* 128–133. doi:10.1080/10510978309368131

Berdahl, J. L., & Anderson, C. (2005). Men, women, and leadership centralization in groups over time. *Group Dynamics, 9,* 45–57. doi:10.1037/1089-2699.9.1.45

Berry, G. R. (2011). Enhancing effectiveness on virtual teams: Understanding why traditional team skills are insufficient. *Journal of Business Communication, 48,* 186–206. doi: 10.1177/0021943610397270

Bettenhausen, K. L. (1991). Five years of groups research: What we have learned and what needs to be addressed. *Journal of Management, 17,* 345–381. doi:10.1177/014920639101700205

Bonito, J. A. (2002). The analysis of participation in small groups: Methodological and conceptual issues related to interdependence. *Small Group Research, 33,* 412–438. doi:10.1177/104649640203300402

Bradley, P. H. (1978). Power, status, and upward communication in small decision-making groups. *Communication Monographs, 45,* 33–43.

Brown, T. M., & Miller, C. E. (2000). Communication networks in task-performing groups: Effects of task complexity, time pressure, and interpersonal dominance. *Small Group Research, 31,* 131–157. doi:10.1177/104649640003100201

Burgoon, J. K., & Dunbar, N. E. (2006). Nonverbal expressions of dominance and power in human relationships. In V. Manusov & M. L. Patterson (Eds.), *The SAGE handbook of nonverbal communication* (pp. 279–297). Thousand Oaks, CA: SAGE.

Byrd, J. T., & Luthy, M. R. (2010). Improving group dynamics: Creating a *Academy of Educational Leadership Journal, 14,* 13–26.

Canary, D. J., & Spitzberg, B. H. (1989). A model of the perceived competence of conflict strategies. *Human Communication Research, 15*, 630–649. doi:10.1111/j.1468-2958.1989.tb00202.x

Carron, A. V., Brawley, L. R., Bray, S. R., Eys, M. A., Dorsch, K. D., ... Terry, P. C. (2004). Using consensus as a criterion for groupness: Implications for the cohesion-group success relationship. *Small Group Research, 35*, 466-491. doi:10.1177/1046496404263923

Carter, D. R., Seely, P. Q., Dagosta, J., DeChurch, L. A., & Zaccaro, S. J. (2015). Leading for global virtual teams: Facilitating teamwork processes. In J. L. Wildman & R. L. Griffith (Eds.), *Leading global teams: Translating multidisciplinary science to practice* (pp. 225-252). New York, NY; Springer.

Cartwright, D. (1968). The nature of group cohesiveness. In D. Cartwright & A. Zander (Eds.), *Group dynamics: Research and theory* (3rd ed., pp. 91–109). New York, NY: Harper & Row.

Chen, C. C., Wu, J., Ma, M., & Knight, M. B. (2011). Enhancing virtual learning team performance: A leadership perspective. *Human Systems Management, 30*, 215–228. doi: 10.3233/HSM-2011-0750

Clegg, S. (1989). *Frameworks of power*. Newbury Park, CA: SAGE.

Clifton, J. (2006). A conversation analytical approach to business communication. *Journal of Business Communication, 43*, 202–219. doi:10.1177/0021943606288190

Clifton, J. (2012). A discursive approach to leadership: Doing assessments and managing organizational meanings. *Journal of Business Communication, 49*, 148–168. doi:10.1177/0021943612437762

Cohen, S. G. (1990). Hilltop Hospital top management group. In J. R. Hackman (Exl.), *Groups that work (and those that don't): Creating conditions for effective teamwork* (pp. 56–77). San Francisco, CA: Jossey-Bass.

Cohen, S. G., & Bailey, D. E. (1997). What makes teams work: Group effectiveness research from the shop floor to the executive suite. *Journal of Management, 23*, 239–290. doi:10.1177/014920639702300303

Comer, D. R. (1995). A model of social loafing in real work groups. *Human Relations, 48*, 647–667. doi:10.1177/001872679504800603

Contractor, N. S., DeChurch, L. A., Carson, J., Carter, D. R., & Keegan, B. (2012). The topology of collective leadership. *Leadership Quarterly, 23*, 994–1011. doi:10.1016/j.leaqua.2012.10.010

Cruz, M. G., Henningsen, D. D., & Smith, B. A. (1999). The impact of directive leadership on group information sampling, decisions, and perceptions of the leader. *Communication Research, 26*, 349–369. doi:10.1177/0093650099026003004

Darics, E. (2014). The blurring boundaries between synchronicity and asynchronicity: New communicative situations in work-related instant message. *International Journal of Business Communication, 5*, 337–358. doi:10.1177/2329488414525440

Davis, C. S. (2008). Dueling narratives: How peer leaders use narrative to frame meaning in community mental health care teams. *Small Group Research, 39*, 706–727. doi:10.1177/1046496408323068

De Dreu, C. K., & Van Vianen, A. E. (2001). Managing relationship conflict and the effectiveness of organizational teams. *Journal of Organizational Behavior, 22*, 300–328. doi:10.1002/job.71

Delbecq, A. L., Van de Ven, A. H., & Gustafson, D. H. (1975). *Group techniques for program planning: A guide to nominal group and delphi processes*. Glenview, IL: Scott, Foresman.

DeStephen, R. S., & Hirokawa, R. Y. (1988). Small group consensus: Stability of group support of the decision, task process, and group relationships. *Small Group Behavior, 19*, 227–239. doi:10.1177/104649648801900204

De Dreu, C. K. W., & Weingart, L. R. (2003). Task versus relationship conflict, team performance, and team member satisfaction: A meta-analysis. *Journal of Applied Psychology, 88*, 741–749. doi:10.1037/0021-9010.88.4.741

de Vries, R. E., Van den Hooff, B., & de Ridder, J. A. (2006). Explaining knowledge sharing: The role of team communication styles, job satisfaction, and performance beliefs. *Communication Research, 33*, 115–135. doi:10.1177/0093650205285366

Druskat, V. U., & Pescosolido, A. T. (2006). The impact of emergent leader's emotionally competent behavior on team trust, communication, engagement, and effectiveness. In N. Ashkanasy, W. J. Zerbe, & C. Hartel (Eds.), *Research on emotion in organizations* (Vol. 2, pp. 27–58). Oxford, England: Elsevier.

Du-Babcock, B., & Tamaka, H. (2013). A comparison of the communication behaviors of Hong Kong Chinese and Japanese business professionals in intracultural and intercultural decision-making meetings. *Journal of Business and Technical Communication, 43,* 21–42. doi:10.1177/1050651913479918

Ellingson, L. L. (2003). Interdisciplinary health care teamwork in the clinic backstage. *Journal of Applied Communication Research, 31,* 93–117. doi:10.1080/00909880032000064579

Emery, C., Daniloski, K., & Hamby, A. (2011). The reciprocal effects of self-view as a leader and leader emergence. *Small Group Research, 42,* 199–224. doi:10.1177/1046496410389494

Fairhurst, G. T. (2007). *Discursive leadership.* Los Angeles, CA: SAGE.

Farmer, S. M., & Roth, J. (1998). Conflict-handling behavior in work groups: Effects of group structure, decision processes, and time. *Small Group Research, 29,* 669–713. doi:10.1177/1046496498296002

Firestien, R. L. (1990). Effects of creative problem solving training on communication behavior in small groups. *Small Group Research, 21,* 507–521. doi:10.1177/1046496490214005

Fisher, B. A. (1971). Communication research and the task-oriented group. *Journal of Communication, 21,* 136–149. doi:10.1111/j.1460-2466.1971.tb00911.x

Fletcher, J. K., & Kaufer, K. (2003). Shared leadership: Paradox and possibility. In C. L. Pearce & J. A. Conger (Eds.), *Shared leadership: Reframing the hows and whys of leadership* (pp. 21–47). Thousand Oaks, CA: SAGE

French, J. R. P., & Raven, B. (1968). The bases of social power. In D. Cartwright & A. Zander (Eds.), *Group dynamics: Research and theory* (pp. 259–269). New York, NY: Harper & Row.

Frey, L., & SunWolf. (2005). The symbolic-interpretive perspective of group life. In M. S. Poole & A. B. Hollingshead (Eds.), *Theories of small groups: Interdisciplinary perspectives* (pp. 185–239). Thousand Oaks, CA: SAGE.

Galanes, G. J. (2003). In their own words: An exploratory study of bona fide group leaders. *Small Group Research, 34,* 741–770. doi:10.1177/1046496403257649

Gardner, W. L., & Avolio, B. J. (1998). The charismatic relationship: A dramaturgical perspective. *Academy of Management Review, 23,* 32–58. doi: 10.5465/AMR.1998.192958

Gastil, J. (1993). Identifying obstacles to small group democracy. *Small Group Research, 24,* 5–27. doi:10.1177/1046496493241002

Gastil, J. (2010). *The group in society.* Los Angeles, CA: SAGE.

Gersick, C. J. G., & Hackman, J. R. (1990). Habitual routines in task-performing groups. *Organizational Behavior and Human Decision Processes, 41,* 65–97. doi:10.1016/0749-5978(90)90047-D

Gibb, J. R. (1961). Defensive communication. *Journal of Communication, 11,* 141–148. doi:10.1016/0749-5978(90)90047-D

Gilstrap, C., & Hendershot, B. (2015). E-leaders and uncertainty management: A computer-supported qualitative investigation. *Qualitative Research Reports in Communication, 18,* 86–96. doi:10.1080/17459435.2015.1086424

Golembiewski, R. T. (1962). *Making decisions in groups.* Glenview, IL: Scott, Foresman.

Gonzales, A. L., Hancock, J. T., & Pennebaker, J. W. (2010). Language style matching as a predictor of social dynamics in small groups. Communication *Research, 37,* 3–19. doi:10.1177/0093650209351468

Gouran, D. S., & Hirokawa, R. Y. (1983). The role of communication in decision making groups: A functional perspective. In M. S. Mander (Ed.), *Communications in transition* (pp. 168–185). New York, NY: Praeger.

Gouran, D. S., Hirokawa, R. Y., Julian, K. M., & Leatham, G. B. (1993). The evolution and current status of the functional perspective on communication in decision-making and problem-solving groups. In S. A. Deetz (Ed.), *Communication yearbook 16* (pp. 573–600). Newbury Park, CA: SAGE.

Graham, E. E., Papa, M. J., & McPherson, M. B. (1997). An applied test of the functional communication perspective of small group decision-making. *Southern Communication Journal, 62*, 269–279. doi:10.1080/10417949709373064

Green, S. G., & Taber, T. D. (1980). The effects of three social decision schemes on decision group process. *Organizational Behavior and Human Performance, 25*, 97–106. doi:10.1016/0030-5073(80)90027-6

Green, T. B. (1975). An empirical analysis of nominal and interacting groups. *Academy of Management Journal, 18*, 63–73. doi:10.2307/255625

Greene, C. N. (1989). Cohesion and productivity in work groups. *Small Group Behavior, 20*, 70–86. doi:10.1177/104649648902000106

Grice, H. P. (1999). Logic and conversation. In A. Jaworski & N. Coupland (Eds.), *The discourse reader* (pp. 76–88). London, England: Routledge.

Guenter, H., van Emmerik, H., Schreurs, B., Kuypers, T., van Iterson, A., & Notelaers, G. (2016). When task conflict becomes personal: The impact of perceived team performance. *Small Group Research, 47*, 569–604. doi:10.1177/1046496416667816

Guetzkow, H., & Gyr, J. (1954). An analysis of conflict in decision-making groups. *Human Relations, 1*, 367–382. doi:10.1177/001872675400700307

Gully, S. M., Devine, D. J., & Whitney, D. J. (1995). A meta-analysis of cohesion and performance: Effects of level of analysis and task interdependence. *Small Group Research, 26*, 497–520. doi:10.1177/1046496412468069

Hackman, J. R. (1992). Group influences on individuals in organizations. In M. D. Dunnet & L. M. Hough (Eds.), *Handbook of industrial and organizational psychology* (Vol. 3, pp. 199–267). Palo Alto, CA: Consulting Psychologists Press.

Hardy, J., Eys, M. A., & Carron, A. V. (2005). Exploring the potential disadvantages of high cohesion in sports teams. *Small Group Research, 36*, 166–187. doi:10.1177/1046496404266715

Harrison, D. A., Price, K. H., & Bell, M. P. (1998). Beyond relational demography: Time and the effects of surface- and deep-level diversity on work group cohesion. *Academy of Management Journal, 41*, 96–107. doi:10.2307/256901

Heath, R. G., & Sias, P. M. (1999). Communicating spirit in a collaborative alliance. *Journal of Applied Communication Research, 21*, 356–376. doi:10.1080/00909889909365545

Henningsen, D. D., & Henningsen, M. L. (2004). The effect of individual difference variables on information sharing in decision-making groups. *Human Communication Research, 30*, 540–555. doi:10.1111/j.1468-2958.2004.tb00744.x

Henry, K. B., Arrow, H., & Carini, B. (1999). A tripartite model of group identification: Theory and measurement. *Small Group Research, 30*, 558–581. doi:10.1177/104649649903000504

Hill, R. P., & Rapp, J. M. (2014). Codes of ethical conduct: A bottom-up approach. *Journal of Business Ethics, 123*, 621–630. doi:10.1007/s10551-013-2013-7

Hill, S. E. K. (2013). Team leadership. In P. G. Northouse (Ed.), *Leadership: Theory and practice* (6th ed., pp. 287–318). Los Angeles, CA: SAGE.

Hirokawa, R. Y. (1982). Group communication and problem-solving effectiveness I: A critical review of inconsistent findings. *Communication Quarterly, 30*, 134–141. doi:10.1080/01463378209369440

Hirokawa, R. Y. (1983). Group communication and problem-solving effectiveness: An investigation of group phases. *Human Communication Research, 9*, 291–305. doi:10.1111/j.1468-2958.1983.tb00700.x

Hirokawa, R. Y. (1988). Group communication and decision-making performance: A continued test of the functional perspective. *Human Communication Research, 14*, 487–515. doi:10.1111/j.1468-2958.1988.tb00165.x

Hirokawa, R. Y., & Johnston, D. D. (1989) Toward a general theory of group decision-making: Development of an integrated model. *Small Group Behavior, 20*, 500-523. doi:10.1177/104649648902000408

Hirokawa, R. Y., Erbert, L., & Hurst, A. (1996). Communication and group decision-making effectiveness. In R. Y. Hirokawa & M. S. Poole (Eds.), *Communication and group decision making* (pp. 269–300) Thousand Oaks, CA: SAGE.

Hirokawa, R. Y., & Johnston, D. D. (1989). Toward a general theory of group decision making: Development of an integrated model. *Small Group Behavior, 20*, 500–523. doi:10.1177/104649648902000408

Hirokawa, R. Y., & Keyton, J. (1995). Perceived facilitators and inhibitors of effectiveness in organizational work teams. *Management Communication Quarterly, 8*, 424–446. doi:10.1177/0893318995008004002

Hoffman, L. R., & Kleinman, G. B. (1994). Individual and group in problem solving: The valence model redressed. *Human Communication Research, 21*, 36–59. doi:10.1111/j.1468-2958.1994.tb00338.x

Hollander, E. P. (1978). *Leadership dynamics A practical guide to effective relationships.* New York, NY: Macmillan.

Hollander, E. P. (1985). Leadership and power. In G. Lindzey & E. Aronson (Eds.), *The handbook of social psychology* (Vol. 2, pp. 485–537). New York, NY: Random House.

Hollingshead, A. B. (1996). The rank-order effect in group decision making. *Organizational Behavior and Human Decision Processes, 68*, 181–193. doi:10.1006/obhd.1996.0098

Hollingshead, A. B., Jacobsohn, G. C., & Beck, S. J. (2007). Motives and goals in context: A strategic analysis of information-sharing groups. In K. Fiedler (Ed.), *Social communication* (pp. 257–280). New York, NY: Psychology Press.

Jacobs, T. O. (1970). *Leadership and exchange informal organizations.* Alexandria, VA: Human Resources Research Organization.

Jackson, J. W. (2008). Reactions to social dilemmas as a function of group identity: Rational calculations, and social context. *Small Group Research, 39*, 673–705. doi:10.1177/1046496408322761

Janssens, M., & Brett, J. (2006). Cultural intelligence in global teams: A fusion model of collaboration. *Group & Organization Management, 31*, 124–153. doi:10.1177/1059601105275268

Jarboe, S. (1996). Procedures for enhancing group decision making. In R. Y. Hirokawa & M. S. Poole (Eds.), *Communication and group decision making* (pp. 345–383). Thousand Oaks, CA: SAGE.

Jarboe, S. C., & Witteman, H. R. (1996). Intragroup conflict management in task-oriented groups: The influence of problem sources and problem analyses. *Small Group Research, 27*, 316–338. doi:10.1177/1046496496272007

Jehn, K. (1995). A multimethod examination of the benefits and detriments of intragroup conflict. *Administrative Science Quarterly, 40*, 256–282. doi:10.2307/2393638

Jehn, K. A. (1997). A qualitative analysis of conflict types and dimensions in organizational groups. *Administrative Science Quarterly, 42*, 530–557. doi:10.2307/2393737

Jehn, K. A., & Mannix, E. A. (2001). The dynamic nature of conflict: A longitudinal study of intragroup conflict and group performance. *Academy of Management Journal, 44*, 238–251. doi:10.13189/ujm.2016.040204

Johansson, C., Miller, V. C., & Hamrin, S. (2014). Conceptualizing communicative leadership: A framework for analysing and developing leaders' communication competence. *Corporate Communications: An International Journal, 10*, 147–165. doi:10.1108/CCIJ-02-2013-0007

Jones, T. S. (2000). Emotional communication in conflict: Essence and impact. In W.F. Eadie & P.E. Nelson (Eds.), *The language of conflict and resolution* (pp. 81–104). Thousand Oaks, CA: SAGE.

Jones, E. E., & Kelly, J. R. (2007). Contributions to a group discussion and perceptions of leadership: Does quantity always count more than quality? *Group Dynamics: Theory, Research, and Practice, 11*, 15–30. doi:10.1037/1089-2699.11.1.15

Katz, N., Lazer, D., Arrow, H., & Contractor, N. (2005). The network perspective on small group: Theory and research. In M. S. Poole & A. B. Hollingshead (Eds.), *Theories of small groups: Interdisciplinary perspectives* (pp. 277–312). Thousand Oaks, CA: SAGE.

Katzenbach, J. R., & Smith, D. K. (1993). *The wisdom of teams: Creating the high-performance organization.* New York, NY: HarperBusiness.

Kauffeld, S., & Lehmann-Willenbrock, N. (2012). Meetings matter: Effects of team meetings on team and organizational success. *Small Group Research, 43,* 130–158. doi:10.1177/1046496411429599

Kellermann, K. (1992). Communication: Inherently strategic and primarily automatic. *Communication Monographs, 59,* 288–300. doi:10.1080/03637759209376270

Ketrow, S. M. (1991). Communication role specializations and perceptions of leadership. *Small Group Research, 22,* 492–514. doi:10.1177/1046496491224005

Keyton, J. (1991). Evaluating individual group member satisfaction as a situational variable. *Small Group Research, 22,* 200–219. doi:10.1177/1046496491222004

Keyton, J. (1999). Relational communication in groups. In L. R. Frey, D. S. Gouran, & M. S. Poole (Eds.), *The handbook of group communication theory and research* (pp. 192–222). Thousand Oaks, CA: SAGE.

Keyton, J., & Beck, S. J. (2010). Examining emotional communication: Laughter in jury deliberations. *Small Group Research, 41,* 386–407. doi:10.1177/1046496410366311

Keyton, J., Beck, S. J., & Asbury, M. B. (2010). Macrocognition: A communication perspective. *Theoretical Issues in Ergonomics Science, 11,* 272–286. doi:10.1080/14639221003729136

Kiesler, S., & Cummings, J. N. (2002). What do we know about proximity and distance in work groups? A legacy of research. In P. J. Hinds & S. Kiesler (Eds.), *Distributed work* (pp. 57–80). Cambridge, MA: MIT Press.

Kirchmeyer, C., & Cohen, A. (1992). Multicultural groups: Their performance and reactions with constructive conflict. *Group and Organization Management, 17,* 153–170. doi:10.1177/1059601192172004

Kirschbaum, K. A., Rask, J. P., Fortner, S. A., Kulesher, R., Nelson, M. T., Yen, T., & Brennan, M. (2014). Physician communication in the operating room. *Health Communication, 30,* 317-327. doi:10.1080/10410236.2013.856741

Knutson, T. J., & Kowitz, A. C. (1977). Effects of information type and level of orientation on consensus-achievement in substantive and affective small group conflict. *Central States Speech Journal, 28,* 54–63. doi:10.1080/10510977709367919

Kramer, M. W. (2005). Communication and social exchange processes in community theater groups. *Journal of Applied Communication, 33,* 159–182. doi:10.1080/00909880500045049

Kramer, M. W. (2006). Shared leadership in a community theater group: Filling the leadership role. *Journal of Applied Communication Research, 34,* 141–162. doi:10.1080/00909880600574039

Kramer, M. W. (2011). Toward a communication model for the socialization of voluntary members. *Communication Monographs, 78,* 233–255. doi:10.1080/03637751.2011.564640

Kramer, M. W., Kuo, C. L., & Dailey, J. C. (1997). The impact of brainstorming techniques on subsequent group processes: Beyond generating ideas. *Small Group Research, 28,* 218–242. doi:10.1177/1046496497282003

Kramer, T. J., Fleming, G. P., & Mannis, S. M. (2001). Improving face-to-face brainstorming through modeling and facilitation. *Small Group Research, 32,* 533–557. doi:10.1177/104649640103200502

Krauss, R. M., & Morsella, E. (2000). Communication and conflict. In M. Deutsch & P. T. Coleman (Eds.), *The handbook of conflict resolution: Theory and practice* (pp. 131–143). San Francisco, CA: Jossey-Bass.

Kuhn, T., & Poole, M. S. (2000). Do conflict management styles affect group decision making? Evidence from a longitudinal field study. *Human Communication Research, 26,* 558–590. doi:10.1111/j.1468-2958.2000.tb00769.x

Kunze, F., & Bruch, H. (2010). Age-based faultlines and perceived productive energy: The moderation of transformational leadership. *Small Group Research, 41,* 593–620. doi:10.1177/1046496410366307

Lafond, D., Jobidon, M-E., Aube, C., & Tremblay, S. (2011). Evidence of structure-specific teamwork requirements and implications for team design. *Small Group Research, 42,* 507–535. doi:10.1177/1046496410397617

Lam, C. (2013). The efficacy of text message to improve social connectedness and team attitude in student technical communication projects: An experimental study. *Journal of Business and Technical Communication, 27,* 180–208. doi:10.1177/1050651912468888

Lam, C. (2015). The role of communication and cohesion in reducing social loafing in group projects. *Business and Professional Communication Quarterly, 78*, 454–475. doi:10.1177/2329490615596417

Lambertz-Berndt, M., M., & Blight, M. G. (2016). "You don't have to like me, but you have to respect me": The impacts of assertiveness, cooperativeness, and group satisfaction in collaborative assignments. *Business and Professional Communication Quarterly, 79*, 180–199. doi:10.1177/2329490615604749

Lammers, J. C., & Krikorian, D. H. (1997). Theoretical extension and operationalization of the bona fide group construct with an application to surgical teams. *Journal of Applied Communication Research, 25*, 17–38. doi:org/ 10.1080/00909889709365463

Larkey, L. K. (1996). Toward a theory of communicative interactions in culturally diverse workgroups. *Academy of Management Review, 21*, 463–491. doi:10.5465/AMR.1996.9605060219

Larson, C. E., & LaFasto, F. M. J. (1989). *TeamWork. What must go right/what can go wrong*. Newbury Park, CA: SAGE.

Lau, D., & Murnighan, J. K. (1998). Demographic diversity and faultlines: The compositional dynamics of organizational groups. *Academy of Management Review, 23*, 325–340. doi:10.5465/AMR.1998.533229

Leach, D. J., Rogelberg, S. G., Warr, P. B., & Burnfield, J. L. (2009). Perceived meeting effectiveness: The role of design characteristics. *Journal of Business and Psychology, 24*, 65–76. doi:10.1007/s10869-009-9092-6

Levine, K. J., Muenchen, R. A., & Brooks, A. (2010). Measuring transformational and charismatic leadership: Why isn't charisma measured? *Communication Monographs, 77*, 576–591. doi:10.1080/03637751.2010.499368

Lewicki, R. J., McAllister, D. J., & Bies, D. J. (1998). Trust and distrust: New relationships and realities. *Academy of Management Review, 23*, 438–458. doi:10.2307/259288

Lia, J., & Robertson, T. (2011). Physical space and information space: Studies of collaboration in distributed multi-disciplinary medical team meetings. *Behaviour & Information Technology, 30*, 443–454. doi:10.1080/0144929X.2011.577194

Lira, E. M., Ripoll, P., Piero, J. M., & Zornoza, A. M. (2008). The role of information and communication technologies in the relationship between group effectiveness and group potency: A longitudinal study. *Small Group Research, 39*, 728–745. doi:10.1177/1046496408323481

Liu, X-Y., Hartel, C. E. J., & Sun, J. J-M. (2014). The workgroup emotional climate scale. *Group & Organization Management, 39*, 626–663. doi:10.1177/1059601114554453

Lovaglia, M., Mannix, E. A., Samuelson, C. D., Sell, J., & Wilson, R. K. (2005). Conflict, power, and status in groups. In M. S. Poole & A. B. Hollingshead (Eds.), *Theories of small groups: Interdisciplinary perspectives* (pp. 139–184). Thousand Oaks, CA: SAGE.

Martin, A. (2007). *What's next? The 2007 changing nature of leadership survey*. Greensboro, NC: Center for Creative Leadership. Retrieved from http://www.ccl.org/wp-content/uploads/2015/04/WhatsNext.pdf

Mathieu, J. E., & Rapp, T. L. (2009). Laying the foundation for successful team performance trajectories: The roles of team charters and performance strategies. *Journal of Applied Psychology, 94*, 90–103. doi:10.1037/a0013257

McComas, K. A. (2003). Citizen satisfaction with public meetings used for risk communication. *Journal of Applied Communication Research, 31*, 164-184. doi: 0.1080/0090988032000064605

McDowell, W. C., Herdman, A. O., & Aaron, J. (2011). Charting the course: The effects of team charters on emergent behavioral norms. *Organization Development Journal, 29*, 79–88.

McGrath, J. E. (1984). *Groups: Interaction and performance*. Englewood Cliffs, NJ: Prentice-Hall.

McGrath, J. E., Berdahl, J. L., & Arrow, H. (1995). Traits, expectations, culture, and clout: The dynamics of diversity in work groups. In S. E. Jackson & M. N. Ruderman (Eds.), *Diversity in work teams: Research paradigms for a changing workplace* (pp. 17–45). Washington, DC: American Psychological Association.

McKinney, B. C., Kelly, L., & Duran, R. L. (1997). The relationship between conflict message styles and dimensions of communication competence. *Communication Reports, 10*, 185–196. doi:10.1080/08934219709367674

Meng, J., Fulk, J., & Yuan, Y. C. (2015). The roles and interplay of intragroup conflict and team emotion management of information seeking behaviors in team contexts. *Communication Research, 42*, 675–700. doi:10.1177/0093650213476294

Merolla, A. J. (2010). Relational maintenance and noncopresence reconsidered: Conceptualizing geographic separation in close relationships. Communication *Theory, 20*, 169–193. doi:10.1111/j.1468-2885.2010.01359.x

Messersmith, A. S. (2015). Preparing students for 21st century teamwork: Effective collaboration in the online group communication course. *Communication Teacher, 29*, 219-226. doi:10.1080/17404622.2015.1046188.

Meyers, R. A. (1989). Persuasive arguments theory A test of assumptions. *Human Communication Research, 15*, 357–381. doi:10.1111/j.1468-2958.1989.tb00189.x

Meyers, R. A., Brashers, D. E., & Hanner, J. (2000). Majority-minority influences: Identifying argumentative patterns and predicting argument-outcome links. *Journal of Communication, 50*(4), 3–30. doi:10.1111/j.1460-2466.2000.tb02861.x

Mirivel, J. C., & Tracy, K. (2005). Premeeting talk: An organizationally crucial form of talk. *Research on Language and Social Interaction, 38*, 1-34. doi:10.1207/s15327973rlsi3801

Moscovici, S. (1976). *Social influence and social change*. London, England: Academic Press.

Moye, N. E., & Langfred, D. W. (2004). Information sharing and group conflict: Going beyond decision making to understand the effects of information sharing on group performance. *International Journal of Conflict Management, 15*, 381–410.

Mullen, B., Anthony, T., Salas, E., & Driskell, J. E. (1994). Group cohesiveness and quality of decision making: An integration of tests of the groupthink hypothesis. *Small Group Research, 25*, 189–204. doi:10.1177/1046496494252003

Mullen, B., Johnson, C., & Salas, E. (1991). Productivity loss in brainstorming groups: A meta-analytical integration. *Basic and Applied Social Psychology, 12*, 3–23. doi:10.1207/s15324834basp1201_1

Myers, S. A., Shimotsu, S., Byrnes, K., Frisby, B. N., Durbin, J., & Loy, B. N. (2010). Assessing the role of peer relationships in the small group communication course. *Communication Teacher, 24*, 43–57. doi:10.1080/17404620903468214

Myrsiades, L. (2000). Meeting sabotage: Met and conquered. *Journal of Management Development, 19*, 879–884. doi:org/10.1108/02621710010379182

Nemeth, C. J. (1986). Differential contributions of majority and minority influence. *Psychological Review, 93*, 23-32. doi:10.1037/0033-295x.93.1.23

Nemeth, C., Swedlund, M., & Kanki, B. (1974). Patterning of the minority's responses and their influence on the majority. *European Journal of Social Psychology, 4*, 437-439. doi:10.1002/ejsp.2420040104

Nicotera, A. M. (1994). The use of multiple approaches to conflict: A study of sequences. *Human Communication Research, 20*, 592–621. doi:10.1111/j.1468-2958.1994.tb00336.x

Northouse, P. G. (2017). *Leadership: Theory and practice* (7th ed.). Los Angeles, CA: SAGE.

Nye, J. L. (2002). The eye of the follower: Information processing effects on attributions regarding leaders of small groups. *Small Group Research, 33*, 337–360. doi:10.1177/104964020330003003

Odermatt, I., Konig, C. J., & Kleinmann, J. (2015). Meeting preparation and design characteristics. In J. A. Allen, N. Lehmann-Willenbrock, & S. G. Rogelberg (Eds.), *The Cambridge handbook of meeting science* (pp. 49–68). New York, NY: Cambridge University Press.

Oetzel, J. G. (2002).The effects of culture and cultural diversity on communication in work groups. In L. R. Frey (Ed.), *New directions in group communication* (pp. 121–137). Thousand Oaks, CA: SAGE.

Oetzel, J. G. (2005). Effective intercultural workgroup communication theory. In W. B. Gudykunst (Ed.), *Theorizing about intercultural communication* (pp. 351–371). Thousand Oaks, CA: SAGE.

Oetzel, J. G., Burns, T. B., Sanchez, M. I., & Perez, F. G. (2001). Investigating the role of communication in culturally diverse work groups: A review and synthesis. In W. B. Gudykunst (Ed.), *Communication yearbook* 25 (pp. 237–269). Mahwah, NJ: Erlbaum.

Oetzel, J. G., McDermott, V. M., Torres, A., & Sanchez, C. (2012). The impact of individual differences and group diversity on group interaction climate and satisfaction: A test of the effective intercultural workgroup communication theory. *Journal of International & Intercultural Communication*, 5, 144–167. doi:10.1080/17513057.2011.640754

Orlitzky, M., & Hirokawa, R. Y. (2001). To err is human, to correct for it divine: A meta-analysis of research testing the functional theory of group decision-making effectiveness. *Small Group Research*, 32, 313–341. doi:10.1177/104649640103200303

Osborn, A. F. (1963). *Applied imagination* (3rd ed.). New York, NY: Scribner.

Pace, R. C. (1990). Personalized and depersonalized conflict in small group discussions: An examination of differentiation. *Small Group Research*, 21, 79–96. doi:10.1177/1046496490211006

Paulus, P. B., Nakui, T., Putman, V. L., & Brown, V. R. (2006). Effects of task instructions and brief breaks on brainstorming. *Group Dynamics: Theory, Research, and Practice*, 10, 206–219. doi:10.1037/1089-2699.10.3.206

Pavitt, C. (1993). What (little) we know about formal group discussion procedures: A review of relevant research. *Small Group Research*, 24, 217–235. doi:10.1177/1046496493242004

Pavitt, C. (1999). Theorizing about the group communication-leadership relationship: Input-process-output and functional models. In L. R. Frey, D. S. Gouran, & M. S. Poole (Eds.), *The handbook of group communication theory and research* (pp. 313–334). Thousand Oaks, CA: SAGE.

Pavitt, C., High, S. C., Tressler, K. E., & Winslow, J. K. (2007). Leadership communication during group resource dilemmas. *Small Group Research*, 38, 509–531. doi:10.1177/1046496407304333

Pavitt, C., & Sackaroff, P. (1990). Implicit theories of leadership and judgments of leadership among group members. *Small Group Research*, 21, 374–392. doi:10.1177/1046496490213006

Pescolido, A. T. (2003). Group efficacy and group effectiveness: The effects of group efficacy over time on group performance and development. *Small Group Research*, 34, 20–42. doi:10.1177/1046496402239576

Pescosolido, A. T., & Saavedra, R. (2012). Cohesion and sports teams: A review. *Small Group Research*, 43, 744–758. doi:10.1177/1046496412465020

Poncini, G. (2002). Investigating discourse at business meetings with multicultural participation. *International Review of Applied Linguistics in Language Teaching*, 40, 345–373. doi:10.1515/iral.2002.017

Pondy, L. R. (1967). Organizational conflict: Concepts and models. *Administrative Science Quarterly*, 12, 296–320. doi:10.1002/job.4030130304

Poole, M. S. (1991). Procedures for managing meetings: Social and technological innovation. In R. A. Swanson & B. O. Knapp (Eds.), *Innovative meeting management* (pp. 53–110). Austin, TX: 3M Meeting Management Institute.

Poole, M. S., & Garner, J. T. (2006). Perspectives on workgroup conflict and communication. In J. G. Oetzel & S. Ting-Toomey (Eds.), *The SAGE handbook of conflict communication: Integrating theory, research, and practice* (pp. 267–292). Thousand Oaks, CA: SAGE.

Poole, M. S., & Dobosh, M. (2010). Exploring conflict management processes in jury deliberation through interaction analysis. *Small Group Research*, 41, 408–426. doi:10.1177/1046496410366310

Poole, M. S., & Zhang, H. (2005). Virtual teams. In S. A. Wheelan (Ed.), *The handbook of group research and practice* (pp. 363–384). Thousand Oaks, CA: SAGE.

Prapavessis, H., & Carron, A. V. (1997). Cohesion and work output. *Small Group Research*, 28, 294–301. doi:10.1177/1046496497282006

Putnam, L. L. (2006). Definitions and approaches to conflict and communication. In J. G. Oetzel & S. Ting-Toomey (Eds.), *The SAGE handbook of conflict communication: Integrating theory, research, and practice* (pp. 1–32). Thousand Oaks, CA: SAGE.

Putnam, L. L., & Stohl, C. (1990). Bona fide groups: A reconceptualization of groups in context. *Communication Studies, 41*, 248–265.

Putnam, L. L., & Stohl, C. (1996). Bona fide groups: An alternative perspective for communication and small group decision making. In R. Y. Hirokawa & M. S. Poole (Eds.), *Communication and group decision making* (2nd ed., pp. 147–178). Thousand Oaks, CA: SAGE.

Putnam, L. L., & Wilson, C. E. (1983). Communicative strategies in organizational conflicts: Reliability and validity of a measurement scale. In M. Burgoon (Ed.), *Communication yearbook 6* (pp. 629–652). Beverly Hills, CA: SAGE.

Raven, B. H. (1993). The bases of power: Origins and recent developments. *Journal of Social Issues, 49*, 227-251. doi:10.1111/j.1540-4560.1993.tb01191.x

Reid, F. J. M., & Reid, D. J. (2010). The expressive and conversational affordances of mobile messaging. *Behavior & Information Technology, 29*, 3–22. doi:10.1080/01449290701497079

Reimer, T., Reimer, A., & Czienskowski, U. (2010). Decision-making groups attenuate the discussion bias in favor of shared information: A meta-analysis. *Communication Monographs, 77*, 121–142. doi:10.1080/03637750903514318

Reimer, T., Reimer, A., & Hinsz, V. (2010). Naïve groups can solve the hidden-profile problem. *Human Communication Research, 36*, 443–467. doi:10.1111/j.1468-2958.2010.01383.x

Renz, M. A. (2006). The meaning of consensus and blocking for cohousing groups. *Small Group Research, 37*, 351–376. doi:10.1177/1046496406291184

Riddle, B. L., Anderson, C. M., & Martin, M. M. (2000). Small group socialization scale: Development and validity. *Small Group Research, 31*, 554–572. doi:10.1177/104649640003100503

Ridgeway, C. L. (2001). Gender, status, and leadership. *Journal of Social Issues, 57*, 637–655. doi:10.1111/0022–4537.00233

Rosenthal, S. B., & Buchholz, R. A. (1995). Leadership: Toward new philosophical foundations. *Business and Professional Ethics Journal, 14*, 25–41. doi:10.5840/bpej199514315

Rozell, E. J., & Gundersen, D. E. (2003). The effects of leader impression management on group perceptions of cohesion, consensus, and communication. *Small Group Research, 34*, 197–222. doi:10.1177/1046496402250431

Ruback, R. B., Dabbs, J. M., & Hopper, C. H. (1984). The process of brainstorming: An analysis with individual and group vocal parameters. *Journal of Personality and Social Psychology, 47*, 558–567. doi:10.1037/0022-3514.47.3.558

Sargent, L. D., & Sue-Chan, C. (2001). Does diversity affect group efficacy? The intervening role of cohesion and task interdependence. *Small Group Research, 32*, 426–450. doi:10.1177/104649640103200403

Scheerhorn, D., Geist, P., & Teboul, J. C. B. (1994). Beyond decision making in decision-making groups: Implications for the study of group communication. In L. R. Frey (Ed.), *Group communication in context: Studies of natural groups* (pp. 247–262). Hillsdale, NJ: Erlbaum.

Schultz, B. (1986). Communication correlates of perceived leaders in the small group. *Small Group Research, 17*, 51-65. doi:10.1177/104649648601700105

Schwarz, R. M. (1994). *The skilled facilitator: Practical wisdom for developing effective groups.* San Francisco, CA: Jossey-Bass.

Schwartzman, H. B. (1989). *The meeting: Gatherings in organizations and communities.* New York, NY: Plenum Press.

Schweiger, D. M., & Leana, C. R. (1986). Participation in decision making. In E. A. Locke (Ed.), *Generalizing from laboratory to field settings: Research findings from industrial-organizational psychology, organizational behavior, and human resource management* (pp. 147–166). Lexington, MA: Lexington Books.

Sharf, B. F. (1978). A rhetorical analysis of leadership emergence in small groups. *Communication Monographs, 45,* 156–172. doi:10.1080/03637757809375960

Shaw, M. E. (1981). *Group dynamics: The psychology of small group behavior.* New York, NY: McGraw-Hill.

Shin, Y. (2014). Positive group affect and team creativity: Mediation of team reflexivity and promotion focus. *Small Group Research, 45,* 337–364. doi:10.1177/1046496414533618

Shuffler, M. L., Burke, C. S., Kramer, W. S., & Salas, E. (2013). Leading teams: Past, present, and future perspectives. In M. G. Rumsey (Ed.), *The Oxford handbook of leadership* (pp. 144–166). Oxford, England: Oxford University Press.

Sillars, A. L., & Wilmot, W. W. (1994). Communication strategies in conflict and mediation. In J. A. Daly & J. M. Wiemann (Eds.), *Strategic interpersonal communication* (pp. 163–190). Hillsdale, NJ: Erlbaum.

Sillince, J. A. A. (2000). Rhetorical power, accountability and conflict in committees: An argumentation approach. *Journal of Management Studies, 37,* 1125–1156. doi:10.1111/1467-6486.00219

Smith, K. K., & Berg, D. N. (1987). *Paradoxes of group life.* San Francisco, CA: Jossey-Bass.

Socha, T. J. (1999). Communication in family units: Studying the first group. In L. Frey (Ed.), *The handbook of group communication theory and research* (pp. 475–492). Thousand Oaks, CA: SAGE.

Solansky, S. T. (2008). Leadership style and team processes in self-managed teams. *Journal of Leadership & Organizational Studies, 14,* 4332–341. doi:10.1177/1548051808315549

Somech, A., & Drach-Zahavy, A. (2007). Schools as team-based organizations: A structure-process-outcomes approach. *Group Dynamics: Theory, Research, and Practice, 11,* 305–320. doi:10.1037/1089-2699.11.4.305

Spink, K. S., & Carron, A. V. (1994). Group cohesion effects in exercise classes. *Small Group Research, 25,* 26–42. doi:10.1177/1046496494251003

Stasser, G., & Titus, W. (1985). Pooling of unshared information in group decision making: Biased information sampling during discussion. *Journal of Personality and Social Psychology, 48,* 1467–1478. doi:org/10.1037/0022-3514.48.6.1467

Stasser, G., & Titus, W. (1987). Effects of information load and percentage of shared information on the dissemination of unshared information during group discussion. *Journal of Personality and Social Psychology, 53,* 81–92. doi:org/10.1037/0022-3514.53.1.81

Stohl, C., & Putnam, L. L. (2003). Communication in bona fide groups: A retrospective and prospective account. In L. R. Frey (Ed.), *Group communication tn context: Studies of bona fide groups* (pp. 399–414). Mahwah, NJ: Erlbaum.

Sung, C. C. M. (2011). Doing gender and leadership: A discursive analysis of media representations in a reality TV show. *English Text Construction, 4,* 85–111. doi:10.1075/etc.4.1.05sun

SunWolf. (2008). *Peer groups: Expanding our study of small group communication.* Los Angeles, CA: SAGE.

SunWolf. (2010). Investigating jury deliberation in a capital murder case. *Small Group Research, 41,* 380–385. doi:10.1177/1046496410366484

SunWolf, & Seibold, D. R. (1999). The impact of formal procedures on group processes, members, and task outcomes. In L. R. Frey, D. S. Gouran, & M. S. Poole (Eds.), *The handbook of group communication theory and research* (pp. 395–431). Thousand Oaks, CA: SAGE.

Thomas, K. W. (1977). Toward multi-dimensional values in teaching: The examples of conflict behaviors. *Academy of Management Review, 2,* 484–490. doi:10.5465/AMR.1977.4281851

Thomas, K. W. (1992). Conflict and negotiation processes in organizations. In M. D. Dunnette & L. M. Hough (Eds.), *Handbook of industrial and organizational psychology* (pp. 651–717). Palo Alto, CA: Consulting Psychologists Press.

Timmerman, C. E., & Scott, C. R. (2006). Virtually working: Communicative and structural predictors of media use and key outcomes in virtual work teams. *Communication Monographs, 73*, 108–136. doi:10.1080/03637750500534396

Tropman, J. E. (2003). *Making meetings work: Achieving high quality group decisions* (2nd ed.). Thousand Oaks, CA: SAGE.

Tse, H. H. M., & Dasborough, M. T. (2008). A study of exchange and emotions in team member relationships. *Group & Organization Management, 33*, 194–215. doi:10.1177/1059601106293779

Türetgen, I. O., Unsal, P., & Erdem, I. (2008). The effects of sex, gender role, and personality traits on leader emergence: Does culture make a difference? *Small Group Research, 39*, 588–615. doi:10.1177/1046496408319884

Turman, P. D. (2008). Coaches' immediacy behaviors as predictors of athletes' perceptions of satisfaction and team cohesion. *Western Journal of Communication, 72*, 162–179. doi: 10.1080/10570310802038424

Van den Hooff, B., & de Ridder, J. A. (2004). Knowledge sharing in context: The influence of organizational commitment, communication climate and CMC use on knowledge sharing. *Journal of Knowledge Management, 8*, 117–130. doi:10.1108/ 13673270410567675

Van de Ven, A. H., & Delbecq, A. L. (1974). The effectiveness of nominal, delphi, and interacting group decision-making processes. *Academy of Management Journal, 17*, 605–621. doi:10.2307/255641

Van Mierlo, H., & Kleingeld, A. (2010). Goals, strategies, and group performance: Some limits of goal setting in groups. *Small Group Research, 41*, 524–555. doi:10.1177/1046496410373628

Van Swol, L. M. (2009). Discussion and perception of information in groups and judge-advisor systems. *Communication Monographs, 76*, 99–120. doi:10.1080/03637750802378781

Waldeck, J. H., Shepard, C. A., Teitelbaum, J., Farrar, W. J., & Seibold, D. R. (2002). New directions for functional, symbolic convergence, structuration, and bona fide group perspectives of group communication. In L. R. Frey (Ed.), *New directions in group communication* (pp. 3–23). Thousand Oaks, CA: SAGE.

Walker, R. C., & Aritz, J. (2015). Women doing leadership: Leadership styles and organizational culture. *International Journal of Business Communication, 52*, 452–478. doi:10.1177/2329488415598429

Wall, V. D., Jr., & Galanes, G. J. (1986). The SYMLOG dimensions and small group conflict. *Communication Studies, 37*, 61–78. doi:10.1080/10510978609368206

Wall, V. D., Jr., Galanes, G. J., & Love, S. B. (1987). Small, task-oriented groups: Conflict, conflict management, satisfaction, and decision quality. *Small Group Behavior, 18*, 31–55. doi:10.1177/104649648701800102

Wall, V. D., & Nolan, L. L. (1986). Perceptions of inequity, satisfaction, and conflict in task-oriented groups. *Human Relations, 39*, 1033–1052. doi:10.1177/001872678603901106

Walther, J. B. (2002). Time effects in computer-mediated groups: Past, present, and future. In P. Hinds & S. Kiesler (Eds.), *Distributed work* (pp. 235–257) Cambridge, MA: MIT Press.

Wang, L., Han, J., Fisher, C. M., & Pan, Y. (2017). Learning to share: Exploring temporality in shared leadership and team learning. *Small Group Research, 48*, 165–189. doi:10.1177/1046496417690027

Warfield, J. N. (1993). Complexity and cognitive equilibrium: Experimental results and their implications. In D. J. D. Sandole & H. van der Merwe (Eds.), *Conflict resolution theory and practice: Integration and application* (pp. 65–77). New York, NY: Manchester University Press.

Watson, W. E., Johnson, L., & Merntt, D. (1998). Team orientation, self-orientation, and diversity in task groups: Their connection to team performance over time. *Group and Organization Management, 23*, 161–188. doi:10.1177/1059601198232005

Watson, W. E., Kumar, K., & Michaelsen, L. K. (1993). Cultural diversity's impact on interaction process and performance: Comparing homogeneous and diverse task groups. *Academy of Management Journal, 36*, 590–602. doi:10.2307/256593

Watzlawick, P., Beavin, J. H., & Jackson, D. D. (1967). *Pragmatics of human communication: A study of interactional patterns, pathologies, and paradoxes.* New York, NY: Norton.

Webber, S. S. (2008). Development of cognitive and affective trust in teams: A longitudinal study. *Small Group Research, 6,* 746–769. doi:10.1177/1046496408323569

Weick, K. E. (1969). Laboratory organizations and unnoticed causes. *Administrative Science Quarterly, 14,* 294-303. doi:10.2307/2391107

Wellman, B. (1988). Structural analysis: From method and metaphor to theory and substance. In B. Wellman & S. Berkowitz (Eds.), *Social structures: A network approach* (pp. 19–61). Cambridge, England: Cambridge University Press.

Wheelan, S. A. (2009). Group size, group development, and group productivity. *Small Group Research, 40,* 247–262. doi:10.1177/1046496408328703

Wheelan, S. A., & McKeage, R. L. (1993). Developmental patterns in small and large groups. *Small Group Research, 24,* 60–83. doi:10.1177/1046496493241005

Whitford, T., & Moss, S. A. (2009). Transformational leadership in distributed work groups: The moderating role of follower regulatory focus and goal orientation. *Communication Research, 36,* 810–837. doi:10.1177/0093650209346800

Whitton, S. M., & Fletcher, R. B. (2014). The Group Environment Questionnaire: A multilevel confirmatory factor analysis. *Small Group Research, 45,* 68–88. doi:10.1177/1046496413511121

Witteman, H. (1991). Group member satisfaction: A conflict-related account. *Small Group Research, 22,* 24–58. doi:10.1177/1046496491221003

Wittenbaum, G. M., Hollingshead, A. B., & Botero, I. C. (2004). From cooperative to motivated information sharing in groups: Moving beyond the hidden profile paradigm. *Communication Monographs, 71,* 286–310. doi:10.1080/0363452042000299894

Wittenbaum, G. M., Hollingshead, A. B., Paulus, P. B., Hirokawa, R. Y. Ancona, D. G., Peterson, R. S., Jehn, K. A., & Yoon, K. (2004). The functional perspective as a lens for understanding groups. *Small Group Research, 35,* 17–43. doi:10.1177/1046496403259459

Wittenberg-Lyles, E., Oliver, D. P., Kruse, R. L., Demiris, G., Gage, L. A., & Wagner, K. (2013). Family caregiver participation in hospice interdisciplinary team meetings: How does it affect the nature and content of communication. *Health Communication, 28,* 110–118. doi:10.1080/10410236.2011.652935

Wolfram, H-J., & Gratton, L. (2014). Gender role self-concept, categorical gender, and transactional-transformational leadership: Implications for perceived workgroup performance. *Journal of Leadership & Organizational Studies, 21,* 338–353. doi: 10.1177/1548051813498421

Wu, J. B., Tsui, A. S., & Kinicki, A. J. (2010). Consequences of differentiated leadership in groups. *Academy of Management Journal, 53,* 90–106. doi:10.5465/AMJ.2010.48037079

Yong, K., Sauer, S. J., & Mannix, E. A. (2014). Conflict and creativity in interdisciplinary teams. *Small Group Research, 45,* 266–289. doi:10.1177/1046496414530789

Yuan, Y. C., Fulk, J., Monge, P. R., & Contractor, N. (2010). Expertise director development, shared task interdependence, and strength of communication network ties as multilevel predictors of expertise exchange in transactive memory work groups. *Communication Research, 37,* 20–47. doi:10.1177/009365020351469

Zanin, A. C., Hoelscher, C., & Kramer, M. W. (2016). Extending symbolic convergence theory: A shared identity perspective of a team's culture. *Small Group Research, 47,* 438–472. doi:10.1177/1046496416658554

AUTHOR INDEX

A

Aaron, J. R., 167
Allison, B. B., 180
Alper, S., 132
Ancona, D. G., 71
Anderson, C., 156
Anderson, C. M., 107, 109
Anthony, T., 101
Aritz, J., 150, 152, 154
Arrow, H., 19, 60, 107, 108
Asbury, M. B., 59
Aube, C., 21
Avolio, B. J., 153

B

Bailey, D. E., 99
Bakar, H. A., 99
Baker, D. C., 146
Bales, R. F., 25, 119
Ballard, D. I., 38
Baran, B. E., 164
Barge, J. K., 140
Barki, H., 79
Baron, N. S., 49
Baruah, J., 79
Bass, B. M., 143, 152
Baxter, J., 152
Bazarova, N. N., 76
Beavin, J. H., 25
Beck, S. B., 59, 97, 164, 185
Behfar, K. J., 103, 121
Bell, B. S., 49, 50, 51
Bell, M. A., 199
Bell, M. P., 44
Berdahl, J. L., 107, 108, 156
Berg, D. N., 119, 120
Berry, G. R., 49, 51
Bettenhausen, K. L., 16
Bies, D. J., 104
Blight, M. G., 141, 142
Bonito, J. A., 17

Botero, I. C., 56, 57, 78
Bradley, P. H., 125
Brashers, D. E., 76
Brawley, L. R., 97
Brennan, M., 41
Brooks, A., 154
Brown, T. M., 63
Bruch, H., 153
Buchholz, R. A., 154
Burgoon, J. K., 123
Burke, C. S., 139
Burnfield, J. L., 172
Burtis, T. B., 43
Byrnes, K., 104

C

Canary, D. J., 127
Carini, B., 19
Carron, A. V., 97, 99, 101
Carson, J., 148
Carter, D. R., 140
Cartwright, D., 100
Chen, C. C., 157
Clegg, S., 123, 125
Clifton, J., 142, 152
Cohen, A., 128
Cohen, S. G., 99, 107
Cohen, S. P., 119
Comer, D. R., 16
Contractor, N., 60, 65, 148
Cruz, M. G., 158
Cummings, J. N., 50, 51
Czienskowski, U., 57

D

Dabbs, J. M. Jr., 77
Dagosta, J., 140
Dailey, J. C., 78, 81
Dagosta, J., 140
Daniloski, K., 146
Darics, E., 49

Dasborough, M. T., 26
Davis, C. S., 165
De Dreu, C. K., 118, 121, 130
de Ridder, J. A., 59, 96
de Vries, R. E., 59
DeChurch, L. A., 140, 148
Delbecq, A. L., 79, 81
Demiris, G., 165
DeStephen, R. S., 82
Deutsch, M., 115, 118, 1129
Devine, D. J., 99
Dobosh, M., 132
Drach-Zahavy, A., 163
Driskell, J. E., 101
Druskat, V. U., 146
Dunbar, N. E., 123
Duran, R. L., 132
Durbin, J., 104

E

Einstein, W. O., 153
Ellingson, L. L., 36
Emery, C., 146
Erbert, L., 66, 88
Erdem, I., 155
Eys, M. A., 101

F

Fairhurst, G. T., 139, 150
Farmer, S. M., 128
Farrar, W. J., 40
Firestien, R. L., 79
Fisher, B. A., 25
Fisher, C. M., 150
Fleming, G. P., 77
Fletcher, J. K., 148
Fletcher, R. B., 97
Floridi, L., 49
Fortner, S. A., 41
French, J. R. P., 124
Friedman, R., 103
Frisby, B. N., 104
Fulk, J., 65, 121

G

Gage, L. A., 165
Galanes, G. J., 117, 119, 128, 129, 157
Gardner, W. L., 153
Garner, J. T., 116, 131
Gastil, J., 182

Geist, P., 169
Gersick, C. J. G., 108
Gibb, J. R., 94
Gilstrap, C., 156
Golembiewski, R. T., 99
Gonzales, A. L., 98
Gouran, D. S., 71, 74
Graham, E. E., 73
Gratton, L., 154
Green, S. G., 81, 87
Grice, H. P., 26
Guenter, H., 121
Guetzkow, H., 121, 122
Gully, S. M., 99
Gundersen, D. E., 157
Gustafson, D. H., 79
Gyr, J., 121, 122

H

Hackman, J. R., 99, 108
Hamby, A., 146
Hamrin, S., 140
Han, J., 150
Hancock, J. T., 198
Hanner, J., 76
Hardy, J., 101
Hare, A. P., 16, 182
Harrison, D. A., 44
Hartel, C. E. J., 96
Heath, R. G., 166
Hendershot, B., 156
Henningsen, D. D., 57, 158
Henningsen, M. L., 57
Henry, K.B., 19,
Herdman, A. O., 167
High, S. C., 139, 141, 142
Hill, R. P., 167
Hill, S. E. K., 146
Hinsz, V., 57
Hirokawa, R. Y., 66, 71, 73, 74, 75, 82, 88, 140, 157
Hoelscher, C., 20
Hoffman, L. R., 82
Hollander, E. P., 145, 146
Hollingshead, A. B., 56, 57, 58, 71, 78, 86
Hopper, C. H., 77
Hurst, A., 66, 88

J

Jackson, D. D., 25
Jackson, J. W., 19

Jacobs, T. O., 145
Jacobsohn, G. C., 58
Jarboe, S. C., 87, 123
Jehn, K. A., 71, 116, 117, 119, 133
Jobidon, M-E., 21
Johansson, C., 140
Johnson, C., 78
Johnson, L., 42
Jones, E. E., 146
Jones, T. S., 117
Julian, K. M., 74

K

Katzenbach, J. R., 109
Kaufer, K., 148
Kauffeld, S., 169
Keegan, B., 148
Kelly, J. R., 146
Kelly, L., 132
Ketrow, S. M., 140
Keyton, J., 28, 58, 59, 97, 101, 102, 157, 164, 1085
Kiesler, S., 50, 51
Kinicki, A. J., 153
Kirchmeyer, C., 128
Kirschbaum, K. A., 41
Kleingeld, A., 20
Kleinmann, J., 82, 172, 173
Knight, M. B., 157
Knutson, T. J., 122
Konig, C. J., 172, 173
Kowitz, A. C., 122
Kozlowski, S. W., 49, 50, 51
Kramer, M. W., 20, 26, 78, 107, 149
Kramer, T. J., 77
Kramer, W. S., 139
Kraus, R. M., 119
Krikorian, D. H., 36, 37
Kruse, R. L., 105
Kulesher, R., 41
Kumar, K., 42
Kunze, F., 153
Kuo, C. L., 78, 81
Kuypers, T., 121

L

LaFasto, F. M. J., 20, 105, 157
Lafond, D., 21
Lam, C.,16
Lambertz-Berndt, M. M., 141, 142

Lammers, J. C., 36, 37
Langfred, D. W., 121
Larkey, L. K., 40, 43
Larson, C. E., 105, 157
Law, K. S., 132
Lazer, D., 60
Leach, D. J., 172
Leana, C. R., 77
Leatham, G. B., 74
Lehmann-Willenbrock, N., 169
Levine, K. J., 154
Lewicki, R. J., 104
Lia, J., 39
Lira, E. M., 45
Littlefield, R. S., 164
Liu, X-Y., 96
Lovaglia, M., 123
Love, S. B., 117, 128, 129

M

Ma, M., 157
Mannis, S. M., 77
Mannix, E. A., 117, 121, 123
Martin, A., 139
Martin, M. M., 107, 109
Martins, L. L., 43
Mathieu, J. E., 166
McAllister, D. J., 104
McComas, K. A., 164
McDermott, V. M., 41
McDowell, W. C.,137
McGrath, J. E., 103, 107, 108
McKeage, R. L., 16
McKinney, B. C.,132
McLeod, P. L., 76
McPherson, M. B., 73
Meier, C., 50, 51
Meng, J.,
Merolla, A. J., 51
Merritt, D., 52
Messersmith, A. S., 180
Meyers, R. A., 76, 78
Michaelsen, L. K., 42
Miller, C. E., 63
Miller, V. C., 140
Milliken, F. J., 43
Mirivel, J. C., 164
Monge, P. R., 65
Morsella, E., 119
Moscovici, S., 76
Moye, N. E., 119

Muenchen, R. A., 154
Mullen, B., 78, 101
Myers, S. A., 104
Myrsiades, L., 164

N

Nelson, M. T., 41
Nicotera, A. M., 132
Nolan, L. L., 177
Northouse, P. G., 139
Notelaers, G., 121
Nye, J. L., 157

O

Odermatt, I., 172, 173
Oetzel, J. G., 41, 42, 43
Oh, S. H., 1032
Oliver, D. P., 165
Orlitzky, M., 75
Osborn, A. F., 77

P

Pace, R. C., 121
Pan, Y., 150
Papa, M. J., 73
Park, H. S., 59
Paskewitz. E. 185A.,
Paulus, P. B.,
Pavitt, C., 78, 87, 1398, 141, 142, 145, 146
Peiro, J. M., 45
Pennebaker, J. W., 98
Perez, F. G., 43
Pescosolido, A. T., 50, 51, 98, 146
Peterson, R. A., 121
Peterson, R. S., 71
Pinsonneault, A., 79
Poncini, G., 43
Pondy, L. R., 43
Poole, M. S., 44, 87, 89, 116, 131, 132
Prapavessis, H., 99
Price, K. H., 44
Putnam, L. L., 33, 116, 130

R

Rapp, T. L., 165, 167
Rask, J. P., 40
Raven, B., 124
Reid, D. J., 44

Reid, F. J. M., 44
Reimer, A., 57
Reimer, T., 57
Renz, M. A., 82
Riddle, B. L., 107, 109
Ridgeway, C. L., 154
Ripoll, P., 45
Robertson, T., 39
Rogelberg, S. G., 164, 172
Rosenthal, S. B., 154
Roth, J., 128
Rozell, E. J., 157
Ruback, R. B., 77

S

Saavedra, R., 98
Sackaroff, P., 145
Salas, E., 98, 101, 139
Samuelson, C. D., 123
Sanchez, C., 41
Sanchez, M. I., 43
Sargent, L. D., 44
Sauer, N. C., 121
Scheerhorn, D. R., 169
Schreurs, B., 121
Schultz, B., 147
Schwartzman, H. B., 163
Schwarz, R. M., 177
Schweiger, D. M., 77
Scott, C. R., 45
Scott, C. W., 164
Seely, P. W., 140
Seibold, D. R., 38, 40, 77, 87, 89
Sell, J., 123
Shanock, L. R., 164
Sharf, B. F., 147
Sheer, V. C., 99
Shepard, C. A., 40
Shimotsu, S., 104
Shin, Y., 96
Shuffler, M. L., 139, 180
Sias, P. M., 166
Sillince, J. A. A., 126
Smith, B. A., 157
Smith, D. K., 109
Smith, K. K., 119, 120
Solansky, S. T., 149
Somech, A., 163
Spink, K. S., 99
Spitzberg, B. H., 127
Stasser, G., 56

Stohl, C., 33
Sue-Chan, C., 44
Sun, J. J-M., 96
Sung, C. C. M.,155
SunWolf, 33, 39, 77, 87, 89

T

Taber, T. D., 87
Tamaka, H.,40
Tannebaum, S. I., 40
Teboul, J. C. B., 169
Thomas, K. W., 116, 122, 128, 130, 135
Timmerman, C. E., 45
Titus, W., 56
Tjosvold, D., 132
Torres, A., 41
Tracy, K.,164
Tremblay, S., 21
Tressler, K. E., 139, 141, 142
Trochim, W. M., 121
Tropman, J. E., 172
Tse, H. H. M., 26
Tsui, A. S., 1053
Türetgen, I. O., 155
Turman, P. D.,102

U

Unsal, P., 155
Urban, D. M.,

V

Van de Ven, A. H., 79, 81
Van den Hooff, B., 59, 96
van Emmerik, H., 121
van Iterson, A., 121
Van Mierlo, H., 20
Van Swol, L. M.,57
Van Vianen, A. E., 130

W

Waldeck, J., 40
Wagner, K., 165
Waldman, D. A., 153
Walker, R. C., 150, 152, 154
Wall, V. D., Jr.,117, 119, 128, 129
Wallace, A. M.,180
Walther, J. B., 50, 51, 76
Wang, L., 150
Warfield, J. N., 120
Warr, P. B., 172
Watson, K. W., 42
Webber, S. S.,
Weber, A., 104
Weick, K. W., 99
Weingart, L. R. 121,
Wellinan, B., 61
Wheelan, S. A., 16, 17
Whitney, D. J., 99
Whitton, S. M., 97
Wilson, C. E., 50
Winslow, J. K., 139, 141, 142
Witteman, H. R., 102, 128
Wittenbaum, G. M., 56, 71, 78
Wittenberg-Lyles, E., 165
Wolfram, H-J., 154
Wu, J. B., 153
Wu, J., 157

Y

Yen, T., 41
Yong, K., 121
Yoon, K.,71
Yuan, Y. C., 65, 121

Z

Zaccaro, S. J., 140
Zanin, A. C., 20
Zhang, H.,
Zornoza, A. M., 45

SUBJECT INDEX

A

Accommodating, 129-130
Affective ties, 60
Agenda, 170-173

B

Bona fide group, 33-37
Brainstorming, 77-79

C

Coalition formation, 15
Code of conduct, 167
Coercive power, 124
Cognitive conflict, 122
Cognitive ties, 60
Cohesiveness, 97-101
Collaboration, 127-128
Communication:
    climate, 93-96
    defined, 6
    overload, 63
    underload, 63
Communication networks, 60-66
    centralized, 62-63
    decentralized, 61-63
    evaluating, 64-66
Concurrency, 38
Conflict:
    accommodating, 129-130
aftermath, 117
avoiding, 130
    collaborating, 127-128
competing, 129
competitive, 121
compromising, 130-131
power, 124-126
cooperative, 121-122
    defined, 115-116
    management strategies, 126-133
    strategy selection, 131-133
    types of, 120-122

Connectivity, 35
Consensus, 82-84
Cooperative information sharing paradigm, 56-57
Copresence, 49

D

Decision-making principles, 88-89
Decision-making procedures:
brainstorming, 77-79
    comparing, 87-88
    consensus, 82-84
    nominal group technique, 79-82
    ranking, 86-87
    voting, 84-86
Delay, 38
Developmental perspective on conflict, 116
Distributive conflict management strategies, 128-129
Diversity, influence of, 40-44

E

Embeddedness, 36
Expert power, 124

F

False consensus, 65
Faultlines, 65
Flexibility, 38
Formal ties, 60
Functional theory of decision making, 71, 73-75

G

Group, defined, 15
Group charter, 165-167
Group climate, 93-96
Group goal, 20, 81, 149
Group identity, 19-20, 93, 182
Groupings, 19
Group polarization, 80
Group presentations, 31, 53, 67-68, 91, 111, 135, 160, 187, 189-197

Group roles, 21
Group size, 15-17, 126
Group structure, 21, 93, 126, 149, 182
Groupthink, 101

H

Heterogeneity, 14
Hidden communication, 15
Hidden profile, 57

I

Information sharing, 55-58
Informational power, 124
Instrumental perspective on conflict, 116
Integrative conflict management strategy, 127
Interdependence, 17, 24-26, 28, 34, 65, 81, 149, 182

L

Law of inherent conflict, 120
Leader selection:
appointed, 145
    elected, 145-146
    emergent, 146-148
Leadership:
    communication competencies, 139-145
    defined, 139
    discursive approach, 150-152
    enhancing, 157-158
    gender assumptions, 154-156
    shared, 148-150
    transformational, 152-154
    virtual, 156-157
Legitimate power, 124
Linearity, 38

M

Majority/minority influence, 76-77
Material ties, 60
Meaning, 7
Meeting citizenship behaviors, 164
Meeting management:
agenda, 171
    code of conduct, 167
    conducting meetings, 174-181
    group charter, 165-167
    importance of, 167-169
    managing relationships, 177
    meeting obstacles, 182-185
    minutes, 175-176
    postmeeting follow-up, 181-182
    premeeting planning, 169-174
    space, 177-178
    virtual meetings, 180-181
    visuals, 178-180
Minutes, 175-176

N

Nominal group technique, 79-82
Normative conflict, 122
Norms defined, 21
Number of group members *See* Group size

P

Pace, 38
Performative act, 26
Persuasive arguments theory, 78
Political perspective on conflict, 116
Process conflict, 121
Power, 122-126
bases of, 124-126
    defined, 123
Presentations. *See* Group presentations
Problem solving. *See* Decision making
Procedural leadership competency, 140-142
Proximity ties, 60
Punctuality, 38

R

Ranking, 86-87
Referent power, 124
Relational conflict, 121
Relational dimension 24-28
Relational group, 4
Relational leadership competency, 142-143
Reward power, 124
Roles, 21

S

Satisfaction, 101-103
Scarcity, 38
Shared leadership, 148-150
Socializing members, 107-109
Social loafing, 16
Sociogram, 61
Space, influence of, 39-40
Superordinate goal, 17, 19

T

Task, *See* Group goal
Task conflict, 121,
Task dimension, 24-28
Task-oriented team, 4
Team cognition, 58-59
Technical leadership competency, 143
Technology:
influence of, 44-49
trends, 49-51
Theory of effective intercultural workgroup communication, 42-43
Time:
influence of, 37-39

perspective, 39
Transformational leadership, 152-154
Trust, 104-106
affective trust, 104
cognitive trust, 104

U

Urgency, 38

V

Voting, 84-86